CONTRIBUTORS TO VOLUME 14

S. Bonotto, *Department of Radiobiology, C.E.N.–S.C.K., 2400 Mol, Belgium.*

D. H. Cushing, *Fisheries Laboratory, Lowestoft, Suffolk, England.*

R. R. Dickson, *Fisheries Laboratory, Lowestoft, Suffolk, England.*

C. Edwards, *Dunstaffnage Marine Research Laboratory, Oban, Argyll, Scotland.*

P. Lurquin, *Department of Radiobiology, C.E.N.–S.C.K., 2400 Mol, Belgium.*

A. Mazza, *International Institute of Genetics and Biophysics, C.N.R., Naples, Italy.*

J. S. Ryland, *Department of Zoology, University College of Swansea, Wales.*

Advances in MARINE BIOLOGY

VOLUME 14

Edited by

SIR FREDERICK S. RUSSELL

Plymouth, England

and

SIR MAURICE YONGE

Edinburgh, Scotland

Academic Press

London New York San Francisco 1976

A Subsidiary of Harcourt Brace Jovanovich, Publishers

ACADEMIC PRESS INC. (LONDON) LTD.
24–28 OVAL ROAD
LONDON NW1 7DX

U.S. Edition published by
ACADEMIC PRESS INC.
111 FIFTH AVENUE
NEW YORK, NEW YORK 10003

Library of Congress Catalog Card Number: 63–14040
ISBN: 0–12–026114–6

PRINTED IN GREAT BRITAIN BY THE WHITEFRIARS PRESS LTD.
LONDON AND TONBRIDGE

CONTENTS

The Biological Response in the Sea to Climatic Changes

D. H. CUSHING AND R. R. DICKSON

Recent Advances in Research on the Marine Alga *Acetabularia*

S. Bonotto, P. Lurquin and A. Mazza

A Study in Erratic Distribution: The Occurrence of the Medusa *Gonionemus* in Relation to the Distribution of Oysters

C. Edwards

Physiology and Ecology of Marine Bryozoans

J. S. Ryland

Adv. mar. Biol., Vol. 14, 1976, pp. 1–122

THE BIOLOGICAL RESPONSE IN THE SEA TO CLIMATIC CHANGES

D. H. Cushing and R. R. Dickson

Fisheries Laboratory, Lowestoft, Suffolk, England

I. Introduction

For a long time events in the sea have appeared to be periodic. Indeed Ljungman (1882) imposed a fifty-five-year cycle upon the long-term records of catches of the Bohuslån herring fishery. Much of the work by Pettersson (for example, 1921–2) during the first two decades of the life of the International Council for the Exploration of the Sea was concerned with physical periodicities as observed in tidal records.

Although records of the presence or absence of some fisheries go back into the Middle Ages, detailed records of catches and stock densities from the fisheries go back to the twenties of the present century. During the thirties and early forties there was a general warming in the northern hemisphere and the climatic change occurred at a time when fisheries records were being properly established and when the quantitative study of marine biology was being developed. As a consequence, this period of warming (or climatic amelioration) was documented in changes in the fish stocks, in the spread of organisms to the north and in profound changes throughout the ecosystem noticed in the western English Channel between 1926 and 1935. These trends of change reversed between 1966 and 1972 and we propose to name this period the Russell cycle (after its discoverer, Sir Frederick Russell F.R.S.). The most remarkable point is that the ecosystem appears to show sharp step-like changes at the beginning and at the end of the cycle so that a relatively smooth turnabout in environmental conditions is accompanied by a " rectified " cycle in marine populations.

If fishes and other animals in the sea can " detect " climatic change so well, the mechanisms of adaptation must involve the stabilization of their populations. Where numbers change by three or more orders of magnitude, such mechanisms must have been disturbed.

The lengthening of biological records and the increasing adequacy of statistical returns from commercial fisheries have enabled us to explain and (to some extent) quantify the effects of exploitation on fish stock abundance; by contrast the effects of the environment are seen largely as the unexplained residual variation when the calculable effects of man's activity have been removed. In the case of species less well documented than those in the major commercial fisheries the influence of the environment may only be noticed when major extremes of abundance or distribution force themselves to our attention, and, of course, over wide areas of the global ocean major changes in the ecosystem may go entirely unrecorded.

Thus our estimates of the environmental contribution to the abundance of various populations in the sea will differ markedly in quality, but even when this contribution appears to be well assessed statistically, the actual environmental processes at work remain to be described in terms of cause and effect. To what environmental parameter is the stock responding? What time scales of change are involved? Over what geographical area can we expect a species to show coherent trends of abundance? At what stage in the life cycle of a species does the environmental control take effect? Are the environmental influences acting through direct or indirect mechanisms? What inbuilt mechanisms does a species possess to resist the influence of climatic change? Stated in this way it is clear that we are far from achieving an adequate knowledge of the processes involved. As Larkin (1970, p. 9) points out " virtually any set of stock-recruit data is sufficiently variable to inspire hypotheses about the effects of trends in environments, especially with the wealth of meteorological and oceanographic data that can be mined for real and fortuitous correlations ". Certainly the danger of fortuitous correlation is present whenever individual case histories are considered in isolation but, viewing the complex record of climatic, hydrographic and biological variation as a whole, it is nevertheless possible to find signs of an underlying order in these events. It is this approach which will be followed here. Section II will attempt to describe the principal climatic " change makers " which are at work and will identify some characteristic time and space scales of climatic variation. In Section III some recent trends of climatic and hydrographic variation will be described. Set against this background of environmental change Sections IV and V will review those changes in marine populations which are thought to be attributable to some variation in the physical environment. Finally in Section VI these various elements of change will be pieced together in an attempt to cast light on the processes involved.

II. Climatic Controls and Characteristics

Although a description of the full complexity of climatic processes is beyond the scope of this work, certain dominant climatic processes appear to be of particular relevance in determining the time and space scales of variations observed in the marine ecosystem. Thus this section does not seek to describe global climate in its entirety, but aims at a description of selected climatic elements which appear to

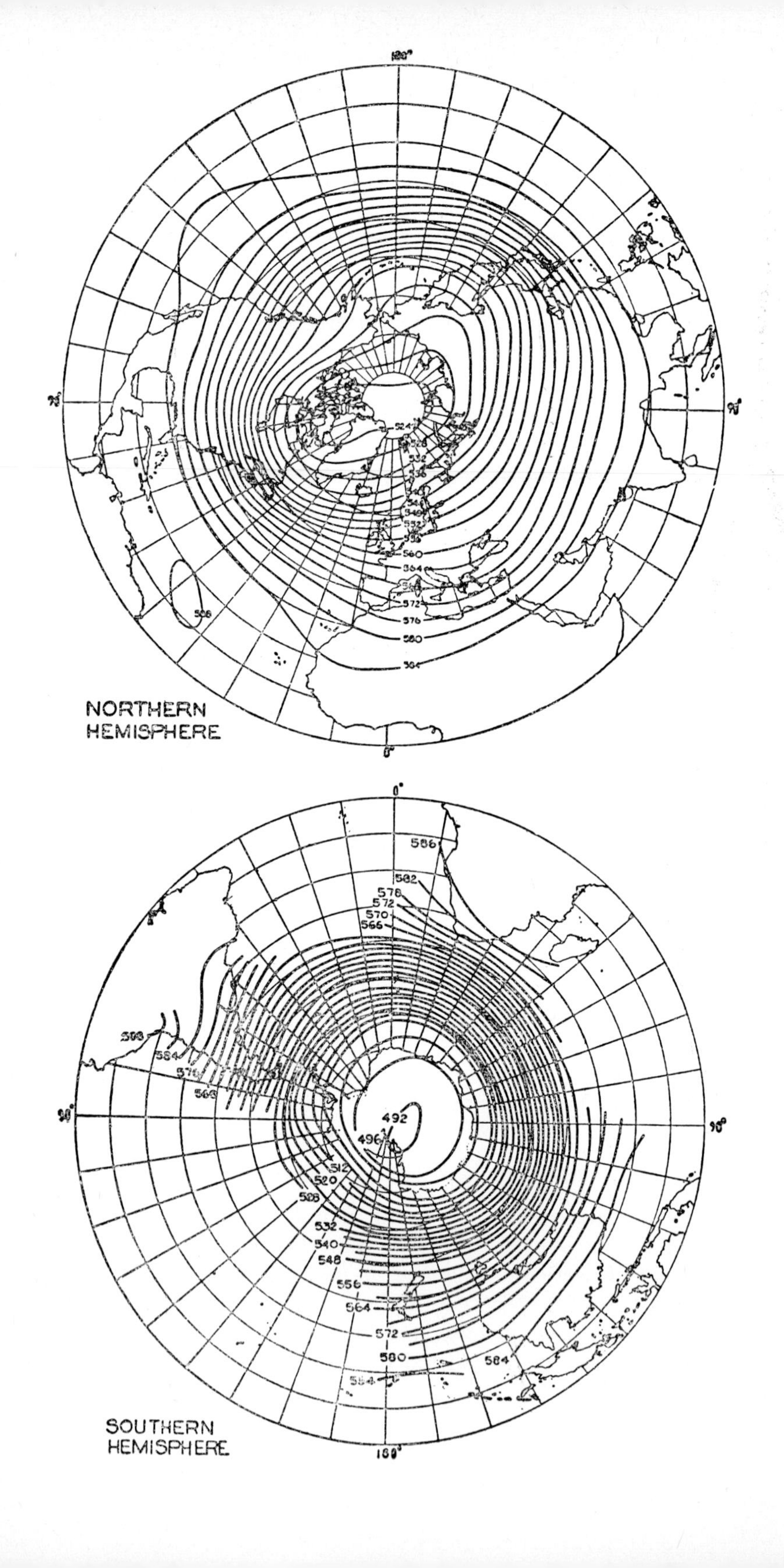

180°
90°
90°
524
528
532
536
544
548
552
556
560
564
568
572
576
580
584
588
0°
NORTHERN
HEMISPHERE
0°
586
582
578
572
570
566
588
584
576
568
90°
90°
492
496
512
520
528
532
540
548
556
564
572
580
584
588
SOUTHERN
HEMISPHERE
180°

have maximum relevance to the observed changes in marine populations. (For a full global account of climatic processes the reader is referred to the classic work "Climate, Present Past and Future" by H. H. Lamb.) Inevitably perhaps, because of the geographical bias of our biological records towards the northern temperate and subarctic regions, the climatic controls described below are predominantly those which prevail in these latitudes.

A. *The temperate westerlies*

In either hemisphere the main momentum of the atmospheric circulation is carried in a single broad belt of westerly winds known as the circumpolar vortex. Essentially these wind belts arise as a result of the gross differences in heating between equator and poles, which finds expression in a greater vertical expansion of the atmosphere over the warm zone. This differential lifting of atmospheric mass against gravity introduces potential energy to the atmosphere and results in the establishment of a pressure gradient between low and high latitudes; the main belts of westerlies occur in middle latitudes where the pressure gradients are steepest (Fig. 1). Since the pressure differences between the warm and cold zones increase with height, the maximum wind speeds (including the jet streams) are observed towards the top of the troposphere.* In the southern hemisphere the greater preponderance of smooth ocean surface results in a reduced surface friction while the absence of warm ocean currents equivalent to the Gulf Stream and Kuro Shio maintains a greater thermal contrast between the equator and the pole. For these reasons, the winds and pressure gradients are generally stronger over the southern ocean. Due to the greater angular momentum of southern hemisphere westerlies, their axis is also displaced slightly equatorward compared with those in the northern hemisphere (see Lamb, 1972, p. 81).

* The troposphere is the lowermost layer of the atmosphere containing the active weather-producing systems. It extends to a height of 11 km over mid and high latitudes and to 16–17 km at low latitudes.

Fig. 1. Average height of the 500 mb pressure level in dekametres (mean of all seasons of the year): (a) northern hemisphere 1949–53 (*b*) southern hemisphere 1952–4. The contours and gradients of the 500 mb pressure level have nearly the same significance as isobars and pressure gradients, at the 5–6 km level. The wind tends to blow along the contours. In accordance with Buy's Ballots law it goes counter-clockwise around the low-pressure regions over the northern hemisphere and clockwise around the low-pressure regions over the southern hemisphere. Note that both hemispheres show a single circumpolar whirl or "vortex" of generally westerly winds at this height. (From Lamb, 1972.)

B. *Planetary waves in the upper westerlies* (*Rossby waves*)

The great westerly wind streams of temperate latitudes do not flow directly around the world from west to east but meander about this track in a series of long waves (Fig. 2). The existence of these waves

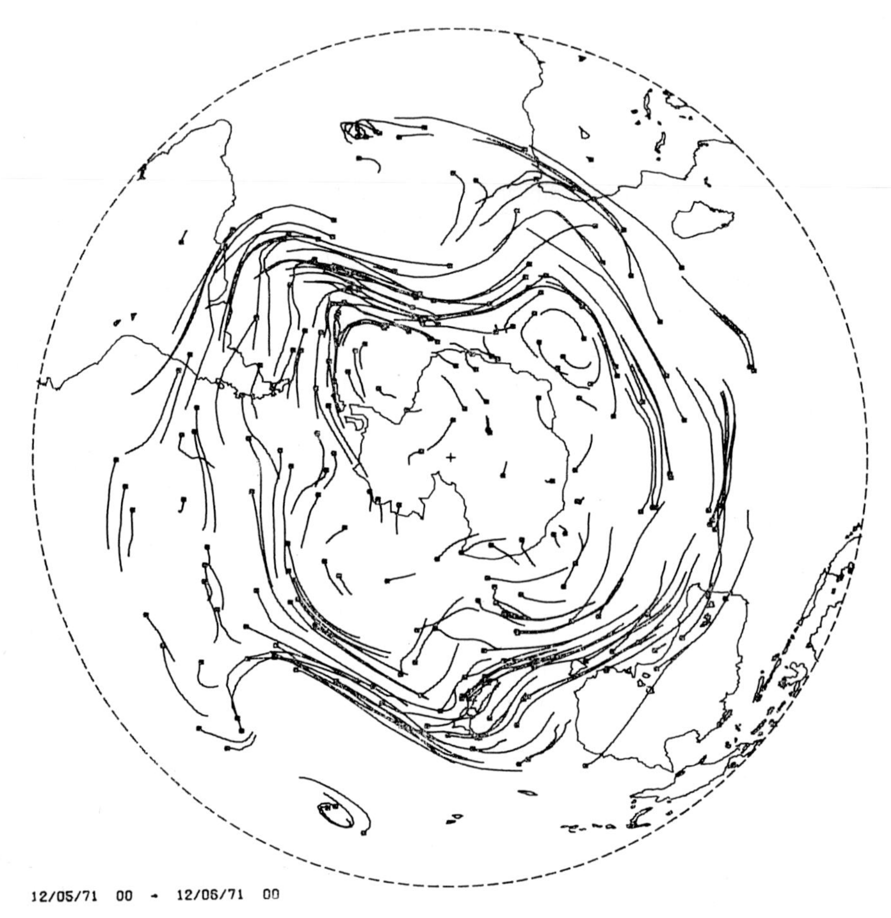

Fig. 2. Rossby waves in the southern hemisphere as described by the trajectories of meteorological balloons flying at a constant pressure level (200 mb) during 5–6 December 1971. (EOLE experiment, courtesy P. Morel.)

was anticipated by Rossby (1939) from vorticity considerations. At any latitude on the earth's surface there is a vorticity or spin about the earth's vertical axis that is appropriate to that latitude. The spin is zero at the equator and greatest at the poles. A parcel of air displaced equatorwards or polewards from its "proper" latitude will tend to retain its original vorticity, and as it moves into latitudes where the vorticity is greater or less than its own, its misfit spin will tend to curve it back towards its original latitude. Each equatorward excursion of the westerlies generated in this way forms a cold trough while each poleward wave takes the form of a warm ridge (Fig. 3). As we will see

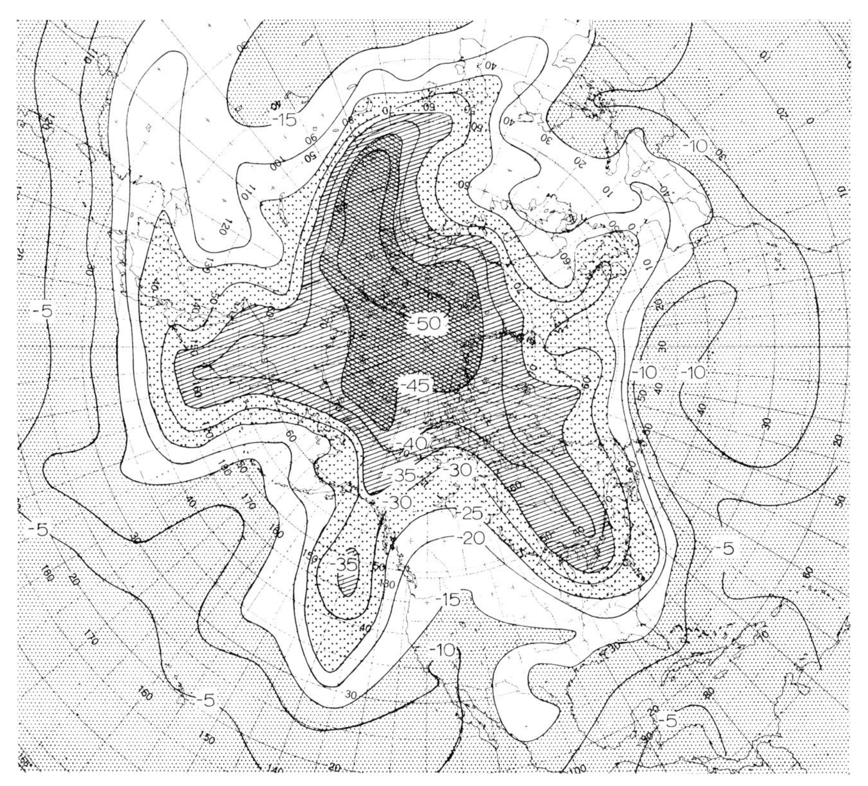

Fig. 3. The temperature distribution at the 500 mb level during a period when the circumpolar circulation was characterized by a well-developed five-wave pattern. Isotherms are drawn at intervals of 5°C, and shaded at intervals of 10°C. (From Rossby, 1959.)

below, there are geographical features on the earth's surface which are capable of deflecting the westerlies from their direct course and which are therefore capable of setting up waves in rather fixed locations, with other secondary waves taking up position at appropriate distances downstream. Though the wave train may slowly progress around the world, the more important condition, climatically, occurs when stationary waves are encouraged to form. Under these circumstances the underlying surface is rather persistently affected by the presence of a standing wave aloft with surface winds and weather systems tending to be steered along the wave-like track followed by the upper winds. The number of waves in the wave train and their latitudinal extent (amplitude) depend upon the strength of the westerlies. During periods of strong westerlies the waves are few and shallow, while five or more large amplitude waves may be observed when the westerlies are weak. In fact for standing waves, the wave length increases as the square root of the velocity of the upper westerlies. Since six or more waves are only occasionally observed, the waves are characteristically large-scale features, perhaps 5 000 km crest-to-crest on average.

C. *Geographical controls on planetary waves*

The importance of these waves as climatic controls arises from the fact that they are not wholly variable in location but tend to be formed and maintained at certain preferred sites.

First, wave formation tends to occur where the underlying surface exhibits a sharp thermal contrast; analogous to the " equator-to-pole " thermal contrast described earlier, ridges tend to form over warm areas of surface while troughs develop over cold zones, so that a pressure gradient tends to develop across the zone of maximum thermal contrast. In the northern hemisphere from early autumn onwards, radiational cooling over an extending snow cover results in the intensification of two great troughs in the upper flow over north-east Asia and north-east Canada and these features will be persistently present until they weaken with the spring thaw. In summer these troughs will retreat to the Arctic or in the case of the Asiatic trough, move out over the cold ocean currents of the north-west Pacific. In effect, they follow the coldest surface.

Second, wave formation occurs where the great mountain ranges act as barriers to the airflow, piling up the air into ridges around which the upper westerlies meander in great poleward excursions. By this mechanism, the Rocky Mountains create a semi-permanent ridge over western North America, and since the conservation of vorticity requires the formation of a trough at some distance downstream, this has the effect of reinforcing the trough over north-east Canada.

The two troughs and the ridge just described represent three great " certainties " of climatic behaviour in the northern hemisphere. They will show year-to-year variations in intensity and in configuration, but on most upper air charts (including the mean annual chart shown in Fig. 1) they will be present in predictable locations. By contrast the secondary waves which may form in the lee of these features are much less predictable, and their presence or absence will be governed by more changeable factors such as the strength of the upper westerlies. If we cannot predict the latter we cannot predict the former. However, the presence of predictable primary waves in the wave train does give us a valuable indication as to where these secondary waves will occur when conditions suit their formation. To give one example from the Pacific sector, White and Clark (1975) show that the Asian trough and the Rocky Mountain ridge are connected by a primarily zonal flow when the westerlies aloft are strong (Fig. 4*a*), but that an additional ridge will rather consistently form in mid-Pacific when these westerlies are weak and when certain conditions of surface

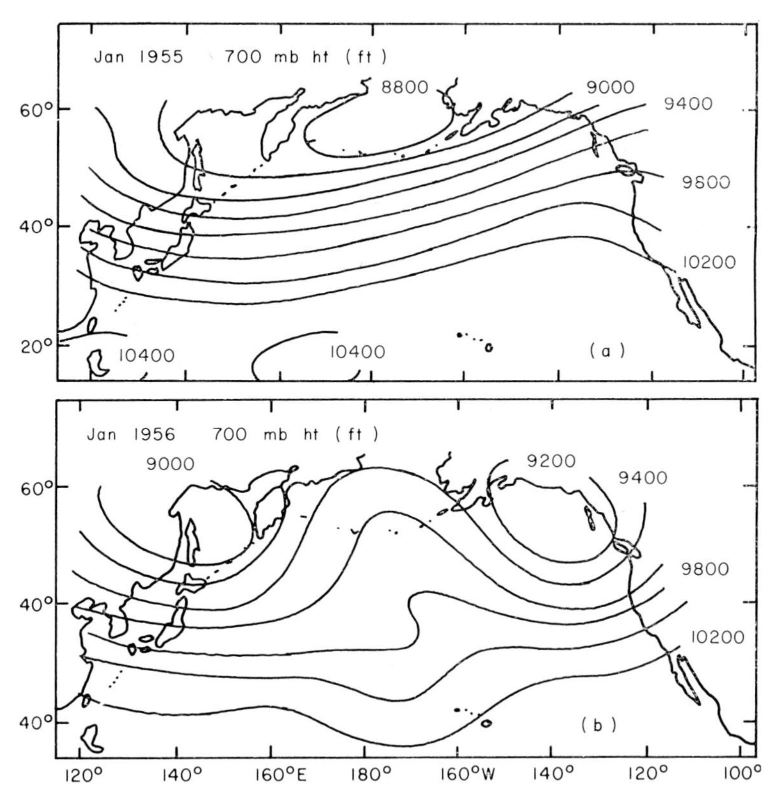

FIG. 4. Charts of 700 mb height (feet) in January 1955 and January 1956 demonstrating the two preferred wave states shown by the mid-latitude westerlies over the North Pacific during the autumn and winter months. (*a*) Primarily zonal flow with a trough off the east coast of Asia and a ridge off the west coast of North America. (*b*) Blocking ridge over the central ocean with flanking troughs. (From White and Clark, 1975.)

heating are favourable (Fig. 4*b*). This example is given, not to show that any simple process is at work, or to give the impression that only two modes of circulation behaviour are possible, but to demonstrate that the location of the secondary waves is not wholly random but is subject to some underlying organization and control.

In the southern hemisphere the effects of geography are less pronounced and the waves in the westerlies are less marked and less consistent as a result. Ridges with their accompanying lee troughs are observed over the Andes and the southern Alps of New Zealand, while the main trough occurs in the Indian Ocean sector where the cold Antarctic continent bulges equatorwards.

D. *Effects of planetary waves on surface weather systems*

The complex mechanisms whereby the long waves in the upper westerlies are able to influence the development and motion of weather systems at surface level is well described by Lamb (1972, pp. 97–107) and will be discussed here only in the most general terms. Essentially these surface cells are encouraged to form at those points in the westerly wave train where flow lines, for any reason, show a tendency to converge or diverge. The westerlies are moving around the waves at a much greater speed than that of the waves themselves and, as they do so, they encounter parts of the wave where the pressure gradient is locally steeper or shallower than average. Entering these areas, the speed of the westerlies is temporarily out of balance with the local pressure gradient and they adjust to this imbalance by turning away from their normal path along the pressure contours to run up- or down-slope towards higher or lower pressure. This in turn causes a redistribution of atmospheric mass which is reflected in the pressure at surface level, bringing enhanced conditions for cyclogenesis or anticyclogenesis at certain characteristic locations along the wave. Frequently these surface systems act in such a way as to reinforce the upper waves which gave rise to them. Cyclogenesis (generation of cyclones) and anticyclogenesis may be stimulated on either side of the axis of westerlies but as Lamb points out (1972, p. 103), " It is observed that anticyclogenic effects prevail along most of the warm side of the jet stream (i.e. of the ' frontal zone ' associated with the main thermal gradient and generally strong upper winds) and cyclogenic effects along most of the cold side. Contrary developments are more limited . . . Along the warm side of the jet is a belt of generally high pressure, the *subtropical anticyclones* . . . The cold side of the jet is where the *subpolar low-pressure belt* is formed."

E. *Anomalies of atmospheric behaviour in the temperate zone*

Thus far we have described some processes which give a degree of continuity to atmospheric behaviour in the zones occupied by the great westerly windbelts. However, we have also identified a number of possible sources of variability in these zones including changes in westerly windstrength, changes in the latitude of strongest flow, and changes in the amplitude, wavelength and position of the planetary waves aloft; Lamb (1972, p. 92) ascribes a further cause of variation to the changes in the eccentricity of the circumpolar vortex around the geographical pole. The following sections will attempt to explore some characteristics and causes of these climatic variations.

An anomaly of atmospheric behaviour is recognized by comparing the average climatic state over any period of interest with the long-term mean condition. Typically we will be concerned below with climatic anomalies which relate to periods of a month or a season, and it is relevant to explain what these time-averaged patterns of behaviour represent. When we mask the influence of the rapidly changing short-term weather systems by averaging meteorological data over such long periods of time, we do not merely arrive at the long-term average or " normal " state. Instead large-scale deviations from normal are brought to light. Of course, these patterns of anomaly are statistical features: the presence of a high pressure anomaly cell on a seasonal pressure anomaly chart does not imply that the cell was continuously present in that location during each day of that particular season. Instead, it may be thought of as reflecting the persistent reappearance of short-lived systems in relatively fixed locations. In the example just given it is probable that the high pressure anomaly cell reflects the passage of a number of anticyclones which merely reached their greatest development or were abnormally delayed in that location. Nevertheless, when one considers a sequence of these time-averaged anomaly charts, the main cells of each pattern may be traced from one chart to the next, and are clearly developing and moving in a slow but systematic manner. These time-averaged charts have therefore a validity in themselves in that they are bringing to light the existence of some agency which is consistently forcing the recurrence or development of fast-moving weather systems in certain preferred locations, and whatever agency is responsible, a sequence of these charts will show how that agency is evolving or progressing with time. These points are best illustrated by reference to Namias' (1951) study of " The Great Pacific Anticyclone of Winter 1949–50 ". Figure 5*a* shows

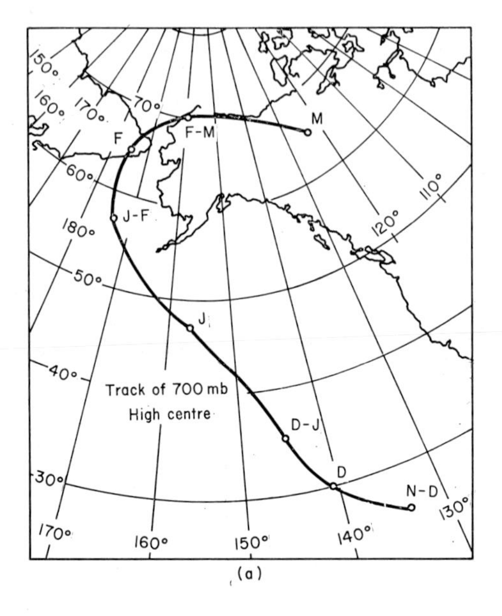

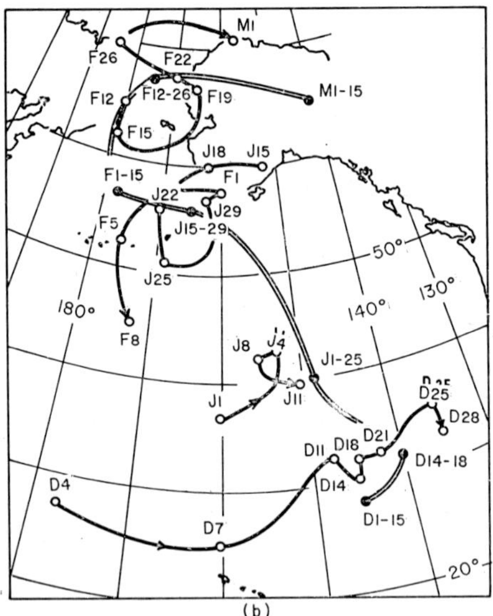

FIG. 5. (*a*) Track of the great Pacific anticyclone of winter 1949–50 as determined from 30-day mean charts of 700 mb height. D stands for December, D-J for mid-December to mid-January, etc. (*b*) Track of the great Pacific anticyclone of winter 1949–50 as determined from 15-day mean charts (solid circles) and 5-day mean charts (open circles) of 700 mb height. For the former J 1-15 refers to period 1–15 January etc; for the latter J1 refers to the 5-day period ending 1 January, etc. (From Namias, 1951.)

the track of this great anticyclonic cell as shown by a sequence of 30-day mean charts. Clearly the progress of the cell is consistent and systematic, reflecting some underlying control. Figure 5*b* shows the increasing complexity which resulted when Namias attempted to track this cell using sequences of 15-day mean and 5-day mean charts. From the latter it is clear that we are in fact dealing with a number of distinct anticyclonic cells, but if (like Namias) we are attempting to explain the factor responsible for encouraging anomalous anticyclogenesis in that sector and with a view to predicting future climatic events, the use of 30-day mean charts will simplify the day-to-day complexity of climate to the point where the underlying processes may be distinguished.

F. *The causes of long-term weather anomalies*

As in the case of the Great Pacific Anticyclone just described, it frequently seems that some external agency is at work during periods when anomalous patterns of atmospheric behaviour are seen to persist or recur over extended periods of time (weeks or months). Much thought has been given in recent years to what the agency or agencies might be, and the subject has been conveniently reviewed by Sawyer (1965). First, he points out that since friction and viscosity appear capable of wiping out the total circulation energy of the atmosphere over a period of about a week, " . . . this fact would make it seem unlikely that one should seek the causes of regularities and anomalies of the atmospheric circulation which persist or develop over much longer periods in the dynamical inertia of the circulation. Rather, one is encouraged to seek factors external to the atmosphere which may affect the rate at which energy is generated in the atmospheric system " (Sawyer, 1965, p. 228). More specifically Sawyer suggests that to be effective as climatic controls, these external factors should fulfill three criteria: they must be comparable in scale with the observed climatic anomalies (i.e. more than 1 000 km across); they must persist for periods of at least one month; and they must be capable of giving up heat at a rate of at least 50 langleys per day during this period (1 langley representing an energy transfer of one calorie per square centimetre of surface). While a number of factors are theoretically capable of influencing the atmospheric circulation, many fail to meet these three criteria and cannot be regarded as the prime cause of climatic anomalies at the observed time and space scales. Variations in the amount of sea ice for example rarely involve extensions of more than 100 km from the " normal " ice front. Soil temperature anomalies are shallow features and are dissipated rather too quickly.

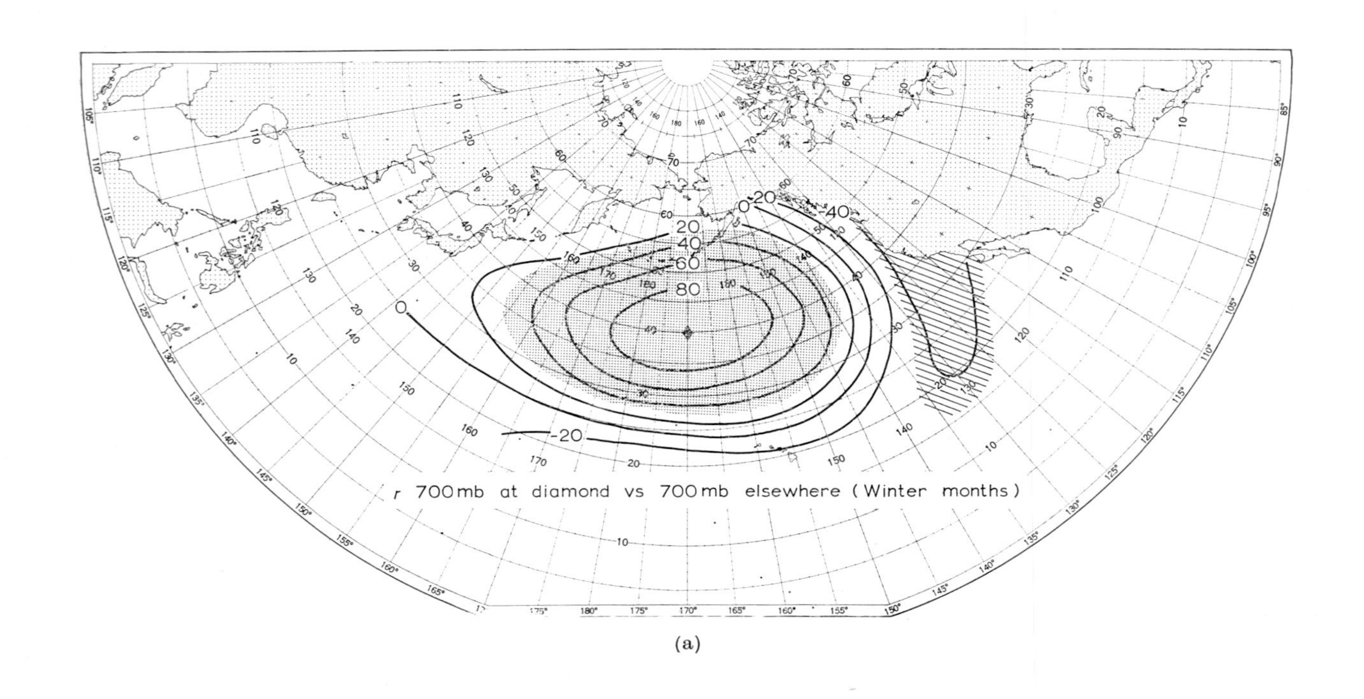

(a)

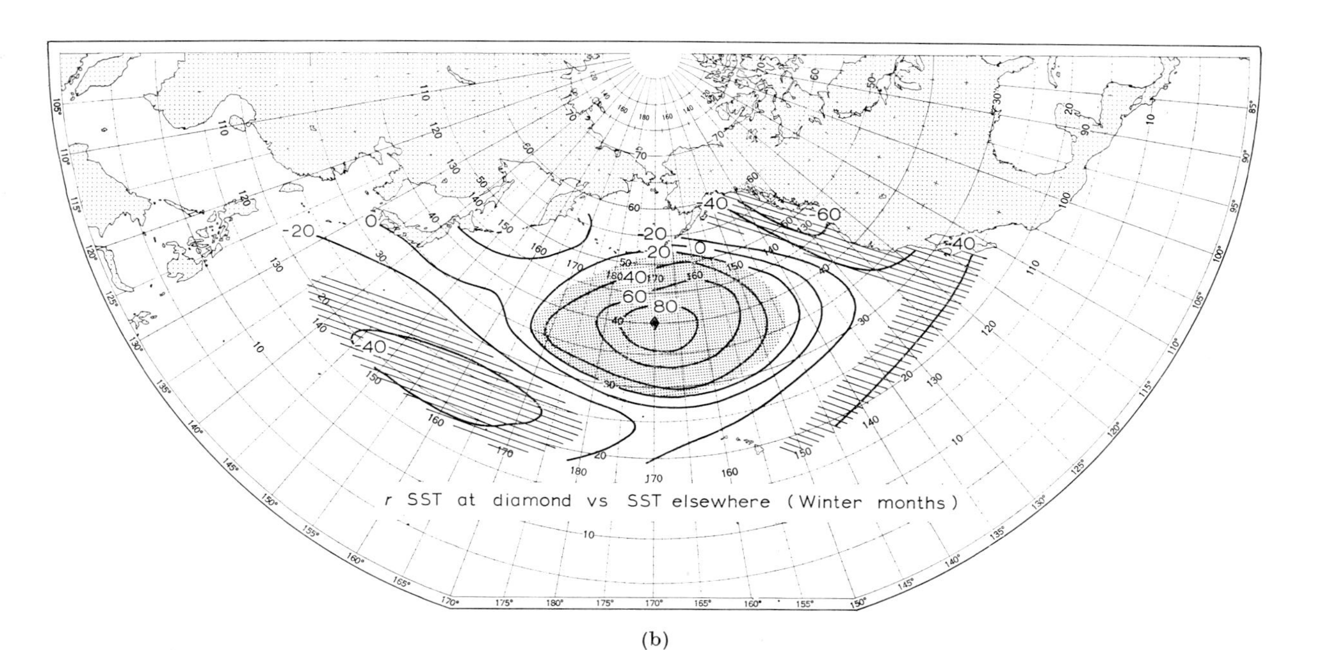

(b)

Fig. 6. Contemporaneous correlations (winter months) between lat. 40°N long 170°W (diamond) and elsewhere for (*a*) 700 mb heights and (*b*) sea surface temperature. Shaded areas represent correlations exceeding 1% level of significance. Positive correlations stippled, negative hatched. (From Namias, 1972.)

Variations in snow cover (with an increase in reflected radiation) have a potential importance, but this is probably confined to limited areas of middle latitudes in spring and autumn. Reviewing these and other factors, Sawyer concludes that only sea surface temperature (SST) anomalies meet all three requisite conditions. SST anomalies of over 1°C are of widespread occurrence, are slow moving and may extend to relatively great depths within the active layer of the ocean [see p. 19] so that they are fully capable of sustaining a significant loss of heat to the atmosphere over periods of a month or even a season.

G. *Time and space scales of SST anomalies*

For the reasons just described, SST anomalies are now regarded as having an important moderating influence on climate at time scales of a few months to a few years in duration and much effort is currently being expended (most notably by Jerome Namias at Scripps Institution of Oceanography) in attempts to characterize these features and to illuminate the coupling mechanisms between ocean and atmosphere.

Figure 6*b* illustrates one important feature of SST anomalies within the temperature zone, the fact that they are typically of large geographical extent. In this illustration Namias (1972) has merely correlated the 20-year record of SST anomaly from one particular grid point intersection in the North Pacific (indicated) with similar records from 154 surrounding points to reveal that these records tend to be coherent over a distance of 3 000–4 000 km, on approximately the same scale as the planetary waves aloft. This point is underlined in Fig. 6*a* which shows the results of similar cross-correlations using time series of 700 mb height anomaly at 5° intersection points.

A second important characteristic of SST anomalies—their persistence—is illustrated in Fig. 7 (from Namias and Born, 1970). Here, a series of monthly *patterns* of SST anomaly covering the North Pacific over the period 1947–66 has been autocorrelated at discrete lags of 1–12 months (each of the monthly patterns was based on a grid of points at 5° intersections from 20°–60°N). The resulting curve traces out the average correlation which is found when the SST anomaly pattern of any particular month is correlated with those of succeeding months. It is clear that SST patterns remain reasonably highly correlated even when they are separated in time by as much as one year; certainly this persistence of SST patterns is very much greater than that which is shown when the fields of 700 mb height anomaly or sea-level pressure anomaly are subjected to the same statistical treatment (see Fig. 7). This SST persistence is ascribed to the large heat capacity of

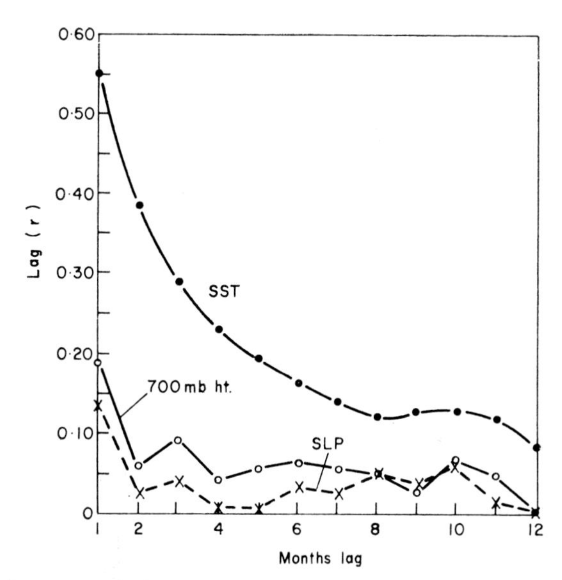

Fig. 7. Overall autocorrelation of standardized values of monthly mean sea surface temperature (SST) 700 mb height, and sea-level pressure (SLP) determined from a 5° grid of points covering the North Pacific (north of 20°N) during the 20-year period 1947–66. (From Namias and Born, 1970.)

the ocean which enables it to store vast amounts of heat for extended periods of time (Namias and Born, 1970, 1974), and this characteristic raises the possibility that the ocean may under certain circumstances act as a " memory " for the atmosphere, forcing or encouraging the repetition of anomalous climatic events.

Needless to say, conditions suitable for the formation of major SST anomalies are not uniformly distributed over an ocean the size of the North Pacific or North Atlantic, but tend to develop in certain preferred locations and at certain times of year. In general, as Namias (1973b) has shown for the North Pacific, SST anomalies tend to reach their extreme development during the cold season when thermal communication between the sea surface and the lower troposphere is maximal, and Fig. 8 (from Dickson, in preparation) identifies the sites where the Pacific has tended to show its greatest surface temperature variability in the winter season over the period 1959–74. As shown, the standard deviation of mean winter surface temperature is maximal in three distinct locations—off Japan, in the east-central Pacific and along the North American seaboard. The specific processes responsible for the generation of extreme SST variability at these sites are at present under investigation, but are likely to be varied in nature. The site off

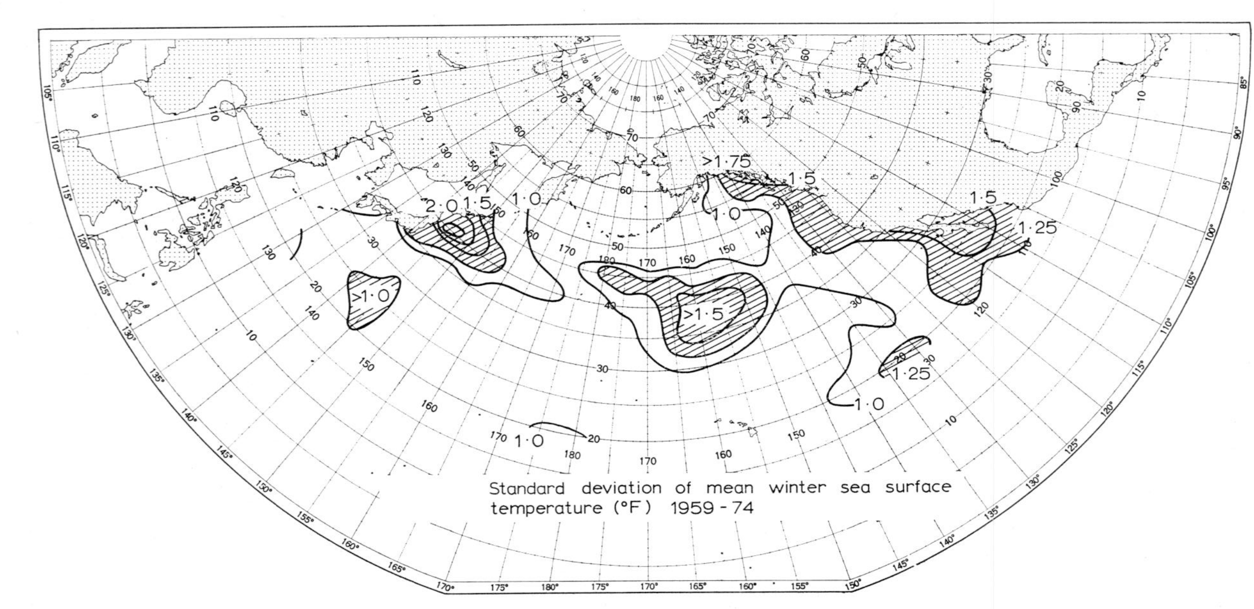

Fig. 8. Standard deviation of mean winter sea surface temperature (°F), 1959–74.

Japan for example lies in the preferred area for warm eddy generation in the confluence zone of the warm Kuroshio and cold Oyashio currents (Kitano, 1975); the site in the east-central Pacific underlies the zone of maximum storm and pressure variability; while that off the American littoral must be at least partly explained by the changing effectiveness of coastal upwelling processes or direct heat exchange with the atmosphere. At each of the two eastern Pacific sites, Namias (1973b) finds a close correlation between the thermal fields of ocean and atmosphere in winter so that in the absence of adequately detailed heat exchange measurements, these sites are held to represent regions of strong thermal interaction with the atmosphere. This conclusion appears to be supported by known atmospheric behaviour. In winter, at the site to the south of the Alaska Peninsula, outbreaks of Polar air are subjected to rapid warming from below, resulting in intense convection instability and in short-term heat exchange rates of up to 2 000 ly/day (Winston, 1955). Within the second zone, curving south-westward along the North American coast from the west coast of Canada the destruction or reinforcement of the west coast inversion in winter provides the potential for major variations in heat exchange.

The three-dimensional (depth) structure of ocean temperature anomalies is difficult to assess. Subsurface ocean temperature records tend to be sparse and too short to describe the " normal " situation and its deviations. There are indications, however, that intense surface anomalies reach deeper into the water column than those which are relatively weak, and show a greater persistence partially as a result of this deep " root ". In the area 35–45°N 155–160°W, at one of the sites of extreme SST variability in the North Pacific and during a period (1968–70) when an intense negative temperature anomaly developed at the ocean surface, investigations show that a cold pool gradually developed down to a depth of over 150 m before being wiped out in the winter of 1970–71 (Dickson, in preparation). Namias (1968, p. 347) reports that in 1968, at one of the principal heat exchange sites in the North Atlantic (SE of Newfoundland) a major and persistent cold pool developed to a depth of " at least 100 m ". This again was an extreme development, with August temperatures as cold as those normally found during the winter minimum. Of course temperature variations may occur at very much greater depths than 150 m and over very much longer time scales, and larger short-term variations may occur at fronts and at discontinuities within the water column. However, at the scales of variation described above (ranging from a few months to a few seasons in duration) those depths of 100–150 m may be regarded as rather extreme for surface-generated ocean temperature anomalies in temperate latitudes. Over the greater part of the open ocean, it is likely that temperature anomalies will be considerably less deep and much more susceptible to change.

H. *Feedback*

Thus far we have seen that large-scale atmospheric flow patterns may be at least partly determined by thermal anomalies and gradients at the earth's surface. While these effects may develop over land (e.g. at the interfaces between snow cover and bare earth or cold land and warm sea) the thermal anomalies of the ocean surface represent the most widespread and persistent controls on atmospheric behaviour. Perhaps the most clear-cut examples of this control are observed when an inherently restless atmosphere is forced into persistent or recurrent patterns of behaviour. One assumes that the thermal reservoirs of the ocean are (through heat exchange) supplying a memory to the atmosphere, guiding the growth and movement of surface weather systems into more or less repetitive patterns and leading, in the aggregate, to the development of short-period climatic fluctuations. One must bear in mind, however, that the ocean temperature anomalies are themselves brought about by abnormal heat exchange under an anomalous atmospheric circulation, so that ocean and atmosphere are in fact capable of mutual adjustment and control (" feedback "). All feedback situations must eventually break down since remote influences will sooner or later break in to disturb the local balance of interaction. However, climatic events of great persistence can result when the feedback loop is set up in such a way as to continuously restore some anomalous initial state. As one example we may cite the case (not infrequently observed) where a warm SST anomaly becomes established to the east of a cold water pool. The intervening sharp thermal gradient is transmitted to the overlying air masses and depressions entering the area linger and develop strongly over this zone. The rapid removal of heat and moisture from the ocean under northerlies running along the western flank of the depression will tend to maintain the cold pool, while the warm southerlies to its east will reduce the heat loss from the warm pool. In this way the thermal gradient on which the depressions are " feeding " will tend to be maintained. A recurrent feedback loop of almost exactly this type appears (through its great persistence) to be a principal cause of inter-annual hydrographic variation in the North Atlantic sector (Dickson, 1971). Figures 9*a*, *b* and *c* illustrate the particular atmospheric anomaly pattern which characterizes this important feedback situation. The principal features are the deep

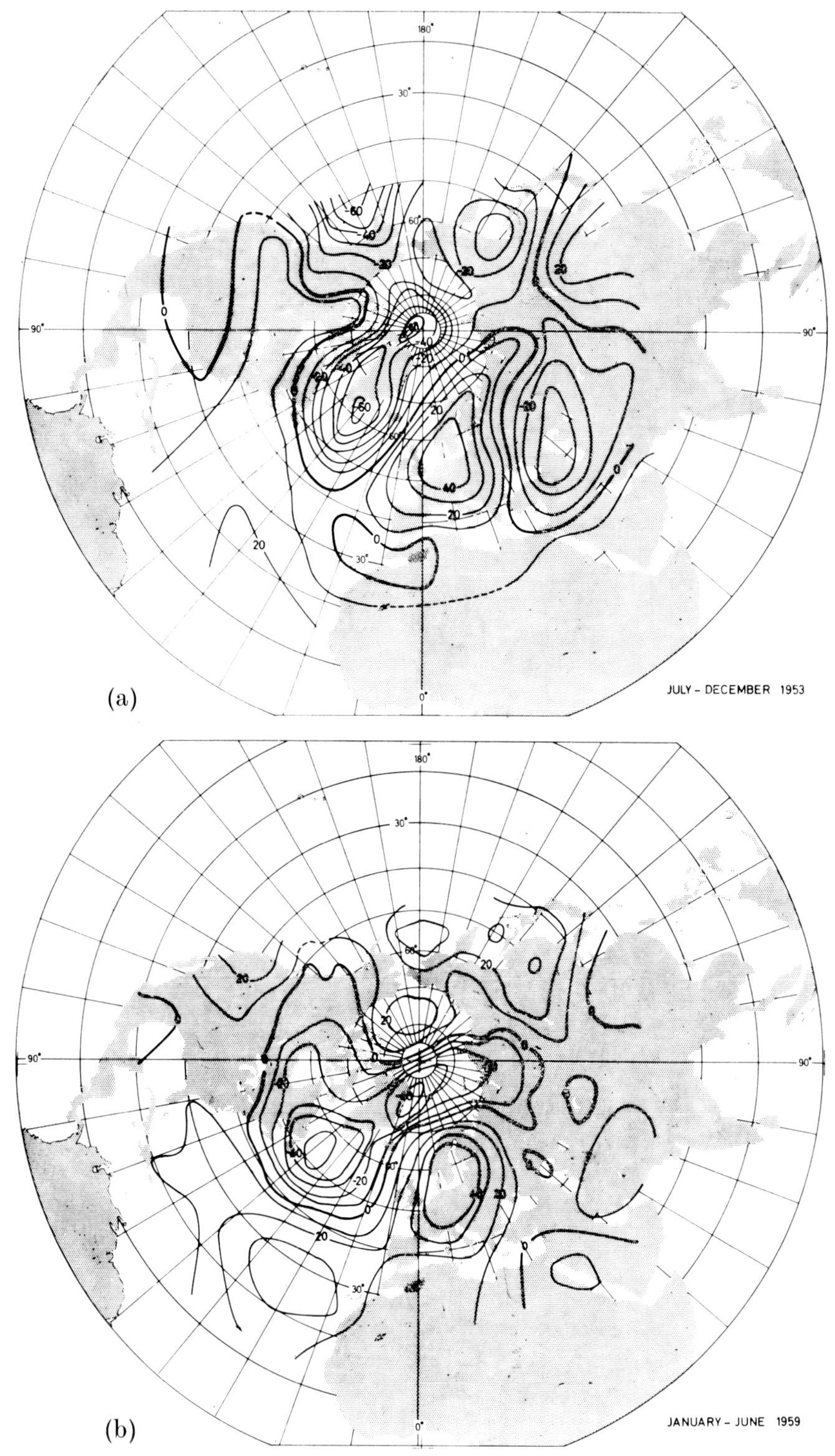

FIG. 9. See caption p. 22

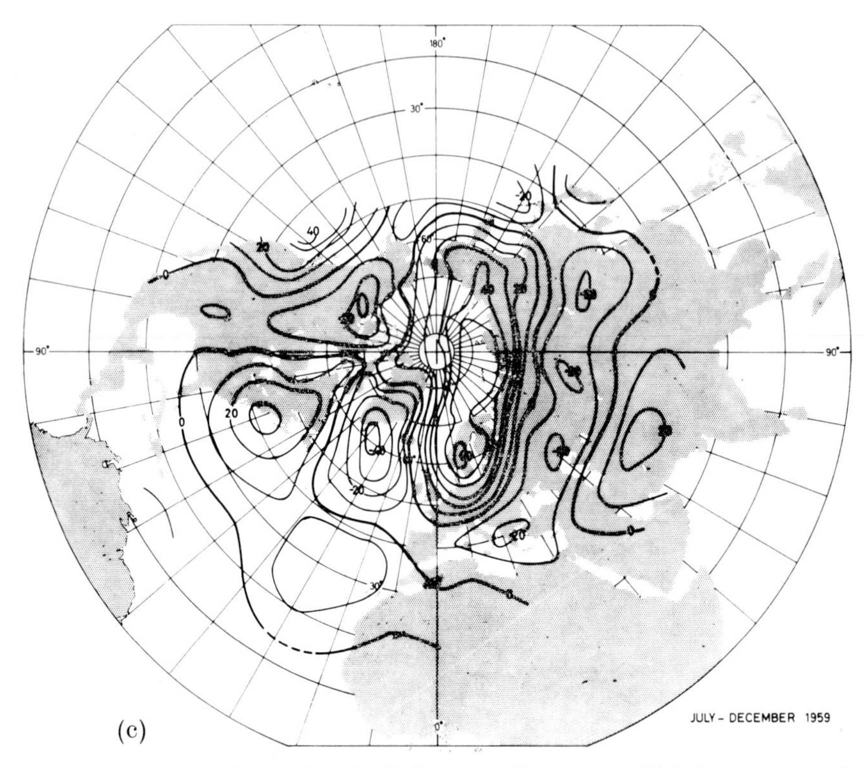

Fig. 9. Three examples of the circulation anomaly pattern which is most strongly associated with ocean/atmosphere " feedback " in the North Atlantic sector. 500 mb height anomaly pattern (m) for (*a*) July–December 1953. (*b*) January–June 1959 (*c*) July–December 1959. (From Dickson, 1971.)

meridional trough which develops over the western Atlantic and the strong meridional ridge over north-west Europe. In response, a cold pool develops in the west Atlantic, with a warm pool in the east, and this surface temperature distribution then encourages the build-up of the initiating circulation anomaly aloft. The 1959 case of feedback was a particularly tenacious event (see Figs 9*b* and *c*), and in fact the characteristic circulation anomaly was present in almost every season between the autumn of 1958 and the spring of 1960. The hydrographic effects (perhaps amplified in the shallow waters of the Continental Shelf) are seen most clearly in the salinity record of the European shelf and in the temperature record from the Kola Meridian of the Barents Sea, but are also clearly marked in the surface temperature and salinity variations at Ocean Weather Ship *Mike* (66°N 2°E) in the open Norwegian Sea. To give an example, Fig. 10 shows the post-war salinity

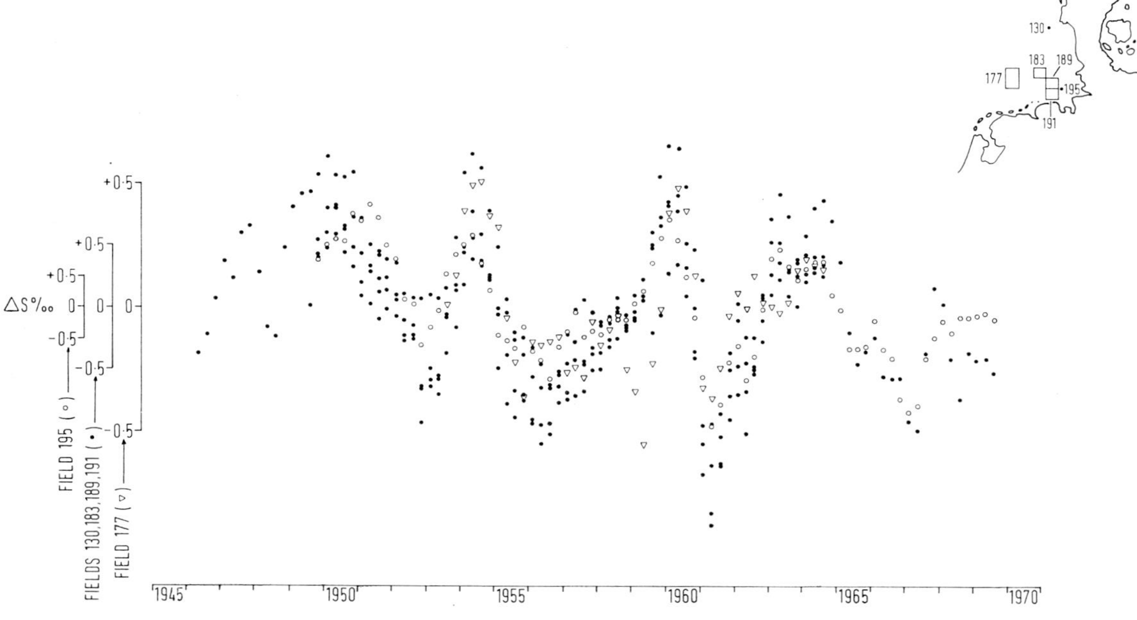

FIG. 10. Running three-quarterly means of salinity anomaly for the post-war period—German Bight (North Sea). This diagram was obtained by superimposing three graphs with different scales, each graph applying to a different set of fields in the German Bight. Though the amplitude of salinity anomaly varied greatly between the different sets of fields each field exhibited the same trend. (From Dickson, 1971.)

variation at various fields in the German Bight of the North Sea. Essentially similar changes may be found in any other part of this shelf and each salinification is associated with the type of sustained feedback described above.

I. *Teleconnections*

The concept of the atmosphere as a continuous fluid carries the implication that a disturbance in one area will tend to be associated with disturbances in other more remote sectors. As we have already seen, the atmospheric fluid does not behave in a wholly chaotic fashion but tends to show elements of an underlying control, often as a result of regularities in the topography or thermal state of the underlying surface. Consequently these remote links, or " teleconnections ", between anomalous events will tend to reflect these controls. The characteristic behaviour of the Rossby waves constitute one form of control: for example, the formation or amplification of a stationary long wave trough at a given point in the wave train will lead to the formation or amplification of a response wave immediately downstream at a distance compatible with the strength of the westerlies. Thus, in the zone of westerlies, teleconnected centres tend to be spaced according to the typical scales of Rossby waves, lying one-half or one wavelength apart.

The principal teleconnections in the northern hemisphere have been identified statistically by O'Connor (1969) from analyses of a 17-year series of 5-day mean 700 mb height anomaly patterns. The resulting charts show the probability of *sign* of an anomaly at other 10° grid intersections when a positive or negative centre lies at a given grid point. Bearing on the discussion of the previous section for example, Fig. 11 shows that with a negative centre located at 60°N 40°W in winter there is a 98% probability that it will be accompanied by a positive anomaly of 700 mb height over north-west Europe. While this type of downstream response is expected from vorticity considerations, the consistent occurrence of this teleconnection in the climatic record must also reflect the persistent feedback situation described above.

Strictly speaking, as Namias and Born (1972) point out, these teleconnections are modal in character, reflecting what usually takes place, and are liable to break down in individual instances. Nevertheless these statistical links have frequently proved of great value in giving some initial indication as to the remote " cause " of some anomalous atmospheric event.

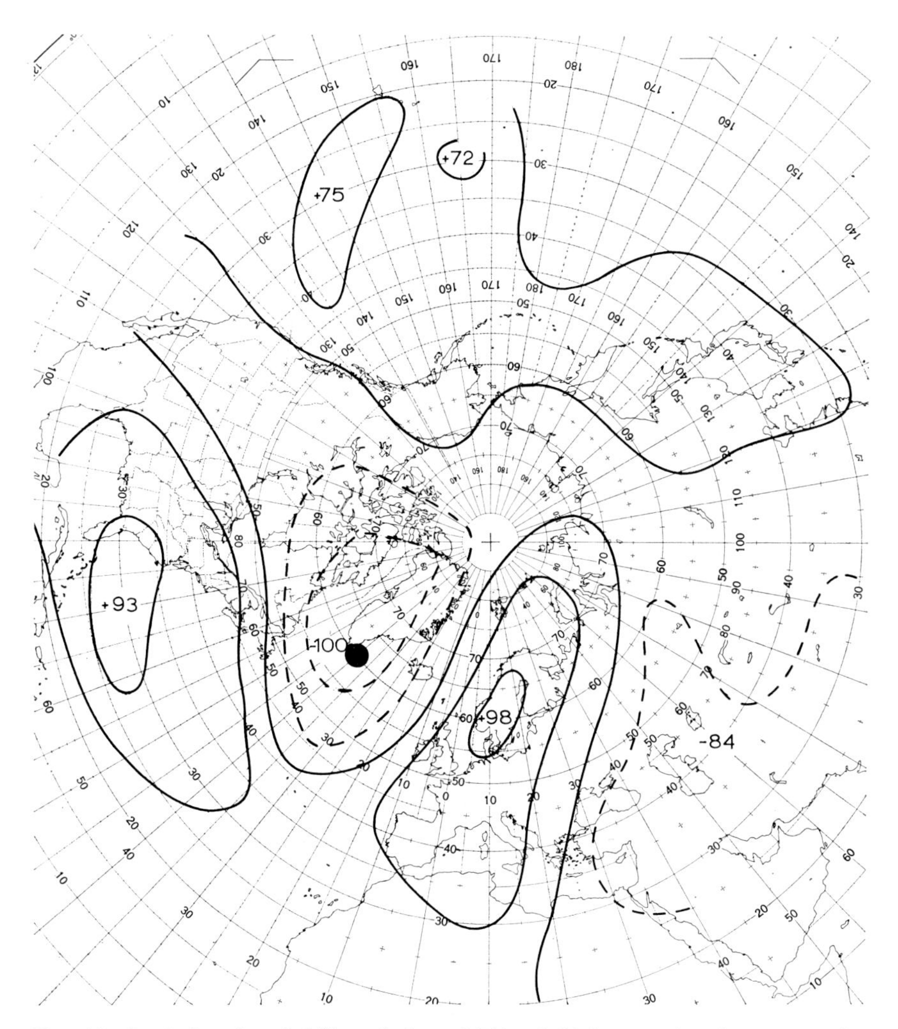

Fig. 11. Isopleths of probability of sign of 700 mb high anomaly when a negative anomaly centre lies within the 10° square centred at 60°N 40°W in winter. (From O'Connor, 1969.)

J. *Tropical and subtropical wind systems*

In the eighteenth century, Hadley postulated that the gross differences in heating between equator and poles should result in a simple one-celled circulation in each hemisphere, with warm air rising and moving meridionally from the equator to descend at the poles and return as a surface flow. In practice as the air rises at the equator and moves poleward, the conservation of its great angular momentum about the earth's axis causes it to turn from a meridional to a zonal track so that at the latitude of 30° it is travelling as a fast upper westerly flow—the subtropical jetstream—overrunning the earth's surface at a speed of over 100 knots. As it continues eastward, overlying the subtropical anticyclone belt, it participates in the general sinking motions which characterize these great cells and finally forms part of a broad easterly return flow towards the equator at surface level.

While the latitudinal limits of this circulation cell may be more restricted than Hadley envisaged, Bjerknes (1972) points out that the equatorial belt may nevertheless have far-reaching influences on the atmospheric circulation. First he notes that the equatorial ocean (notably the eastern equatorial Pacific) represents the greatest " heat engine " available to the atmosphere. Second he points out that " the atmosphere at the equator is in its average state close to the threshold of vertical instability, so that even a small positive anomaly of heating and evaporation, distributed over the large interface area along the equator, can release vast amounts of atmospheric potential energy " (Bjerknes 1972, p. 108). And finally, as already noted, the great angular momentum of equatorial air masses means that any latitudinal exchange of air is tantamount to an export of westerly angular momentum from the equator.

K. *Upwelling*

In temperate latitudes the space scales of short-term climatic variation are linked to those of the surface pressure cells or the associated Rossby waves aloft. Within the subtropics, however, pressure variability is small and SST anomalies are principally controlled by variations in the strength of the tradewind systems which flow from the north-east (northern hemisphere) and south-east (southern hemisphere) to converge at the meteorological equator. As a result of these large-scale variations within the broad belt of the trades, " . . sea surface temperature anomalies in the equatorial Pacific may vary as much as 3°C from one January to another and extend over an area perhaps 10 000 km long in a zone about 1 000 km wide " (Namias, 1974, p. 170). More specifically, the variations in tradewind strength appear to be translated into hydrographic variation *via* the mechanism of upwelling.

Upwelling of cold subsurface water occurs as a result of horizontal divergence in the surface of the ocean. As Smith (1968) points out, a similar surface-cooling effect might appear to be produced by wind mixing and other factors, but these factors cannot account for the *persistent* ascent of subsurface water which is what we mean by the term " upwelling ". While upwelling can occur throughout the world ocean it is of greatest biological importance within the subtropical and tropical zones.

Along the eastern boundaries of the subtropical oceans the easterly trade winds tend to blow parallel with the coast as they flow towards the equator. Due to the Ekman effect, the wind driven surface layer is directed 90° to the right of the wind in the northern hemisphere (90° to the left in the southern hemisphere), and as this surface layer moves offshore, it is replaced along the coast by water upwelling from the deeper layers. Coastal upwellings of this type are especially well developed (and well described) along the coasts of Southern California/Baja California (most intense off Cape Mendocino), Peru and North-west and South-west Africa. Above 200 m depth (approximately) a slow coastal current moves offshore and equatorward; below this level a counterflow moves poleward. The offshore boundary of the coastal upwelling is marked rather imprecisely by a cell or " roller bearing " of convergence and divergence situated some 50–100 km from the coast.

Further offshore, open ocean upwelling may occur in response to spatial variations of the windstress on the sea. This is normally of secondary importance except in the special case of equatorial upwelling. Along the equator the easterlies exert a stress on the sea towards the west, and since the Coriolis parameter is zero at the equator, the surface water flow is also towards the west. However, a few degrees to the north and south of the equator, the Ekman transport under this easterly wind stress becomes increasingly poleward in direction as the Coriolis force increases, so that the surface flow within the equatorial zone is strongly divergent, and upwelling results (Smith, 1968).

Thus in the classic picture, upwelling tends to develop within the tradewind belt along narrow coastal strips which run equatorward along the eastern boundaries of the oceans; these tend towards, but do not necessarily merge with, a second broad zone of upwelling which runs along the equatorial belt, strong in the east but weakening towards the west. In this idealized geographical pattern no upwelling is expected along the western ocean boundaries since, although the wind systems at the western margins of the subtropical high pressure cells have a poleward component, parallel with the coast, the western boundary currents are too strong and too stable for upwelling to develop. Needless to say, this idealized pattern may be considerably distorted in practice; the situation in the Pacific sector approximates fairly closely to it, and the principal elements of the pattern can be recognized in the Atlantic sector, but the Indian Ocean is certainly anomalous, with upwelling being unexpectedly weak or absent off Western Australia and with strong upwelling off the Somali coast—in a western boundary current—during the south-west monsoon. The latter is not due to local wind stress, but appears instead to be due to the tilting of isotherms as the density field adjusts to the acceleration and offshore swing of the Somali Current. Thus, as in the case of the western boundary current itself, this " geostrophic upwelling " is associated with wind stress over the whole ocean as well as locally.

L. *Seasonal variations of upwelling*

The great productivity of upwelling areas in the tropics and subtropics stems not so much from any " fertilization " of the surface layers by nutrient-rich subsurface water (as widely suggested), but rather from the fact that the upwelling system is capable of supporting a temperate type of production cycle within the subtropical zone (Cushing, 1969a, 1971a). The seasonal shifts in the intensity and position of upwelling provide the opportunity to spread the areas of production in the eastern boundary currents and elsewhere. Figure 12, from Bakun (1973) provides one illustration of seasonal variation in upwelling for the west coast of North America, and details of seasonal variations in all other upwelling centres are given in Cushing (1969a, 1971a). In general, there is a poleward movement of the great coastal upwelling belts from spring to autumn as the subtropical high pressure cells intensify, while in the rather special case of the Indian Ocean, the

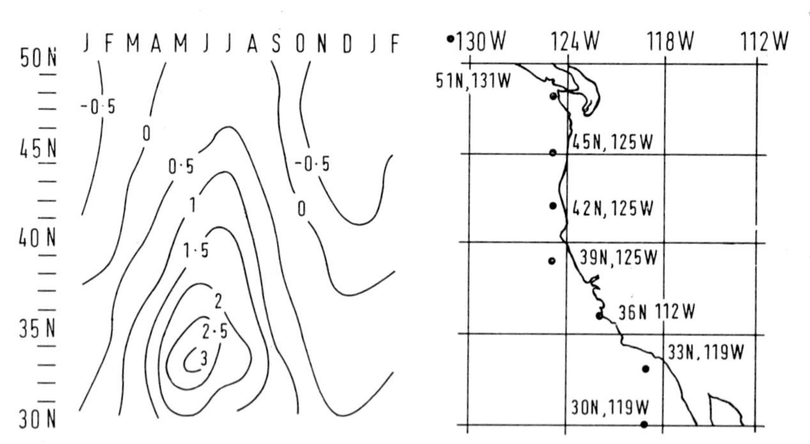

FIG. 12. Long-term mean annual cycle of offshore Ekman transport formed from upwelling indices computed from monthly mean atmospheric pressure data at points indicated by dots on the coastline drawing. Units are cubic metres per second per metre of coastline. (From Bakun, 1973.)

upwellings off SW Arabia, in the Somali Current and off the Malabar coast are tied to the timing of the SW monsoon (April to September). The great Equatorial upwellings appear to reach their greatest development during the northern winter and spring.

III. Climatic Variation

The realization during the first four decades of this century that global climates were subject to change has been followed by a large and increasing scientific activity aimed at describing the magnitude and extent of these changes. It is now apparent that the atmosphere is in a state of considerable unrest, exhibiting significant variations and shifts over a wide range of time and space scales. Recently, as our observational network has achieved a wider global coverage and as our existing meteorological records have lengthened, the tendencies, global interconnections and (sometimes) the causes of prominent climatic events have emerged from this large accumulation of local climatic records. At the same time, fundamental economic and social effects have been identified so that, increasingly, there is a demand for advancing our knowledge of the causes of climatic variation to the point of prediction. However, the underlying secular patterns of global climatic change have a tendency to affect local climate in a radical and unpredictable way, often altering the local climate beyond the range of past experience, and frequently we are confronted by global climatic events which are acyclic in character and may last so long that no analogues are available within our climatic records. In these cases we rely on monitoring climatic events as they occur, and determining the likely cause and effect relationships which will come into play following the establishment of particular large-scale climatic anomalies.

This chapter seeks to describe the variations of a limited range of climatic elements during the present century and these will later be related to changes in marine populations during the same period and over the same time scales.

A. *Ultra long-term changes in the general atmospheric circulation*

The first four decades of this century were characterized by a sustained and remarkably uniform tendency towards high latitude warming in the northern hemisphere with peak warmth developing in the European sector of the Arctic. In the vicinity of Spitzbergen mean annual air temperatures increased by $>7°F$ between the periods 1900–19 and 1920–39, though during this period increases of 2–5°F were more generally observed in the European Arctic (Mitchell, 1963). These changes found expression in a widespread reduction of sea ice in this sector, to a general minimum in 1938, and in a poleward retreat of the terrestrial permafrost limit. In parts of Siberia the permafrost boundary retreated northwards by some 40 km between the 1830s and the 1930s.

This protracted high latitude warming was linked by Defant (1924), Wagner (1940) and Scherhag (1950) with an increase in the strength of the general circulation over the north Atlantic sector, and the surface and upper-air zonal circulation indices of Lamb (1965, Fig. 13),

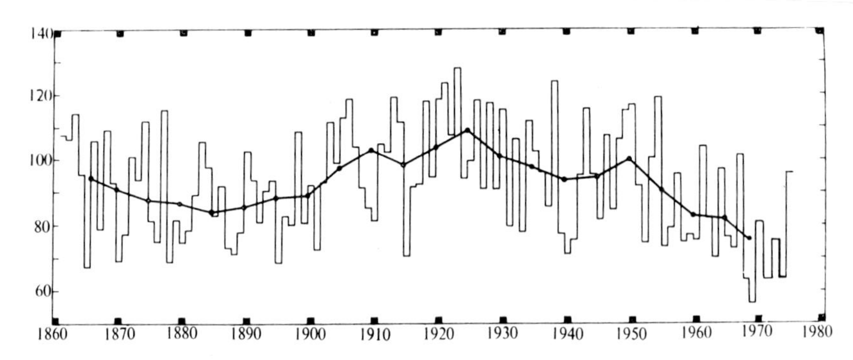

Fig. 13. Number of days per year of general westerly circulation type over the British Isles (1861–1974). 10-year means plotted at 5-year intervals are shown by the bold line. (From Lamb, 1972.)

Mironovitch (1960), Girs (1963) and Trenkle (1956) were later able to confirm the occurrence of this steady strengthening of the westerlies from the 1880s (at least) until the mid-1920s. From early analyses of cyclone and anticyclone tracks it also became evident that this increased zonality was accompanied by a contraction of the circumpolar westerly vortex, accompanied by a lengthening of wavelengths in the upper westerlies and by a northward shift of the zonal wind at surface level and aloft. By the early 1930's the Iceland Low (and indeed the whole of the subpolar minimum pressure belt) lay farther north on average than at any other time in this century, and the more frequent invasion of the Arctic by depressions is held at least partly responsible for the retreat of Arctic sea ice referred to above (Dickson and Lamb, 1972).

Further studies showed that this increase in circulation strength from the nineteenth century to the mid-1920s and its subsequent weakening were not merely "local" North Atlantic events, but involved the major momentum-carrying winds of both hemispheres. As Lamb (1964a, p. 497) notes: " We find that the average circulation (including all the main windstreams of the world—e.g. North Atlantic westerlies, trade winds, southern hemisphere westerlies and the monsoon currents of south and east Asia) was intensifying, attaining a maximum strength about 1920 $\pm$ 15 years and since shows signs of decreasing. Along with this the strength and breadth of the subtropical anticyclone belt waxed and has begun to wane. " In keeping with this finding we are now aware from Mitchell's studies that the warming of the European Arctic was merely a particularly striking component of a warming tendency which was global in extent (Fig. 14).

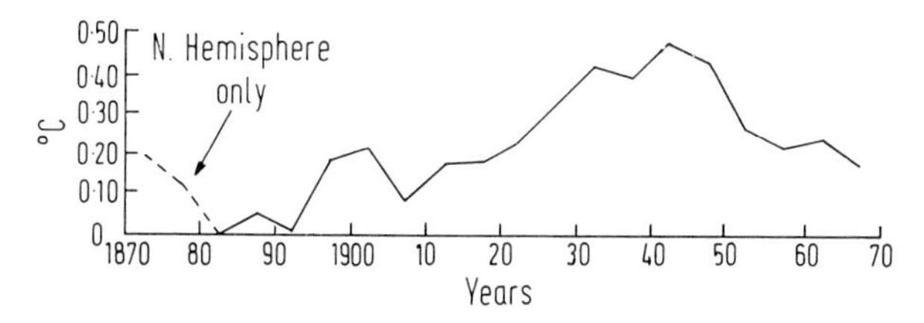

FIG. 14. The variation of global mean surface air temperature, expressed as deviation from the mean, in successive 5-year averages from 1880–84 to 1965–69, after computations by J. M. Mitchell updated at the National Centre for Atmospheric Research (NCAR), Boulder, Colorado.

As regards the cause, Lamb (1964b, p. 8) notes that " it is hard to avoid the suggestion that the world-wide access of energy to the general circulation implied an increase in the supply of solar energy effectively available ", and there is some evidence that such a change did take place, partly through an increase in the strength of the direct solar beam, and partly through a gradual reduction in the volcanic dust load of the upper atmosphere which took place from the 1820s onwards (Dickson and Lamb, 1972).

Since the 1940s the trends described above have been reversed with remarkable suddenness, again on a global scale. As already seen, the principal windstreams of the world reached peak strength as early as the 1920s and the subsequent rapid reduction in atmospheric vigour has been accompanied by four main events. First there appears to be some evidence for a slight equatorward shift of the axes of winds and pressure belts, resulting in important latitudinal shifts of the zonal rain belts within the tropics (Winstanley, 1973a, b; Lamb, in press; Wahl and Bryson, 1975). Second as the main westerly windbelts have weakened in strength, their path around the globe has become less direct, and the upper westerlies have shown an increasing tendency to meander about the west-to-east track which they formerly occupied. Wavelengths in the upper westerlies have shortened, and the north-south amplitude of the waves has increased. Put differently, the recent climatic history of middle and high latitudes has been determined less by any latitudinal shift of the westerly axis itself than by changes in the amplitude and position of waves about this axis. For example, Lamb (1973), points out that before 1960 the circulation over the European Arctic and Subarctic was controlled by two persistent cold troughs in high latitudes, generally over north-east Canada and north-east Siberia. Since then the Canadian trough has tended to regress westward to the position which it occupied in the middle decades of the nineteenth century (Wahl and Lawson, 1970; Eichenlaub, 1971) and an extra trough has tended to develop in the Franz Josef Land-Novaya Zemlya sector of the European Arctic (Lamb, 1973). This single rearrangement of waves has had important climatic repercussions in the Atlantic sector as will be demonstrated below (p. 34).

The third major event to accompany the recent decline in atmospheric vigour has been the cooling of the global atmosphere. Integrating over a world-wide network of stations Mitchell shows a cooling of some 0·3°C in pentade mean surface temperatures since the warmest years of this century (Fig. 14), but this decline in temperature has shown temporal and geographical variations and is not confined merely to the lower layers of the atmosphere. Dronia (personal communication) shows that the mean temperature of the lowest 5·5 km of the atmosphere has declined by 0·25°C at 65°N, and by 0·85°C in the polar regions of the northern hemisphere, over the last 25 years. Over 5 years of this period, between 1958 and 1963, Starr and Oort (1973) find that the mean temperature of the bulk (92%) of atmospheric mass in the northern hemisphere declined by around 0·60°C. In general the cooling at surface level was most pronounced at high latitudes, with maximum cooling taking place in the Arctic, where temperatures had earlier shown their greatest increase. The resulting changes in the extent of snow and ice cover throughout the Arctic have been dramatic (Dickson and Lamb, 1972; Kukla and Kukla, 1974; Lamb, in press) and since these changes represent a southward extension of a highly reflective surface, they may in themselves have helped to intensify the cooling of the Arctic in the most recent years.

A fourth and final major event which has accompanied the long-term turnabout in atmospheric vigour during the present century has been an apparent reduction in available solar energy. The principal evidence for such a change is derived from actinometric measurements of the strength of the direct solar beam on clear days from the middle latitudes of the northern hemisphere (30–60°N, observations from America, Europe, North Africa, Asia and Japan). An overall decline in incoming radiation since the 1940s amounting to approximately 4% has been reported (Budyko, 1968; Pivovarova, 1968) and this has been suggested as a contributory cause of the observed weakening of the general circulation.

The ultra long-term global changes just described must appear to be very far removed from the short-term local changes in environment to which individual marine populations must respond. Nevertheless, they form an essential and relevant background to the events (at progressively smaller time and space scales) which will be described in the remainder of this section.

B. *Recent inter-decadal changes of climate*

This subsection will deal with events at slightly shorter time scales than those just described; to do so we will restrict our attention to a single sector of the northern hemisphere—the European Arctic and Subarctic—and to the short period from the early 1950s to the present, in order to identify in greater detail the types of environmental change which have occurred as the mid-latitude westerlies have continued to weaken from a strongly zonal to a relatively meridional circulation type. While the events to be described are specific to the European sector, equally radical climatic shifts have been experienced in other mid-latitude sectors during this change of the westerly circulation towards increasing meridionality.

Over the European Arctic and Subarctic seas, the early years of this period were characterized by northerly meridionality and climatic deterioration, with direct northerly outbreaks of increasing frequency sweeping the Norwegian–Greenland Sea and adding almost every year to the severity of climate in areas as far south as the British Isles. This change was associated with the build-up of an intense and persistent pressure anomaly ridge over Greenland in the early 1950s, and its subsequent maintenance (on average) through the late 1950s and 1960s. Compared with the climatic "normal period" 1900–39, this cell represented an increase of over 3 mb in mean annual pressure at Greenland over the period 1956–65, but was very much more prominent during the winter half year (November to March). Figure 15 shows that in the winter quarter, December to February, the increase in mean sea-level pressure at Greenland amounted to over 7 mb between 1900–39 and 1956–65, and this pressure rise continued in the late sixties.

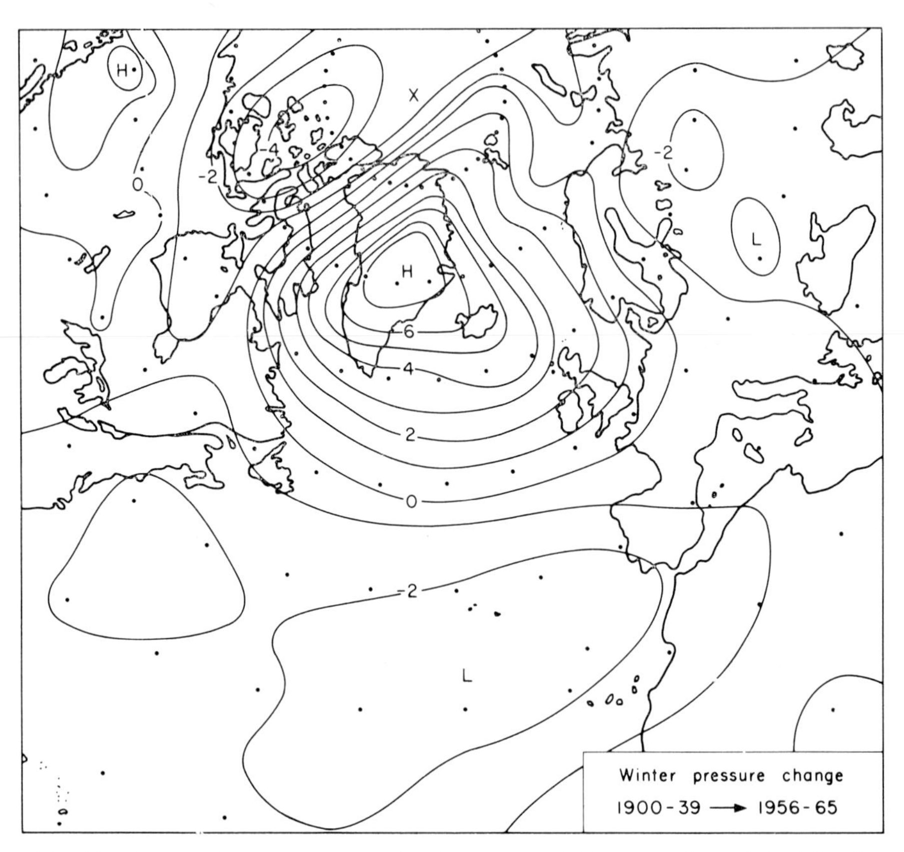

Fig. 15. Change of mean winter sea-level pressure (mb) between the periods 1900–39 and 1956–65.

Figure 16 shows that a further increase of over 5 mb occurred at Greenland between the winters of 1956–65 and 1966–70. Coupled with a slight decrease of pressure over the eastern Norwegian and Barents Seas this change has resulted in " a remarkable difference of pressure between Greenland and the eastern Norwegian Sea; this has increased since 1950 in every month of the year, and the annual mean value of the difference between 40° and 0°W at 70°N for the 1950s decade was 4·2 mb compared with 1·2 mb for 1900–39 " (Lamb, 1965, p. 10).

The resulting increase in northerlies and its physical repercussions have been reviewed by Dickson and Lamb (1972) and by Rodewald (1972). The great boosting of northerlies during the winter months resulted in a steep decline in mean winter air temperature throughout this sector, especially in those areas lying to the north and west of the

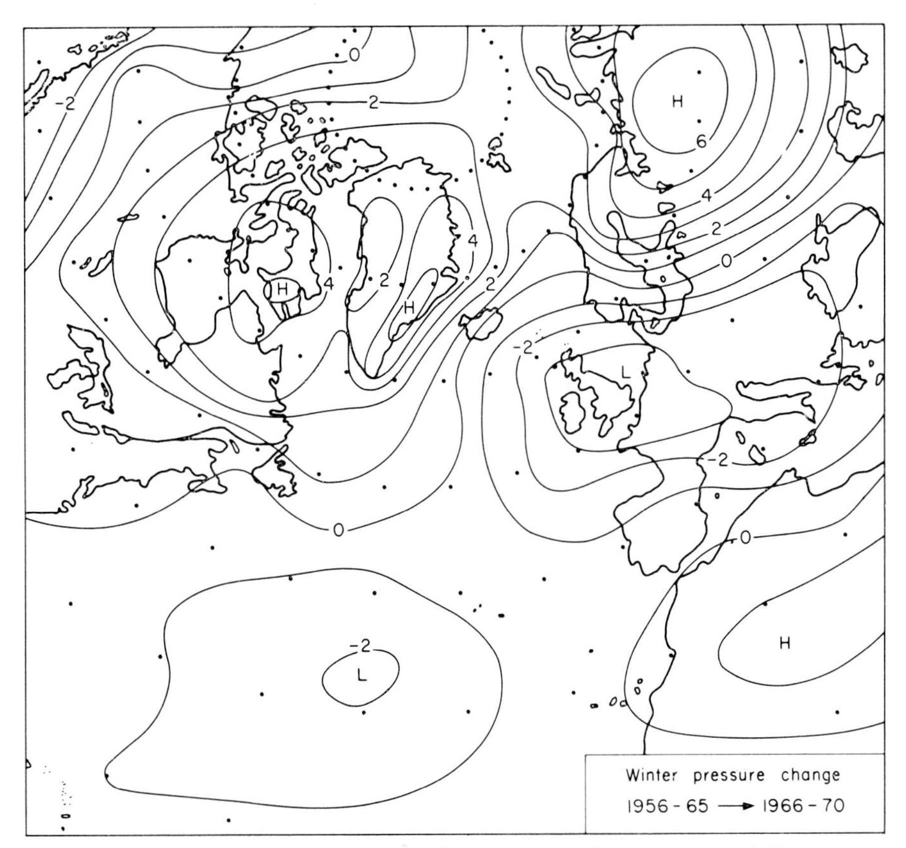

Fig. 16. Change of mean winter sea-level pressure (mb) between the periods 1956–65 and 1966–70.

atmospheric and oceanic polar fronts. Between the winters of 1949/50 and 1967/68 mean winter temperatures declined by 5·54°C at Isfjord (Spitsbergen), 5·60°C at Bear Island and 4·06°C at Jan Mayen. At Franz Josef Land the cooling was even more abrupt; decade mean air temperatures from the 1920s to the 1950s had varied only slightly, averaging between −10·8°C and −10·0°C, but in the 1960s the *decade mean* temperature dropped sharply to −14·5°C (Rodewald, 1972). Moving south from these high-Arctic stations towards the British Isles the winter cooling becomes rapidly less marked, but there is evidence that since the early sixties the resulting decline in sea temperature on the European shelf in the critical spring months March to May has been sufficient to affect (favourably) the survival of North Sea cod, which in these latitudes are approaching the equatorward limits of the species range (Dickson *et al.*, 1973).

Apart from this effect on air temperatures, the strengthening of the northerly airstream in this sector has been responsible for a progressive southward extension of sea ice across the Greenland Sea to a maximum extent in the spring of 1968 (Fig. 17). Icelandic observations (e.g. Malmberg, 1969) suggest that this change took place in two stages. First, as the northerlies increased in strength, the hydrographic character of the cold East Greenland and East Icelandic Currents was altered, so that the currents became cooler and less saline as the proportion of polar water increased. The East Icelandic Current changed from being an ice-free arctic current in 1948–63 to a polar current in 1964–71, transporting drift ice and preserving it. Second, the fresh polar water component has been so great in the most extreme years (1965, 1967, 1968) that the surface layers will not reach a sufficiently high density, even at the freezing point, to start deep convective mixing with the warmer (but denser) layers below. For this reason Malmberg concludes that active ice formation has contributed to the extension of sea ice north of Iceland in these years. The biological implications of this south-eastward extension of the oceanic polar front are discussed by Dickson and Lamb (1972). Of these, the chief event has been the change in the migration route of Atlanto-Scandian herring, identified by Jakobsson (1969).

A third event of importance in this sector has been an apparent change in the timing and duration of oceanic production in recent years. In summary it appears that between 1948 and 1965 a change in cloud cover associated with the progressive increase in northerlies may have brought about a reduction in solar radiation of over 30% in waters to the north of the British Isles, not in every month of the year certainly, but in the months of April and May which are critical to the initiation and development of the spring phytoplankton bloom (Dickson and Lamb, 1972). Our evidence for this change is rather poorly based, relying on solar radiation observation at Lerwick (Shetlands) in the absence of any more general coverage of radiation measurements, but it appears to be confirmed by the independent finding of Robinson (1969) that the peak of the spring phytoplankton outburst at the southern approaches to the Norwegian Sea has been progressively delayed by a total of almost one month over the period 1948–68 (see also p. 100 and Fig. 36). The importance of a delay in oceanic production arises from the fact that certain economically important fish stocks appear to spawn at relatively fixed times of year and are unable to adapt their spawning time to rapid changes in their environment (Cushing, 1969b); this thesis is discussed more fully below.

The reduction in solar radiation just described should not be

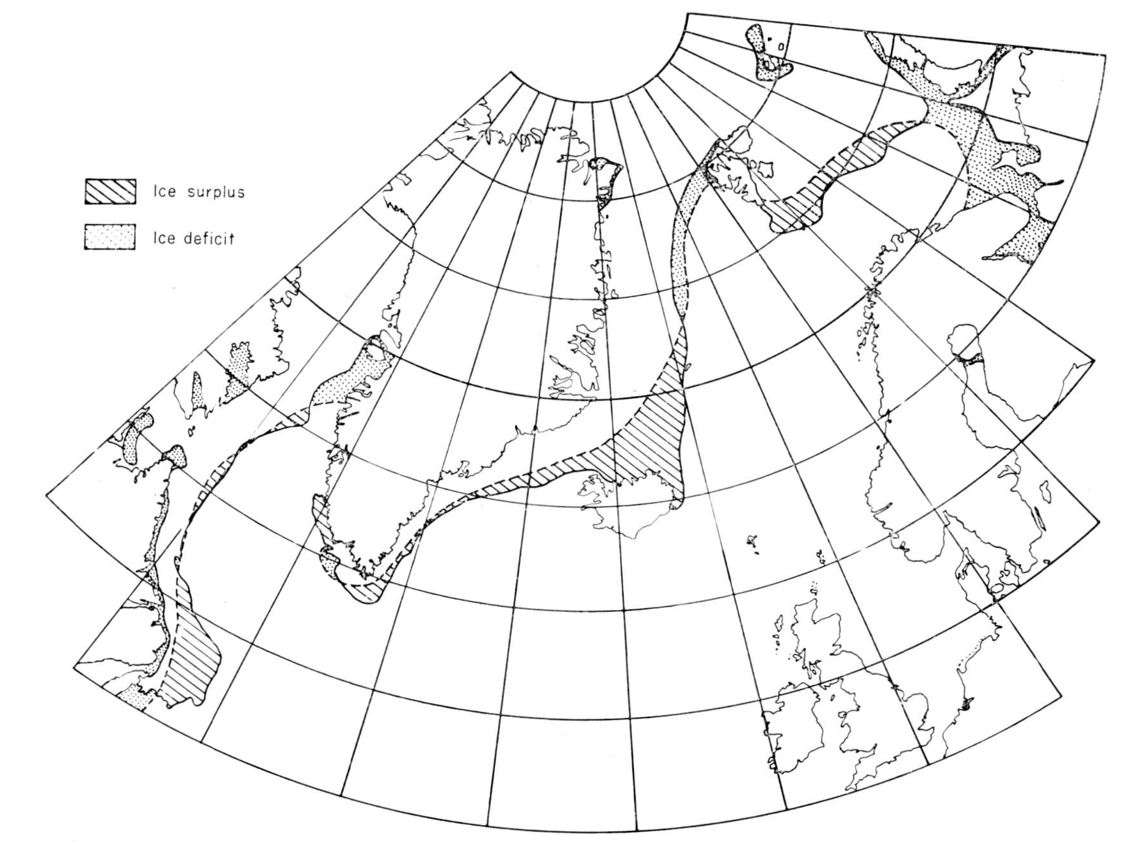

Fig. 17. Ice extent at 8 May 1968 (solid line) compared with normal (dashed line). The normal is a composite one based on an American average concentration of ice 1911–1950. (From Dickson and Lamb, 1972.)

confused with long-term global changes in incoming radiation. As mentioned earlier (p. 33) such a change has apparently taken place since the 1940s, and has been associated with radical changes in the global atmospheric circulation, but this global change of around 4% in incoming radiation is insignificant compared with the *local* effect of a change in cloud cover in certain months resulting from a *local* change in the atmospheric circulation. Equally the two week reduction in the length of the growing season in England, reported in the 1950s is attributable more to the prevailing decline in air temperature than to any local change in solar radiation (see Lamb, 1969). It is relevant to add that, during this period of climatic deterioration over the Norwegian–Greenland Sea, the waters off south and west Greenland were locally experiencing the opposite tendency. As shown in Fig. 15

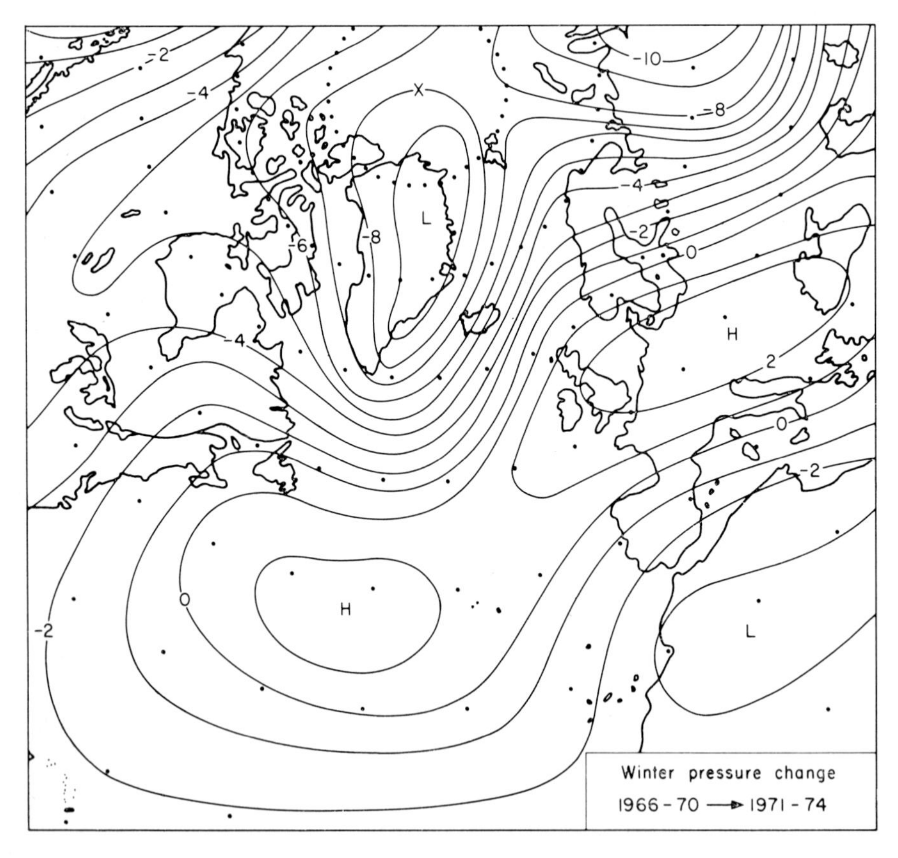

FIG. 18. Change of mean winter sea-level pressure (mb) between the periods 1966–70 and 1971–74.

the dominant ridge at Greenland was responsible for maintaining an anomaly wind running in a cyclonic sense over the eastern Atlantic to the Irminger Sea (off SE Greenland). These conditions appear to have been responsible for boosting the underlying warm branch of the Atlantic current system (the Irminger Current) resulting in a wave of deep warming along the West Greenland banks (Hermann, 1967; Blindheim, 1967; Dickson and Lamb, 1972) and in improved environmental conditions for *adult* cod which here lie close to the poleward limits of the species range (see p. 58). Conditions remained cold in the surface layer, however, resulting in poor survival of eggs and larvae.

In the winter of 1970–71, these striking, quasi-linear climatic trends in the Norwegian, Greenland and Irminger Seas came to an abrupt halt. The high pressure anomaly cell over Greenland which had

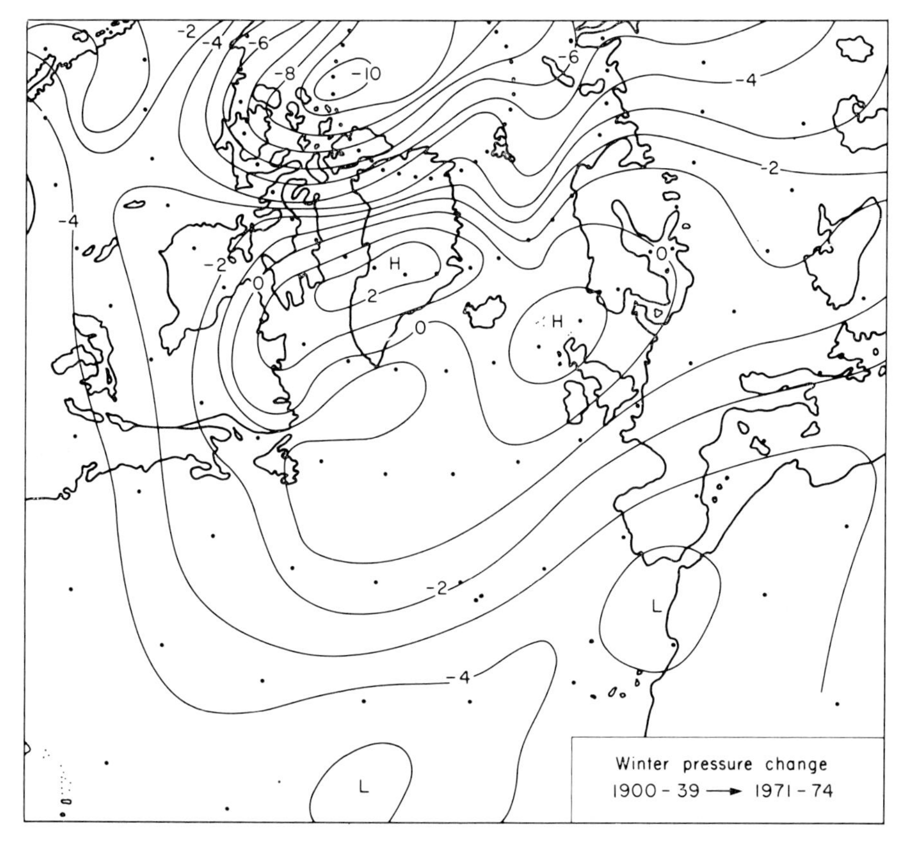

FIG. 19. Change of mean winter sea-level pressure (mb) between the periods 1900–39 and 1971–74.

for so long dominated the patterns of climatic behaviour in this sector showed an almost total collapse, with mean winter sea-level pressure falling by over 9·5 mb at Greenland between the periods 1966–70 and 1971–74 (Fig. 18). The ridge has not yet been totally eradicated (Fig. 19), since in the earlier of these two periods the mean winter sea-level pressure had attained a peak of 12 mb above the 1900–39 normal, yet the northerlies of the Norwegian–Greenland Sea have been weakened drastically, a rapid amelioration of the marine climate at Iceland has taken place and there are signs that earlier trends in marine production cycles have been reversed (Dickson *et al.*, in press).

C. *Inter-annual variations*

Examining climatic behaviour at an even larger "magnification" we encounter striking year-to-year variations in ocean and atmosphere which are superimposed on the longer-term trends just described. These short-term changes contribute to, but are not recognizable in, the decade averages of climatic parameters which have been discussed hitherto. Nevertheless, when they are brought to light by a more appropriate averaging of the data they may be shown to be of great amplitude and geographical extent.

As regards the Atlantic sector, the dominant inter-annual hydrographic variation has already been described (Section II , p. 22) as a response to a recurrent ocean/atmosphere feedback situation. Put differently, one particular pattern of the Atlantic windfield is peculiarly able to set up the specific oceanic conditions suitable for feedback, and through its own persistence this circulation pattern is capable of inducing an unusually sustained trend of change in the ocean which stands out as the dominant variation in the hydrographic record.

The causative circulation anomaly is illustrated in Fig. 9*a*, *b* and *c* and it is clear from this that we can expect the resultant hydrographic variations to be coherent in trend over the scale of a Rossby wave. Figure 20 confirms one aspect of this coherence by comparing the temperature trends for the Kola Section of the Barents Sea (70°30′N 33°00′E–72°30′N 33°00′E) with the contemporaneous trend of temperature and salinity anomaly at a group of fields in the German Bight of the North Sea. It is evident that there is a considerable similarity of trend between Barents Sea temperatures and German Bight salinities, and although this agreement is shown here for only a short period of years for ease of presentation, a similar conformity between these two parameters could have been shown during any other part of the inter-war and post-war periods. The link with North Sea temperatures is much less clear and although some common tendency can perhaps be detected in the temperature records at Kola and in the German Bight, it remains true that these inter-annual fluctuations in ocean temperature become progressively less clear-cut with distance from the high Arctic, while time series of surface salinity throughout the European shelf (North Sea, Irish Sea, English Channel) and as far north as Ocean Weather Station " Metro " (66°N 2°E) show these variations rather clearly. (No suitable salinity time series are available from the Arctic.)

We also have some evidence to suggest that ocean temperature trends of the western and eastern Atlantic are related, again confirming a large-scale coherence of hydrographic change. However, the relationship is in an inverse sense as might be expected from the atmospheric circulation pattern responsible for these changes (see Fig. 9). Strengthened northerlies running along the western flank of the meridional trough in the west Atlantic create cooling along the North American coast (by advection and by an increased exchange of sensible and latent heat with the ocean), while the southerly anomaly wind in the eastern Atlantic creates conditions of anomalous warmth along the European seaboard and in the Barents Sea. Figure 21 illustrates the resultant inverse trend of temperature in the western and eastern Atlantic by comparing the mean annual temperature of the Kola Section (integrated over the 0–200 m depth layer) with similar data from the cold component of the Labrador current (data from Elizarov, 1962; Izhevskii, 1964; see also Burmakin, 1971).

Figures 10 and 20 show that these dominant inter-annual variations of temperature and salinity are regular in periodicity, but are not truly cyclic. Generally, our records are too " gappy " to be suitable for spectral analysis but in the case of an unbroken 70-year salinity record from Anholt Lightvessel in the Kattegat this type of analysis has been carried out (Dickson, 1973). It showed a dominant 5-year periodicity in the record over the period 1947–70, but confirmed that the dominant periodicity had been slowly changing over the present century. (A similar result may be inferred for the European shelf seas as a whole since they can be shown to share a similar trend of salinity variation to that at Anholt.) This result may be confirmed by autocorrelating the inter-war and post-war temperature records from the Kola Meridian Section of the Barents Sea. As shown in Fig. 22 the mean temperature in the 0–200 m layer showed a peak autocorrelation after 51 months lag during the period 1921–40 while the corresponding peak is found after 64 months lag in the period 1951–71. (The inter-war and post-war data series were made up of 240 and 252 consecutive monthly means

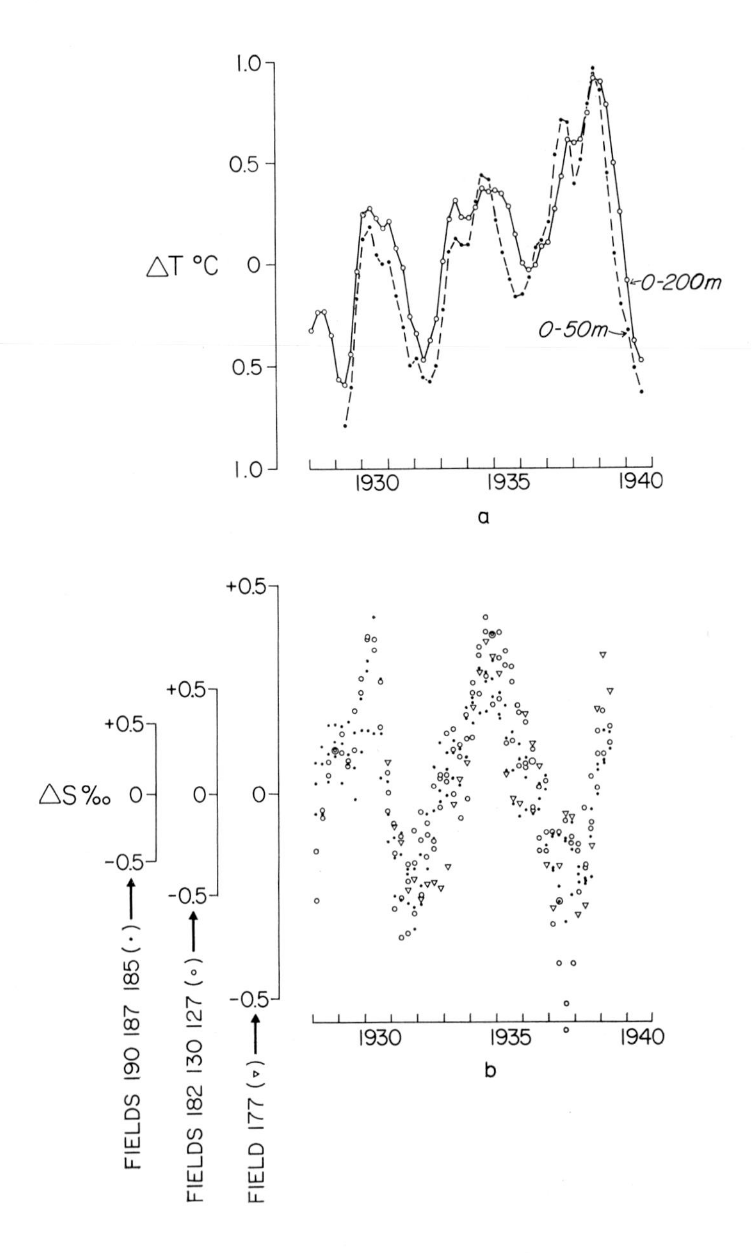

1.0
0.5
ΔT °C
0
0.5
1.0
0-200m
0-50m
1930
1935
1940
a
+0.5
+0.5
+0.5
ΔS ‰
0
0
0
-0.5
-0.5
-0.5
FIELDS 190 187 185 (•)
FIELDS 182 130 127 (○)
FIELD 177 (▽)
1930
1935
1940
b

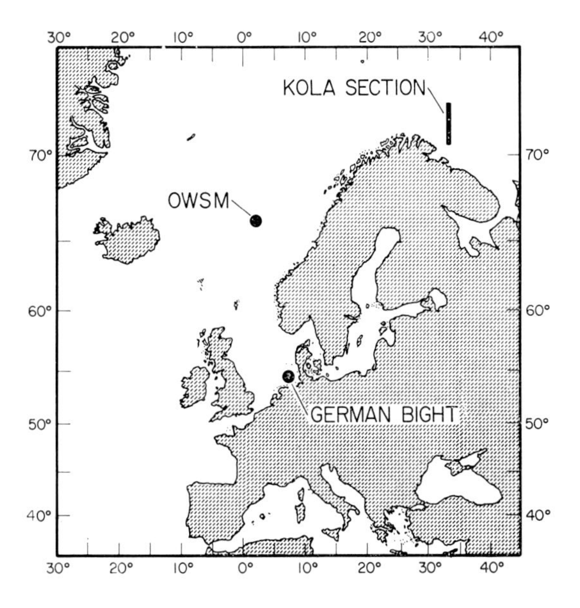

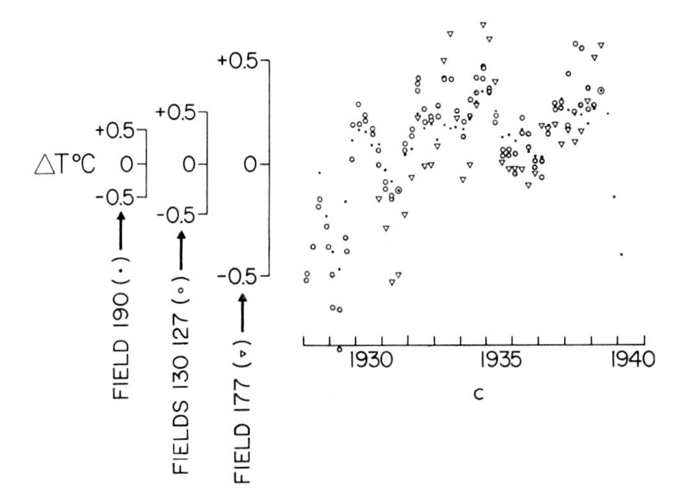

FIG. 20. Comparison of hydrographic trends in the Barents Sea and North Sea during a part of the inter-war period. (*a*) Running three-quarter year means of temperature integrated over the 0–5 m and 0–200 m depth layers on the Kola Section (70°30′N, 33°00′E – 72°30′N, 33°00′E. (*b*) and (*c*) Running three-quarter year means of salinity and temperature anomaly for seven fields in the German Bight.

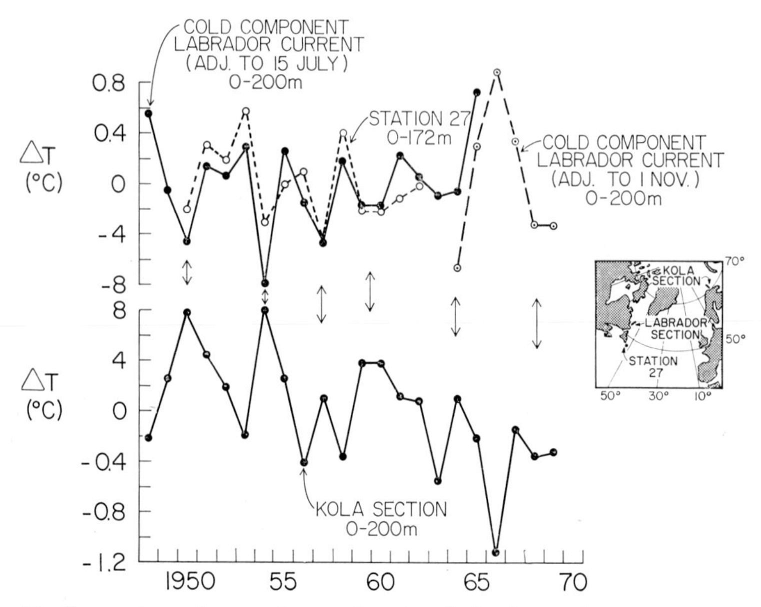

FIG. 21. Comparison of ocean temperature trends in the north-west and north-east Atlantic. (Adapted from Burmakin, 1971.)

of temperature anomaly, respectively.) Again, this result can be regarded as applying rather generally to the north-east Atlantic since Midttun (1969) shows that inter-annual temperature trends along the Norwegian coast, are essentially similar to those of the Kola Section.

Though this inter-annual hydrographic variation can be shown to dominate the hydrographic record over wide areas of the northern North Atlantic, its amplitude is difficult to quantify with any great precision. Our hydrographic time series are few in number and variable in quality, their duration is frequently inadequate to provide an accurate determination of the normal seasonal cycle of events and, of course, the inter-annual fluctuations about this seasonal trend may differ considerably in amplitude. However, some approximate indication of the amplitude of these inter-annual variations can be derived; in the following table the mean range of inter-annual temperature variation (seasonal trend removed) is expressed as a percentage of the seasonal range of temperature at a variety of North Atlantic locations. In the case of the Kola Section the data are reported as depth averages over the 0–50 m and 0–200 m layers; each of these depth layers shows approximately the same amplitude of inter-annual temperature change (see, for example, Fig. 20) suggesting that these temperature

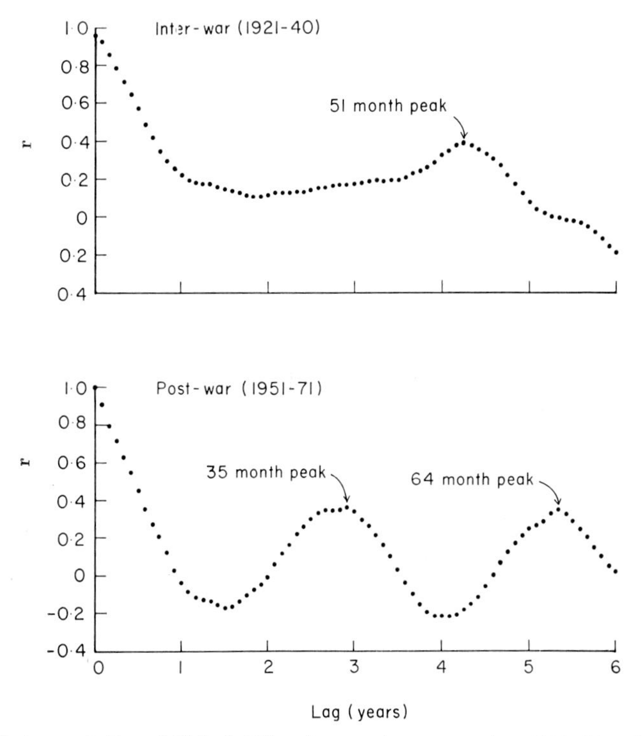

FIG. 22. Autocorrelation of Kola 0–200 m temperature anomaly. 1921–40 and 1951–71.

variations have a considerable depth of influence, but since the *seasonal* changes of temperature are greatest in the near-surface layers, the ratio of inter-annual : annual temperature range is greater in the case of the 0–200 m depth layer. In all cases, since the seasonal change of temperature is a smoothed long-term average, the inter-annual (anomalous) temperature variations with which they are compared have also been smoothed (i.e. as running three-quarter year or annual means).

Each of these estimates conceals a considerable range of variation about the mean, but even allowing for a considerable error about these values it is plain that these 4–6 year variations represent a very much greater rate of temperature change than any longer—or shorter—period variation (with the exception of the seasonal cycle). It is even more difficult to provide similar estimates of the amplitude of salinity variation at this time scale. Adequate time series of salinity data are even more restricted than those of temperature, and the seasonal change of salinity tends to be relatively ill defined. It is fairly clear however that the inter-annual variation of salinity may be considerably

TABLE I. MEAN RANGE OF INTER-ANNUAL TEMPERATURE VARIATION (SEASONAL TREND REMOVED) EXPRESSED AS A PERCENTAGE OF MEAN SEASONAL RANGE OF TEMPERATURE AT 4 ATLANTIC STATIONS

Station	*Depth or depth range considered* 0 m	0–50 m	0–200 m
Kola Section (Inter-war)	—	22%	45%
Kola Section (Post-war)	—	20%	52%
O.W.S. M	17%	—	—
Labrador Section 8-A	—	—	38–50%
Station 27 (Cape Spear)	—	—	20% (0–172 m layer)

greater than the seasonal change. In the German Bight of the North Sea the mean 1905–54 seasonal salinity range amounts to 0·46‰ and 0·91‰ at fields 130 and 177 respectively. As shown in Fig. 10 the mean range of the inter-annual variation (seasonal trend removed) amounted to 1·60‰ and 0·78‰ respectively at the two fields when the data are smoothed in the form of running three-quarterly means. To provide an example from the open ocean rather than from the shelf, the inter-annual range of salinity anomaly at Ocean Weather Station " METRO " has varied between 68% and 153% of the seasonal range in the period 1949–68 (again using running three-quarterly means of salinity anomaly).

D. *Within-year variations* (*non-seasonal*)

While the inter-annual hydrographic variations just described are of considerable geographical extent, their effect on fisheries is not uniform but will vary in space and time; geographically the more important effects will occur where the range limits of some species are locally altered, or temporarily where some environmental threshold to which the fish respond is exceeded. In this subsection we will be chiefly concerned with the latter effect, a short-lived excursion through some significant environmental threshold during the working-out of some longer-term fluctuation.

The hydrographic situation of the Baltic provides a case in point. In the Baltic a light fresh surface layer of water deriving from precipitation and river discharge overlies a dense saline water mass which derives from the North Sea and which enters the Baltic as a deep undercurrent through the Danish Belts and Sound. The intense density gradient between the surface and deep water masses generally prohibits the transfer of oxygen to the bottom layers, and aeration of the chain of deep basins along the Baltic floor relies on renewal of the bottom water by inflow from the North Sea. This renewal is not a constant process but takes the form of sudden irruptions during which an intense bottom current penetrates deeply into the Baltic, filling each of the deep basins in turn and spilling from one basin to the next across the intervening sills (Fonselius, 1962, 1967, 1969). These surges are separated by years of stagnation, during which the oxygen is progressively mined from the bottom water by marine organisms and by aerobic decomposition. If the stagnation is sufficiently prolonged, the oxygen disappears completely and H_2S may be detected in the bottom layers; in this situation the spawning areas of economically important fish stocks are drastically reduced in extent and the benthos may totally disappear over wide areas of the innermost Baltic deeps (Lablaika, 1961; Shurin, 1961, 1962; see also p. 73 and Fig. 28).

As regards the cause of Baltic inflows, it appears that although local meteorological conditions over the Baltic are capable of boosting or retarding the speed and/or volume of inflow (so that the inflowing stream may take anything from 6 months to 2 years to pass through the chain of basins), these local effects seem incapable of producing a major influx by themselves. Instead, as Wyrtki (1954, p. 24) has observed in connection with the exceptionally powerful inflow of 1951, " The condition for the large scale of the salt surge was the high salinity of water in the deep part of the Kattegat and its further penetration into the Belts. This latter effect surely does not depend so much on meteorological happenings as on a more or less strong influx of oceanic water into the northern North Sea and Skagerrak ". Indeed, it can be shown that in the post-war period a major Baltic inflow has accompanied each inter-annual salinity maximum in the deep water of the Skagerrak (Dickson, 1971, 1973).

This is important from the point of view of the previous discussion since the periodic recurrence of high salinity conditions in the bottom layers of the Skagerrak (and in the European shelf seas as a whole) is linked to the broad-scale 4–6 year hydrographic variation described in the preceding subsection. In short, a recurrent and persistent pan-Atlantic circulation anomaly is responsible for a periodic salinification of the north-east Atlantic and European shelf; when the density of the deeper layers of the Skagerrak and Kattegat becomes greater than that of the resident stagnant water-mass of the Baltic deep (and when local meteorological conditions are favourable) a renewal of the Baltic deep water occurs as a sudden and relatively short-lived event.

Figure 23 illustrates several cycles of this cause-and-effect chain

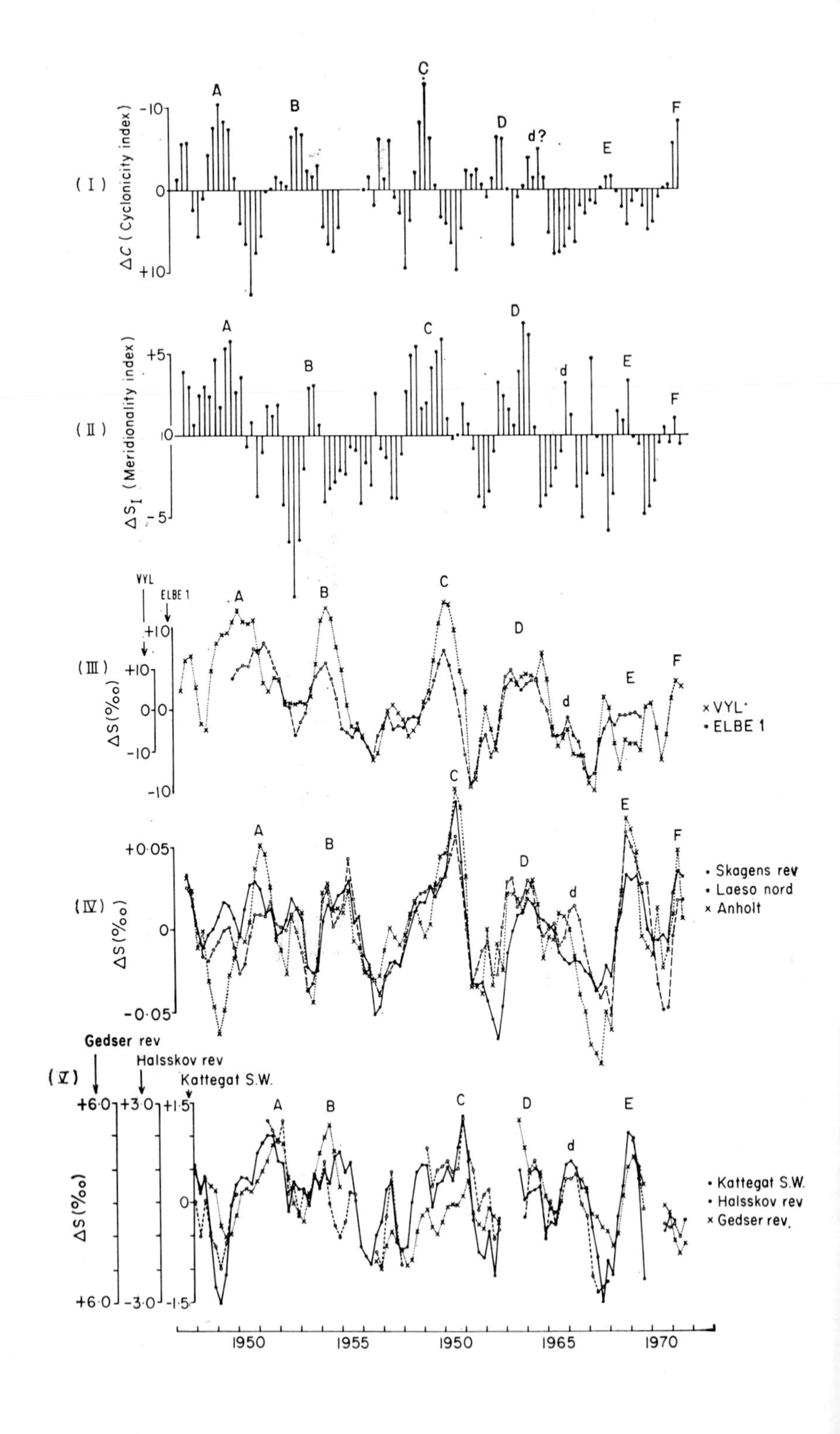
(I)
ΔC (Cyclonicity index)
-10
0
+10
A
B
C
D
d?
E
F
(II)
ΔS_I (Meridionality index)
+5
0
-5
A
B
C
D
d
E
F
(III)
VYL
ELBE 1
+10
+10
0
0
-10
-10
ΔS(‰)
A
B
C
D
d
E
F
× VYL
• ELBE 1
(IV)
ΔS(‰)
+0.05
0
-0.05
A
B
C
D
d
E
F
• Skagens rev
• Laeso nord
× Anholt
Gedser rev
Halsskov rev
Kattegat S.W.
(V)
+6.0
+3.0
+1.5
+6.0
-3.0
-1.5
ΔS(‰)
A
B
C
D
d
E
• Kattegat S.W.
• Halsskov rev
× Gedser rev.
1950
1955
1950
1965
1970

from the post-war period. The atmospheric indices shown in Fig. 23 (i) and (ii) show the periodic re-establishment of abnormal anticyclonicity over north-west Europe, and the simultaneous increase in southerly airflow off the European seaboard (see also Fig. 9), while the lower graphs illustrate the resultant salinification in the surface waters of the eastern North Sea (iii), in the deeper layers of the Skagerrak and Kattegat (iv) and in the Belts at the entrance to the Baltic (v). Major Baltic inflows occurred during each salinity maximum.

E. *Preliminary discussion*

The four preceding subsections have considered variations in the atmospheric and marine climates ranging from changes of over a century in duration to changes occurring within a single year. Their separation is in one sense artificial since in nature these changes are superimposed to form a unified complex of climatic variation. Nevertheless, it is relevant to add that marine populations can be shown to respond separately to climatic variation at each of these time scales. Indeed this variable response to climate can be shown for one species in one ocean area.

Figure 24 shows the change in the mean date of capture of spawning Arcto-Norwegian cod at Lofoten between 1894 and 1974, assumed here to reflect a change in the mean date of spawning over this period. The data are smoothed in the form of running 10-year means and interpolated values were used for the years 1899, 1902 and 1906. It is clear, as shown by Cushing (1969b) that cod do not adapt their spawning time to rapid changes in their environment, yet in the longer term they appear to be responding to some ultra long-term shift in their environment, perhaps to some effect of the change in westerly wind strength, discussed earlier (p. 31). Equally the survival of North Sea cod has been linked to the recent inter-decade cooling trend in the North Sea (see p. 89) at the equatorward limit of the species range, while at the poleward limits of range the survival of Arcto-Norwegian cod appears to be linked to the large-amplitude inter-annual warming

---

Fig. 23. Running three-quarterly means (post-war period) of: (i) Cyclonicity Index anomaly, British Isles (anticyclonic tendency negative, but graph inverted); (ii) Meridional Index anomaly (southerlies positive), British Isles; (iii) Surface salinity anomaly at lightvessels VYL and ELBE I, German Bight (for explanation of scales see text); (iv) Near-bottom salinity anomaly at lightvessels SKAGENS REV, LAESO NORD and ANHOLT, Skagerrak and Kattegat; (v) Near-bottom salinity anomaly at lightvessels KATTEGAT SW, HALSSKOV REV and GEDSER REV, Danish Belts. (From Dickson, 1973.)

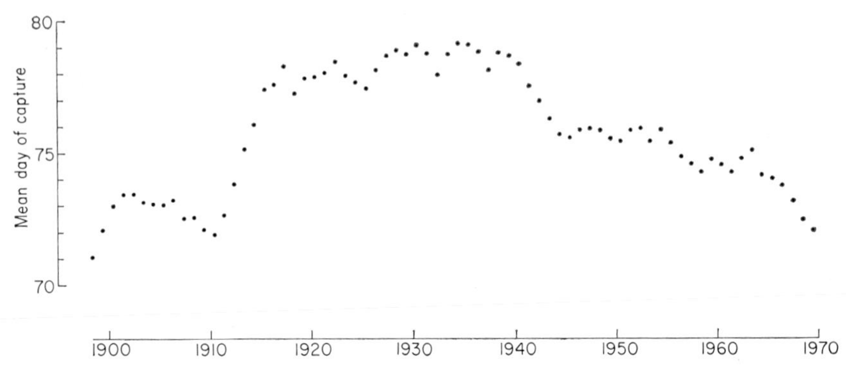

FIG. 24. 10-year running means of the mean date of capture of the Lofoten spawning cod fishery 1894–1974 (1 January = day 1 etc.).

and cooling trends along the coast of Arctic Norway and in the nursery grounds of the Barents Sea (Garrod, 1970). Finally, as regards the Baltic cod, the available spawning area may be reduced almost to zero in the innermost Baltic deeps during the long periods of anoxic stagnation (Lablaika, 1961), but following renewed inflow from the North Sea this situation is very rapidly relieved. The response of marine populations to their changing environment is dealt with more fully in subsequent chapters, but these four examples serve to show the range of response, and hence the range of climatic variation, which must be considered if hydrobiological interactions are to be adequately described.

F. *Climatic variation in mid- to low-latitudes*

Thus far our discussion of climatic variation has centred on events in middle and high latitudes of the Atlantic sector where the response of the ocean to changes in the atmospheric circulation has been relatively clear-cut. This is not necessarily the case in the lower latitudes of the North Atlantic, between 20° and 45°N.

Certainly, large-scale, long-term changes in ocean temperature may be detected in this zone. Lamb (1964a) shows clearly enough from the changes in surface temperature distribution that the Gulf Stream/North Atlantic Drift before 1850 crossed the Atlantic in a lower latitude than since, a displacement parallel with that of the winds. However, according to Worthington (personal communication) the interpretation of shorter-term changes in surface temperature in this zone is more ambiguous, due to the nature of the two water masses which make up two-thirds of the warm water (> 17°C) in the North Atlantic.

These two water masses may be termed " 18° water " and " salinity maximum " water and each represents a volume of approximately 1·8 million km^3. " 18° water " is formed south of the Gulf Stream by outbreaks of polar continental air. These outbreaks form a deep mixed layer in the northern Sargasso Sea whose temperature is always very close to 18°C, according to a great number of sections dating back to the CHALLENGER section of 1873. When very cold conditions prevail the water is *not* cooled to temperatures below 18°C, but instead excessive amounts of water at this temperature are formed and flow southwards towards the Caribbean Sea at a core depth of 300 m. Although there are indications that production of 18° water does vary from year to year in this manner, no quantitative estimates of the amounts formed can be made at present. However, it can be seen from this that surface temperature is of no great value as an index of climate in this region.

" Salinity maximum water " is formed in mid-Atlantic close to 23°N at the northern limit of the tradewind belt. When the trades are strong they extract water vapour from the sea surface and transport it equatorward to the zone where the trades of both hemispheres converge. Water underlying the tradewind belt flows slowly northward, gradually increasing in salinity until it sinks at the salinity maximum zone and returns to low latitudes. It can be shown that this process stops abruptly when the trade winds weaken drastically.

Thus in the case of both these water masses, the water formed at the surface sinks and moves to lower latitudes while surface water is drawn from more southerly latitudes to replace it, so that there is not necessarily a direct link between surface hydrographic conditions and climate.

G. *Aperiodic variations of equatorial upwelling*

The study of equatorial upwelling and its variations in many parts of the world ocean is a very recent event, and though a considerable research effort has been devoted to the Canary Current and the Somali Current since 1968, these records are probably too short to describe the year-to-year variations which must undoubtedly exist. The longer-term changes in equatorial upwelling in the Pacific sector are very much better known, since long time series of surface temperature data exist at a number of equatorial stations, yet as the adequacy of hydrographic and meteorological records has improved, the analysis of these records has identified an increasingly complex situation, apparently linked to a range of events over a wide area of the North and South Pacific, so that a comprehensive understanding of these changes has proved elusive. A brief description of what *is* known will perhaps act as an appropriate antidote to sections which have (inevitably) been concerned with identifying, in an over-simplified way, a limited number of processes which lend a degree of ordered regularity to the complexity of climate!

From time to time during the southern summer, a flow of warm tropical water extends over the cooler waters along the coast of Peru. This " El Niño " condition causes a catastrophic *absence* (though not necessarily a mortality) of anchoveta, with severe economic repercussions. Recent estimates (1966–70) show that around 11 million tons of anchoveta can be removed by birds and man from this coastal strip in the course of a year, equivalent to about 16% of the annual world landings of fish in these years from 0·02% of the world ocean.

From early studies by Bjerknes (1966, 1972) the basic cause of El Niño appears to lie in a weakening of the south-east and north-east trade winds which flank the equatorial zone. As a result, heat loss from sea to atmosphere is lessened over a large part of the equatorial ocean, but more important, the weakening of these easterlies brings a reduction of surface water divergence along the equator so that upwelling ceases. This event, coupled with the annual insolation maximum, leads to a vast accumulation of warm water along the eastern tropical Pacific. By the classic theory the weakening of the trades also brings an end to upwelling at the Peruvian coast and the resulting warmth offshore is apparently heightened by invasions of warm water from the equatorial belt.

Wyrtki (1973, 1974) has extended this chain of events to include changes in the North Equatorial countercurrent which flows eastward between latitudes 4° and 10°N. Using tide gauge records from islands which straddle this flow, he suggests that with the weakening of easterly wind stress from the weakening of the trades, the countercurrent increases in strength to carry an abnormally large volume of warm water from the western to the eastern tropical Pacific. A peak in ocean temperature off Central America follows a peak in countercurrent transport by some 3 months. Wyrtki also notes, however, that these changes in countercurrent transport and equatorial warmth may be associated with some larger-scale anomaly of the atmospheric circulation over a large part of the Pacific, and in a later paper Namias (1973a) supplies evidence to that effect.

In essence Namias shows that the changes described by Wyrtki develop a few months after a change in the zonal windflow in the remote subtropical atmosphere. Computing the strength of the upper westerlies (at 700 mb level) in a Pacific-wide strip between 20°N and 35°N, Namias suggests that each strengthening of these westerlies aloft brings a weakening in the underlying trades with the remaining teleconnections following Wyrtki's scheme. In the short term (days and weeks) the warmer equatorial ocean that results from these changes is apparently able to feed energy back into this subtropical upper westerly airflow via the Hadley circulation (see Bjerknes, 1972) but in the longer term (months and seasons) there is some evidence that the periodic intensification of these upper westerlies is a response to events in the extra-tropical North Pacific, through interaction with slowly evolving Rossby wave patterns in the temperate zone (Namias, 1972).

A full description of this complex interplay of events has not yet been achieved, however, and other important contributory factors continue to be identified. Quinn (1974) has developed an El Niño prediction technique involving the strength of the south-east trades and based on the difference in sea-level atmospheric pressure between Easter Island and Darwin, Australia (27–12°S). A further very recent study by Wyrtki (El Niño Workshop, Guayaquil, December 1974) suggests that *prior* to El Niño the south-east trades in the eastern tropical Pacific do not appear to weaken or disappear but may in fact intensify. He holds this factor responsible for the accumulation of an anomalously large body of water in the western Equatorial Pacific causing a large-scale zonal slope of the sea surface. The subsequent relaxation of these south-east trades from an abnormally strong to an abnormally weak intensity (or even to their normal intensity) causes an increased transport in the eastward-flowing Equatorial counter-currents (North and South) and undercurrent during readjustment of the ocean's density field. By this argument the flooding of the eastern tropical Pacific and Peru coast with an abnormally large body of warm water would not necessarily require a weakening of trade winds below their normal intensity and it is even conceivable that an El Nino-type condition could occur off Peru with a *normal* intensity of coastal upwelling, if the upwelling concerned the warm invading water mass rather than the normal cool subsurface water. This " hot upwelling " would certainly be capable of effecting the major observed changes in oceanic production offshore.

While the full complexity of El Niño has yet to be described it is clear that the space scales involved are enormous, spanning the entire equatorial Pacific and extending at least 35 degrees of latitude from the equator in either hemisphere. Within this vast area the climatic and hydrographic records are modulated with the same dominant inter-annual rhythm, but although this unique " signal " must have a far-reaching influence on world climate, the details of this influence are as yet almost wholly unknown.

IV. The Warming in the Thirties and Early Forties

A. *Introduction*

The warming in the northern North Atlantic ocean during the thirties and early forties is well documented, as indicated in an earlier section. In recent years, the physical changes in the north-east Atlantic have become well established and in the last decade the biological ones have started to emerge. The changes described below in the western English Channel or in the populations off West Greenland are profound ones that involve the survival of populations either in their existing environments or in new ones. The invasion of an area by an animal can represent an enormous change in its numbers and often in the ecosystem as a whole.

The history is interesting from two standpoints. First, the colonization of West Greenland by the cod (and other animals) has been described in considerable detail and the increments in year class strength in successive years must have been considerable. The northward movement of the herring on the Murman coast, or the albacore off Oregon, may indicate similar events. Second, some of the ecosystem structure in the western Channel has been revealed by the events that have occurred there in the last half century, which may imply considerable changes in numbers in many populations.

B. *The rise of the Greenland cod fishery*

Between 1810 and 1850, cod (*Gadus morhua* L.) were numerous at West Greenland, and British vessels caught them as far north as Disko Bay in summertime (Tåning, 1953); they were absent in the year 1820 and between 1845 and 1849. Although there were always some cod in the fjords of southern Greenland, none were found on the banks that were visited from time to time by North American and Faroese cutters (Jensen and Hansen, 1931). Experimental fishing by such vessels in 1906 and 1914 yielded few fish (Hansen, 1949) and the Tjalfe expedition in 1908 and 1909 caught cod in only a few places (Jensen, 1925). Between 1912 and 1923 annual catches gradually increased from 23·5 tons in 1912 to 243·5 tons in 1917 and about 600 tons between 1919 to 1923 (Jensen, 1925), but two cod were found on the offshore banks for the first time in 1921. Hence the first increase must have occurred in the fjord population. Figure 25 shows the international catches between 1930 and 1972 in thousands of tons; from 1930 to 1952, the information is taken from Beverton and Hodder (1962) and subsequently from the *ICNAF Statistical Record*. Following the first strong year class of

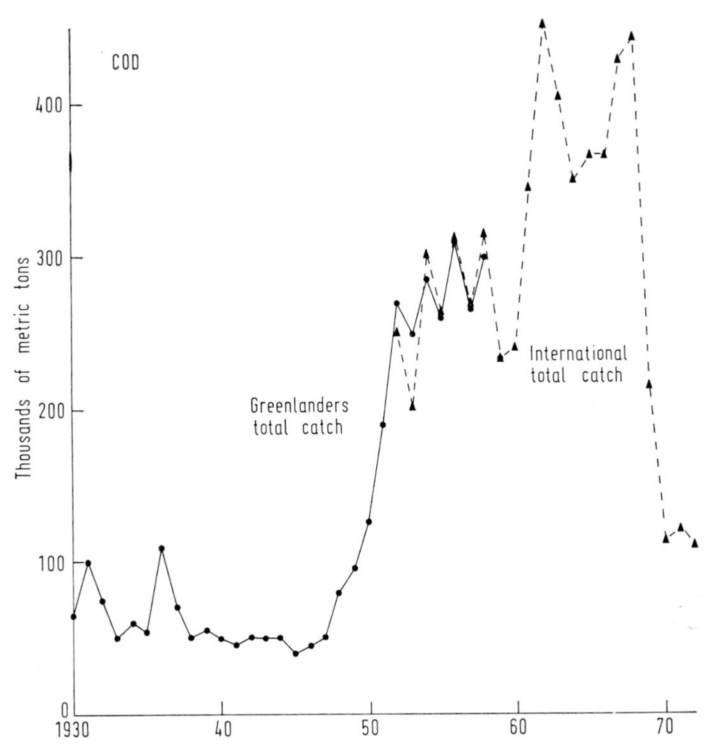

FIG. 25. The increase in catch of cod at West Greenland in thousands of tons between 1930 and 1972.

1917, there was a peak in catch in the thirties that comprised the three year classes 1922, 1924 and 1926, and during the forties a second peak comprised those of 1934 and 1936 (Hansen, 1954). The positions of catches progressed northwards along the coast of West Greenland in the following order: Julianehaab (1917), Godthaab (1919), Sukkertoppen (1922), Holsteinberg (1927), Disko (1931) and 72°45′N in 1936 (Jensen, 1939). The fishery at East Greenland started when German fishermen began to work east of Cape Farewell in 1952. The German and Icelandic fishery in East Greenland started in 1955 after the Dohrn Bank had been discovered by Dr Arno Meyer. The 1945 year class of the East Greenland stock sustained the Cape Farewell fishery and in 1953 the Icelandic spawning fishery.

By 1925, eggs and fry were recorded on Fiskennes Bank, Storehellefiske Bank and Godthaab fjord, all presumably from the 1917 year class (incidentally in 1936, as many as 12 000 eggs/haul were recorded from the mouth of Angsmagssalik fjord in East Greenland

(Jensen, 1939)) but the development in this region may have been later than on the west coast. The Norwestlant expedition in 1963 (Hansen, 1968) showed the presence of stage I cod eggs at West Greenland, East Greenland, Iceland and right across the Denmark Strait

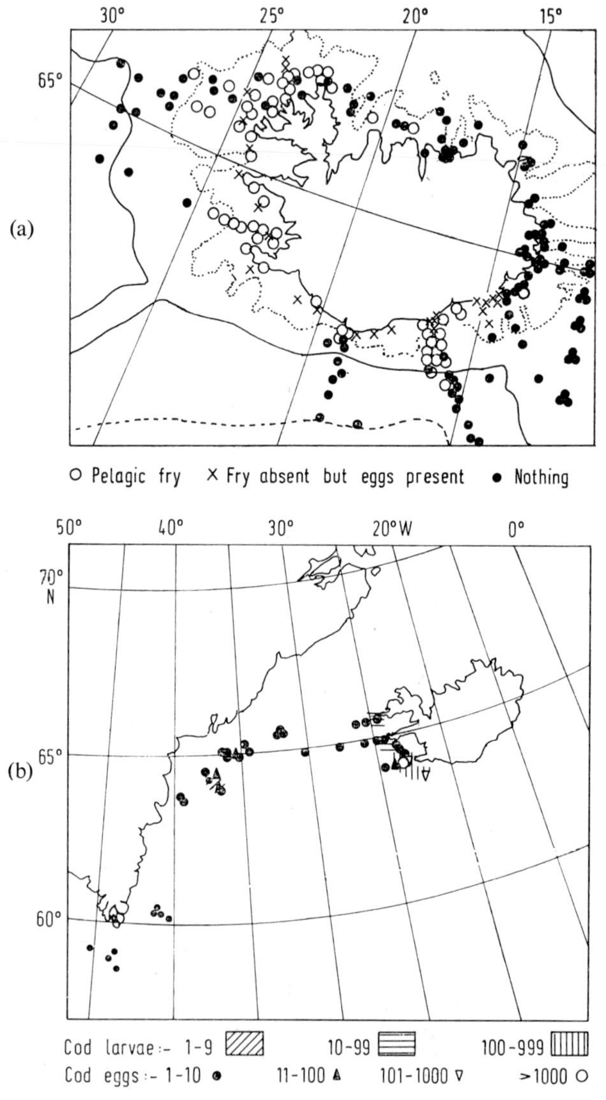

FIG. 26. (*a*) Distribution of cod eggs in April, May and June from Schmidt's surveys, limited by 25°W (Schmidt, 1909); (*b*) Distribution of cod eggs on Norwestlant I, extending beyond 25°W. (Hansen, 1968.)

(Fig. 26*b*), a more extensive distribution than that given in Tåning (1937), and very much more extensive than Schmidt's (1909) observations in 1903–06 at Iceland alone (Fig. 26*a*). Schmidt found no cod eggs on the north coast of Iceland at that time, but showed that most spawning occurred in March and April on the south coast of Iceland; some spawning occurred on the west coast in May, which he called an "afterspawning". Subsequent surveys between 1924 and 1938, showed small quantities of eggs and fry on the north and north-east coasts and shoals of mature cod were also found there (Jensen, 1939).

The question arose whether the stocks at Iceland and the new stock at Greenland were connected and a considerable series of tagging experiments was carried out. Mature and maturing cod were tagged in central West Greenland between 1924 and 1936 and in central and southern West Greenland between 1929 and 1936; a proportion was recaptured on the spawning ground south and south-west of Iceland during the spawning season (Hansen *et al.*, 1935) which indicated a migration of adults from Greenland to Iceland. None were recaptured during the period October–January (Tåning, 1937). The recaptures in numbers at Iceland started in 1930 (except one in 1927) and a fair proportion came from northern and central areas of West Greenland (Tåning, 1937; Hansen *et al.*, 1935). During the same period, 4 939 fish were tagged at Iceland on the spawning ground; 443 were recaptured there and 17 at West Greenland (Tåning, 1937). Whether these animals were of Icelandic or Greenland origin is unknown. After the Second World War, when such tagging experiments were resumed, there were no recaptures at West Greenland from the Icelandic spawning ground. Rasmussen (1959) showed that of cod tagged during the fifties in the northern part of West Greenland, none migrated to Iceland. The proportion recovered from taggings at central and southern West Greenland declined during the same period; from 1929 to 1944 the proportion of recaptures taken on the Icelandic spawning ground was 45·1%, whereas between 1945 and 1952 it fell to 4·4%. Most were tagged in the southern part of West Greenland (Hansen, 1954). In 1968–9, the proportion remained at about the same low level, the greater part coming from the central area off West Greenland (Horsted, 1973). Thus during the thirties and early forties some mature fish migrated from Iceland to Greenland, but not since. The Icelandic recaptures of mature West Greenland fish started in 1930, but the proportion decreased after 1945, during the period when no more Greenland recaptures of Icelandic cod were made. Since 1945, the West Greenland fish that spawn at Iceland never return to Greenland, but continue to spawn in Icelandic waters.

The southern, south-western and western coasts of Iceland are washed by a warm Atlantic stream, the Irminger Current. West of Iceland it splits into westerly and easterly branches. The westerly one moves across the Denmark Strait to flow alongside and below the cool and low salinity East Greenland Current. The two currents tend to mix south of Cape Farewell and to move northward along the coast of West Greenland. Detailed charts of the spring transport are given in Dietrich (1957) and Lee (1968) and both show a rather complex structure west and north-west of Iceland. However, a drift bottle released by Tåning (1931) off the Westmann Islands was recovered off West Greenland, providing some evidence that there is a residual current structure by which eggs and larvae and adult fish can drift from Iceland to West Greenland and mature fish travel from Greenland to Iceland.

By 1912 cod appeared at West Greenland and the year classes appearing in three successive groups: 1917, and 1922, 24, 26; and 1934, 36. It is likely (but cannot be shown) that this colonization was fed by the drift of eggs and larvae and perhaps of O-group fish from Iceland via the Irminger Current as West Greenland waters became warmer. By 1930 adult fish started to return to Iceland to spawn and between that date and 1944 the proportion was fairly high and indeed a few fish made the reverse migration. After 1945 no reverse migrants were recorded and the proportion migrating from Greenland to Iceland was reduced. Very roughly the period 1930–44 corresponds to the warmest period of the present century (it peaked in 1945, Beverton and Lee, 1965) and since then the climate has become cooler and indeed there has been little recruitment to the West Greenland stock since 1968 (but see p. 39 for a description of effects on the adult stock at West Greenland). A very rough comparison of Schmidt's survey of eggs and larvae at Iceland in the cool period with those of Tåning (1967) and Hansen (1968) during warmer years suggests that the cod spawned at Iceland farther offshore and farther to the north and east than during the cool period. The suggestion is that this put them in the westerly branch of the Irminger, and that transport via this current must have been enough to maintain the stocks at West Greenland (see p. 40).

This interpretation suggests that the Icelandic stock itself might have increased with the warming; perhaps Schmidt's "after-spawning" component in May became more important. There is a northern group off West Greenland which appears now to be isolated and there is a southern group which, at least up to 1968, recruited partly to the West Greenland group in the south and partly to Iceland. The process of colonization has established one stock in northern West Greenland, where no eggs and larvae were observed during the first decade of the century, and the other remains loosely in contact with the Icelandic one. Perhaps it was Schmidt's " after-spawning " stock that generated the colonization, being later in the season. As the climate deteriorates one might imagine a reversal of these processes of colonization. Indeed the decline of catches since the early sixties and the failure of recruitment to the West Greenland stock since 1968 suggests that the period of colonization has already come to an end. Figure 16 shows that in recent years the predominantly westerly winds from Iceland to Greenland in the fifties in spring have been replaced by northerly ones which might prevent the transport of Icelandic larvae to West Greenland.

The good East Greenland year classes were born in 1945, 1956, 1958, 1961, 1962, 1963, 1964 and 1968. Eight years after the hatching of a good year class, a good spawning fishery occurs off East Greenland, followed by emigration to Iceland. Dr Arno Meyer has pointed out to us that the decline of good year classes at West Greenland has been followed by a succession of good ones at East Greenland during the sixties. The questions arise firstly whether migration from Iceland is sustaining the East Greenland stock but is no longer penetrating to West Greenland, and secondly whether the East Greenland stock existed during the early period of absence of that at West Greenland (1900s).

C. *The penetration of high latitudes in the Atlantic in the thirties*

Some of the earliest evidences of climatic amelioration were the appearance of southern species off Western Europe and the penetration of boreal species into high latitudes or arctic regions. Up to 1935, the total number of the trigger fish, *Balistes*, ever recorded off the French coast was 25 (Desbrosses, 1935) and in 1930–1, 17 specimens were taken; a turtle, *Dermatochelys coriacea* L. appeared in 1932 (Desbrosses, 1932) together with the medusa, *Callanthias ruber* (Rafinesque 1810) (Desbrosses, 1936). In 1932, the siphonophore *Velella*, the planktonic mollusc *Ianthina* and the tunicate *Salpa fusiformis* were recorded at Plymouth, the tunicate *Cyclosalpa bakeri* and the shark *Oxynotus paradoxus* Frade at Fastnet and twenty whales were stranded on the Scottish coasts (Stephen, 1938); in 1935 *Ianthina*, *Velella* and *Lepas*, the goose barnacle, were recorded off Skye and in 1933 and 1934 albacores were found in the Firth of Clyde and in the Holy Loch, on the west coast of Scotland.

In the Faroe Islands, the twaite shad *Alosa finta* (Cuvier), the swordfish *Xiphias* and the pollack *Pollachius pollachius* L. were recorded (Tåning, 1953). In the Barents Sea, cod and haddock, *Melanogrammus aeglefinus* (L.), were plentiful in 1925, having probably been absent since 1883; some trial fishings had been carried out at Bear Island with no result between 1898 and 1914 (Jensen, 1939). By 1929–30, cod had appeared off Novaya Zemlaya and in 1931 haddock appeared in the White Sea, where it had not been previously recorded. Two gastropods *Gibbula tumida* (Montagu) and *Akera bullata* (Müller), a radiolarian *Collozoon*, the hermit crab *Eupagurus bernhardus* (L.) and the cockle *Cerastoderma edule* (L.) appeared on the Murman coast for the first time. Catches of the Murman coast herring, *Clupea harengus pallasi* Valenciennes, during the twenties amounted to a few hundred tons, but by 1933 they reached nearly 70 000 tons (from 1 000 tons in 1930) (Berg, 1935). Blacker (1957, 1965) contrasted the distribution of Atlantic and Arctic benthic species in the western Barents Sea between 1878 and 1931 and that between 1949 and 1959. He found that the Atlantic species had extended from Bear Island right up the west coast of Spitzbergen, even to the Norske Bank; north west of the islands, they had also extended eastwards towards Hope Island. The sponge *Geodia barreti* Bowerbank, the spider crab *Lithodes maia* (L.), and the starfish *Hippasteria phrygiana* (Parelius) were found at 79°30′N and the prawn *Sabinea sarsi* Smith at 80°09′N. The same sort of picture emerges from the distribution of boreal (or Atlantic) forms along the Kola meridian; in 1921 and 1925 there were 42% and 44% respectively of boreal forms, by 1955 it was 72% and the proportion remained high until 1956–7 (86%). It fell to 73% in 1958 (Nesis, 1960).

In 1930 and 1931, cod and herring appeared off Jan Mayen and a few of the cod caught there had Icelandic hooks in them (Jensen, 1939). Off Iceland, large shoals of mature cod appeared off the north coast and during the period up to 1938 there were small quantities of eggs and larvae there. Herring spawned on the east and north-west coasts of Iceland, witch *Glyptocephalus cynoglossus* (L.), and turbot *Scophthalmus maximus* (L.), spread to the northern and eastern coasts and capelin *Mallotus villosus* (Müller) tended to disappear from the southern coasts and to spawn later on the northern and eastern coasts (Jensen, 1939). Between 1924 and 1949 new records at Iceland included swordfish, *Xiphias gladeus* L., pollack *Pollachius pollachius* (L.), twaite shad *Alosa finta* (Cuvier), the dragonet *Callionymus maculatus* Rafinesque and the ray *Raja nidarosiensis* Collett. Animals that became more frequent included mackerel *Scomber scombrus* (L.), tunny *Thunnus thynnus* (L.,) horse mackerel *Trachurus trachurus* (L.), conger *Conger conger* (L.), basking shark *Cetorhinus maximus* (Gunner), thornback ray *Raja clavata* (L.), the mullet *Crenimugil labrosus* Risso, the fork-beard, *Phycis blennoides* Brunnich, *Paralepis rissoi kroyerei* Lütken, *Paralepis brevis* Zugmayer, *Raja lintea* Fries, the saury pike *Scombresox saurus* (Walbaum), the rudderfish *Centrolophus britannicus* Günther, and the myctophid *Scopelus elongatus*. The great silver smelt *Argentina silus* Ascanius and the Greenland shark *Somniosus microcephalus* (Bloch and Schneider) extended their distribution (Fridriksson, 1949).

During the twenties, there were sharks, rays, catfish *Anarrhichas minor* Olafsen, Greenland halibut *Reinhardtius hippoglossoides* (Walbaum), and Norway haddock *Sebastes marinus* (L.), in small shoals off East Greenland, but by the early thirties they were found everywhere (Jensen, 1939). The colonization of West Greenland by the cod has already been referred to. Between 1923 and 1934, there were seven records of the North Atlantic dealfish *Trachypterus arcticus* (Brünnich) where there was only one previous record. In 1925, the pelagic fry of the redfish *Sebastes marinus* L. were found all over the Davis Strait. The Ca'ing whale *Globicephala melaena* Traill increased in numbers in 1926. In 1928 the ling and the craspedote medusa *Halopsis ocellata* A. Agassiz were found in September at Frederikshaab. In 1929, the haddock was found south of Julianehaab and in 1932 at Nanortalik and it had not been previously recorded in West Greenland. At the end of the twenties the salmon appeared off Sukkertoppen, whereas it had been previously limited to the Godthaab and Amerdlockfjords only. The coalfish spawned off West Greenland for the first time in the present century in 1930–2 and between 1931–6 the piked dogfish appeared where it had been very rare before. The halibut moved north from Store Hellefiske Bank and in the south the capelin, Greenland halibut and fjord cod all decreased in numbers (Jensen, 1939). By the early forties, fishermen were catching cod in the Davis Strait (Hansen, 1949).

In the Bay of Fundy and off Nova Scotia, new records of both northerly and southerly species were obtained. In 1932 and 1936, the sea tadpole *Careproctus longipinnis* Burke was recorded and in 1935 the catfish, both of which are arctic species. Southern species recorded in 1932 and 1936 included the Greater silver smelt *Leptagonus decagonus* Black and Schneider and the common sea snail *Neoliparis atlanticus* Jordan and Evermann. The sucker fish *Remora remora* L. was observed in 1933 and 1934 the tunicates *Salpa zonaria* (Pallas) in 1932 and 1935, *Salpa vagina* Tilesius in 1936 and the mousefish *Histrio histrio* L. in 1937 (McKenzie and Homans, 1937). The mixture of northern and southern species is probably due to the intensification of the Labrador current which appears to occur during increased transport by the North Atlantic drift.

Two trends are discernible, first, the northerly movement of sub-tropical forms into British waters, into the Bay of Fundy and to the Faroe Islands and Iceland. Secondly, temperate, boreal or Atlantic forms appeared on the Murman coast off Spitzbergen, north and east of Iceland and in West Greenland. Both trends are very widespread across the whole of the North Atlantic. The spread northwards includes many phyla, although most of the animals are large and rather obvious ones. The important point of evidence is that this widespread northerly movement occurred during the period of Atlantic warming.

D. *Northward movements in the Pacific*

The evidence for such movements in the Pacific during the initial period of warming is not very great, but the years in which they are recorded were 1926 and 1931, which are significant dates in the north-east Atlantic. In 1926, the siphonophore *Velella* and the sunfish *Mola mola* L. were recorded off British Columbia and the anchovy *Engraulis mordax* Girard was recorded north of the Columbia river. The albacore was seen to the north of Oregon for the first time and *Pneumatophorus planiceps* (a mackerel) and *Atherinopsis californiensis* Girard (a jack smelt) were recorded in Coos Bay in Oregon (Hubbs, 1948). In 1931 *Seriola dorsalis* (Gill), the yellowtail, was recorded off Washington State, *Pleuroncodes planiceps* Stimpson, a pelagic crab, was seen north of S. Catalina Island and *Palometa simillimma* (Ayres) (Californian pompano) was caught in the north (Hubbs, 1948). During the " 1959 event ", when the waters off Southern California were distinctly warmer, *Pleuroncodes* moved north for a considerable distance (Longhurst, 1967). Off California in 1931 a number of southern species were recorded, *Sphyrna zygaena* (L.), (hammerhead shark), *Etrumeus micropus* Temminck and Schlegel (Japanese herring), *Manta birostrus* (Walbaum) (giant ray), *Fodiator acutus* Cuvier et Valenciennes (sharp nosed flying fish), *Verrunculus polylepis* Steindachner (a trigger fish) and *Chaetodipterus zonatus* (Girard) (a spadefish) (Walford, 1931). Such northward extensions to range were of the same scale as that reported from the Atlantic and occurred at the same period; as will become clear, the very dates 1926 and 1931 are significant in this context.

E. *Events in the western English Channel*

Having dealt in general terms with the broadscale distributional changes of the present century in the Atlantic and Pacific Oceans let us return to events in the English Channel where a more complete biological record permits us to investigate in more detail the dynamics of changes in the ecosystem during this period.

The events in the western English Channel were recorded in the main in weekly samples with a stramin trawl and in phosphate observations at surface and bottom at International Station E1, not very far south of Plymouth. In later years, samples from other stations in the western English Channel were examined. The first event was the decline in recruitment to the Plymouth herring stock, *Clupea harengus harengus* L., which started with the 1925 year class. In 1931, the last recorded year class entered the fishery, which itself collapsed in 1936 or 1937 (Cushing, 1961; Ford, 1933). In the summer of 1926, pilchard eggs, (*Sardina pilchardus* Walbaum), were recorded in considerable numbers in July, and in 1934 and 1935 they were abundant in summertime and remained so until 1960 (Cushing, 1961; Southward, 1963); in 1950, the stock of pilchards was estimated at 10^{10} (Cushing, 1957) but very recent evidence suggests that this figure was too high by 50% (Macer, 1974). Between 1925 and 1935, the winter phosphorus at International Station E1 declined by one third and the sharpest decrease appears to have occurred in 1929 or in 1930 (Cooper, 1938).

In the autumn of 1931, the macroplankton declined by a factor of four and the numbers of summer-spawned fish larvae decreased. The indicator species *Sagitta elegans* Verrill, an arrow worm, was replaced by *Sagitta setosa* J. Müller (Russell, 1935) and between 1931 and 1934 the total number of the arrow worms declined. More generally, northwesterly forms, the medusa *Aglantha digitalis* (O. F. Müller), the pelagic polychaete *Tomopteris helgolandica* Greef, the pelagic mollusc *Spiratella retroversa* (Fleming) and the euphausids *Meganyctiphanes norvegica* (M. Sars), *Thysanoessa inermis* (Krøyer) in the macroplankton were replaced by south-westerly ones [the copepod *Euchaeta hebes* Giesbrecht, the medusa *Liriope tetraphylla* (Chamisso and Eysenhardt), the amphipod *Apherusa* spp. and the euphausid *Nyctiphanes couchi* (Bell)] particularly in autumn and winter (Southward, 1963). The numbers of spring-spawned larvae decreased sharply in 1935. Later, it was shown that the stock densities of skates and rays increased after 1929/30, but those of spurdogs (*Squalus acanthias* L.) decreased and some species of shell-gravel molluscs disappeared between the twenties and fifties, *Myrtea spinifera* (Montagu), *Gouldia minima* (Montagu), *Tellina pygmaea* Loven, *Thracia villosiuscula* (Macgillivray) (Holme, 1961). Southward (1963) compared research vessel catches in 1919–22 with those made in 1944–52. The ling *Molva molva* L. disappeared, the spurdog decreased by three orders of magnitude, the boarfish *Capros aper* L. and the gurnard *Trigla gurnardus* L. by one order of magnitude and the whiting *Odontogadus merlangus merlangus* (L.), the dab *Limanda limanda* L., the streaked gurnard *Trigla lineata* Gmelin and the dory *Zeus faber* L. by a factor of three. The conger *Conger conger* L., the angler fish *Lophius piscatorius* L., the flounder *Platichthys flesus* (L.) and the nursehound *Scyliorhinus stellaris* (L.) increased slightly. The brill *Scophthalmus rhombus* (L.) increased by a factor of three and the horse mackerel *Trachurus trachurus* L., the hake *Merluccius merluccius* L., the red mullet *Mullus surmuletus* L. and the red bandfish *Cepola rubescens* L. all increased by an order of magnitude. The sea bream *Pagellus centrodontus* (de la Roche) was not recorded in the earlier period. All those that increased, except the flounder, were southern species; three of those that decreased were northern species, but three were southern. Not only had a considerable ecological change taken place, but also a distributional one. A much more striking example of invasion was shown in the *Octopus* plague of 1950, when adults reproduced in warmer water in many parts of the western Channel, particularly off the French coast (Rees, 1951; Rees and Lumby, 1954). Dr G. T. Boalch (of the Plymouth Laboratory) has told us that the seaweed *Laminaria ochroleuca* de la Pylaie was recorded in the Plymouth area in the late 1940s and that it spread over the south-west by the early sixties. In the last ten years, however, it has decreased in abundance, markedly. Dr Boalch also writes that the phytoplankton in the twenties resembled that at the present time in species composition, a northerly distribution, whereas that in the intermediate years comprised a number of southerly species.

In 1965, the numbers of spring-spawned fish larvae (generally northerly species) increased again by an order of magnitude (Russell, 1973). In October of that year the Scyphomedusan *Pelagia noctiluca* (Forskål) appeared in the Channel with ephyra larvae. In January 1966 small individuals were found near the Eddystone Lighthouse, and in the following summer *Pelagia* were found on the north coast of Cornwall; a widespread invasion had arisen from the breeding in the autumn before (Russell, 1967). In 1970, the macroplankton increased for the first time since the autumn of 1930 and in the following winter, the phosphorus rose to the pre-1930 level. The whole cycle is summarized in Fig. 27, in Russell (1935) and Fig. 1 in Russell *et al.* (1971). In the summer of 1972, *Sagitta elegans* returned for the first time since the early thirties and by 1973 it had reached the levels of abundance experienced before 1931 (Southward, 1974).

There are three events common to the two changes in the thirties and the seventies. In 1930, macroplankton and winter phosphorus

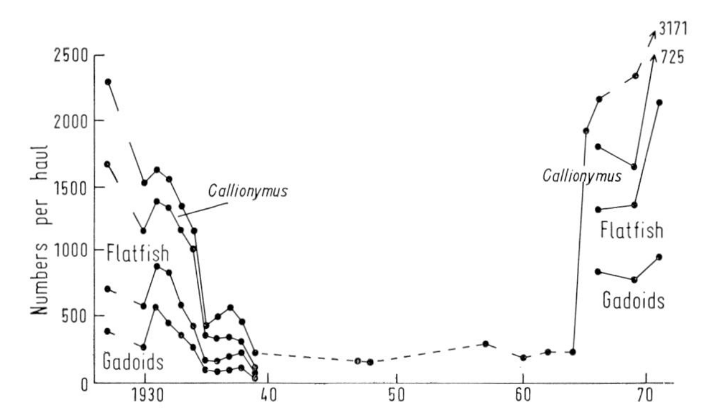

Fig. 27. The Russell cycle in monthly averages of the planktonic stages of teleostean fish, excluding clupeids, between 1924 and 1971. (Russell, 1973.)

declined and the spring spawners decreased 5 years later. In 1965 the spring spawners increased in numbers to be followed by macroplankton and winter phosporus in 1970. Thus, the later events are the mirror image of the earlier ones; if the decline of spring spawners in 1935 indicated the end of the earlier sequence of events, then recovery signalled the start of the later ones.

The interpretation of the changes has been discussed by Cushing (1961) and Southward (1963). Cushing showed that the magnitude of herring recruitment was directly correlated with the quantity of winter phosphorus one year after hatching and that the numbers of pilchard eggs (or stock) were inversely correlated with winter phosphorus, six months after hatching. Using the winter phosphorus as a link, herring hatched in winter were inversely related to pilchards hatched in the following summer (but not in the same years). It was suggested that the decrement of phosphorus represented an increment of larval pilchards, in three stages, 1926, 1930/1, and 1935. The quantity of winter phosphorus was directly correlated with the numbers of pilchard eggs, or stock $3\frac{1}{2}$ years later, which would imply that the larval pilchards represented recruits which were themselves inversely related to the size of their parent stock. If this interpretation were right, a competitive change occurred between herring and pilchard through three generations. That the sequence of events in reversal mirrored the original ones implies that the changes did take place during a period of 5 or 10 years.

Southward (1974) has examined earlier evidence and believes that an alternation between herring and pilchard has occurred before; the summer pilchard period, like that just finished, occurred in the first half of the seventeenth century, the early years of the eighteenth century and in the first thirty years of the nineteenth century.

Southward's interpretation is a more general one which states that with an increase in temperature in the waters of the western English Channel, considerable changes in distribution occurred. One of the distributional changes perhaps due to the lessened extension of the Atlantic drift into the north was the appearance of *Pelagia* in 1965 at the start of the reversal of the Russell cycle. However, the two explanations are not exclusive and the profound changes in the ecosystem that occurred between 1925 and 1935 and reversed between 1965 and 1973 (which may be more competitive than distributional) might have an origin common to both. Recently, Southward *et al.* (1975) have analysed sea surface temperatures at E1 between 1924 and 1972 and have shown a short-term periodicity of about eleven years, like the sunspot cycle. They have also detected a secular trend of warming which peaked about 1945. Catches of hake and of summer pilchard eggs were correlated positively with the short-term trends, whereas catches of cod and of autumn pilchard eggs were correlated negatively. We have witnessed a change from warm water species to cold water ones and a shift in pilchard spawning from summer to autumn as the warmer period gave way to the colder.

The changes shown in Fig. 27 are part of a more widespread climatic change, a warming in the thirties and early forties to be followed by a deterioration. During the warming, the general picture in the North Atlantic and even perhaps in the North Pacific is of the northward movement of animals. The two interpretations of the events given above are part of a more general pattern. The one describes the general distributional change that fits into the pattern of events throughout the North Atlantic and the other describes the dynamic changes that might have been consequential, perhaps due to changes in wind strength and direction as will be indicated below. It is only to be expected that the gross changes are accompanied by subtle and dynamic ones in populations.

F. *Intertidal organisms*

Southward and Crisp (1956) compared the distributions of the barnacles *Balanus balanoides* (L.) and *Chthamalus stellatus* (Poli) in the British waters in the thirties and in the fifties. During the fifties, *Chthamalus* advanced eastward from St Alban's Head, Swanage, to the Isle of Wight, whereas *Balanus* retreated on the south-west of Cornwall. At the same time, *Chthamalus* advanced south, down the east coast of

Scotland from Nybster to Wick, Lybster and Latheron. Temperature for the cirrhal beat ranges from 0 to 17°C for *Balanus* and from 5 to 30°C for *Chthamalus* and so there is a physiological basis for the temperature preference. *Chthamalus* also increased in the Gulf of St Malo and *Balanus perforatus* (Bruguière) increased between the Scillies and St Catherine's Head (Isle of Wight).

Crisp and Southward (1958) established a geographical boundary to the southern forms between Cherbourg and the Isle of Wight. On the English coast, the boundary varies slightly, being near St Alban's Head for *Chthamalus*, *B. perforatus* (some of which were recorded at Poole), the limpets *Patella aspera* Lamarck and *P. intermedia* Jeffreys; for the gastropoda *Gibbula umbilicalis* da Costa, *Monodonta lineata* da Costa and *Littorina neritoides* (L.), the boundary lies much farther west, in the centre of Lyme Bay. The boundary is a well-known one between the shallow eastern Channel well mixed by tidal streams and the somewhat deeper western area that becomes stratified in summer. Crisp (1965) has noticed that the distribution of *B. balanoides* is bounded around the world by the mean annual isotherm of 45°F. The two boundaries obviously do not correspond in the Channel, which suggests that the control of numbers at any position is more complex than a simple and direct response to temperature. Yet Southward (1967) has reported that between 1960 and 1966 *Balanus* has moved eastward and has increased in numbers on the south-west coast of England, in accordance with the deterioration of climate in the sixties.

G. *Discussion*

The changes that became noticeable in the thirties are now seen to be part of a widespread climatic change expressed as a periodic warming in the north-east Atlantic. The Greenland cod increased in stages 1912, 1917, 1922, 1924 and 1926, and 1934, 1936, presumably as generations succeeded each other. It is also possible, but not shown, that the Icelandic cod stock either increased in numbers or altered its spawning distribution. Between 1930 and 1944 fish tagged at West Greenland were recovered in Iceland and during the same period a few fish tagged at Iceland were recaptured in West Greenland. Since 1945 the proportion of Iceland recoveries from West Greenland declined and no Iceland tags were recaptured at West Greenland. Thus, the climatic deterioration since the mid-forties can perhaps be detected in the sequence of tag returns. During the period of warming, the West Greenland stock was established probably from Iceland and it has separated into two parts, the northern group from which no tags were recovered in Iceland since the fifties and a southern or central one from which tags are still being recovered in Iceland. Presumably if climatic deterioration continues, the isolation will be reinforced and then the question will be whether spawning can continue in West Greenland waters; indeed, there has been no recruitment to the West Greenland cod stock since 1968. However, as the recruitment at West Greenland has declined since the fifties, that at East Greenland during the sixties has increased. As climatic deterioration continues, the stock at Iceland might continue to be sustained from East Greenland, if not West Greenland.

During the same period, the thirties, invading animals were noticed all over the north-east Atlantic, in the Bay of Fundy and to some degree off the west coast of North America. Boreal forms increased in high latitudes in West Greenland, Iceland, Jan Mayen and in the northern and eastern Barents Sea. Subtropical forms appeared in British waters and off the Faroes and one or two off Iceland. A mixture of northern and southern forms appeared in the Bay of Fundy and off Nova Scotia perhaps because increased transport in the Gulf Stream may be accompanied by an increase in the Labrador current. Events off the west coast of North America must be under the control of other systems if only because the Californian current is a southbound one; in itself, this suggests that the upwelling system varies in intensity and in position as suggested earlier and animals can be carried north on the countercurrent. The important point is that the invaders were noticed in the thirties in three different oceanographic structures (a) off California, (b) off Nova Scotia and (c) in the north-east Atlantic. Most were invaders from the south, although some appeared from the north off Nova Scotia.

Events in the western Channel were studied at the time and since in considerable detail. They occurred first during the years 1925/6, 1930/1 and 1934/5 and they reversed in 1965/6, 1970/1 and 1973. Phosphorus, macroplankton and summer spawners were reduced in 1930/1 and they recovered forty years later in 1970/1. The spring spawners declined in 1935, five years after the initial change and they recovered in 1965, five years before the reversal. There were extensive distributional changes in 1930/1 and 1970/1. There were consequential changes involving the benthos. It is likely that the changeover from herring to pilchard and probably back again were associated with the changes that were spread over about a decade. The distributional changes occurred at a critical time in the sequence and the whole Russell cycle exemplifies the rectification of climatic periodicity that appears to take place in the sea. It is interesting that the intertidal organisms had altered their distributions in south-west England by 1966 back to those characteristic of the thirties.

There are two general conclusions. First, the climatic perodicity is widespread and extends across oceanographical structures and may even be common to the Pacific and the Atlantic. The second is that an apparent distributional change may involve considerable changes in the ecosystem as occurred in the western Channel and off West Greenland during the thirties, all of which were responses to climatic change. It is likely that the climatic changes were gradual throughout the time period and that in the biological systems, some were rectified as in the Russell cycle.

V. The Biological Effects of Climatic Changes

A. *Introduction*

In the last section, the warming in the thirties and early forties was described in the general biological terms of colonization and extension of distribution, although possible mechanisms were indicated, particularly in recruitment to the cod stocks at West Greenland and in the ecosystem changes associated with the Russell cycle. In the following section an attempt will be made to analyse the mechanisms involved. The effects of cold winters in the British Isles will be studied to show how extreme conditions can modify distribution and the population mechanisms. The changes that have occurred in the Baltic will be described partly to show how relatively minor changes in temperature and salinity may be associated with considerable distributional changes, but also to illustrate yet again the profound modifications to ecosystem structure that appear to be associated with the large-scale changes to fish stocks.

The most important part of the following section is the analysis of changes in recruitment to fish stocks. First are considered the long-term changes across the centuries, the rectification of periodicity and the association of periodicities in different parts of the northern hemisphere. Second, the variability of recruitment is described, the magnitudes of year class strength that are common to different stocks of different species throughout the North Atlantic, the description of the warming in the thirties in terms of year class strength only and the two sorts of recruitment variability (i.e. the low level type that sustains a stock in approximately steady conditions and the large change of two or more orders of magnitude that rectifies periodicities such as the Russell cycle). Third, the dependence of year class strength on climatic factors is described particularly in terms of the hypothetical match of the production of larvae to that of their food during the production cycle.

If climatic factors affect the recruitment to fish stocks by a mechanism such as that indicated in the match/mismatch hypothesis (to be described below, p. 91), it would be reasonable to suppose that similar principles apply to the ecosystem as a whole and some attempt is made to analyse the ecosystem changes in such a light.

B. *The effect of cold winters*

Once every ten or fifteen years an exceptionally cold winter occurs in north-west Europe. Typically an anticyclone settles over Scandinavia bringing very cold easterly winds to the southern North Sea and in these conditions ice may form off the Dutch and German coasts. Lamb (1966) has recorded the incidence of such cold winters since 1400 and lists the number per decade as an index of climatic change; for example there were six in the decade 1770–80 and none in the decade 1930–40. In recent years the most notable cold winters were 1928–9, 1946–7 and 1962–3.

Two groups of biological effect of the cold winter have been recorded in British waters: damage to the intertidal fauna and death of fish in the southern North Sea. Crisp (1964) collated reports from inshore areas in 1963. In the River Crouch and the River Roach, oysters, crabs, cockles, and the polychaete *Sabella pavonina* Savigny were killed. In the south-west, in addition to these animals, the gastropod *Ocenebra erinacea* (L.), the bivalves *Anomia ephippium* L., *Pecten maximus* (L.) and *Venerupis pullastra* (Montagu) were killed. The polychaete, *Lanice conchilega* (Pallas) and the bivalve *Scrobicularia* were killed at Whitstable. Dead razorshells *Ensis ensis* L., were found in Studland Bay, empty tubes of piddocks *Pholas dactylus* L. in Lyme Bay and dead shipworms *Teredo norvegica* Spengler at Exmouth. At the Mumbles near Swansea, dead fish, echinoderms, crustacea, annelids and coelenterates were recorded and large numbers of the barnacle *Elminius modestus* Darwin, the crabs *Corystes cassivellaunus* (Pennant), *Portunus puber* (L.) and *Pilumnus hirtellus* (L.) were killed. The southern forms suffered greater mortality in the south-west, probably because the air temperature anomalies were greater in that region.

In all three of these extreme cold winters, a number of fish species were killed; these were brill, conger, cod, whiting, dabs *Limanda limanda* (L.), plaice *Pleuronectes platessa* L., soles *Solea solea* (L.), miller's thumbs *Cottus gobio* L., gurnards, pout whiting *Trisopterus luscus* L., eels *Anguilla anguilla* (L.), mullets, wrasses, skates and rays

(Lumby and Atkinson, 1929; Simpson, 1953; Woodhead, 1964). In 1947 up to half the catch of soles was dead (13·4 cwt/landing; Woodhead, 1964). In the Silver Pit, in the southern North Sea, where the cooled water collected, the stock densities of sole increased by a factor of 15; thus, not only were they killed in the cold water (3°C), but they were caught more readily in it (Woodhead, 1964). Simpson (1953) showed that there were no cod, plaice or dab eggs in the water of 0°C in 1947 and so recruitment might have been modified, yet the 1963 year class of plaice was the highest recorded (Bannister *et al.*, 1974) perhaps because the development of the eggs and larvae was delayed and the production cycle was advanced to meet it as indicated in a later section.

Because the rate of larval development is an inverse power function of temperature, on the match/mismatch hypothesis one might expect better year classes in cooler water. However, Uda (1952) has correlated the poor catches of the Hokkaido herring with the occurrence of cold winters four years earlier (in 1866, 1869, 1884, 1902, 1905, 1913, 1926, 1935, 1945). The Hokkaido herring *Clupea harengus pallasi* Valenciennes lays its eggs along the shoreline and is probably as vulnerable to cold winters as the intertidal organisms referred to above and is not a case to be explained by the match/mismatch hypothesis. The specific dates of cold winters in Japanese waters are not of course the same as those listed by Lamb for Europe, but in general terms the decades of high cold winter frequency appear to be similar in both Europe and Japan (Lamb, 1972). The biological effects are likely to be common; the death of intertidal organisms and low year classes of such animals (for example, pallasi herring), the death of fish at sea and in some instances high year classes from the survivors.

C. *Changes in the Baltic*

The description of Baltic hydrography given on pp. 47–8 has already emphasized the importance of periodic deep inbursts of dense water from the Skagerrak in aerating the chain of deep basins along the Baltic floor, in the absence of any effective exchange of oxygen with the fresher superficial layers. The cycles of inflow and stagnation in the post-war period are detailed in Fig. 23 and a similar 4–6 year periodicity of change could be shown in the inter-war period. These short-term (inter-annual) variations in Baltic hydrography are superimposed on an underlying trend of much longer period since, after a period of strong dilution in the 1920s (Fig. 28), there was a progressive and general rise in the salinity of the Baltic basin from the early 1930s, levelling off in the 1940s and 1950s at a level unprecedented since the start of

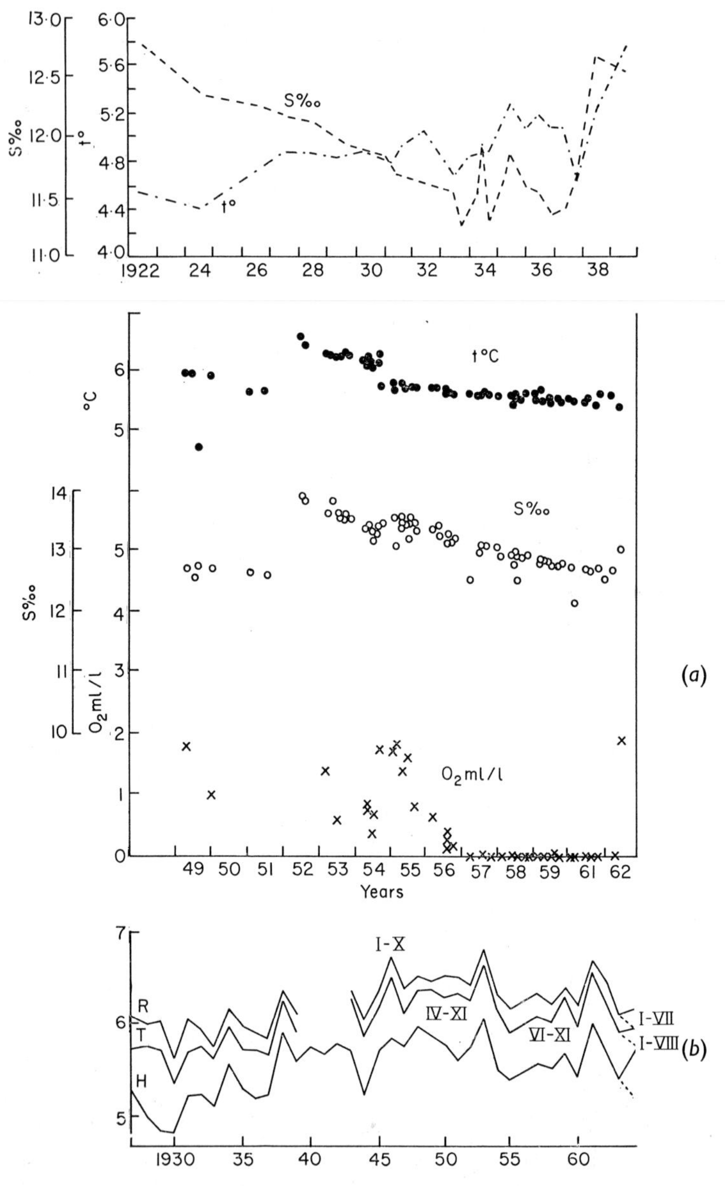

FIG. 28. (*a*) The changes in temperature and salinity at 200 m in the Gotland Deep between 1922 and 1939 and 1949 and 1961 (Dietrich, 1963, and Segerstråle, 1969). (*b*) The changes in surface salinity off the Gulf of Finland between 1927 and 1964.

regular recordings (though with some slight decline since then). This long term "drift" in Baltic salinity and the short-term variations superimposed on it have both been associated with changes in the ecosystem.

The general increase in salinity at the surface up to the middle fifties was accompanied by a decline in the brackish water copepod *Limnocalanus grimaldii* de Guerne and an increase in the stenohaline animals in the Gulf of Bothnia and the Gulf of Finland, for example, the copepods *Acartia longiremis* Lilljeborg, *Temora longicornis* (O. F. Müller), *Centropages hamatus* (Lilljeborg), *Pseudocalanus elongatus* (Boeck), the ctenophore *Pleurobrachia pileus* (O. F. Müller), the medusa *Aurelia aurita* (L.), the barnacle *Balanus improvisus* Darwin, however, the influx of this animal might have been associated with temperature directly as it is a subtropical species at its northern limit, (personal communication, Dr A. J. Southward), the mackerel, the garfish *Belone belone* L., and the medusa *Cyanea capillata* L. In the south the appendicularian *Oikopleura dioica* Fol and the copepod *Oithona similis* Claus became common and the horse mackerel *Trachurus trachurus* (L.), made sporadic appearances (Segestråle, 1969).

With an inflow in 1923, haddock larvae appeared and in 1925–6 a fishery developed, but the fish did not spawn (Johansen, 1926). There are Baltic stocks of cod and whiting that increased during the thirties, the cod catches increasing from 5 000 tons/year to 70 000 tons/year by the early forties. Flounder catches increased from about 1 000 tons/year in 1921 to 5 000 tons/year in 1927, but they declined to very low levels indeed by 1945, presumably as the river discharges decreased, and as the salinity rose.

Some ecological effects of the shorter-term cycles of saline inflow and anoxic stagnation in the Baltic deeps have already been described (p. 47) but there is evidence that these changes also have an effect on the fertilization system in the innermost deeps, most notably in the Gotland Basin to the east of Bornholm. During periods of peak stagnation, such as that observed in 1959 in this basin, phosphate is released in large amounts since it appears that the low pH of the bottom water favours a release of phosphate from the sediments (Fonselius, 1967). Fonselius estimates that 40 000 tons of phosphate-phosphorus were released in this way in the Gotland Basin in 1959 and were later mixed in winter through the primary halocline to the surface layers. A similar massive release of phosphate occurred in the Gotland Basin in the early 1930s, during the stagnation period which preceded the major inflow of 1933–4 (Kalle, 1943). Meyer and Kalle (1950) associated the increase in cod catches by a factor of about 16 during the late thirties with this fertilization. If so nutrients must have been at a low level during the spring outburst before fertilization and the recruitment to the cod stock must have been markedly limited then.

Thus there are two types of change that have occurred in the Baltic since the twenties. The first is the general and slow salinification up to the mid-fifties that is associated with the spread of stenohaline animals and the input of invaders like *Oikopleura* and haddock larvae from the North Sea. The second one, the stagnation and fertilization in the Gotland Deep, is a much more profound one that could well provide a basis for experimental studies on the possible mechanisms of eutrophication in the sea.

D. *Recruitment to the fish stocks*

1. *Long-term changes in catches*

Long-term changes are those that extend across the centuries and are based on the presence or absence of fisheries in historical records, upon judgements of good and bad periods, on records of occurrence and, by the nineteenth century, upon catch records. In the earlier material there is always some doubt about the quality of evidence, but if the change recorded is a large one, for example the extinction of a fishery and men's livelihoods, it is probably well recorded. Seven series of records are available, some of which extend back into the fourteenth century, and some are illustrated in Fig. 29.

(i) The fishery for overwintering North Sea herring (Johansen, 1924; Høglund, 1972) on the western coast of Sweden, the Bohuslån coast, has appeared and disappeared for centuries. The documentation is very complete and is given in Ljungman (1882). He fitted a 55-year period to the cycle, assuming that the fisheries depended upon sunspot and tidal variation. The sequence of fisheries since 1400 is shown in Fig. 29, together with Ljungman's periods; on four occasions out of five, the fishery failed at the end of the 55-year cycle, but the start of the fishery is not related to it at all. The peak catches, shown by black circles, were given by Pettersson (1921–2) who also attempted to fit tidal periodicities to the sequence of fisheries.

The alternation between the Norwegian spring fishery for Atlanto-Scandian herring and the Bohuslån fishery has been well known for a long time and the sequence of events during the fishery was described by Boeck (1871) and Devold (1963); towards the end of a Norwegian period, the fishery shifts from Utsira, northwards to Stadt and finally to the Lofoten Islands and the date of the fishery occurs later in the year. There are spawning groups at each of the three positions (Runnstrøm,

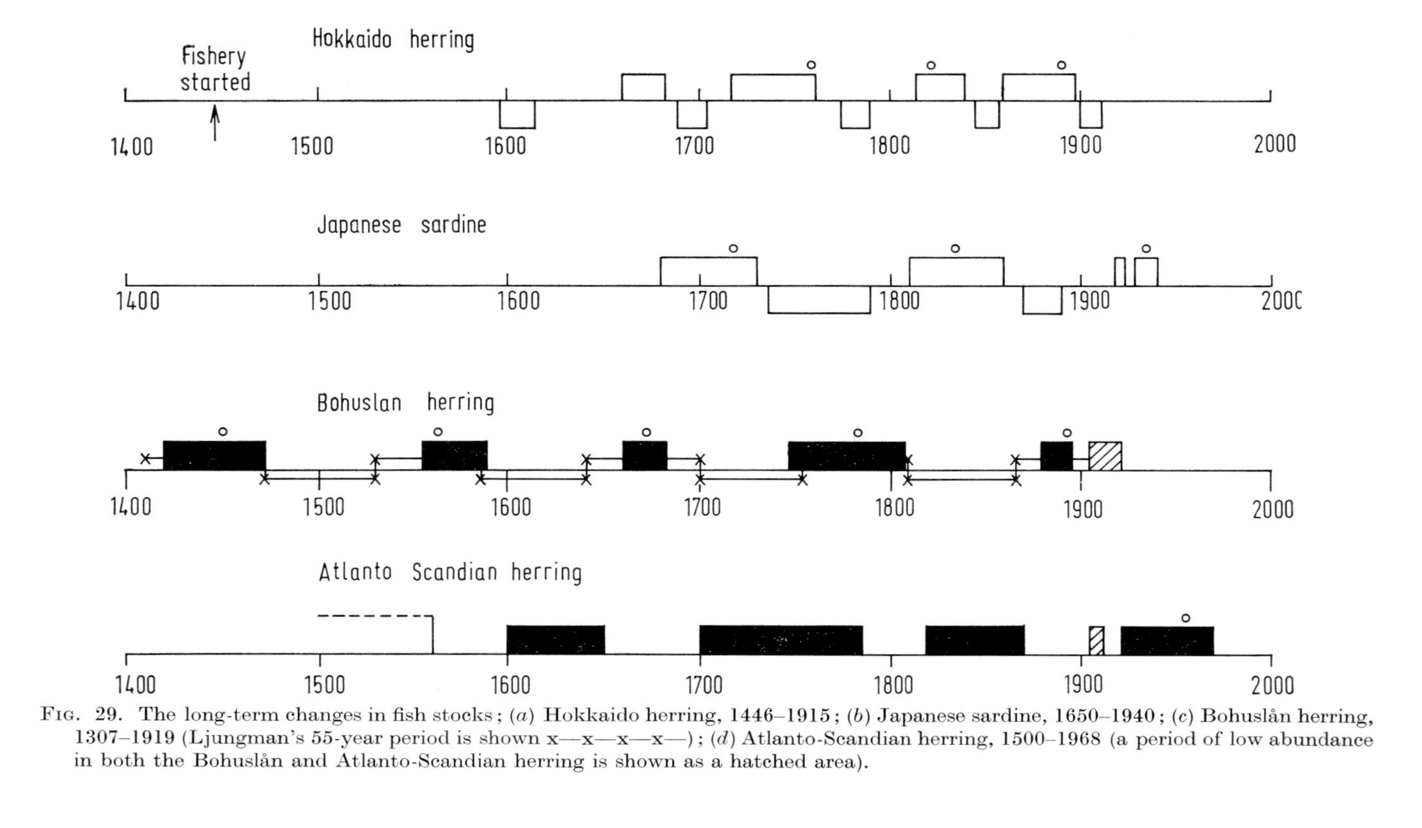

FIG. 29. The long-term changes in fish stocks; (*a*) Hokkaido herring, 1446–1915; (*b*) Japanese sardine, 1650–1940; (*c*) Bohuslån herring, 1307–1919 (Ljungman's 55-year period is shown x—x—x—x—); (*d*) Atlanto-Scandian herring, 1500–1968 (a period of low abundance in both the Bohuslån and Atlanto-Scandian herring is shown as a hatched area).

1933) and as the period progresses there may be a change in relative quantities from stocklet to stocklet. The dates of the periods of the Norwegian spring herring fishery are given in Storrow (1947), Rollefsen (1949) and Devold (1963), but they are less well established than those of the Bohuslån fishery. In particular, the period 1700–84 is described as " variable ". Again the period 1904–19 is low, but the fishery is present and this is also true for the period 1904–19 in the Bohuslån fishery. In other words, in the distant past, the periods appear clearer than the more recent ones with the better information. The periods of the Atlanto-Scandian herring were correlated by Beverton and Lee (1965) with the periodicity of ice cover north of Iceland, which links the Atlanto-Scandian periods with climatic ones.

The alternation between the two fisheries is fairly clear, particularly when it is recalled that the period 1700–84 in Norway is a variable one. There is overlap between the fisheries between 1745 and 1784 and again between 1904 and 1919 and hence the alternation is not the precise one that Ljungman and Pettersson suggested on the basis of the Swedish material alone. The nature of a possible climatic periodicity is described elsewhere in this paper, and a possible explanation is also given. The only point that need be made here is that a somewhat irregular periodicity in the two fisheries is much more explicable when the cause is attributed to terrestrial climatic fluctuations (e.g. changes in the planetary wave system aloft) rather than directly to the rigid periodicities of sunspots or tides. Further, the detailed explanation of the alternation in the fisheries depends on periodic shifts in wind strength and direction that must allow for some degree of overlap between periods, as will be explained below.

(ii) The fishery for the Japanese sardine *Sardinops melanosticta* (Temminck and Schlegel) has fluctuated since the eighteenth century. The dates of the periods of good fishery, as distinct from poor fishery, are given in Uda (1957) and the peak dates, shown by black circles, are given in Uda (1952). For our present purpose the judgement of " good " and " poor " is regarded as being equivalent to the presence and absence used in the description of the Swedish and the Norwegian fisheries. During the recent period of high sardine catches in the thirties, the stock was very heavily fished and may indeed have suffered from recruitment overfishing (Cushing, 1971b) and the period of good catches may have been artificially shortened as a result. It is obvious that the periodicity of the Japanese sardine corresponds with that of the Norwegian herring and that it alternates with the Bohuslån herring, suggesting that these stocks are linked through some globally coherent climatic change.

(iii) Records of the Hokkaido herring fishery extend back to 1447. The dates of the period of " good " and " poor " fishery are given in Uda (1957) and the dates of peak catches are given in Uda (1952). He believes that the Hokkaido herring period occurred at the same time as that of the Bohuslån herring. The heaviest catches from the stock of Hokkaido herring were taken towards the end of the nineteenth century after which catches declined slowly but steadily during a time of heavy fishing effort. This may explain why there has been no recovery during the twentieth century. There is no relationship between the Hokkaido herring periods and those of the other three fisheries in Fig. 29 although some likenesses might be perceived with an optimistic eye. If the peak catches are examined, it will be seen that they are 111 or 112 years apart in the Bohuslån fishery, 100–119 years apart in Japanese sardine fishery and 65–70 years apart in the Hokkaido herring fishery. Ljungman and Pettersson would have noted the 55-year cycle in the first two and its absence in the third. There might be some connection between these three fisheries as shown in Fig. 29, but it cannot be established from such short data series.

(iv) Material since the sixteenth century is available in the records of the Adriatic sardine (Zupanovitch, 1968), which are also classified as " good " and " poor " year classes. Table II summarizes a comparison made by Zupanovitch between the period of the Japanese sardine and the Adriatic one:

TABLE II.

Japan	*Yugoslavia*
1500–1600 (start)	1553^{++}, 1588^{++}
$1680–1730^{+}$	$1670–1730^{+}$
($1716–1724^{+++}$	$1718–1725^{+++}$)
$1736–1789^{-}$	$1730–1780^{-}$
($1760–1780^{---}$	1775^{---})
1830^{+++}	$1830–40^{+++}$
($1884–1880^{---}$	$1882–1878^{---}$)
$1917–21^{+}$	$1919–21^{+}$
$1929–39^{+++}$	$1929–40^{++}$
$1941–57^{-}$	$1946–55^{-}$

The association is based on the judgement of good or bad year classes in the two fisheries. If it can be sustained, a periodicity common to the Atlantic, the Pacific and the Mediterranean can be established between the two north-east Atlantic herring stocks, the Japanese sardine and the Adriatic sardine.

(v) Soutar and Isaacs (1974) have examined the records of fish scales in the cores of anoxic sediments off California. They established high correlations between numbers of scales/cm²/year and billions of sardines and millions of tons of anchovies for recent decades. Figure 30 shows

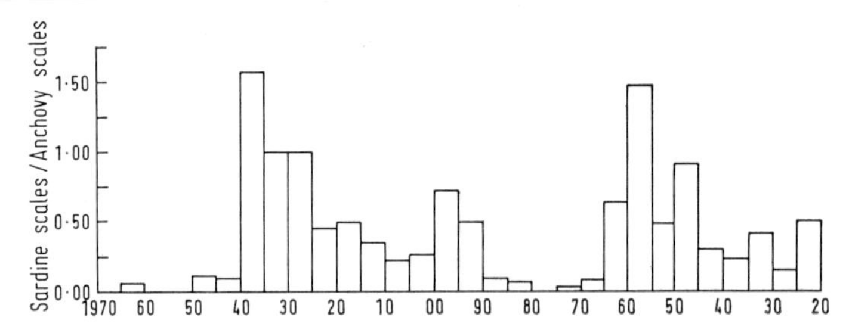

FIG. 30. Ratio of numbers of sardine to anchovy scales in anoxic sediments off California. (Soutar and Isaacs, 1974.)

the ratio of numbers of sardine scales to those of anchovies from 1810 onwards. There are two distinct periods each of which is about fifty years or so in duration. The second corresponds roughly with the period of warming in the first half of the present century; the end cannot be properly specified because the stock of sardines might have been reduced by fishing.

(vi) and (vii) The last two series of records are restricted to the nineteenth century. The first gives a summary of occurrences of bluefin tuna in the North Sea and north-east Atlantic to the north of the British Isles. The second summarizes records of northerly and southerly species of rare fish taken in British waters (Storrow, 1947). Both northerly and southerly records increased toward mid-century, as did those of the bluefin tuna, which is of course a southerly species in British waters. This situation resembles that of the Bay of Fundy and Nova Scotia where records of both southerly and northerly species increased during the thirties.

(viii) Storrow (1947) also summarizes events on the Eastern Seaboard of the U.S. during the nineteenth century and distinguished bluefish *Pomatomus saltatrix* (L.) periods from those of the weakfish *Cynoscion regalis* (Bloch and Schneider). They are in opposite phase, the former being associated with seals, capelin and basking shark and the latter with herring, mackerel and menhaden. The bluefish period was a warm period in European waters and the weakfish one was a cool one.

Thus in Figs. 29 and 30, which show the only long time series of periodicities in fisheries available to us, a periodicity appears to exist between warm and cold periods, whether the estimate of abundance is shown in terms of the presence or absence of a fishery, in the sequences of " good " or " bad " years, or in the occurrence of peak years of catch. The periods are common to the Atlantic and the Pacific; those of the Hokkaido herring do not appear to fit the warm and cold periods.

In Section IV the events in the thirties of the present century have been described in some detail. Figure 31 shows the catches of the

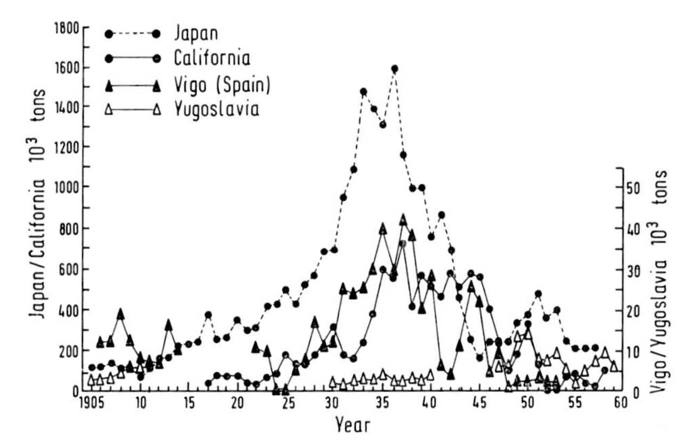

FIG. 31. Catches of sardines off Japan, off California, off Spain and in the Adriatic between 1905 and 1960. (Zupanovitch, 1968.)

Japanese sardine, the Californian sardine, the Spanish sardine and the Adriatic sardine all of which reached a peak during the thirties (Zupanovitch, 1968). Two of these stocks may have suffered subsequently from recruitment overfishing, but each fishery must have been based on strong early year classes in the stocks at about the same time. Roughly this is the period of warming in the north, the period of the Norwegian herring, and of course it was that of a general warming in the North Atlantic.

In conclusion, two points emerge from this brief study. (a) The periodicity exists on a widespread scale and is probably associated with climatic periods that may be described as warming or cooling the north-east Atlantic. Such periods exist in the biological material, but they overlap and are not precisely defined as Ljungman and Pettersson suggested on rather different grounds. Indeed a fuzzy periodicity that might vary in wavelength and amplitude might seem to be much more likely on climatological grounds. (b) The variation in stock that accompanies the extinction or emergence of a fishery must be of several orders of magnitude. The ordinary variation in recruitment to a fish stock ranges from 0·3 to 2·0 orders of magnitude in a stable situation (Cushing, 1973). The differences when a fishery collapses or expands rapidly must be very much greater. The recent Norwegian period was initiated with the famous 1904 year class which in 1910 raised the stock (of many age groups) by one order of magnitude; this year class must have been very much larger than its predecessors. There is some evidence that for some decades from emergence to collapse the stock remains fairly stable and it is in this sense that periods shown in Fig. 29 are rectified ones, like the $\approx$ 40-year Russell cycle in the western Channel (see Fig. 27).

2. *Variations in recruitment to stocks of fish*

The gross long-term changes in fish stocks are indicated by considerable increments or decrements to stock in the form of changes in year class strength or recruitment. If fish stocks rectify the climatic periodicities, it is of considerable interest to know whether there are temporal trends in recruitment and whether there are changes in the variability of recruitment with time.

One of the first studies of such temporal changes was given by Beverton (1962) who compared the catches of some demersal stocks in the North Sea between 1905 and 1962. Catches of cod and lemon sole *Microstomus kitt* (Walbaum) remained steady during the period; those of plaice rose after the Second World War following the war-time decrease in fishing intensity (Gulland, 1968) and those of turbot rose slightly. However, catches of sole increased by an order of magnitude, reaching a steady level in the fifties. Catches of haddock fell by an order of magnitude and Gulland (personal communication) later showed that catches in the southern North Sea fell by two orders. In the year that Beverton's paper was published, the 1962 year class of haddock was hatched, which was twenty-five times larger than any of its predecessors. Thus, changes in stock, probably independently of changes due to fishing, occur from decade to decade and must indicate larger variations in year class strength.

A comprehensive summary of the year class strengths for ten stocks of fish in the North Atlantic between 1902 and 1962 was published by Templeman (1965); the stocks were of cod, haddock and herring. The year classes were ranked as " no star ", and " one " to " five star ". The classification is somewhat subjective because in the early years, the age distributions were not expressed in stock density as they are now. Table III summarizes Templeman's results:

TABLE III. COMMON YEAR CLASSES IN NORTH ATLANTIC 1902–62

1904	Atlanto-Scandian herring^{+++++}; Arcto-Norwegian cod^{++}; Icelandic haddock^{+}
1922	W. Greenland cod^{+++}; Iceland cod^{++++}; Arcto-Norwegian cod^{+}; Icelandic haddock^{++}
1934	W. Greenland cod^{+++}; Iceland cod^{++}; Arcto-Norwegian cod^{+}; Atlanto-Scandian herring
1943	All cod stocks and Grand Bank haddock
1945	Six stocks (+, ++)
1949	Seven stocks (+, ++)
1950	All except Grand Bank haddock; Arctic haddock^{++++}; Arcto-Norwegian cod^{++++}; Atlanto-Scandian herring^{+++++}
1955	Iceland cod; Grand Bank haddock
1956	All except Atlanto-Scandian herring (+, ++)

There are two points of evidence from the table, (a) that quite good year classes are common to most of the stocks cited throughout the North Atlantic in some years, 1949, 1950, 1956, and (b) that some outstanding year classes were common to different species in the same region, for example, haddock, cod and herring in 1950. Even the 1904 year class which may have initiated the recent Atlanto-Scandian herring period was quite good for cod and haddock. This is in keeping with the pan-Atlantic nature of inter-annual hydrographic change (see pp. 33–40) and illustrates the fact that different species may respond in a common way within a given region.

Templeman (1972) repeated this work on cod and haddock stocks in the North Atlantic between 1941 and 1970. He based his classification on observations in stock density and so they should be intrinsically more reliable. It is possible to add the stars by years for different regions on the assumption that the ratios of recruitment to stock remain constant between stocks. Figure 32(*a*) shows the distributions of " total recruitment " to the gadoid stocks in North Atlantic between 1941 and 1970. There is a build-up during the forties and possibly a slow decline during the fifties. There are low values in the late sixties, but this is partly because the rankings were not totally established by 1972 and because the cod does not recruit to the adult stock until it is five to nine years of age; but there is some evidence of decline in

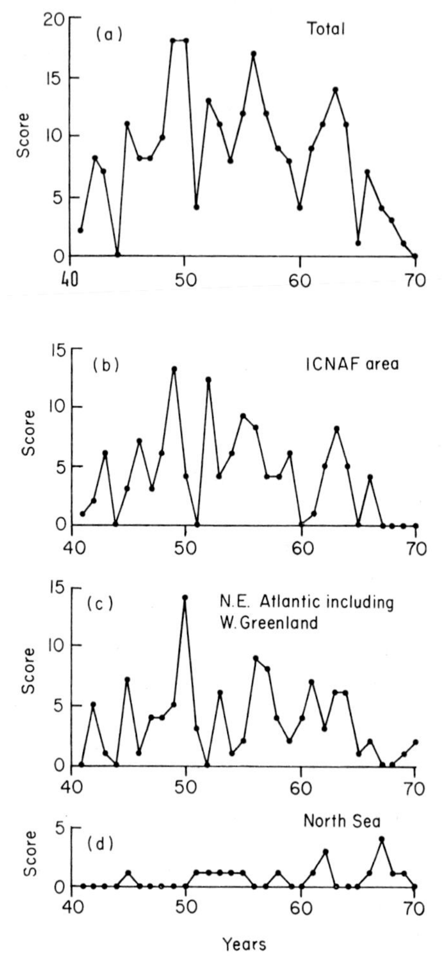

Fig. 32. Indices of gadoid year classes derived Templeman's (1972) estimates, themselves based on stock densities: (*a*) all species and all stocks; (*b*) ICNAF area, excluding West Greenland; (*c*) north-east Atlantic, including West Greenland; (d) North Sea. Note that estimates from 1967 onward are of not fully represented year classes.

the North Atlantic, both in the east and in the west. Figure 32(*b*) for the north-west Atlantic (excluding West Greenland) is not very different from the distribution of totals. Figure 32(*c*) for the north-east Atlantic (including West Greenland) shows the same trends, but in a rather less pronounced way. Our colleague Mr D. J. Garrod believes that a good year class in the Atlantic gadoids tends to appear in the north-west a year before it does so in the north-east; the stronger year classes in the north-west were those of 1949, 1952 and 1955 and they were followed by the strong year classes of 1950, 1953 and 1956 in the north-east. Figure 32 (*d*) shows the distribution for the North Sea, where the augmentation of gadoid stocks during the sixties is now well established. Thus the gadoid stocks in the North Atlantic reached a peak of production in the early fifties to be followed by an outburst in the North Sea during the sixties and there is a little evidence of any one-year delay between the western ocean and the eastern. Dow (1969) has correlated the abundance of Maine lobsters with warmer sea temperature in the fifties. There is no evidence that the gadoid outburst in the sixties has decreased; indeed the 1972 year class of haddock in the northern North Sea was a strong one. It should be pointed out that if the gadoid stock/recruitment curve is a dome shaped one, with low recruitment at both high and low stock, as it is (Cushing and Harris, 1973), an increase in fishing mortality through the North Atlantic could produce such an effect. Such a relation has been established for the Arcto-Norwegian cod stock (Garrod, 1968; Cushing, 1973; Garrod and Jones, 1974) and the pattern of recruitment may follow this course. It is, however, unlikely that other gadoid stocks in the North Atlantic are in the same condition as that in the Arcto-Norwegian cod except perhaps Georges Bank haddock (for which the evidence is equivocal).

The variation in recruitment may be examined from two sources; first, Kasahara (1961, 1964) has published three series of year class strengths (as simulated catches), for a mackerel *Pneumatophorus diego* Ayres, *Sardinops caerulea* Girard (Californian sardine) and for Hokkaido herring. Between 1938 and 1953, the year class strength of the mackerel varied by one and a half orders of magnitude (i.e. a factor of seventy) and possibly showed an overall slow decrease with time. Between 1930 and 1950 the recruitment of the Californian sardine varied by a factor of seven or eight with little overall trend (we have excluded the year classes 1949 and 1950 which were probably reduced by recruitment overfishing (Cushing, 1971)). Between 1907 and 1948, there was a variation by a factor of about 55 in the recruitment of the Hokkaido herring and a downward trend of one order of magnitude in each ten years.

The second source is presented in Cushing (1973), and shown here as Fig. 33. Between 1924 and 1962 the recruitment to the Downs herring fluctuated by about a factor of three to five and declined in the later years due to recruitment overfishing. The year classes in the British Columbia herring increased between 1914 and 1952 but were very low in 1917 and 1924, by a factor of 50 or so, but apart from these, they

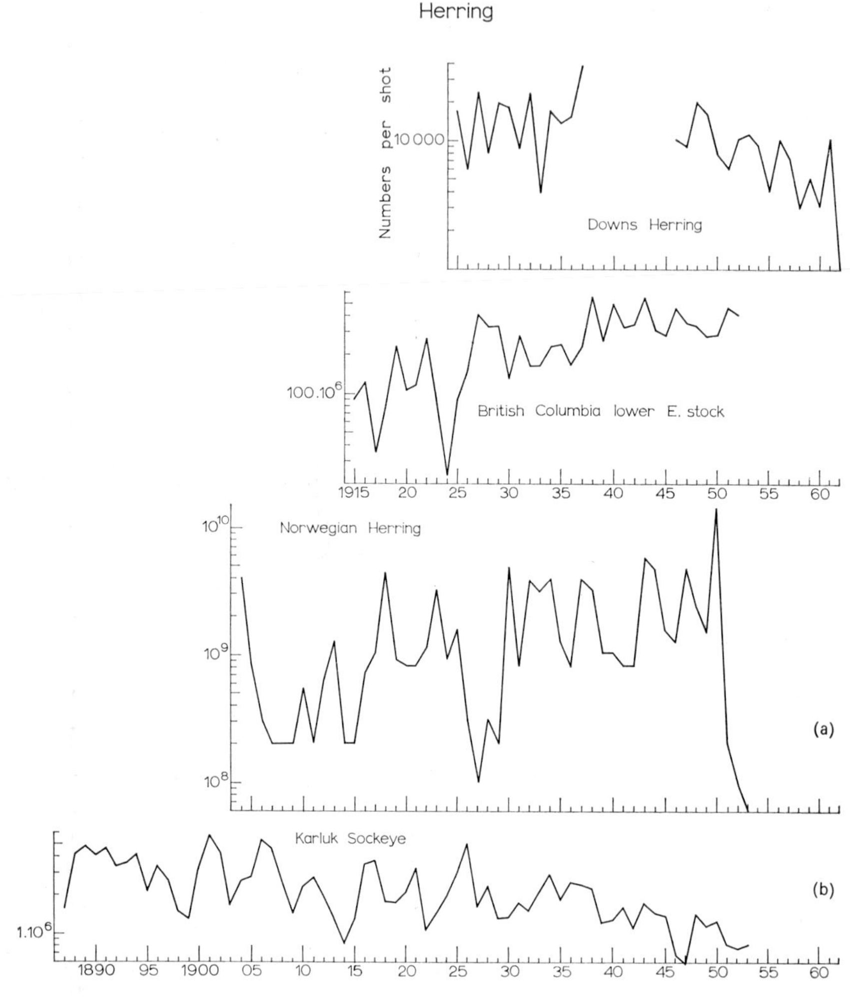

Fig. 33. Recruitment to a number of fish stocks: (*a*) Downs herring, British Columbian herring (Lower East stock) and Norwegian herring; (*b*) Karluk sockeye salmon; (*c*) West Greenland cod, Iceland cod, Arcto-Norwegian cod and Georges Bank haddock (Cushing, 1973); (*d*) North Sea sole.

varied by a factor three or five, about a rising trend which amounted to a factor of 12 in each decade. The variability of the Norwegian or Atlanto-Scandian herring is considerable, of between one and two orders of magnitude, and the year class strength roughly doubled itself in each decade. The variation of the Karluk salmon *Onchorhynchus nerka* (Walbaum) between 1887 and 1953 was low, between $\times$ 3 and $\times$

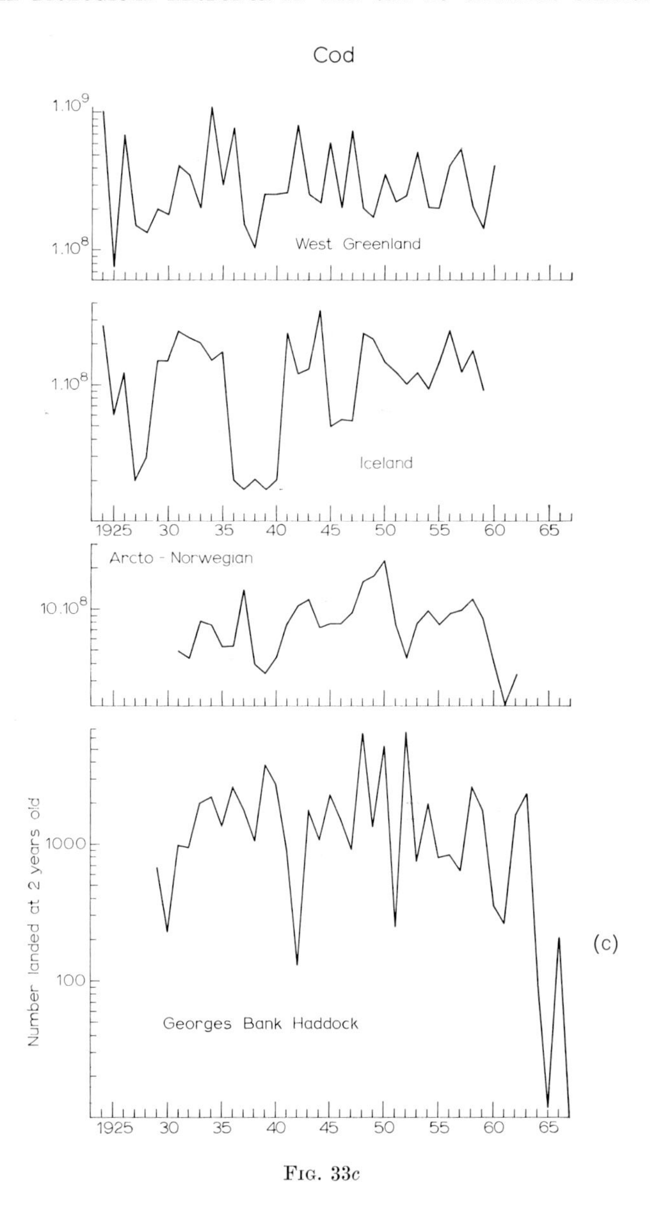

Fig. 33*c*

10 and there was a slow downward trend by a factor of five per decade. Very roughly there was no trend in the cod stocks, except that in the Arcto-Norwegian stock which might have been due to recruitment overfishing in the sixties. In the West Greenland cod, the Iceland cod and the Arcto-Norwegian cod, year classes vary by about an order of

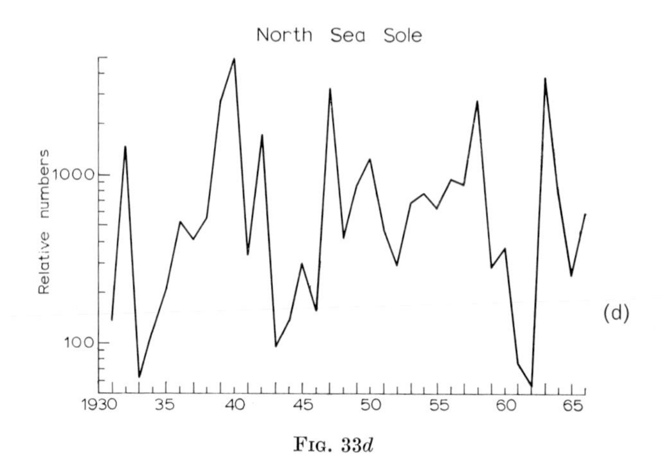

FIG. 33*d*

magnitude or less, whereas those of the Georges Bank haddock range over a factor of thirty to forty (Fig. 33*c*). The sole year classes in the fifties and sixties showed no upward trend (in contrast to the catches of earlier years), but they varied by two orders of magnitude (Fig. 33*a*).

Fluctuations in the catches of the Malabar oil sardine *Sardinella longiceps* Valenciennes have been considerable. The animal spawns in July and August towards the end of the south-west monsoon; after twelve months they form the basis of an inshore canoe fishery. They spawn twice and by the age of two and a half years they are dead. Their biology and the fluctuations in the fishery are well described by Raja (1969) and he has shown that failures in recruitment were associated with lack of rainfall; when the June rainfall is less than 10 mm/day the ovaries suffer from a breakdown of eggs or atresia. However, the dramatic rises in catch are not accounted for. In a four-year period, let the catches by C_1, C_2, C_3 and C_4. Then the stock is represented by $(C_3/E) + (C_2/E)\exp - Z + (C_3/E)\exp - 2Z$ and recruitment by (C_4/E), where E is the exploitation rate and Z is the total instantaneous mortality rate. Let $F = 0{\cdot}5$ where F is the instantaneous fishing mortality rate, $Z = 1{\cdot}0$, and with a series of catches going back to 1940 a stock/recruitment curve can be constructed. Because we are concerned with differences in recruitment the value of the exploitation ratio (F/Z) does not matter too much. A Ricker curve was fitted (Cushing and Harris, 1973), $R = AP \exp - BP$, where R is recruitment, P is stock, A the coefficient of density independent mortality and B that of density dependent mortality; $A = 0{\cdot}774 \pm 0{\cdot}13$ and $B = 0{\cdot}0014 \pm 0{\cdot}003$. The curve is a near linear one, but it is remarkable for

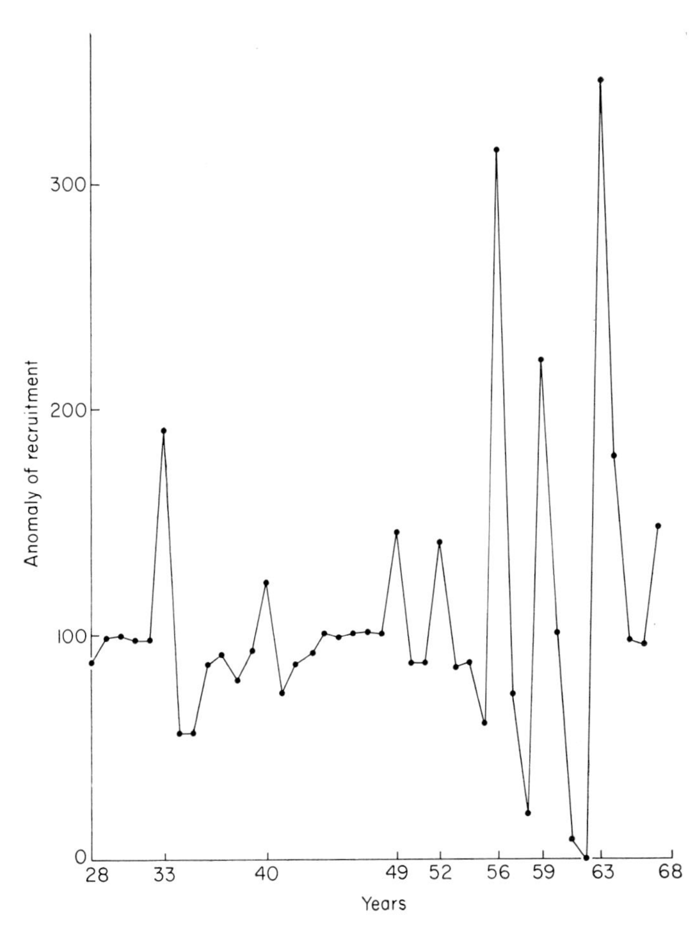

FIG. 34. Malabar sardine. Recruitment, 1928–68, plotted as deviations from the stock/recruitment curve.

a number of positive anomalies outside the expected range of error. In Fig. 34, recruitment is plotted as deviations from the stock/recruitment curve in a time series. Such high recruitments were those responsible for the rise in the fishery. The period is not long enough to draw firm conclusions and the present decline in the fishery since 1968 is probably due to the lack of rainfall as Raja has suggested. In broad terms the period of the fishery resembles that of the cod off West Greenland and some of the outstanding year classes are contemporaneous with those in North Atlantic gadoid stocks.

From the study of recruitment to the fish stocks, a number of conclusions emerge. The first is that there are considerable differences between stocks of the same species; for example, the variability in recruitment to the Atlanto-Scandian stock of herring is much greater than that of the Downs stock of herring, or the British Columbian stock of herring (a pallasi herring), both of which live in relatively enclosed waters as compared with the open Norwegian Sea. The haddock year classes appear to be more variable than those of the cod and those of the North Sea sole are very variable indeed; the latter might be expected, as it is on the northern edge of its distribution. Such differences in variability may be related to local variation in the production cycles, but the information is not yet available to confirm this. It is interesting to note that the abundance of Dungeness crab off Oregon is correlated with an upwelling index six months to one year earlier (Bostford and Wickham, 1975).

The second conclusion is that the clupeid or salmonid year classes may trend slowly upward or downward in decades, whereas those of the gadoids do not appear to do so in the data presented. Because the gadoids are more fecund, it has been suggested that they have greater capacity for stabilization (Cushing, 1971b; Cushing and Harris, 1973), that is, they respond more quickly to a change in stock. Hence they might be expected to resist the small changes from decade to decade, to which the herring stocks appear to respond. On a longer time scale, the gadoids do not exhibit the dramatic exits and entrances of the clupeids, although there might be secular variations in year class strength on longer time scales than shown in Fig. 33; Ottestad (1942) showed that towards the end of the last century, the Arcto-Norwegian cod stock was lower in magnitude than in later years. The clupeids appear to rectify the effects of climatic change with a dramatic rise or catastrophic collapse every fifty to seventy years, whereas the gadoids tend to adapt to such effects more successfully. Such a result might be expected by comparing the shapes of the stock/recruitment curves of the two groups of fishes.

The third conclusion is that outstanding year classes might be common across broad areas of the ocean. Templeman's work cited above shows that it may well be true for gadoid and clupeid stocks in the North Atlantic. Some of the correlations to be described below between events distant from each other are explicable if the climatic factors that generate outstanding year classes are systematic and coherent at these space scales.

The fourth conclusion is that since 1941, the gadoid stocks were very productive in the North Atlantic during the fifties but were less so during the sixties, but in the North Sea a gadoid outburst started during the sixties and continues today. The index of production is the sum of outstanding year classes, of the magnitude that sustain the stock for a number of years. Very roughly with the warming by the thirties, the cod increased in the north, colonized West Greenland, and reached its zenith of production during the fifties. Since 1968, recruitment has failed at West Greenland and during the sixties, gadoid year classes have declined (but not at East Greenland). In a sense, the cod has retreated into the North Sea although there has been no migration; all that has happened is that the North Sea stocks have become more relatively productive than those elsewhere in the North Atlantic. Just as the end of the period of warming coincided with the start of the Russell cycle, the events described in the gadoid stocks appear to be linked with it.

3. *The dependence of year class strength on climatic factors*

The high variability of year class strength due to environmental factors (as shown in any stock and recruitment curve) has led fisheries biologists to search for relationships with particular variables. For example, Jensen (1952) showed a correlation between the stock density of small plaice in the Transition Area and the surface temperature and salinity at Schultz's Grund L.V. in the Transition Area between 1905 and 1935. Johansen (1928) demonstrated a similar effect for small plaice between 1904 and 1926 in the Southern Kattegat and the Belt Seas. The area is one of considerable differences in salinity from year to year and they must equally well represent a whole gamut of related climatic factors.

There are positive relationships between year class strength and temperature, for example in the West Greenland cod (Hermann *et al.*, 1965), in the Arcto-Norwegian cod (Kislyakov, 1961), in the Adriatic sardine (Zupanovitch, 1968) and in the Giant scallop *Placopecten magellanicus* (Gmelin) in the Bay of Fundy (Dickie, 1955). A number of inverse relationships have been established between year class strengths and temperature, in the Californian sardine (Marr, 1960), in the Hokkaido herring (Uda, 1952), in the English sole *Parophrys vetulus* Girard (Ketchen, 1956) and in the New Brunswick cod (Martin and Kohler, 1965). Recently, Dickson *et al.* (1973) have shown that during the fifties and sixties in the North Sea the year class strengths of cod were inversely related to temperature. The gadoid outburst in the North Sea during the sixties has already been referred to and the correlation might be extended to haddock and whiting.

From the very long-term changes and from the events in the thirties, fish stocks are able to respond to the variability in climate through variation in year class strength. There are two correlations between recruitment and climatic change in general terms, one by Ottestad (1942) on the Arcto-Norwegian cod and the other by Zupanovitch (1968) on the Adriatic sardine, and both use the same methods. Ottestad extracted four periodicities from the growth of pine trees in the Lofoten area by smoothing the time series in a form of harmonic analysis and so phases could be allocated to each year; each fraction of a year class in each year was raised by the sine of the phase and the four curves were added and fitted by least squares to the annual catches ($r = 0{\cdot}84$; $n = 55$). A similar technique was used by Zupanovitch who extracted four periods in climatic factors and correlated the sum with catches of the Adriatic sardine. The technique was used successfully to predict catches up till 1970. The method common to the two disparate fisheries depends upon a periodicity in climatic events which is at first sight a little remote from events in the sea. However, it is its general character that is important rather than any particular relationship, because all climatic events are pervasive on rather a broad scale.

In temperate waters some fish (cod, plaice, herring, salmon) appear to spawn at fixed seasons (Cushing, 1969b). Although the period of spawning may last as long as three months, the peak date of spawning has a standard deviation of about a week or less. This conclusion was based on seasonal egg surveys of the plaice in the Southern Bight of the North Sea since 1911, on grab surveys for herring eggs during the thirties in the region off the island of Utsira in Norway, on records of catches of cod by weeks in the Lofoten fishery since 1888 and upon observations of salmon spawning on the redds by date in about fifty streams in the Fraser river system. In contrast, there appeared to be no regularity in the spawning dates of the Californian sardine in one of the major upwelling regions, and as tuna larvae are spread all over the North Pacific anticyclone in nearly all months of the year the spawning of their parents is not restricted to any season. The distinction between the apparently fixed season of spawning in temperate waters and the variable season in low latitudes probably lies in the nature of the production cycle. In tropical and subtropical waters ($< 40°$ latitude) the cycle is continuous all through the year, i.e. food is always available. In temperate waters the cycle stops in winter and most food is available during the spring outburst of plankton. Most fish live for a time in the plankton as larvae and the period of larval drift is probably quite crucial in the population processes, i.e. in establishing the magnitude of recruitment, the control of numbers and the competition with other species.

The production cycle in temperate waters varies in time of onset, amplitude and spread (Colebrook, 1965) and although there is very little precise information on the point, it is likely that it can shift by weeks and perhaps even by a month between years. The evidence on variability in peak date comes from Davidson (1934) in the Bay of Fundy, Corlett (1953) in the north-east Atlantic and Lund (1964) in Windermere. The production of eggs occurs at about the same time each year, but the production of food for fish larvae must occur at different times from year to year; the production of larvae may be said to be matched or mismatched to that of their food (Fig. 35). The

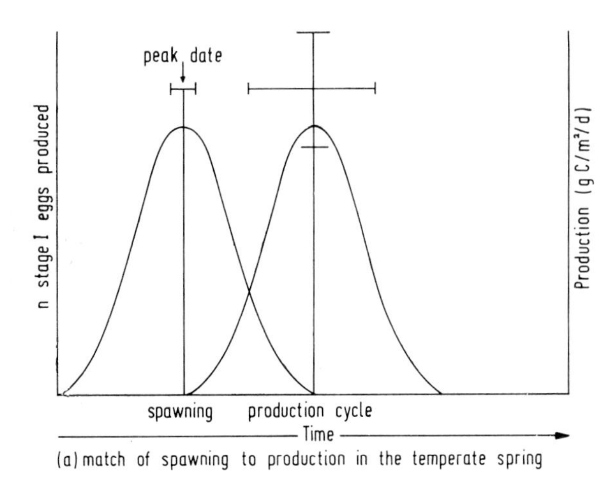

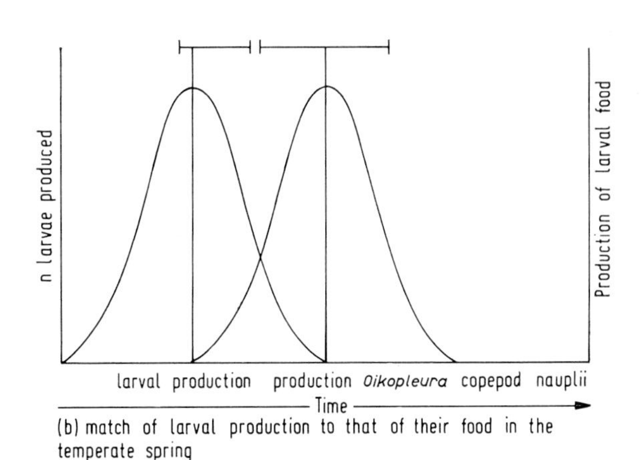

FIG. 35. The match/mismatch hypothesis.

larvae need food within ten days or so of yolk sac exhaustion and the degree of match could be estimated by measuring the food per larva within critical periods. Figure 35 shows that the food per larva may be very low at an early stage because of mismatch and very high at a later stage when the cycles are matched. In the first case, the larvae may starve and in the second they may reach the point of no return when they must die after starvation, even if food is presented before the superabundant food becomes available (Blaxter and Hempel, 1963). The best chance of obtaining food occurs neither too soon nor too late. Differences in year class strength at constant stock values might be accounted for by the match or mismatch of larval production to that of their food.

The 1963 year class of plaice in the southern North Sea was the largest recorded and it was hatched in the cold winter of 1962–3. The effect of cold water is to delay development and hence perhaps improve the chance of a good match between larval production and that of their food. The cold winter occurs when an anticyclone is centred over Scandinavia and cold easterly winds persist for a long period in southern and eastern England. Under such conditions the production cycle in the southern North Sea is likely to be advanced because the strong northerlies are absent; Cushing (1972) suggested that the delay of the production cycle in the Southern Bight was related directly to wind strength, and the northerlies are the stronger. Two factors combined to obtain a best match and a very strong year class was the result.

One of the important points is that the dependence of development on temperature is an inverse power function and it follows that cold water delays development to a much greater degree than any acceleration in warm water. Hence we would expect that recruitment should be well correlated with temperature inversely, but poorly correlated in a positive manner, merely because the differences in development rate are very much less in warm water than they are in cold. Recently, Dickson *et al.* (1973) have shown that the year class strengths of cod in the central North Sea (Dogger area) are correlated with water temperature in April; in a spatio-temporal matrix of correlation coefficients, the highest were found in April and between the Dogger and the Spurn, which corresponds to the spawning on the Flamborough Off ground. During the sixties, the temperature was significantly cooler than a decade earlier and the increment in cod year classes might well be a consequence of cooler springs. It is an important point because it bears on the more general rise of the gadoids (haddock, cod and whiting) in the North Sea during the sixties.

More generally the negative correlations noted above between year class strength would not be unexpected on the match/mismatch thesis. Fish may spawn far enough ahead of the production cycle so that the average match is of two overlapping distributions in time, larvae before that of food. Then an exact match where the distributions coincide would be unusual and the condition in which the larval distribution succeeds that of their food would be rare enough to be impossible. Within these constraints a positive correlation with temperature would be the product of specialized conditions, but the negative correlation would support the thesis. It is possible that the survival of cod at Labrador or at West Greenland is a special case that depends directly upon temperature.

The effect of temperature is a special case of the general thesis of match and mismatch. The correlations established by Ottestad and Zupanovitch demonstrated the general effect of climate that can only be effective during the larval drift. Because the onset of the production cycle is determined by wind strength and direction together with irradiance, the match/mismatch thesis is sufficient to account for the correlation. Then the correlations of haddock year classes with wind strength, but not direction (Carruthers, 1938; Rae, 1957) which subsequently failed become explicable.

One of the most spectacular series of changes is the alternation between the Norwegian and Swedish herring periods already referred to. The Norwegian fishery is based on the Atlanto-Scandian stock of herring of spring spawners that spawn on three distinct grounds, off Utsira, off Stadt and off the Lofoten Islands (Runnstrøm, 1933). The Swedish fishery is based on North Sea autumn spawners (Johansen, 1924; Andersson, 1956; Høglund, 1972) that overwinter on the western edge of the Norwegian deep water and move close to the Bohuslån coast during the Swedish periods. Devold (1963) suggested that at the end of the Norwegian period the Atlanto-Scandian fish changed their spawning time, size and racial characters and became the autumn spawners of the Bohuslån fishery. Although simple enough to explain all the facts, it is a complicated hypothesis in that such profound physiological changes have not been observed in other stocks of fish.

In the north-east Atlantic the wind shifts periodically from southerly to westerly and such periods correspond roughly to those of the Norwegian and Bohuslån herring. During periods of westerly wind, the Norwegian coast would be exposed to onshore winds with considerable fetch. During periods of southerly wind, the fetch would be much less and the more northerly grounds off Stadt and off Lofoten would be protected to some degree because of the angle between the coastline and the southerly wind. In the North Sea, the autumn spawning stocks most likely to enter the Skagerrak are those that spawn off the Scottish coasts. There, a westerly wind would be offshore and a southerly one would have a relatively long fetch for most of the spawning grounds. The period of westerly winds is that of the Bohuslån fishery as shown in Lamb's figure of the periodic change in west wind distribution over the British Isles, and the southerly period is that of the Atlanto-Scandian herring. Hence the alternation can be explained not only in terms of the climatic changes that have occurred, but also in terms of the match/mismatch hypothesis. A southerly wind in the North Sea might delay the production cycle, generating mismatch for the North Sea herring, whereas the same wind in the Norwegian Sea might advance it, generating a match of the production of Atlanto-Scandian herring to that of the larval food.

Hill and Lee (1957) correlated the year class strength of the Arcto-Norwegian cod with the southerly wind component at Bear Island. There is no well-established periodicity in the catches of this cod stock, but the correlation suggests that if present it might follow that of the Atlanto-Scandian herring and the Japanese sardine. Garrod and Dickson (quoted in Cushing, 1972) correlated the year class strengths of cod with temperature on the Kola meridian and also with salinity anomalies in the German Bight of the North Sea. Ishevskii (1964) established a number of distant correlations of this type based on a periodicity of 18–20 years in the heat budget.

Climatic changes are pervasive and world wide and events in one region can be quite properly correlated with those in another. Hence the correlations noted above are sensible ones. However, the only way in which climatic events can be connected causally with events in fish populations is by means of a structure such as is outlined in the match/mismatch hypothesis. It used to be thought that fish larvae were blown by the wind away from the entrance to a nursery ground and that variance in wind strength was correlated with year class strength for that reason. The correlations established by Carruthers and Rae of haddock recruitment on wind strength for a period depended on the assumption that the larvae were drifted by a westerly wind over the Norwegian trench (Rinne) and failed to settle on the seabed. Saville (1959) showed that haddock larvae never reached the deep Norwegian trench from the northern North Sea. The match/mismatch hypothesis, however, can account for such events and a variety of climatic factors at a number of positions may be correlated with the critical ones (wind strength and direction, irradiance and temperature).

There is a very interesting special case of the dependence of recruitment upon environmental factors; indeed it might be an example where the match/mismatch hypothesis should be applied. It is the effect of El Niño upon the recruitment to the Peruvian anchoveta stock. During the late southern summer, warm water replaces the cold upwelling waters off Peru and this is El Niño, so called because it arrives at Christmas or just after. The immediate and obvious effect is that the rich production cycle characteristic of the cool upwelled water is replaced by warmer and less productive waters (see pp. 51 *et seq.* for a climatic description of the phenomenon).

El Niño has been recorded in certain years, for example, 1911–12, 1917, 1925, 1932, 1939–42, 1953, 1957–8, 1965 and 1972–3 (Vildoso, 1974). The recent event was remarkable for the appearance of the pelagic crab *Euphylax dovii* (Stimpson) in the second half of 1971 in the waters of south of Pisco (14°S), where the water was anomalously warm (Wooster and Guillen, 1974) and it may have come from the west; to the north, the water was colder than usual.

The most accurate indicator of the appearance of El Niño is the departure of the guano birds (the cormorant *Phalocrocorax bougainvillii* Lesson 1837; the booby *Sula variegata* Tschudi, 1845; and the pelican *Pelecanus occidentalis thegus* L.) which abandon the guano islands, their nests and their chicks within a few days of the start of El Niño. They fly south, eventually to Chile.

Vildoso (1974) lists a large number of pelagic fishes, rare off Peru, which appeared there during the recent Niño of 1972–3. The frigate mackerel *Auxis rochei* (Risso), the skipjack *Katsuwonus pelamis*, (L. 1758), the yellowfin Tuna *Thunnus albacares* (Bonnaterre, 1778), the dolphin *Coryphaena hippurus*, (L. 1758), the Spanish mackerel *Scomberomorus maculatus* (Mitchill, 1815) and the manta *Manta birostris* (Walbaum) all appeared in the coastal waters off Peru. Vildoso lists a number of other fishes many of which originate from Central America and California. Perhaps the most remarkable migrant was the milkfish *Chanos chanos* (Forsk.) from Indonesia, *recorded for the first time* in the eastern Pacific Ocean. Perhaps it had crossed the equatorial Pacific in the North Equatorial countercurrent; Wyrtki (see p. 52) has correlated the appearance of El Niño with increased transport in the countercurrent.

Economically, the most important effect of the 1972–3 Niño was the collapse of the anchoveta fishery, the largest fishery in the world. The year class hatched in the autumn of 1971 was only one tenth of the previous four or five classes; the subsequent ones were also low (Valdivia, 1974). Because the number of eggs laid in the 1971 year class was not much lower than in its predecessor, there must have been a greater mortality of larvae or of young fish. However, this must have occurred before El Niño started in the following February of 1972. Before El Niño, there is always a period of strong winds and intense upwelling. Under such conditions, production at the surface near the coast is not very high (Cushing, 1971a) and Paul Smith (personal communication) has suggested that the year class failed through food lack in the autumn of 1971.

During 1972 and 1973 the winds at three coastal stations in Peru remained much as they had been in 1971 and presumably upwelling continued. A special effect occurs when the equatorial water spreads poleward; the temperature layering is depressed and there is a strong poleward flow subsurface. The latter is a planetary effect whereas some upwelling of warm water might occur in the surface layers (O'Brien, personal communication). The effect of this might be to maintain some production. The productivity of normally upwelled water is about $1gC/m^2/day$ and that during El Niño is about $0{\cdot}3gC/m^2/day$ (Guillen, personal communication); but that of tropical surface waters is much less. Thus it seems likely that the poor year classes of anchoveta during 1972 and 1973 are related to the reduced productivity of El Niño water; it is not impossible (but cannot be shown) that the failure of the 1971 year class was due to abnormal upwelling in the pre-El Niño phase.

From this information the match/mismatch hypothesis cannot be sustained. However, in more general terms, the possibility is established that the strength of a year class may depend upon the quantity of food available to it. Because El Niño varies the amount of food, the stock of anchoveta responds by reducing its recruitment. Fortunately, it now appears possible to forecast El Niño (see p. 53) and in future some of the ill effects to the fishery will be mitigated. This study emphasizes the point that a stock responds to its environment by modifying its recruitment and that environment includes the effect of fishing. During the decade of heavy exploitation, the recruitment to the anchoveta stock increased until the 1971 year class, probably as the number of guano birds declined (Jordan, 1974). With low succeeding year classes, stock has declined considerably, from the right to the left of a possible stock/recruitment curve. The point here is that the time-honoured distinction between the effects of fishing and the environment is becoming tiresome: it is just possible that an environmental effect has rendered the stock more vulnerable to fishing.

Summarizing, fishermen and fisheries biologists have been aware of a relationship between fish stocks and the climate (but not the weather) for a long time. The correlations with a general climatic trend shown by Ottestad and Zupanovitch show that there is some basis for the legend, and the distant correlations of Ishevskii suggest that many factors over a very broad area might indicate the critical ones. The match/mismatch hypothesis has the great advantage that such critical factors are taken into account. The inverse correlation with temperature, which occurs in a number of stocks, is well explained on the basis of such a hypothesis, as are the correlations with wind from one direction that fail after a period of time.

If the match/mismatch hypothesis is true and if a good model of the production cycle can be constructed, then the information from the daily weather report might be used to estimate the degree of match or mismatch in any one year. In other words the environmental factors that drive the production cycle would be used in the model to estimate the distribution of the cycle. The match or mismatch is estimated by the overlap of the distribution of larvae (which is a function of the production cycle). Year class strength might well be correlated with an index of this overlap.

E. *Ecosystem changes*

The events in the western English Channel during the thirties have already been described. An ecosystem of winter spawning herring, macroplankton (including arrow worms), and spring and summer spawning fish was largely replaced by one dominated by summer spawning pilchards. The winter phosphorus declined by about one third and the abundance of elasmobranchs and the benthos was reduced. The recent reversal of the change indicates a link with climatic events, but whatever the mechanism involved, a large part of the ecosystem changed its character completely. The initial change may have occurred in three stages, 1925, 1930/1 and 1934/5, and the reversal started in 1965/6 and continued in 1970/1. In 1951 and 1952, the Californian sardine collapsed, possibly due to heavy fishing and by 1966 the anchovy stock had replaced the sardine stock in weight (Ahlstrom, 1966). The changeover from one competitor to another had taken about a decade or about three generations. Thus the "rectification" of climatic change observed over very long periods of time in some pelagic stocks may originate in the same sort of competitive change and may equally involve large parts of the ecosystem.

Rather similar changes occurred in the North Sea during the fifties. The first change to be noticed was in the biology of the North Sea herring. By 1950 the copepod *Pseudocalanus* had been replaced by another, *Calanus*, and a sharp increment in the growth of herring in that year was associated with changes in the pattern of recruitment to the herring stock in subsequent years (Cushing and Burd, 1957). Later it was shown that the increments of growth in the second and third summers of the herring's life were correlated with the quantities of *Calanus* in April, May and June, both east and west of the Dogger Bank, where the herring nurseries lie; in the first summer, no such effect was detected because the growth then is markedly density dependent. The increment of growth added more than 10% to the catches of herring in the North Sea during the fifties. Indeed on a somewhat longer time scale from 1924 onwards, the increment in catch due to growth alone was much greater (Burd and Cushing, 1962).

The planktonic change in the North Sea was not confined to copepods. It was a more extensive change from neritic animals (the copepods *Pseudocalanus*, *Paracalanus*, *Centropages hamatus*, *Acartia*) to oceanic ones (the molluscs *Clione*, *Spiratella* and the copepod *Centropages typicus*) (Glover, 1957). Williamson (1961) analysed a matrix of 23 entities from the plankton recorder (e.g. *Temora longicornis*, Cladocera, young fish) for the period 1949–59; rank correlations showed a difference between 1949–52 (and 1959) and 1953–58. Variance was analysed by principal component analyses and two-thirds was correlated with one index of vertical mixing and with another describing the rate of its decline in spring. Bainbridge and Forsyth (1972) have carried out a similar analysis in the restricted area off Buchan Head and the Shetlands (3°W to 1°E; 57°N to 62°N) on the material from the plankton indicator (which is a little plankton net); the largest component was correlated with vertical mixing, but the second component was correlated with advection, which is not surprising in that rather restricted area. This work is most important because it provides a direct link between the climatic factors and the changes that took place in the ecosystem; indeed they were related to the variability of the production cycle, which implies that the match/mismatch hypothesis need not be limited to the generation of year classes in fish stocks but could be applied to the ecosystem as a whole.

Similar changes on a more extensive scale have been described by Glover *et al.* (1972). In the North Sea and in the waters west of the British Isles considerable changes in the planktonic populations have been described since 1948. The numbers of *Calanus*, *Metridia lucens* Boeck, *Candacia armata* Boeck, *Centropages typicus*, *Spiratella retroversa* (Fleming) and of *Pseudocalanus* and *Paracalanus* have declined over the period of twenty years. The numbers of *Pleuromamma borealis* (F. Dahl), *Euchaeta norvegica* Boeck and *Acartia clausi* Giesbrecht have increased, whereas those of *Temora longicornis* and

Clione limacina Phipps have remained more or less steady. There has been a decline in total biomass in the North Sea and the North Atlantic and in the North Sea, at least, the spring out-burst has occurred later by about two weeks, on average. Bainbridge and Cooper (1973) have shown that the main area of blue whiting spawning has shifted to the south from Rockall towards the Porcupine Bank since 1965.

Perhaps the most striking long-term changes in ecosystems are shown in Fig. 36. A systematic decline in copepod numbers and

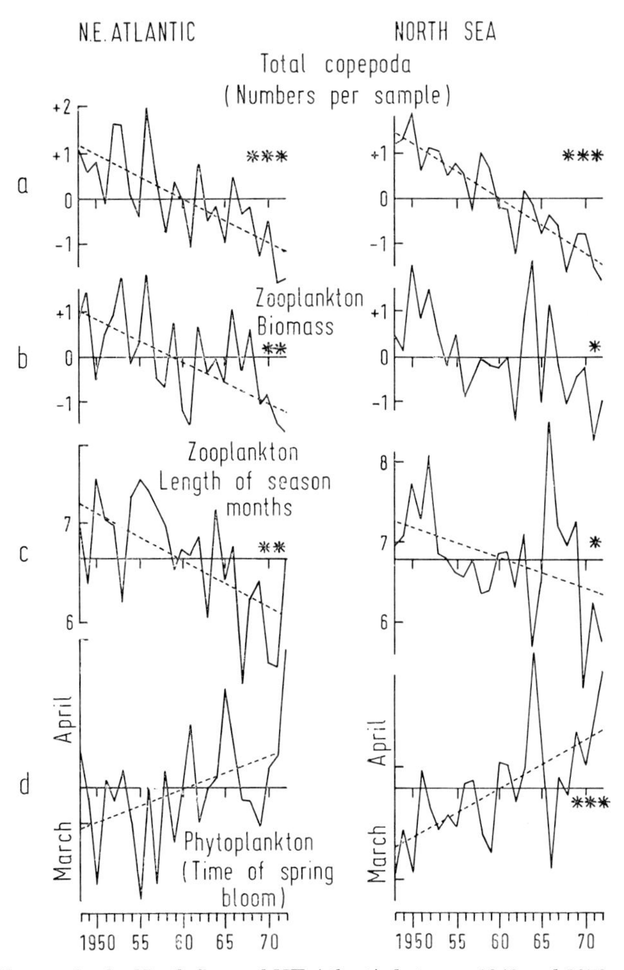

FIG. 36. Changes in the North Sea and NE Atlantic between 1948 and 1972 as shown by the plankton recorder network; (a) numbers of copepods ; (*b*) zooplankton biomass; (*c*) length of zooplankton season; (*d*) timing of spring bloom. (From Glover, Robinson and Colebrook, 1972.)

zooplankton biomass is associated with a contemporaneous delay in the timing of the spring bloom and decline in the length of the zooplankton season in the North Sea and North Atlantic between 1948 and 1972. The average delay to the spring bloom in the two regions is of the order of two to four weeks in 24 years and the decline in numbers and biomass of zooplankton is probably consequential. The mechanism of delay is unknown, but it might well have been associated with long-term changes in wind strength and direction which are part and parcel of the differences in climate recorded across the years.

Such changes recorded in Fig. 36 are of the type that must affect the whole ecosystem profoundly. The Russell cycle in the western English Channel, the gadoid upsurge during the sixties and the effects of the warming during the thirties in the North Atlantic were all serial events in which different parts of the ecosystem appeared and disappeared in an order which we do not yet understand.

Such changes are those that might be expected if the long-term climatic changes were mediated in the recruitment to the fish stocks and plankton populations by the match or mismatch of larval production and that of larval food.

F. *Conclusion*

There is a little more to the match/mismatch hypothesis than just a relation between year class strength and the quantity of food available, however acceptable such a thesis might be on general grounds. The numbers of any population are controlled by three processes, the potential recruitment as given by the number of eggs hatched, the density dependent or compensatory mechanisms, and the competition between species that has to be maintained from generation to generation. The match/mismatch hypothesis merely states that the quantity of food available depends upon the variability of the production cycle. It is likely, however, that the mortality of larvae (and perhaps 0-group fish) also depends upon the food. If larval fish eat well, they grow well, swim well, they tend to avoid predation, whereas if they starve, they will be subject to predation; thus differences in mortality are due to differences in available food. In such a system, differences in numbers of larvae will generate differences in death rate and so the dependence of mortality on food is also density dependent.

By the same reasoning, growth must be density dependent and so mortality rate and growth rate are both linked to the availability of food. Because the animals grow quickly in their early life history, they pass from one predatory field to another and each succeeding predator is larger, but very much less numerous than its predecessor, both growth rate and mortality rate decrease with age. Competition between populations is expressed by an increase in biomass by the successful competitor. In the long term of generations this is an increase in the number of eggs hatched, an increase in stock biomass and increases in the integrated growth throughout a cohort. In the short term of life in the larval drift or on the nursery ground, competition might well be expressed in the ratio of growth to mortality during a period of time. Thus, during the early stages of the life history, the potential recruitment is modified by density-dependent and competitive factors and all three population processes may take place at the same time and may be expressed very generally in the use of the available food by the developing cohort. It is a single process that may persist for a considerable proportion of the life cycle, but the major changes must occur during the early stages. The single process is not a simple one, however well it is described as the dependence of larvae upon their food. The initial numbers of larvae are provided by the stock and their food is produced in the variable production cycle. The density-dependent control may be effected in the number of larvae per food organism and the magnitude of recruitment is the result

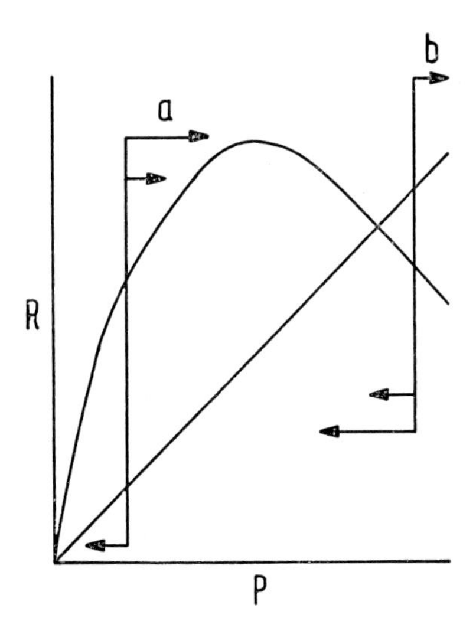

FIG. 37. Variability about a stock and recruitment curve. Normally the stock stabilizes itself about the stabilization point; at low stock any recruitment greater than stock (i.e. above the bisector) returns stock to the stabilization point and at high stock any recruitment less than stock fulfills the same purpose. At very low stock, a succession of poor year classes (each less than stock) must lead to a sharp and dramatic decline of stock; at very high stock, a succession of rich year classes will ensure a dramatic increase (Cushing, 1971).

of the maximization of biomass during the life history, but primarily in the early stages; competition is maintained in the ratio of growth to mortality at particular critical stages of food sharing during their development. The real point is that the fish stock has become adapted to the variability of the production cycle at a particular position, that of the larval drift and nursery ground.

The processes may be summarized in a curve expressing the dependence of recruitment on parent stock (Fig. 37). The curve itself expresses the ratio of density-dependent to density-independent mortality that increases with increasing stock. The variance about the curve expresses the variability of recruitment probably due to differences in available food at one level of stock and is therefore environmentally determined, presumably by the factors that control the production cycle (irradiance, wind strength and direction) and the production of larvae (primarily temperature). The figure shows other possibilities: a succession of extremely low year classes at low stock can cause collapse and a succession of very high ones at high stock can cause an extraordinary increase. Thus, within the mechanisms described by the stock and recruitment curve there exists the possibility of the large changes required by the rectification of climatic periodicity, as for example in the long-term clupeid periodicities and in the Russell cycle. If the changes in the ecosystem expressed in the Russell cycle are rectified in the way suggested, then the stock and recruitment relationship developed for fishes might be applicable to other populations within the system.

In order to achieve the effect of rectification, the variance of recruitment must be fairly high; a stock with very low variance might not be able to survive in temperate waters—although it might do so in tropical or subtropical ones where food is continuously available. Perhaps rectification is absent in tropical and subtropical systems. When recruitment fails dramatically as in the Plymouth herring (Ford, 1933) the mismatch may be considerable, food is very scarce and the larval mortality rate increases. Then stock survives at very low levels by reducing the numbers of larvae. In later decades the reverse process occurs, as if by colonization; the match improves, more food is available, mortality decreases and perhaps a competitor is eliminated.

It is likely that the changes in distribution, for example the colonization by cod of West Greenland, are essentially changes of the same sort. If cod larvae reach West Greenland from Iceland there must be enough available food during their first summer to support the cohort and of course its abundant successors. If our interpretation of the match/mismatch hypothesis is right, then much of the success of colonization must have depended upon production on the long larval drift from Iceland to West Greenland. *Octopus* did not survive the plague years in the western English Channel, but the general northerly movement during the thirties must have depended first on survival and secondly upon colonization.

The dramatic changes in the fisheries depend upon the mechanisms that have been indicated in the rectification of the climatic periodicities. It is often said that a forecast of recruitment to a fish stock should not be attempted because it is a random variable. Although many differences from year to year are small enough not to need forecasting there are two needs for prediction. First, predictions of year class strength on a pragmatic basis are already being used in catch quota regulations. Secondly, the large changes should be predictable; for example, the haddock year class of 1962, the 1904 year class of Atlanto-Scandian herring, the collapse of the Plymouth herring fishery. Such changes concern the very existence of industries, not merely their year to year profitability.

An important scientific point concerns the considerable changes that occur within the ecosystem, for example in the Russell cycle and in the North Sea plankton. There is no information yet on the variation in algal species under such circumstances but every other part of the food web appears to have been affected in such changes. This implies first that the variability of the recruitment to the stocks of plankton animals, herbivores and carnivores, is as variable as that of the fishes or the dramatic changes could not take place. Secondly, it seems likely that the stabilization mechanisms resemble those of fishes and indeed most copepods are relatively fecund. It is obvious that the death of one animal is the food of the other, or that the mortality rate of one is for a short time the growth rate of another, but it is a consequence that the variation in one trophic level can be transferred to another. Thus variation in the herbivore composition of species in the North Sea affected the growth rate and recruitment of the herring. Conversely, the competition between pilchard and herring in the western English Channel may have affected other parts of the ecosystem, if not all.

We tend to imagine the ecosystem as a somewhat stable structure with the carrying capacity of the environment exploited to the utmost. This may or may not be true, but within the total energy transferred to the top predators, there is scope for considerable variability probably at all levels. Just as a fish population may be considered as responding continuously to climatic change through the quantity of food available, so must other populations in the system. The carrying capacity might be expected to vary in the same sort of way and its capacity to respond can be little more than the capacity of its component populations to do so.

VI. The Biological Response to Climatic Change

There has been a long history of the biological indications of climatic change in the sea. Records of the appearances of uncommon animals were kept in the museums and formed the bases of early analyses by Day and later by Storrow. The long memories of fishermen stored oral records of when they saw whales, herring in the waves or strange invaders like the sunfish or Ray's bream. Because they lived within the weather, they had a sense of climate before the science of climatology developed. They knew that the herring appeared or disappeared at particular places capriciously and might guess that the time had come again to make a voyage to Iceland or Bear Island.

Such indications imply that a relationship exists between biological events in the sea and the changes in climate. Marine biologists have tried to establish such links for many years and the purpose of this review is to present a contemporary account of them. In the last decade or so, the part played by the ocean in modifying climatic events has become clarified to some degree, particularly in the Pacific. The scale of such events is large and it is no longer surprising to find links between one side of the world and the other, or even between the northern and southern hemisphere. The biological events are usually of smaller scale and are reported in regions such as the North Sea, Gulf of Maine or the northern and central coasts of Peru. Indeed, the places where some of the crucial biological mechanisms occur might be small in scale and particular in character, such as the western English Channel or the north-west coast of Iceland. This is because the mode of biological response to climatic events tends to be (but is not exclusively) that of a population; the numbers change slowly or sharply in time and such changes are manifest most clearly in restricted areas such as spawning grounds. This final section summarizes the scales of climatic change, the biological events associated with each scale and the forms of biological change that might be expected under different circumstances.

A. *The scales of climatic change*

The shortest scale of climatic change of interest is a seasonal one. Presumably the wind systems persist for rather longer and the temperate spring becomes delayed. Such effects are seen directly in changes in the growing season on land or in the long-term changes in air or sea temperature. In the sea, the most important seasonal effect in temperate waters is the delay or advance in the spring outburst which might determine the recruitment to the fish populations. Such effects are probably the consequence of annual shifts in the local pressure systems. In the north-east Atlantic it is common experience that in one year the depressions may track in one spring across the British Isles and in the next towards the Lofoten Islands, with consequent difference in wind strength and direction. Farther south, as spring advances into summer, the subtropical anticyclone shifts poleward to some degree and the zones of upwelling in the eastern boundary currents shift poleward as well. It is likely that particular points of upwelling, for example, equatorward of a cape, might become more or less important in a particular year as the angle made by the wind to the coast changes. On a larger spatial scale, seasonal differences in the equatorial shift of the subtropical anticyclone might affect the system of equatorial currents, the migration of animals within them, the productivity in the associated divergences and such phenomena as El Niño. The important point is that each seasonal progress is a world-wide event and that its consequences may be shown in an extensive array of apparently disparate biological events across the world ocean.

Seasonal changes are less noticeable in a climatic sense than those that persist for a period of years, sometimes as few as five years and sometimes as long as a century. In the British Isles during the five years up to 1973, the month of October was exceptionally fine and warm and over the years one learns to detect similar patterns. Probably the most important of such spatio-temporal patterns is the appearance of broad areas of subsurface temperature anomaly for a few years at a time in the northern subtropical Pacific. These areas in the sea are linked to the Rossby wave system aloft and with atmospheric cyclones and anticyclones that spin from west to east at surface level across the northern hemisphere.

In the waters of the north-east Atlantic, there is a periodicity of about five years or so in the incursion of high salinity water into the Baltic similar to the periodicity in salinity anomalies in different regions of the North Sea and Western Approaches. Indeed this change in salinity on the European shelf is merely one local expression of a dominant inter-annual hydrographic change which is pan-Atlantic in scale. The dates of recurrence of this trend of hydrographic change appear time and time again in the literature. The 1950 year classes of gadoid fishes, and the 1904 year class of Atlanto-Scandian herring are examples. In 1925/6, 1930/1 and 1934/5, profound changes occurred in the marine ecosystem of the western English Channel, in the appearance of turtles off western France, the appearance of southern fishes off California and Oregon and in the Bay of Fundy. Similarly there are years when certain indicator species appear, *Dolioletta lesueuri* in 1921 in the North Sea, *Pelagia* in the western Channel in 1966. Frequently, such events occur in or close to the years of high salinity inflow to the North Sea. Biologically, each event is a remarkable and noticeable one wherever it occurs.

There are longer periods of climatic importance with a " cycle " of a century or more, for example, the shift in wind direction from north-westerly to south-westerly or even southerly over the British Isles (Fig. 13). Although indications of such periods can be elicited from the data of earlier centuries, it is only during the recent warming since the end of the last century and the subsequent cooling in the last twenty years that the periodicity has been adequately documented. During the thirties the north-east Atlantic became warmer and northern waters were colonized by a variety of animals. The proximate cause of the warming was the weakening of the westerlies as the pressure difference between the Iceland Low and the Azores High slackened; as the wind strength was reduced, more heat was retained in the water. However, many other changes must have occurred; at particular positions at which fish spawn, the strength, direction and persistence of the winds probably changed. The amount of cloud cover would have been reduced in some positions and increased at others. The northern ice boundary retreated and with it the polar front and perhaps the nature and strength of the east Icelandic current. Most dramatically, an increment of warmth in the north-east Atlantic is accompanied by a decrement in the north-west.

At any one position a set of events may change radically in a quasi-periodic manner. Although they may originate in remote events (in the output of the great heat engine in the eastern tropical Pacific, for example), the effects spread in a climatic sense downstream from that remote source. Consequently biological events may be linked across enormous distances, but the local agents of change may differ profoundly. The year classes of cod at local Labrador may be directly affected by differences in temperature, those in the north-east Arctic by differences in wind direction and the invasion of west Greenland by the cod might have been the consequence of structural water mass changes off north-west Iceland. Such are three examples from the same species in the Atlantic. The diverse reactions by the three stocks are in response to climatic events that might have a common or a simpler origin.

The physical events of El Niño can be related convincingly to three sets of climatic events, the strength of the upper westerlies in the Pacific, the difference in sea level between one end of the equatorial Pacific and the other, and the differences in transport in the North Equatorial countercurrent. All three components express facets of the same simple heat engine that straddles the Pacific Ocean. This expresses the climatic diversity from the opposite point of view, that of the climatologist, not the biologist. If we attempt to relate the biological events with the climatological, we must be aware that both are multivariate.

Not only are both sets of processes multivariate, but the climatic events exist at all scales and so are also multivariate in a temporal sense. They are thus linked spatially and temporally. For the biological events, the ultimate climatic causes may be remote but the proximate causes are as diverse as the spread of climatic factors downstream from the same source. To understand the biological events, the proximate causes are more important than the remote and ultimate ones although the latter are useful in linking those biological events remote from each other.

B. *The forms of biological response to climatic change*

The simplest response was probably the earliest to be recorded, the appearance of an exotic invader, the indicator species. Our intuitive picture of events in the sea tends to be one of turbulent and confused mixture and we tend to resist evidence of stable structure; the invader appears capriciously from distant places and indicates an unknown event. Most dramatically, the appearance of the milkfish in the eastern tropical Pacific from the Western Ocean during the most recent El Niño might have been associated with the increased transport in the North Equatorial countercurrent; a distinction between this event and earlier ones was that such augmented transport is now considered as a reliable predictor of El Niño.

An indicator is a noticeable, rather large and long-lived animal. It is, however, an individual and not a population. Of course, groups of indicators appear, but they represent no transfer of population, no colonization and the group itself is lost from the parent stock. There is a disadvantage to the use of indicators and it is that their presence may be interpreted as significant, but their absence cannot; El Niño has occurred on many occasions during the present century but the milkfish never appeared in the eastern tropical Pacific until the last one in 1972 and 1973. However, during a long enough period, an array of indicators, as opposed to a somewhat vulnerable single species, will probably show events that are based upon climatic change. Yet the events are biologically trivial because indicating animals are expendable by the population or by the species.

There is a point in scale at which a group of indicators becomes significant and a colonization takes place. There are two components to a colonization. The first is a stray from the parent population, even if it is as low as that from salmon parent streams in the Pacific. The second is a gap in the niche structure of the colonized area, like that exploited by the small-mouthed bass released in San Francisco Bay which grew to yield catches of 500 tons in twenty years. The two examples have been deliberately chosen to suggest that colonization can proceed in either of two modes, as juveniles or as adults, their parents. Common to the two modes is the generation of colonization through increased recruitment from generation to generation. This is how climatic change can be mediated in biological terms.

The colonization of northern waters in the Atlantic during the period of warming was one of great extent geographically, of enormous biological complexity and of great economic importance. The most striking single event was the appearance of a great cod stock off West Greenland which did not exist in the first decade of the century. The cause of this event is unknown, but it may have been the survival in warmer water off West Greenland of larvae that had travelled from Iceland; it is also possible that a change occurred in the spawning distribution off north-western Iceland. Whatever the cause, a climatic change indicated by rising sea temperatures was mediated by an increase in recruitment; it is significant that as the climatic change regressed, during the fifties and sixties, the year classes at West Greenland declined. Since 1968, they have been very low.

If colonization is the effective transport of populations in response to climatic changes, so the same populations can respond *in situ*. Let us consider the changes in recruitment to fish populations as representing those of animals less well known. The variability of year classes is high and in temperate waters it might well be related to the match of larval production to that of their food. The thesis is illustrated well in Fig. 32 and Fig. 35. It is observed that the variability of year classes is often greater in deeper water, where the variability of the production cycle in temperate waters must be greater. Such variation follows from the fact that the spring outburst must start from nothing as winter passes. In subtropical waters (or perhaps more exactly, equatorward of 40°) production is more or less continuous and there is no high amplitude production cycle; consequently we would not expect the match/mismatch hypothesis to apply. The variability in year class strength in subtropical waters is not very well documented, but the stocks there are not very long-lived. In high latitudes, some fish live very long and the variation of their year classes is high; one might suggest that they live long because there are long intervals between very large year classes. The reverse may be true in low latitudes because the variability of year classes does not appear to be very great.

Because the production cycle in temperate waters may be modulated by variations in wind strength and direction and by changes in radiation, it is the only agent by which variation in recruitment might be affected by climatic changes. The trends in year class strength upward or downward must reflect trends in one of the many facets of climatic change. There are two types of such change, the adaptation of cod stocks by slowly varying their levels of recruitment, and the yielding of herring stocks by dropping their year class strengths by many orders of magnitude leading to periodic extinctions of the fisheries. The difference between the cod and the herring is in the nature of the stock/recruitment curve. The dome-shaped curve of the cod allows the stock to respond quickly, but the near linear curve for the herring allows it to only respond slowly and every hundred years or so it may fail to do so. Thus we may distinguish between slow upward and downward trends in recruitment that occur more or less continuously in cod stocks and the sharp rectifications that take place periodically in herring stocks. The same distinction can be drawn between the slow and continuous changes shown by the plankton biomass in the north-east Atlantic in the last 25 years and the dramatic rectification evinced in the Russell cycle. However, we are no longer examining the stabilization mechanisms of single populations, but the responses of ecosystems. One might imagine that the trends in timing of the production cycle affect the biomass of zooplankton or the duration of its growing season because the production cycle was shown to occur later during the two decades. The same cause might affect the Russell cycle but it must have penetrated all the nooks and crannies of the ecosystem. Perhaps the real point is that the Russell cycle represents the only study of an ecosystem for a long period of years.

C. *The links between the scales of climatic change and the forms of biological response*

We have selected three scales of climatic change as being of greatest relevance: the seasonal ones each year, those that occur every five or ten years, and those that take place with a period of about a century. There are four categories of biological response: the appearance of indicator species, of colonizing populations, of numerical increments or decrements to the fish stocks as year classes trend slowly upward or downward, and the structural changes that ecosystems suffer.

The prime effect of the seasonal changes is upon the generation of recruitment (for example, to the stocks of fish as representing the other populations in the sea). We have put forward the hypothesis that year class strength in temperate waters depends upon the match or mismatch of larval production to that of their food. Because spring-spawning fish appear to spawn at fixed seasons, the match or mismatch is generated by the variation in food production, itself determined by climatic factors. As such factors slowly change in short periods of a decade or less the same mechanism is sufficient to describe the trends in year class strength.

On the longer time scale, the period of which is about a century or so, stocks of herring appear to rectify the periodicity. The simplest explanation may be that the mismatch is great enough to reduce recruitment by many orders of magnitude and then some forty or fifty years later, the reverse process takes place because the match has once more become good. The response of stocks of cod is probably of similar character but of less magnitude. The fishery in the Vestfjord has survived for centuries, never becoming extinct, so the differences in average year class strength between periods cannot be too great. However, in an ecosystem, the more numerous herring-like fishes may play a more controlling part. The changeover from herring to pilchard in the Russell cycle was accompanied by an array of changes which may have been consequential or contemporaneous.

The appearance of an indicator is quite unrelated to the match/mismatch hypothesis, being a more direct consequence of changes in the current systems, themselves mediated by climatic events. A colonization, however, comprises components of oceanic transport and an excess of food available in the area colonized. The match/mismatch hypothesis describes the changes that occur downstream of a spawning ground that is not only constant in time but fixed in position. But a colonization implies a new spawning distant from the parent one. However, the appearance and the potential disappearance of the cod stock at West Greenland rectifies the long-term changes much as does the Atlanto-Scandian stock of herring. Cod as a species can resist environmental changes, but as a colonizer, cannot. Perhaps a more productive way of considering a colonization is as a change or even a distortion of migration. The triangle of migration is based on the larval drift between spawning ground and nursery ground; its apex is the adult feeding ground to which the immature fish move from the nursery ground and from which the adults migrate annually to the spawning ground and back again. A colonization starts as an extension of the larval drift and indeed part of the West Greenland stock continued to recruit to the parent spawning ground in Iceland. Yet at the same time when the new spawning ground was established the result was a completely new migratory circuit.

D. *Conclusion*

We have examined the response of animals to climatic change on a world-wide scale (in so far as we have been able) because there is evidence of coherent trends of climatic change on a hemispheric scale. In terms of populations, the response to climatic change is the prime environmental agent. Individuals are affected by a gamut of environmental factors, but the populations have evolved to resist them in so far as they can. The evolution of migration has resulted in a larval drift between spawning ground and nursery ground, where presumably the opportunities for larval survival, in terms of food, have been maximized for very long periods. Thus the match/mismatch mechanism is really part of the pattern of migration; it should be absent from populations that do not migrate and from those that may not experience lack of food (as in the subtropical and tropical ocean). Because the mechanism is linked with the evolutionary survival of the population it may also play a part in the stabilization mechanism; it is possible that the density-dependent mortality responsible is also a function of available food. Indeed it would be parsimonious to link match/mismatch and stabilization.

VII. References

Ahlstrom, E. H. (1966). Distribution and abundance of sardine and anchovy larvae in the California Current region off California and Baja California, 1951–64: a summary. *Special Scientific Report of the U.S. Department of the Interior Fish and Wildlife Service (Fisheries)*, No. 534. 71 pp.

Andersson, K. A. (1956). De stora sillfiskeperioderna på Sveriges Vastkust ochsillen, some framkallade dem. *Institute of Marine Research, Lysekil Series Biology Report* No. 5 1–28.

Bainbridge, V. and Cooper, G. A. (1973). The distribution and abundance of the larvae of the blue whiting *Micromesistius poutassou* (Risso) in the Northeast Atlantic 1948–70. *Bulletin of Marine Ecology*, **8** (2), 99–114.

Bainbridge, V. and Forsyth, D. C. T. (1972). An ecological survey of a Scottish herring fishery. V. The plankton of the Northwestern North Sea in relation to the physical environment and the distribution of herring. *Bulletin of Marine Ecology*, **8** (1), 21–52.

Bakun, A. (1973). Coastal Upwelling Indices, West Coast of North America 1946–71. *NOAA Technical Report*, NMFS SSRF–671, 103 pp.

Bannister, R. C. A., Harding, D. and Lockwood, S. J. (1974). Larval mortality and subsequent yearclass strength in the Plaice (*Pleuronectes platessa* L.). *In* " The Early Life History of Fish " (ed. J. H. S. Blaxter), pp. 21–38. Springer-Verlag, Heidelberg and New York.

Berg, L. S. (1935). Rezente Klimaschwankungen und ihr einfluss auf die geographische Verbreitung der Seefische. *Zoogeographica*, **3**, 1–15.

Beverton, R. J. H. (1962). Long term dynamics of certain North Sea fish populations. *In* " Exploitation of Animal Populations " (E. D. Le Cren and M. W. Holdgate, eds.), pp. 242–264. Blackwell Scientific Publications, Oxford.

Beverton, R. J. H. and Hodder, V. M. (1962). Report of Working Group of Scientists on Fishery Assessment in relation to regulation problems. *Supplement of the Annual Proceedings of the International Commission on North-west Atlantic Fisheries*, **11,** 81 pp.

Beverton, R. J. H. and Lee, A. J. (1965). Hydrographic fluctuations in the North Atlantic Ocean and some biological consequences. *In* " The Biological Significance of Climatic Changes in Britain " (C. G. Johnson and L. P. Smith, eds.). Institute of Biology Symposia, No. 14, pp. 79–107. London and New York.

Bjerknes, J. (1966). Survey of El Niño 1957–58 in its relation of tropical Pacific Meteorology. *Bulletin of the Inter-American Tropical Tuna Commission*, **12,** (2), 25–86.

Bjerknes, J. (1972). Global Ocean-Atmosphere Interaction. *Rapports et Procès-verbaux des Réunions. Conseil international pour l'Exploration de la Mer*, **162,** 108–119.

Blacker, R. W. (1957). Benthic animals as indicators of hydrographic conditions and climatic change in Svalbard waters. *Fishery Investigations, London*, Series II, **20** (10) 1–49.

Blacker, R. W. (1965). Recent changes in the benthos of the West Spitzbergen fishing grounds. *Special Publication of the International Commission on the North-west Atlantic Fisheries*, No. 6, 791–794.

Blaxter, J. H. S. and Hempel, G. (1963). The influence of egg size on herring larvae (*Clupea harengus* L.). *Journal du Conseil*, **28,** 211–40.

Blindheim, J. (1967). Hydrographic fluctuations off West Greenland during the years 1959–1966. *Red book of the International Commission on the North-west Atlantic Fisheries*, 1967, Pt. IV, 86–105.

Boeck, A. (1871). On silden og silde fiskerierne, naun lig om det norske varsildfisket. Indberetning til den Konglige norske Regierings Department for der Indre am foretagst Praktisch-vidensk abelige Undersøgelser.

Bostford L. W. and Wickham D. E. (1975). Correlation of upwelling index and Dungeness crab catches. *Fisheries Bulletin*. **73,** 901–907.

Budyko, J. I. (1968). On the causes of climatic variations. *Meddelanden från Sveriges meterologiska och hydrologiska institut* Serie B 28, 6–13.

Burd, A. C. and Cushing, D. H. (1962). I. Growth and recruitment in the herring of the North Sea. II. Recruitment to the North Sea herring stocks. *Fishery Investigations, London*, Ser. 2. **23**, (5) 1–71.

Burmakin, V. V. (1971). Seasonal and year-to-year variability of water temperature in the areas of Labrador and Newfoundland. *ICNAF Research Document* 71/96, 18 pp. (mimeo).

Carruthers, J. N. (1938). Fluctuations in the herrings of the East Anglian Autumn fishery, the yield of the Ostend spent herring fishery and the haddock of the North Sea in the light of the relevant wind conditions. *Rapports et Procès-verbaux des Réunions. Conseil international pour l'Exploration de la Mer*, **107,** 1–15.

Colebrook, J. M. (1965). On the analysis of variation in the plankton, the environment and the fisheries. *Special Publication of the International Comission on the North-west Atlantic Fisheries*, No. 6, 291–302.

Cooper, L. H. N. (1938). Phosphate in the English Channel, 1933–8, with a comparison with earlier years, 1916, and 1923–32. *Journal of the Marine Biological Association of the United Kingdom*, **23,** 181–195.

Corlett, J. (1953). Net phytoplankton at Ocean Weather Stations " I " and " J ". *Journal du Conseil*, **19,** 2. 178–190.

Crisp, D. J. (1964). The effects of the severe winter of 1962–3 on marine life in Britain. *Journal of Animal Ecology*, **33,** 165–210.

Crisp, D. J. (1965). Observations on the effects of climate and weather on marine communities. *In* " The Biological Significance of Climatic Changes in Britain " (C. G. Johnson and L. P. Smith, eds.), No. 14, pp. 63–77. Institute of Biology, London and New York. (Synopsis.)

Crisp, D. J. and Southward, A. J. (1958). The distribution of intertidal organisms along the coasts of the English Channel. *Journal of the Marine Biological Association of the United Kingdom*, **37,** 157–208.

Cushing, D. H. (1957). The number of pilchards in the Channel. *Fishery Investigations, London*, Series II, **21,** (5) pp. 27.

Cushing, D. H. (1961). On the failure of the Plymouth herring fishery. *Journal of the Marine Biological Association of the United Kingdom*, **41,** 799–816.

Cushing, D. H. (1969a). Upwelling and fish production. *FAO Fisheries Technical Paper*, No. 84, 40 pp.

Cushing, D. H. (1969b). The regularity of the spawning season of some fishes. *Journal du Conseil international pour l'Exploration de la Mer*, **33,** 81–92.

Cushing, D. H. (1971a). A comparison of production in temperate seas and the upwelling areas. *Transactions of the Royal Society of South Africa*, **40,** (1), 17–33.

Cushing, D. H. (1971b). The dependence of recruitment on parent stock in different groups of fishes. *Journal du Conseil*, **33,** 340–362.

Cushing, D. H. (1972). The production cycle and numbers of marine fish. *Symposium of the Zoological Society, London*, **29,** 213–232.

Cushing, D. H. (1973). The natural regulation of fish populations. *In* " Sea Fisheries Research " (F. R. Harden Jones, ed.), pp. 399–411. Elek Science London.

Cushing, D. H. and Burd, A. C. (1957). On the herring of the Southern North Sea. *Fishery Investigations, London*, Series II, **20,** (11) 1–31.

Cushing, D. H. and Harris, J. G. K. (1973). Stock and recruitment and the problems of density dependence. *Rapports et Procès-verbaux des Réunions. Conseil international pour l'Exploration de la Mer*, **164,** 142–55.

Davidson, V. M. (1934). Fluctuations in the abundance of planktonic diatoms in the Passamoquoddy region, New Brunswick, from 1924 to 1931. *Contributions to Canadian Biology and Fisheries*, **8,** 357–407.

Defant, A. (1924). Die Schwankungen der atmosphärischen Zirkulation. *Geografiska Annaler*, **6,** 131–141.

Desbrosses, P. (1932). Capture d'une Tortue (*Dermatochelys coriacea* L.) dans la baie d'Etel. *Bulletin de la Société zoologique de France*, **57,** 274–277.

Desbrosses, P. (1935). Echouage d'un Tetrodon, *Tetrodon lagocephalus* L. pres de Quiberon et remarques sur la présence de cette éspèce et de *Balistes capriscus* L. au nord du 44°L.N. *Bulletin de la Société zoologique de France*, **60,** 43–48.

Desbrosses, P. (1936). Présence à l'entrée occidentale de la Manche de *Callanthias ruber* (Rafinesque 1810). *Bulletin de la Société zoologique de France*, **61,** 406–407.

Devold, F. (1963). The life history of the Atlanto-Scandian herring. *Rapports et Procès-verbaux des Réunions. Conseil international pour l'Exploration de la Mer*, **154,** 98–108.

Dickie, L. M. (1955). Fluctuations in abundance of the Giant Scallop *Placopecten magellanicus* (Gmelin) in the Digby area of the Bay of Fundy. *Journal of the Fisheries Research Board of Canada*, **12.** (6). 797–856.

Dickson, R. R. (1971). A recurrent and persistent pressure anomaly pattern as the principal cause of intermediate scale hydrographic variation in the European shelf sea. *Deutsche hydrographische Zeitschrift*, **24** (3), 97–119.

Dickson, R. R. (1973). The prediction of major Baltic inflows. *Deutsche hydrographische Zeitschrift*, **26** (3), 97–015.

Dickson, R. R. and Lamb, H. H. (1972). A review of recent Hydro meteorological events in the North Atlantic sector. *ICNAF Special Publication*, No. 8, 35–62.

Dickson, R. R., Pope, J. G. and Holden, M. J. (1973). Environmental influences on the survival of North sea Cod. *In* " The Early Life History of Fish " (J. H. S. Blaxter, ed.), pp. 69–80. Springer-Verlag, Heidelberg and New York.

Dickson, R. R., Lamb, H. H., Malmberg, S. A., and Colebrook, J. M. (1975). Climatic reversal in the northern Atlantic. *Nature, London*, 256 (5517), 479–481.

Dietrich, G. (1957). Schichtung und Zirkulation der Irminger See im Juni 1955. *Bericht der Deutschen wissenschaftlichen Kommission fur Meeresforschung* **14,** (4), 255–312.

Dietrich, G. (1963). New hydrographical aspects of the Northwest Atlantic. *Special Publication of the International Commission on the North-west Atlantic Fisheries*, **6,** 29–51.

Dow, R. L. (1969). Cycle and geographic trends in sea water temperatures and abundance in American lobster catches. *Science*. **164**, 1060–1062.

Eichenlaub, V. L. 1971. Further comments on the climate of the mid nineteenth century United States compared to current normals. *Monthly Weather Review*, **99** (11), 847–850.

Elizarov, A. A. (1965). Long-term variations of oceanographic conditions and stocks of cod observed in the areas of West Greenland, Labrador and Newfoundland. *Special Publicaiton of the International Commission on the North-west Atlantic Fisheries*, No. 6, 827–831.

Fonselius, S. H. (1962). Hydrography of the Baltic deep basins. *Report of the Fisheries Board of Sweden. Series Hydrography*, **13,** 40 pp.

Fonselius, S. H. (1967). Hydrography of the Baltic deep basins II. *Report of the Fisheries Board of Sweden Series Hydrography*, **20,** 31 pp.

Fonselius, S. H. (1969). Hydrography of the Baltic deep basins III. *Report of the Fisheries Board of Sweden, Series Hydrography*, **23,** 97 pp.

Ford, E. (1933). An account of the herring investigations conducted at Plymouth during the years from 1924 to 1933. *Journal of the Marine Biological Association of the United Kingdom*, **19,** 305–84.

Fridriksson, A. (1949). Boreo-tended changes in the marine vertebrate fauna of Iceland during the last twenty five years. *Rapports et Procès-verbaux des Réunions. Conseil international pour l'Exploration de la Mer*, **125,** 30–32.

Garrod, D. J. (1968). Population dynamics of the Arcto-norwegian cod. *Journal of the Fisheries Research Board of Canada*, **24** (1), 145–190.

Garrod, D. J. (1970). Reports of the north-east Arctic Fisheries Working Group. *ICES Cooperative Research Report*, Series A, No. 16, 60 pp.

Garrod, D. J. and Jones, B. W. (1974). Stock and Recruitment relationship in North-east Arctic cod stock and the implications for management of the stock. *Journal du Conseil*, **36** (1), 35–41.

Girs, A. A. (1963). The general distinguishing features of the interaction of various patterns of the atmospheric circulation over the northern hemisphere. *Izvestiya Akademii Nauk SSSR (geography)*, **6,** 15–26.

Glover, R. S. (1957). An ecological survey of the drift net herring survey off the north east coast of Scotland II; the planktonic environment of the herring. *Bulletin of Marine Ecology*, **5,** 1–43.

Glover, R. S., Robinson, G. A. and Colebrook, J. M. (1972). Plankton in the North Atlantic—an example of the problems of analysing variability in the environment. *In* "F.A.O. Marine Pollution and Sea Life", (M. Ruivo, ed.), pp. 439–445. Fishing News (Books), Ltd. West Byfleet, Surrey and London.

Gulland, J. A. (1968). Recent changes in the North Sea plaice fishery. *Journal du Conseil*, **31,** 305–22.

Hansen, P. M. (1949). Studies on the biology of the cod in Greenland waters. *Rapports et Procès-verbaux des Réunions. Conseil international pour l'Exploration de la Mer*, 5–77.

Hansen, P. M. (1954). The stock of cod in Greenland waters during the years 1924–52. *Rapports et Procès-verbaux des Réunions. Conseil international pour l'Exploration de la Mer*, **136,** 65–71.

Hansen, P. M. (1968). Report on cod eggs and larvae. *Special Publication of the International Commission of the North-west Atlantic Fisheries*, No. 7. (Environmental surveys NORWESTLANT 1-3 1963) 127–138.

Hansen, P. M., Jensen, A. S. and Tåning, A. V. (1935). Cod marking experiments in the waters of Greenland, 1924–33. *Meddelelser fra Kommissionen for Danmarks Fiskeri og Havundersøgelser*, **10,** 1, 119 pp.

Hermann, F. (1967). Temperature variations in the West Greenland area since 1950. *Red book of the International Commission on the North-west Atlantic Fisheries*, Part IV, 76–85.

Hermann, F., Hansen, P. M. and Horsted, S. A. (1965). The effect of temperature and currents on the distribution and survival of cod larvae at West Greenland. *Special Publication of the International Commission on the North-west Atlantic Fisheries*, No. 6. 389–409.

Hill, H. W. and Lee, A. J. (1957). The effect of wind on water transport in the region of the Bear Island fishery. *Proceedings of the Royal Society*, B **148,** 104–16.

Höglund, H. (1972). On the Bohuslån herring during the great herring fishery period in the eighteenth century. *Report of the Institute of Marine Research,* **20,** 1972. 1–86.

Holme, N. A. (1961). The bottom fauna of the English Channel. *Journal of the Marine Biological Association of the United Kingdom*, **41,** 397–461.

Horsted, S. (1973). Recent information on landings, age composition and recruitment of Subarea 1 cod, and estimates of yield in 1972–75. *International Commission on the North-west Atlantic Fisheries*, Res. Doc. 73/107. 23 pp.

Hubbs, C. L. (1948). Changes in the fish fauna of Western North America correlated with changes in ocean temperature. *Journal of Marine Research,* 459–482.

Izhevskii, G. K. (1964). Forecasting of Oceanological conditions and the Reproduction of Commercial Fish. Translated by the Israel Program for Scientific Translations, Jerusalem 1966 from Vsesoyuznyi Nauchno-Issledovatel'skii Institute Morskogo Rybnogo Khozyaistra I Okeanografii (VNIRO) 95 pp.

Jakobsson, J. (1969). On herring migrations in relation to changes in sea temperature. *Jokull*, **19,** 134–145.

Jensen, A. J. C. (1952). The influence of Hydrographical factors on fish stocks and Fisheries in the Transition Area, especially on their fluctuations from year to year. *Rapports et Procès-verbaux des Réunions. Conseil international pour l'Exploration de la Mer*, **131,** 51–60.

Jensen, Ad. S. (1925). On the fishery of the Greenlanders. *Meddelelser fra Kommissionen for Danmarks Fiskeri og Havundersøgelser*, **7** (7), 39 pp.

Jensen, Ad. S. (1939). Concerning a change of climate during recent decades in the arctic and subarctic regions, from Greenland in the West to Eurasia in the East and contemporary biological and geophysical changes. *Det Konglike Videnskabernes Selska Biologiske Meddelelser*, **14**, (8), 75 pp.

Jensen, Ad. S. and Hansen, P. M. (1931). Investigations on the Greenland cod *Gadus callarias* (L.). *Rapports et Procès-verbaux des Réunions. Conseil international pour l'Exploration de la Mer*, **72,** 1–41.

Johansen, A. C. (1924). On the summer and autumn spawning herrings of the North Sea. *Meddelelser fra Kommissionen for Denmarks Fiskeri og Havundersøgelser*, **5,** 119 pp.

Johansen, A. C. (1925). On the influence of the currents upon the frequency of the mackerel in the Kattegat and adjacent parts of the Kattegat. *Meddelelser fra Kommissionen for Havundersøgelser*, **7** (8), 26 pp.

Johansen, A. C. (1926). On the remarkable quantities of haddock in the Belt Sea during the winter of 1925–26 and causes leading to the same. *Journal du Conseil*, **1,** 140–156.

Jordan, R. S. (1974). Biology of the anchoveta; summary of present knowledge. 21 pp. IDOE Workshop on the Niño phenomenon, Guayaquil.

Kalle, K. (1943). Die grosse Wasserumschuchtung im Gotland-Tief vom Jahre 1933/4. *Annalen der Hydrographie und maritime Meteorologie, Berlin*, **71,** 142–146.

Kasahara, H. (1962). "Fisheries Resources of the North Pacific Ocean." H. R. MacMillan Lectures on Fisheries, University of British Columbia.

Ketchen, K. S. (1956). Climatic trends and fluctuations in yield of Marine Fisheries in the North Pacific. *Journal of the Fisheries Research Board of Canada*, **13,** 357–374.

Ketchen, K. S. (1956). Factors influencing the survival of the lemon sole (*Parophrys vetulus*) in Hecate Strait, British Columbia. *Journal of the Fisheries Research Board of Canada*, **13,** (5) 647–94.

Kislyakov, A. G. (1961). The relationship between hydrological conditions and variations of cod year-class abundance. *Trudy soveshchani. Ikhtiologich eskaya Kommisiya* **13,** 260–264.

Kitano, K. (1975). Some properties of the warm eddies generated in the confluence zone of the Kuroshio and Oyashio Currents. *Journal of Physical Oceanography*, **5,** (2), 245–252.

Kukla, G. J. and Kukla, H. J. (1974). Increased Surface Albedo in the Northern Hemisphere. *Science, New York* **183** (4126), 709–714.

Lablaika, I. A. (1961). Distribution and age composition of cod in the Gothland Deep in 1959. *Annales biologiques, Copenhagen.* **16** (1959), 146.

Lamb, H. H. (1964a). Climatic changes and variations in the atmospheric and ocean circulations. *Geologische Rundschau*, **54,** 486–504.

Lamb, H. H. (1964b). Neue Forschungen über die Entwicklung van Kleimaanderung. *Metallotechnische Rundschau*, **17,** 63–74.

Lamb, H. H. (1965). Frequency of weather types. *Weather, London*, **20,** 9–12.

Lamb, H. H. (1966). " The changing climate : selected papers," 236 pp. Methuen, London.

Lamb, H. H. 1969. The new look of climatology. *Nature, London*, **223** (5212): 1209–15.

Lamb, H. H. (1972). " Climate, past, present and future. I. Fundamentals and climate now ", 613 pp. Methuen, London.

Lamb, H. H. (1973). " Whither Climate Now? " *Nature, London*, **244** (5416), 395–397.

Lamb, H. H. (in press). The Current trend of world climate. A perspective. *UEA Climatic Research Unit. Research Publication* No. 3.

Larkin, P. A. (1970). Some observations on models of stock and recruitment relationships for fishes. *ICES Symposium on Stock and Recruitment.* Document No. 17, 14 pp. (mimeo).

Lee, A. J. (1968). Norwestlant surveys : Physical oceanography. *Special Publication of the International Commission on the North-west Atlantic Fisheries*, No. 7. Environmental surveys NORWESTLANT, 1–3, 1963, 31–54.

Ljungman, A. (1882). Contribution towards solving the question of the secular periodicity of the great herring fisheries. *United States Commission on Fish and Fisheries*, **7** (7), 497–503.

Longhurst, A. R. (1967). The pelagic phase of *Pleuroncodes planipes* STIMPSON (Crustacea, Galatheidae) in the California current. *CALCOFI Report* 11, 142–154.

Lumby, J. R. and Atkinson, G. T. (1929). On the unusual mortality amongst fish during March and April 1929, in the North Sea. *Journal du Conseil* **4,** 309–32.

Lund, J. W. G. (1964). Primary production and periodicity of phytoplankton. *Verhandlungen der Internationalen Vereinigung für theoretische und angewandte Limnologie*, **15,** 37–56.

Macer, C. T. (1974). Some observations on the fecundity of the pilchard (*Sardina pilchardus* Walbaum) off the south west coast of England. *International Council for the Exploration of the Sea*, CM 1974 J9, 6 pp (mimeo).

Malmberg, S.-A. (1969). Hydrographic changes in the waters between Iceland and Jan Mayen in the last decade. *Jökull*, **19,** 30–43.

Marr, J. C. (1960). The causes of major variations in the catch of the Pacific sardine *Sardinops caerulea* (Girard). *Proceedings of the World Scientific Meetings on the Biology of Sardines and related Species F.A.O.* III, 667–79.

Martin, W. R. and Kohler, A. C. (1965). Variation in recruitment of cod (*Gadus morhua* L.) in southern ICNAF waters as related to environmental changes. *Special Publication of the International Commission on the North-west Atlantic Fisheries*, No. 6, 833–846.

McKenzie, R. A. and Homans, R. E. S. (1937). Rare and interesting fishes and salps in the Bay of Fundy and off Nova Scotia. *Proceedings of the Nova Scotian Institute of Science*, **19,** (3) 277.

Meyer, P. F. and Kalle, K. (1950). Die biologische Umstimmung in der Ostee in den letzten Jahrezenten, eine Folge hydrographischer Wasserumschichtungen. *Archiv für Fischereiwissenschaft*, **2,** 1–9.

Midttun, L. (1969). Variability of temperature and salinity at some localities of the coast of Norway. *Progress in Oceanography* (5), 41–54.

Mironovitch, V. (1960). Sur l'évolution séculaire de l'activité solaire et ses liasons avec la circulation atmosphérique generale. *Meterologische Abhandlungen. Institut fur Meteorologie and Geophysik Freie Universität, Berlin.* **9,** 1–17.

Mitchell, J. M. Jr (1963). On the world-wide pattern of secular temperature change. *In* " Changes of Climate, Proceedings of the UNESCO and WMO Symposium, Rome ", pp. 161–181. UNESCO, Paris.

Namias, J. (1951). The great Pacific anticyclone of winter 1949–50 : a case study in the evolution of climatic anomalies. *Journal of Meteorology*, **8,** 251–261.

Namias, J. (1968). On the causes of the small number of Atlantic hurricanes in 1968. *Monthly Weather Review*, **97** (4), 346–348.

Namias, J. (1972). Large-scale and long-term fluctuations in some atmospheric and oceanic variables. *In* " The Changing Chemistry of the Oceans. Nobel Symposium 20," pp. 27–48. Almqvist and Wiksell, Stockholm.

Namias, J. (1973a). Response of the Equatorial countercurrent to the subtropical Atmosphere. *Science, New York*, **181** (4106), 1244–1245.

Namias, J. (1973b). Thermal communication between the sea surface and the lower troposphere. *Journal of Physical Oceanography*, **3** (4), 373–378.

Namias, J. (1974). Collaboration of Ocean and Atmosphere in weather and climate. *Proceedings of the Marine Technology Society 9th Annual Conference*, pp. 163–178.

Namias, J. and Born, R. M. (1970). Temporal coherence in North Pacific Sea-surface temperature patterns. *Journal of Geophysical Research*, **75** (30): 5952–5955.

Namias, J. and Born, R. M. (1972). Empirical techniques applied to large-scale and long-period air-sea interactions. A preliminary experiment. SIO Reference Series, pp. 72–1.

Namias, J. and Born, R. M. (1974). Further studies of temporal coherence in North Pacific sea surface temperatures. *Journal of Geophysical Research*, **79** (6) : 797–798.

Nesis, K. N. (1960). Variations in the bottom fauna of the Barents Sea under the influence of fluctuations in the hydrological regime. *Soviet Fishing Investigations in North European Seas, Moscow*, 129–138.

O'Connor, J. F. (1969). Hemispheric teleconnections of mean circulation anomalies at 700 mb. *ESSA Technical Report* WB10.

Ottestad, P. (1942). On periodical variations in the yield of the great sea fisheries and the possibility of establishing yield prognoses. *Fiskeridirektoratets-skrifter, Serie Havundersøkelser*, **7,** (5).

Pettersson, O. (1921–2). Kosmiska orsaker till rövelserna uti hafvets och atmosfarens mellanskikt. *Svenska Hydrografisk-biologiska kommissionens fyrskeppsundersökning*, **7,** 23 pp.

Pivovarova, Z. I. (1968). The long-term variation of intensity of solar radiation according to observations of actinometric stations. Leningrad. *Trudy Glavnoi geofizicheskoi observatorii.*, **233,** 17–37.

Quinn, W. H. (1974). Outlook for El Nino-type conditions in 1975. *NORPAX Highlights*, **2** (6) : 2–3.

Rae, K. M. (1957). A relationship between wind, plankton distribution and haddock brood strength. *Bulletin of Marine Ecology*, **4,** 247–69.

Raja, B. T. A. (1969). The Indian oil sardine. *Bulletin of the Central Marine Fisheries Research Institute*, **16,** 128 pp.

Raja, B. T. A. (1971). Fecundity fluctuations in the oil sardine *Sardinella longiceps* Val. *Indian Journal of Fisheries*, **18,** 84–98.

Rasmussen, B. (1959). On the migration pattern of the West Greenland stock of cod. *Annales biologiques*, **14,** 123.

Rees, W. J. (1951). The distribution of *Octopus vulgaris* Lamarck in British waters. *Journal of the Marine Biological Association of the United Kingdom*, **29,** 361–378.

Rees, W. J. and J. R. Lumby (1954). The abundance of *Octopus* in the English Channel. *Journal of the Marine Biological Association of the United Kingdom*, **33,** 515–536.

Robinson, G. A. (1969). Fluctuations in the timing of the spring outbreak of phytoplankton in the north-east Atlantic and the North Sea. ICES CM 1969 Doc. No. L:16 4 pp. and 4 figs. (mimeo).

Rodewald, M. (1972). Einige hydroklimatische Besonderheiten des Jahrezehuts 1961–1970 im Nordatlantik und in Nordpolarmeere. *Deutsche hydrographische Zeitschrift*, **25** (3), 97–117.

Rollefsen, G. (1949). Fluctuations in two of the most important stocks of fish in northern waters, the cod and the herring. *Rapports et Procès-verbaux des Réunions. Conseil international pour l'Exploration de la Mer*, **125,** 33–35.

Rossby, C.-G. (1959). Current problems in meteorology. *In* " The Atmosphere and the Sea in Motion ", pp. 9–50. (B. Bolin, ed.). Oxford University Press, London.

Rossby, C.-G. and *Collaborators* (1939). Relation between variations in the intensity of the zonal circulation of the atmosphere and the displacements of the semi-permanent centres of action. *Journal of Marine Research*, **2,** 38–55.

Runnstrøm, S. (1933). "Sildeundersøkelser." *Årsberetning Norges fiskerier*, 1931, 1, 65–77.

Russell, F. S. (1935). On the value of certain plankton animals as indicators of water movements in the English Channel and North Sea. *Journal of the Marine Biological Association of the United Kingdom*, **20,** 309–332.

Russell, F. S. (1967). On the occurrence of the Scyphomedusan *Pelagia noctiluca* in the English Channel in 1966. *Journal of the Marine Biological Association of the United Kingdom*, **47,** 363–66.

Russell, F. S. (1973). A summary of the observations of the occurrence of planktonic stages of fish off Plymouth, 1924–1972. *Journal of the Marine Biological Association of the United Kingdom*, **53,** 347–355.

Russell, F. S., Southward, A. J., Boalch, G. T., and Butler, E. I. (1971). Changes in biological conditions in the English Channel off Plymouth during the last half century. *Nature, London*, **234,** 468–470.

Saville, A. (1959). The planktonic stages of the haddock in Scottish waters. *Marine Research*, **3.**

Sawyer, J. S. (1965). Notes on the possible physical causes of long-term weather anomalies, *W.M.O. Technical Note*, No. 66, 227–248.

Scherhag, R. (1950). Die Schwankungen der atmosphärischen Zirkulation in den letzten Jahrzehnten. *Bericht des Deutschen Wetter dienstes*, **12,** 40–44.

Schmidt, J. (1909). The distribution of the pelagic fry and the spawning regions of the gadoids in the North Atlantic from Iceland to Spain. *Rapports et Procès-verbaux des Réunions. Conseil international pour l'Exploration de la Mer*, **10,** 4.

Segestråle, S. G. (1969). Biological fluctuations in the Baltic Sea. *Progress in Oceanography*, **5,** 169–184.

Shurin, A. T. (1961). Characteristic features of the bottom fauna in the eastern Baltic in 1959. *Annales biologiques, Copenhagen*, **16** (1959), 86.

Shurin, A. T. (1962). The distribution of bottom fauna in the eastern Baltic in 1960. *Annales biologiques, Copenhagen*, **17** (1960), 93.

Simpson, A. C. (1953). Some observations on the mortality of fish and the distribution of plankton in the southern North Sea during the cold winter, 1946–1947. *Journal du Conseil*, **19,** 150–77.

Smith, R. L. 1968. Upwelling. *Annual Review of Oceanography and Marine Biology*, **6,** 11–46.

Souter, A. and Isaacs, J. D. (1974). Abundance of pelagic fish during the 19th and 20th centuries as recorded in anaerobic sediments off California. *Fisheries Bulletin*, **72,** 257–75.

Southward, A. J. (1963). The distribution of some plankton animals in the English Channel and approaches. III. Theories about long term biological changes including fish. *Journal of the Marine Biological Association of the United Kingdom*, **43,** 1–29.

Southward, A. J. (1967). Recent changes in abundance of intertidal barnacles in southwest England: a possible effect of climatic deterioration. *Journal of the Marine Biological Association of the United Kingdom*, **47,** 81–95.

Southward, A. J. (1974*a*). Changes in the plankton community in the western English Channel. *Nature, London*, **249** (5433), 180–1.

Southward, A. J. (1974*b*). "Long term changes in abundance of eggs of the Cornish pilchard (*Sardina pilchardus* Walbaum) off Plymouth". *Journal of the Marine Biological Association of the United Kingdom*, **54,** 641–649.

Southward, A. J. and D. J. Crisp (1956). Fluctuations in the distribution and abundance of intertidal barnacles. *Journal of the Marine Biological Association of the United Kingdom*, **35,** 211–29.

Southward, A. J. Butler, E. I. and Pennycuick (1975). Recent cyclie changes in climate and in abundance of marine life. *Nature, London*, **253,** 5444. 714–7.

Starr, V. P. and Oort, A. H. (1973). Five-year climatic trend for the Northern Hemisphere. *Nature, London*, **242** (5396), 310–313.

Stephen, A. C. (1938). Temperature and the incidence of certain species in western European waters in 1932–1934. *Journal of Animal Ecology*, **7,** 125–129.

Storrow, B. (1947). Concerning fluctuations and the teaching of ecology. *Report of the Dove Marine Laboratory*, Third Series, No. 9. 7–580.

Tåning, A. V. (1931). Drift Bottle experiments in Icelandic Waters. *Rapports et Procès-verbaux des Réunions. Conseil international pour l'Exploration de la Mer*, **72,** (v).

Tåning, A. V. (1937). Some features in the migration of cod. *Journal du Conseil* **12,** 1–35.

Tåning, A. V. (1953). Long term changes in Hydrography and fluctuations in Fish Stocks. *Annual Proceedings of the International Commission on the North-west Atlantic Fisheries*, **3,** 69–77.

Templeman, W. (1965). Relation of periods of successful year classes of haddock on the Grand Bank to periods of success of year classes for cod, haddock and herring in areas to the north and east. *Special Publication of the International Commission on the North-west Atlantic Fisheries, Environmental Symposium*, No. 6. 523–533.

Templeman, W. (1972). Year class success in some North Atlantic stocks of cod and haddock. *Special Publication of the International Commission on the North-west Atlantic Fisheries*, **8,** 223–239.

Trenkle, H. (1956). Naherungswerte der Zonalgeschwindigkeit im Atlantisch-Europaischen Sektor für den Zeitraum 1881–1955. *Metallotechnische Rundschau*, **9,** 153–158.

Uda, M. (1952). On the relation between the variation of the important fisheries conditions and the oceanographical conditions in the adjacent waters of Japan 1. *Journal of the Tokyo University of Fisheries*, **38,** (3), 364–89.

Uda, M. (1957). A consideration on the Long Years trend of the Fisheries fluctuation in relation to sea conditions. *Bulletin of the Japanese Society of Scientific Fisheries*, **23**, 368–372.

Uda, M. (1960). The fluctuation of the sardine fishery in oriental waters. *Proceedings of the World Scientific Meeting on the Biology of Sardines and Related Species*, III, 937–947.

Valdivia, J. (1974). The anchovy population. IDOE Workshop on the Niño phenomenon, Guayaquil.

Vildoso, A. Ch. (1974). Biological aspects of the 1972–73 El Niño ; I. Distribution of the Fauna. IDOE Workshop on the Nino phenomenon, Guayaquil, 7 pp.

Wagner, A. (1940). Klimaänderungen und Klimaschwankungen. *Die Wissenschaften*, **92,** Braunschweig. 221 pp.

Wahl, E. W. and Bryson, R. A. (1975). Recent changes in Atlantic surface temperatures. *Nature, London*, **254,** 45–46.

Wahl, E. W. and Lawson, T. L. (1970). The climate of the mid-nineteenth century United States compared to current normals. *Monthly Weather Review*, **98,** 259–265.

Walford, L. A. (1931). Northward occurrence of southern fish off San Pedro in 1931. *California Fish and Game*, **17,** 401–405.

White, W. B. and Clark, N. (1975). On the development of blocking ridge activity over the central North Pacific. *Journal of atmospheric Science*, **22,** 490–502.

Williamson, M. H. (1961). An ecological survey of a Scottish herring fishery IV; changes in the plankton during the period 1949–59. *Bulletin of Marine Ecology*, **5,** 207–29.

Winstanley, D. (1973a). Recent rainfall trends in Africa, the Middle East and India. *Nature, London*, **243,** 464–5.

Winstanley, D. (1973b). Rainfall patterns and general atmospheric circulation. *Nature London*, **245,** 190–4.

Winston, J. S. (1955). Physical aspects of rapid cyclogenesis in the Gulf of Alaska. *Tellus*, **7,** 481–500.

Woodhead, P. M. J. (1964). Changes in the behaviour of the sole *Solea vulgaris*, during cold winters and the relation between the winter catch and sea temperatures. *Helgoländer wissenschaftliche Meeresuntersuchungen*, **10,** 328–42.

Wooster, W. S. and O. Guillen (1974). Characteristics of El Niño in 1972. *Journal of Marine Research*, **32,** 387–404.

Wyrtki, K. (1954). Der grosse Salzeinbruch in die Ostsee im November und Dezember 1951. *Kieler Meeresforschungen*, **10.** 19.

Wyrtki, K. (1973). Teleconnections in the Equatorial Pacific Ocean. *Science, New York*, **180,** 66–68.

Wyrtki, K. (1974). Equatorial currents in the Pacific 1950 to 1970 and their relations to the Trade Winds. *Journal of physical Oceanography*, **4,** (3), 372–380.

Zupanovitch, S. (1968). Causes of fluctuations in sardine catches along the eastern coast of the Adriatic Sea. *Anali Jadranskog Instituta*, IV, 401–489.

Adv. mar. Biol., Vol. 14, 1976, pp.123–250

RECENT ADVANCES IN RESEARCH ON THE MARINE ALGA *ACETABULARIA* *

S. Bonotto, P. Lurquin and A. Mazza

Department of Radiobiology, C.E.N-S.C.K., 2400 Mol, Belgium

and

International Institute of Genetics and Biophysics, C.N.R., Naples, Italy

* Work dedicated to Professors Joachim Hämmerling and Jean Brachet.

I. Introduction

The green marine alga *Acetabularia*, because of its peculiar morphology (Fig. 1), long ago attracted the attention of naturalists. As early as 1640 Parkinson published a beautiful drawing of this plant (Parkinson, 1640), which was called at that time *Umbilicus marinus* (see Table I for other synonyms). This " historical " drawing is shown in Fig. 2.

In the eighteenth century, various species of *Acetabularia* and some other members of the order Dasycladales were mentioned or studied by several authors (Tournefort, 1700, 1719; Donati, 1750; Ellis, 1755; Linnaeus, 1758, 1767; Pallas, 1766; Scopoli, 1772; Ellis and Solander, 1786). Curiously enough, these algae were thought to belong to the animal kingdom (Lamouroux, 1812, 1816, 1821). Later on, they were

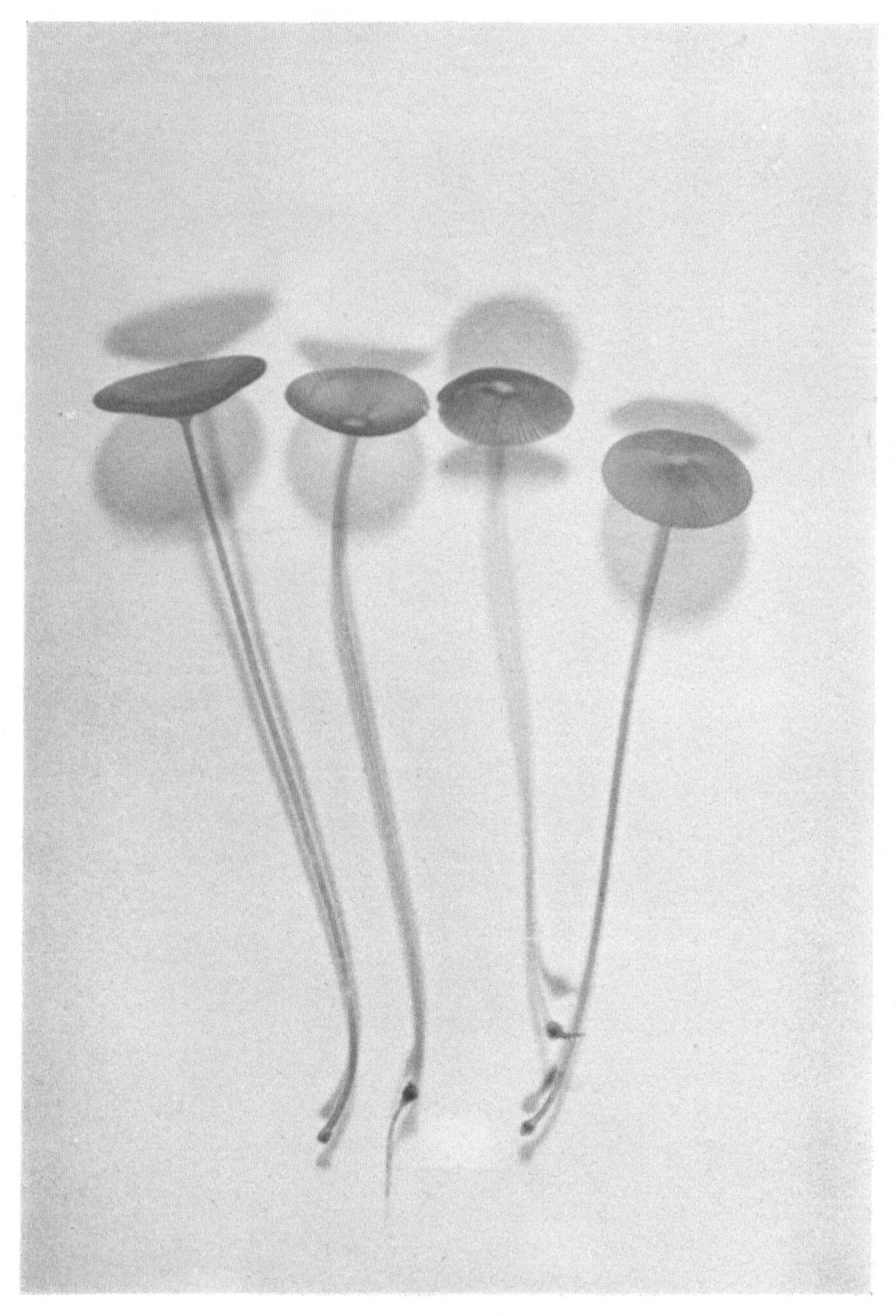

FIG. 1. *Acetabularia mediterranea.* Full grown cells at stage 8 (Bonotto and Kirchmann, 1970), from a laboratory culture obtained according to Lateur and Bonotto (1973). The cells, which contain several million chloroplasts, have a typical green colour.

[*facing p. 124*

placed with certitude under the green algae (Decaisne, 1842; Munier-Chalmas, 1877, 1879). Most of the publications which appeared in the nineteenth century were concerned with the morphology, the systematics and the geographic distribution of these marine algae (references in: Puiseux-Dao, 1962, 1963, 1970b; Valet, 1968a, b, 1969a). A few reports dealt with the biological cycle of *Acetabularia mediterranea**

Fig. 2. *Acetabularia mediterranea*. Line drawing of a group of cells growing on a shell, made by Parkinson in 1640. Parkinson called *Acetabularia* " *Umbilicus marinus* ". (From Crawley, 1970, with permission.)

(Woronine, 1861; de Bary and Strasburger, 1877) or that of *Dasycladus clavaeformis* (Derbes and Solier, 1856; Berthold, 1880).

The importance of *Acetabularia* as a tool for research on cellular biology was fully recognized only during the first decades of this century. The pioneer work of Joachim Hämmerling in Germany, started in 1926 (Schweiger, 1970a), has been extremely important for

* Under the International Code of Botanical Nomenclature the correct name is now *Acetabularia acetabulum* (L.) Silva. As the name *mediterranea* has been used in all the published work referred to in this article it has been retained to avoid confusion.

further research on *Acetabularia*. Hämmerling discovered that this giant unicellular alga (Woronine, 1861) has a single nucleus, located in the basal region of the stalk (Hämmerling, 1931) and that anucleate algae* are capable of differentiating such complex structures as stalk, whorls and species specific reproductive caps (Hämmerling, 1932, 1934c, 1943a). In more elegant experiments, Hämmerling and his collaborators obtained grafts between an anucleate half of one species and a nucleate fragment of another (Hämmerling, 1940, 1943b, 1946b, 1963a, b; Beth, 1943b, c; Maschlanka, 1943b, 1946). These investigations clearly showed that the nucleus produces " morphogenetic substances " (Hämmerling, 1934a, b, 1935), which migrate into the cytoplasm, where they can be stored for several weeks (Hämmerling, 1963a, b; Hämmerling and Zetsche, 1966).

Some twenty years later, a biochemical study of *Acetabularia mediterranea* was undertaken in Belgium by Jean Brachet and his collaborators (Brachet, 1951, 1952, 1957, 1958a, b, 1959, 1960, 1961a, b; Brachet and Chantrenne, 1951, 1952, 1953, 1956; Brachet and Brygier, 1953; Chantrenne *et al.*, 1953; Brachet, Chantrenne and Vanderhaeghe, 1955). Shortly before the discovery of the messenger ribonucleic acid (mRNA), Brachet came to the conclusion that the information present in the genes (nuclear DNA) must be carried to the cytoplasm in the form of stable molecules of RNA (Brachet, 1961a, b, 1962b, 1963a, c, d, 1964a, 1965a, b, c, 1967a, c, 1968, 1970). This conclusion proved to be very important for an approach to the understanding of gene action during cellular differentiation (Brachet, 1971, 1972a, b). Later on, several other laboratories became interested in *Acetabularia*, which received the honour of three international symposia, in 1969, 1972 and 1974 respectively (Brachet and Bonotto, 1970; Schweiger and Berger, 1972; Puiseux-Dao, 1975a).

The increasing interest in *Acetabularia* as a tool for research on cellular and molecular biology, resulted in the creation in 1972 of an " International Research Group on *Acetabularia* " (IRGA), whose main purpose is the coordination and development of studies on *Acetabularia* and other species of the Dasycladales. As a consequence of these initiatives, progress of research on *Acetabularia* has been particularly rapid in the last years (Schweiger *et al.*, 1974a, 1975; Puiseux-Dao, 1975b Brachet, 1975).

The most significant work performed on *Acetabularia* in the past has been reviewed by several authors (Hämmerling, 1953, 1963a;

* Anucleate algae are cells from which the nucleus has been removed by cutting off the basal part of the stalk. They are called also anucleate fragments or anucleate cells. Whole cells from which the nucleus has been dissected out are enucleate.

Puiseux-Dao, 1962, 1963, 1965; Gibor, 1966; Brachet, 1968, 1970, 1972a, b; Schweiger, 1967, 1968, 1969; Werz, 1969a, 1970a, b, 1974). Moreover, an excellent book on *Acetabularia* has been written by Simone Puiseux-Dao (1970b) and reviewed by Shephard (1971). The present article, though referring to previous work, will draw attention to the most significant recent advances in research on *Acetabularia* and related Dasycladales.

II. Ecology and Systematics of *Acetabularia* and the other Dasycladales

The order Dasycladales belongs to the Chlorophyceae. Some species are found as fossils (Reuss, 1861; Munier-Chalmas, 1877, 1879; Pia, 1920, 1926, 1927; Valet and Segonzac, 1969; Gusić, 1970; Raviv and Lorch, 1970; Segonzac, 1970a, b; for further references see Popa and Dragastan, 1973). Most of the living species grow in warm or subtemperate shallow waters. They are fixed to natural solid substrata such as rocks, stones, pebbles or shells, but are able to adhere also to artificial materials (Dao, 1954a; Puiseux-Dao, 1962) (Fig. 3).

The systematic classification of the Dasycladales suffered some confusion, as frequently one species has been called by different names, and the same name (e.g. *Acetabularia caraibica*) has been used for different species. A complete revision of the systematics of the Dasycladales was made by Valet (1967b, 1968a and b, 1969a, b, 1972). This author grouped the species so far described into two main families, Dasycladaceae and Acetabulariaceae. A new species of *Acetabularia* has been found very recently in the Pacific; it belongs to the Acetabulariaceae (according to Valet, 1969a) and was called *A. haemmerlingii* in honour of Professor Hämmerling (Berger, 1975b; Schweiger and Berger, 1975). The systematics of the Dasycladales are summarized in Table I, which also gives the synonyms found in the literature and their respective references. This classification follows essentially that proposed by Valet (1968a, 1969a, 1972) and contains information from publications by Puiseux-Dao (1970b) and Schweiger and Berger (1975). Recently, Berger *et al.* (1974) and Schweiger *et al.* (1974b) have studied the ultrastructure of the morphological markers of several species of *Acetabularia*, with the scanning electron microscope, and have analysed some molecular markers, such as the isozyme pattern of the malic dehydrogenase (MDH). According to Berger *et al.* (1974), and to Schweiger and Berger (1975), three sections should be distinguished in the genus *Acetabularia:* 1. *Acetabularia;* 2. *Acetabuloides;* 3. *Polyphysa.* As the advantages and limitations of scanning electron

TABLE I. SYSTEMATICS OF *Acetabularia* AND THE OTHER DASYCLADALES

Classification	*Synonyms*	*References*
Order DASYCLADALES		
1. Family : *Dasycladaceae*		
A. Sub-family : *Dasycladoïdeae*		
Tribe : *Dasycladeae*		
Genus : *Dasycladus* C. Agardh		Agardh (1827), Köhler (1957)
Species : *Dasycladus vermicularis* (Scopoli) Krasser		Collins (1909), Taylor (1960), Giaccone (1969b, 1971), Cinelli (1971), Giaccone *et al.* (1972)
	Spongia vermicularis	Scopoli (1772)
	Conferva clavaeformis	Roth (1806)
	Myrsidium clavatum	Rafinesque-Schmaltz (1810)
	Fucus vermicularis	Bertoloni (1819)
	Cladostephus clavaeformis	Agardh (1824)
	Dasycladus clavaeformis	Agardh (1828, 1863), De Toni (1889), Giaccone and Pignatti (1967), Pignatti and Giaccone (1967), Pignatti *et al.* (1967), Giaccone (1968 b, 1969 a, 1969 b, 1970), Tolomio (1973)
	Myrsidium bertolonii	Bory de St-Vincent (1832)
	Dasycladus cylindricus	Meneghini, in Kützing (1849)
Dasycladus densus Womersley		Womersley (1955)
Dasycladus ramosus Chamberlain		Chamberlain (1958)
Genus : *Chlorocladus* Sonder		Sonder (1871)
Species : *Chlorocladus australasicus* Sonder		Sonder (1871)
	Dasycladus australasicus	Cramer (1888), De Toni (1889)
Tribe : *Batophoreae*		

TABLE I—*contd.*

Classification		Synonyms	References
Genus	: *Batophora* J. Agardh		Agardh (1854)
Species	: *Batophora oerstedii* J. Agardh		
		Dasycladus conquerantii	Mazé and Schramm (1870)
		Botryophora conquerantii	Cramer (1890)
		Coccocladus occidentalis var. *conquerantii*	Howe (1904)
		Coccocladus occidentalis var. *laxus*	Howe (1904)
		Batophora oerstedii var. *occidentalis*	Howe (1905a, b)
		Dasycladus occidentalis	Harvey (1858)
		Botryophora occidentalis	Agardh (1887)
		Coccocladus occidentalis	Cramer (1888)
B. Sub-family	: *Bornetelloideae*		
Genus	: *Bornetella* Munier-Chalmas		Munier-Chalmas (1877)
Species	: *Bornetella capitata* (Harvey) J. Agardh		Agardh (1887), De Toni (1889), Cramer (1887), 1890), Weber-van Bosse (1913), Gilbert (1943), Valet (1967b).
		Neomeris capitata	Harvey (1857), in Valet (1969a)
		Bornetella capitata var. *brevistylis*	Arnoldi (1912)
	Bornetella sphaerica (Zanardini) Solms-Laubach		Solms-Laubach (1893), Weber-van Bosse (1913), Gilbert (1943), Egerod (1952)
		Neomeris sphaerica	Zanardini (1878)
		Bornetella ovalis	Yamada (1933, 1934)
		Bornetella capitata	Okamura (1912)

	Bornetella nitida (Harvey) Munier-Chalmas		Munier-Chalmas (1877), Cramer (1890), Solms-Laubach (1893), Weber-van Bosse (1913), Gilbert (1943), Valet (1967b) Sartoni (1974)
		Neomeris nitida	Harvey (1857), in Valet (1969a)
		Bornetella nitida var. *minor*	Boergesen (1946)
	Bornetella oligospora Solms-Laubach		Solms-Laubach (1893), Weber-van Bosse (1913), Gilbert (1943), Valet (1967b)
C. Sub-family :	*Neomeridoïdeae*		
Tribe :	*Neomerideae*		
Genus :	*Neomeris* Lamouroux		Lamouroux (1816)
Species :	*Neomeris cokeri* Howe		Howe (1904, 1095, 1909), Collins (1909), Taylor (1960).
	Neomeris annulata Dickie		Dickie (1874), De Toni (1889), Solms-Laubach (1893), Howe (1909), Yamada (1934), Gilbert (1943), Egerod (1952), Taylor (1960).
		Neomeris kelleri	Cramer (1887, 1890), De Toni (1889)
		Neomeris eruca	Cramer (1890)
	Neomeris dumetosa Lamouroux		Lamouroux (1816, 1821), De Toni (1889), Cramer (1887), Howe (1909), Collins (1909), Taylor (1960).
	Neomeris van bosseae Howe		Howe (1909), Setchell (1926), Gilbert (1943), Boergesen (1946), Egerod (1952), Valet (1967b)
		Neomeris dumetosa	Sonder (1871), Agardh (1887), Cramer (1890), Solms-Laubach (1893).
	Neomeris stipitata Howe		Howe (1909)
		Neomeris dumetosa	Church (1895)
	Neomeris mucosa Howe		Howe (1909), Yamada and Tanaka (1938), Dawson (1956), Valet (1967b)

Table I—*contd.*

Classification	Synonyms	References
Neomeris bilimbata Koster		Koster (1937), Dawson (1956), Valet (1967b)
Tribe : *Cymopolieae*		
Genus : *Cymopolia* Lamouroux		Lamouroux (1816)
Species : *Cymopolia barbata* (L.) Lamouroux		Lamouroux (1816), Kützing (1849), Agardh (1886), Cramer (1887), De Toni (1889), Boergesen (1925), Hämmerling (1952), Taylor (1960)
	Corallina barbata	Pallas (1766)
	Corallina rosarium	Ellis and Solander (1786)
	Cymopolia rosarium	Lamouroux (1821), Kützing (1849), De Toni (1889)
	Cymopolia bibarbata	Kützing (1843)
	Cymopolia barbata var. *bibarbata*	Agardh (1886), De Toni (1889)
	Cymopolia barbata var. *rosarium*	Agardh (1886), De Toni (1889)
	Cymopolia mexicana	Agardh (1886, 1887), De Toni (1889)
Cymopolia van bosseae Solms-Laubach		Solms-Laubach (1893), Weber-van Bosse (1913), Yamada (1934), Gilbert (1943)
2. Family : *Acetabulariaceae*		
Tribe : *Halicoryneae*		
Genus : *Halicoryne* Harvey		Harvey (1859)
Species : *Halicoryne wrightii* Harvey		Harvey (1859), Agardh (1886), Cramer (1895), Solms-Laubach (1895), Okamura (1907), Yamada (1934), Gilbert (1943).
	Halicorine spicata	Dickie (1876)

Rank	Taxon	Synonym	References
	Halicoryne spicata (Kützing) Solms-Laubach		Solms-Laubach (1895)
		Polyphysa spicata	Kützing (1863, 1866), De Toni (1889), Solms-Laubach (1895), Valet (1967b)
Tribe	: *Acetabularieae*		Hämmerling (1944)
Genus	: *Acetabularia* Lamouroux		Lamouroux (1816)
A. Section	*Acetabularia sensu stricto*		
Species	: *Acetabularia schenckii* Moebius		Möbius (1889), Egerod (1952)
		Acicularia schenckii	Solms-Laubach (1895), Howe (1901), Collins (1909), Beth (1943a), Taylor (1960)
	Acetabularia mediterranea Lamouroux		Lamouroux (1816), Solms-Laubach (1895), Hämmerling (1931, 1932), Schulze (1939), Stich (1951a), Puiseux-Dao (1962, 1970b), Brachet (1965d), Giaccone and Pignatti (1967), Pignatti and Giaccone (1967), Pignatti *et al.* (1967), Rizzi *et al.* (1967a, b), Giaccone (1968a, 1969), Bonotto (1970), Tolomio (1973)
		Umbilicus marinus	Parkinson (1640)
		Calopilophora mathioli	Donati (1750)
		Madrepora acetabulum	Linnaeus (1758)
		Tubularia acetabulum	Linnaeus (1767)
		Olivia androsace	Bertoloni (1819)
		Acetabulum marinum	Tournefort (1719)
		Acetabularia acetabulum	Silva (1952), Cinelli (1969, 1971), Giaccone (1969b, 1971, 1972), Giaccone *et al.* (1972)
	Acetabularia kilneri J. Agardh		Agardh (1886), Solms-Laubach (1895), Valet (1967b)

TABLE I—*contd.*

Classification	Synonyms	References
Acetabularia major Martens		von Martens (1866), De Toni (1889), Solms-Laubach (1895), Reinbold (1902), Yamada (1925), Howe (1932), Gilbert (1943), Schweiger *et al.* (1974b)
	Acetabularia crenulata var. *major*	Sonder (1871)
	Acetabularia denudata	Zanardini (1878)
Acetabularia major Martens var. *gigas* Solms-Laubach		Valet (1969a)
	Acetabularia gigas	Solms-Laubach (1895)
Acetabularia ryukyuensis Okamura et Yamada		Okamura and Yamada (1932), Okamura (1932)
	Acetabularia mediterranea	Okamura (1916)
Acetabularia ryukyensis Okamura et Yamada var. *philippinensis* Gilbert		Valet (1969a)
	Acetabularia philippinensis	Gilbert (1943)
Acetabularia dentata Solms-Laubach		Solms-Laubach (1895), Yamada (1934), Gilbert (1943), Valet (1967b)
	Acetabularia calyculus	Zanardini (1878), Valet (1969a)
Acetabularia crenulata Lamouroux		Lamouroux (1816), Harvey (1858), Solms-Laubach (1895), Howe (1901), Collins (1909), Taylor (1960)
	Acetabularia caraïbica	Kützing (1856), Solms-Laubach (1895), Howe (1901).
Acetabularia haemmerlingii Schweiger and Berger		Berger *et al.* (1974), Berger (1975b), Schweiger and Berger (1975).

	Acetabularia calyculus Quoy et Gaimard		Quoy et Gaimard (1824), Harvey (1858), Solms-Laubach (1895), Collins (1909), Okamura (1912, 1932), Arasaki (1942), Taylor (1960)
		Acetabularia suhrii	Solms-Laubach (1895)
		Acetabularia caraïbica	Okamura (1912)
	Acetabularia farlowii Solms-Laubach		Solms-Laubach (1895), Howe (1905), Collins (1909), Taylor (1960)
B. Section	*Acetabularia polyphysa*		
	Acetabularia peniculus (R. Brown) Solms-Laubach		Solms-Laubach (1895), Puiseux-Dao (1970b)
		Fucus peniculus	Brown (1819)
		Polyphysa aspergillosa	Lamouroux (1816)
		Polyphysa cliftoni	Harvey (1858), Schweiger (1969), Clauss *et al.* (1970), Zerban *et al.* (1973), Berger *et al.* (1974)
		Acetabularia peniculus var. *cliftoni*	Solms-Laubach (1895)
	Acetabularia pusilla (Howe) Collins		Howe (1909), Collins (1909), Taylor (1960)
	Acetabularia clavata Yamada		Yamada (1934), Egerod (1952), Moorjani (1970)
	Acetabularia exigua Solms-Laubach		Solms-Laubach (1895), Moorjani (1970)
		Acetabularia tsengiana	Egerod (1952)

TABLE I—*contd.*

Classification	Synonyms	References
Acetabularia parvula Solms-Laubach		Solms-Laubach (1895), Taylor (1950)
	Acetabularia parvula var. *americana*	Taylor (1945)
	Acetabularia moebii	Solms-Laubach (1895), Nizamuddin (1964), Boergesen (1940), Egerod (1952), Moorjani (1970), Cinelli (1971)
	Acetabularia minutissima	Okamura (1912)
	Acetabularia polyphysoïdes	Okamura (1913)
	Acetabularia wettsteinii	Schussnig (1930b), Hämmerling (1934c, d, 1944), Schulze (1939), Feldmann and Feldmann (1947)
Acetabularia myriospora Joly *et al.*		Joyly *et al.* (1965)
	Acetabularia polyphysoïdes var. *deltoidea*	Howe (1909), Collins (1909), Taylor (1960)
Acetabularia polyphysoïdes (Crouan) Kuntze		Kuntze (1891), Crouan, in Mazé and Schramm (1870), Solms-Laubach (1895), Howe (1909), Collins (1909), Taylor (1960).
Acetabularia antillana (Solms-Laubach) Egerod		Valet (1969a)
	Acetabularia antillaria	Egerod (1952)
	Chalmasia antillana	Solms-Laubach (1895)

microscopy and molecular markers as tools for taxonomic classification have still to be established, we adopt at present the previous classification (Valet, 1969a, 1972). However, it is likely that a thorough investigation of the morphological and molecular markers of the *Dasycladaceae* could lead to a revision of the present classification.

The question of whether or not *A. major* and *A. gigas* (Solms-Laubach, 1895) are identical species has not yet been resolved (Schweiger *et al.*, 1974b) and some doubts still exist for other species (Berger *et al.*, 1974).

III. Reproductive Cycle of *Acetabularia*

A. Reproduction of normal Acetabularia

At the end of the vegetative phase, the cell develops a reproductive cap, into which the secondary nuclei, originated from the fragmentation of the primary nucleus, migrate later. The primary nucleus and the perinuclear cytoplasm undergo fine structural changes during the reproductive phase (Bouloukhère, 1969, 1970; Woodcock and Miller, 1973b; Berger, 1975a; Berger *et al.*, 1975). After settlement in the cap, the secondary nuclei become anchored by cytoplasmic microtubules (Woodcock, 1970, 1971) and then undergo encystment (Hämmerling, 1939). It has been shown that mitotic divisions of secondary nuclei take place in the rhizoid, during capward translocation through the stalk, as well as in the young cyst (Schulze, 1939; Puiseux-Dao, 1962, 1970b; Woodcock and Miller, 1973b; Berger *et al.*, 1975a). Moreover mitotic figures were observed by Valet (1969a) in the related species *Halicoryne spicata*. Mature cysts release biflagellate gametes (Crawley, 1966, 1970).

A. mediterranea reproduces most frequently by gamete fusion (Fig. 4), but some variations are known to occur (Puiseux-Dao, 1970b; Bonotto, 1975). Figure 5 illustrates the different types of reproduction so far reported in the literature. Gamete formation and/or fusion were studied by several authors (Hämmerling, 1934d; Schulze, 1939; Werz, 1953; Puiseux-Dao, 1962, 1966b; Crawley, 1966, 1970; Woodcock and Miller, 1972b, 1973a; Berger *et al.*, 1975a). The early report by Schulze (1939) that meiotic figures occur in *Acetabularia* has been discussed by Puiseux-Dao (1970b). Moreover, the occurrence of meiosis in *Batophora oerstedii* (a species related to *Acetabularia*) is still uncertain (Puiseux-Dao, 1970b). According to Woodcock and Miller (1973a), if meiosis takes place during gametogenesis, it occurs in the final nuclear division in the cysts. However, according to Green (1973a), meiosis in *Acetabularia* should take place before cyst formation. The very recent

FIG. 4. Most frequently occurring reproductive cycle of *Acetabularia mediterranea*. 1, zygote; 2, zygote showing polar growth; 3, stalk and rhizoids development; 4, whorl formation; 5, cap initiation: the form of the apical region changes; 6, cap growth; 7, maximal cap and cell size; 8, secondary nuclei formation and their migration into the cap's rays; 9, cyst formation; 10, cyst with nuclei and cytoplasmic organelles; 11, release of gametes; 12, two gametes; 13, gamete fusion. The secondary nuclei are haploid. (From Bonotto, 1975, with permission.)

paper by Koop (1975c) shows that the secondary nuclei are indeed haploid. In this regard the recent discovery of lampbrush chromosomes in the primary nucleus of *Acetabularia mediterranea* (Spring *et al.*, 1975b) is of great interest. Lampbrush chromosomes, in fact, have so far been observed with certitude only in specific meiotic stages of animal cells (Gall, 1966; for further references see Spring *et al.*, 1975b).

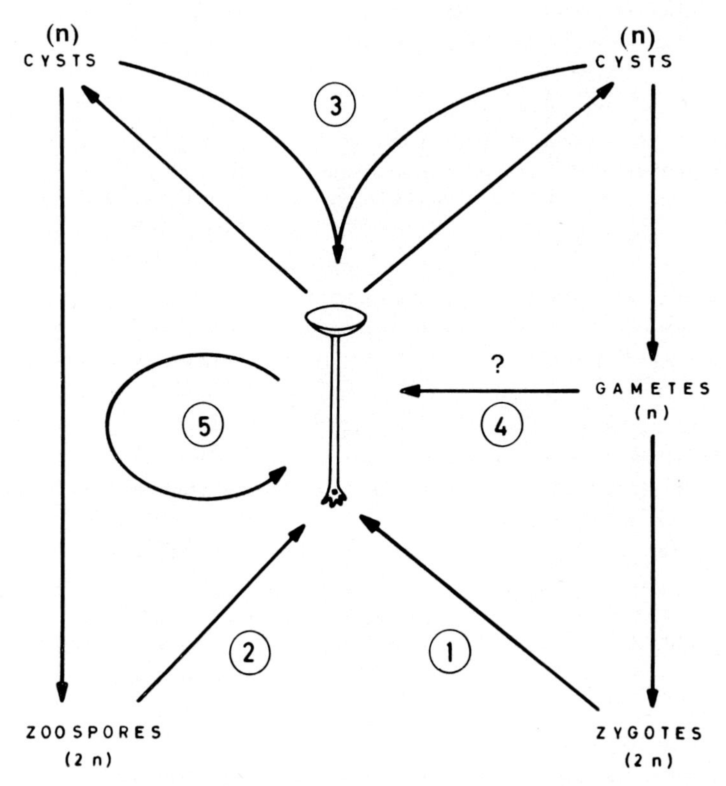

FIG. 5. Scheme showing the different possible types of reproduction of *Acetabularia mediterranea:* 1, by gamete fusion; 2, by zoospore formation; 3, by cyst germination; 4, by gamete parthenogenesis; 5, by germination of basal cytoplasm. (From Bonotto, 1975, with permission.)

Moreover, Herth (1975) and Herth *et al.* (1975) have shown, by X-ray diffraction and by electron microscopy, that crystalline cellulose is the predominant polysaccharide of the cyst wall, thus confirming previous investigations (Werz, 1968a, b, 1969a, b, 1970a; Zetsche, 1967; Zetsche *et al.*, 1970). This finding is particularly important since several authors had previously demonstrated that the stalk and cap walls are constituted of a highly polymeric fibrillar β 1, 4-mannan

(Iriki and Miwa, 1960; Miwa, 1960; Werz, 1963c, 1970a; Frei and Preston, 1961, 1968; Zetsche, 1967; Preston, 1968; Mackie and Preston, 1968; Zetsche *et al.*, 1970).

A similar shift from one structural polysaccharide to another has so far been reported only for the diplohaplophasic cycle (Kornmann, 1938; Zetsche and Wutz, 1975) of *Derbesia marina-Halicystis ovalis*. According to Preston (1968), the diploid form *Derbesia* contains β-1, 4-mannan, whereas the haploid form *Halicystis* possesses β-1,3 xylan plus a small amount of glucan. It appears, therefore, that in *Acetabularia* also the observed shift, from β-1,4-mannan to cellulose, is coupled to a transition from a diploid to a haploid form.

In some conditions fertilization in *Acetabularia* would not seem necessary. Hämmerling (1934d), in fact, provided circumstantial evidence that the gametes of *Acetabularia* might develop parthenogenetically. This possible type of reproduction has been recently discussed by Puiseux-Dao (1970b). However, it would be difficult to make a clear distinction between parthenogenesis, and germination of another kind of zoid, such as, for example, the zoospores (Puiseux-Dao, 1962, 1970b). The production of zoospores by *Acetabularia mediterranea* has been questioned by Hämmerling (1964). In later investigations, Valet (1968a, 1969a) observed direct germination of zoids into the cysts of *Acetabularia clavata.* Moreover, asexual reproduction by quadriflagellate zoospores has been observed in *Acetabularia parvula* by Puiseux-Dao (1965). *Acetabularia* may reproduce also by direct germination of cysts and of basal cytoplasm. Cyst germination is not a rare event (Puiseux-Dao, 1970b; Bonotto, 1975). It has been found to occur also in *Acetabularia dentata* (Valet, 1968a, 1969b) and in *Batophora oerstedii* (Puiseux-Dao, 1970b). The reproduction of *Acetabularia mediterranea* by germination of basal cytoplasm occurs at a low rate (only to 7%), under laboratory conditions (Bonotto, 1975). In this case a new plant develops most probably from a residual body (the " Restkörper " of Schulze, 1939) left in the rhizoids after the migration of secondary nuclei. According to de Bary (de Bary and Strasburger, 1877), in nature, part of the rhizoids (the " Basal-blase ") winters to reproduce a new plant in the next year. This kind of reproduction has been observed by Valet (1969a), under laboratory conditions, not only in *Acetabularia* but also in *Bornetella* and in *Halicoryne*. If one combines our observations (Bonotto, 1975) with those of other authors (de Bary and Strasburger, 1877; Schulze, 1939; Valet, 1969a), one can conclude that the disappearance of the visible thallus (reproductive cap and cylindrical stalk) after cyst formation does not necessarily involve the death of the plant.

In conclusion it should be said that *Acetabularia mediterranea*, although unicellular, may show different types of reproduction, which probably depend on the environmental conditions (Puiseux-Dao, 1970b). This wonderful reproduction capability, however, is not exceptional in the world of marine algae (Dawson, 1966 Levring *et al.*, 1969).

B. *Reproduction of branched* Acetabularia

Spontaneous branching of the stalk appears at a rather low frequency (0·5–2%) in natural populations as well as in laboratory cultures (Woronine, 1861; Hämmerling, 1934a, 1939; Nasr, 1939; Dao, 1954a; Bonotto and Janowski, 1968; Valet, 1968a and b, 1969a;

FIG. 6. *Acetabularia mediterranea*. Spontaneous branching of the stalk. The branches bear ripe caps with normal ovoidal cysts. Scale = 10 mm. (From Bonotto *et al.*, 1970a, with permission.)

Bonotto, 1970; Bonotto *et al.*, 1970a; Loni and Bonotto, 1971). However, branching can be induced to an appreciable rate by puromycin treatment (Brachet, 1963b), by adopting particular culture conditions (Bonotto and Bonnijns-Van Gelder, 1969) or by gamma irradiation (Bonotto *et al.*, 1970b; Bonotto and Kirchmann, 1972a). The peculiar morphology of branched *Acetabularia mediterranea* is illustrated in Fig. 6 (spontaneous branching) and in Fig. 24 (branching induced by gamma irradiation).

These cells are generally capable of forming a reproductive cap at the top of each branch. At maturity, all the reproductive caps born by a single cell become replete with normal ovoidal cysts. Like those of normal plants, the cysts of branched *Acetabularia* are fertile. We have observed also that branched cells may reproduce by germination of basal cytoplasm (Bonotto, 1975). From the reported observations a question arises: is the total number of cysts (and of zoids) formed by a single cell influenced by the number of reproductive caps born by the stalk? If this is so, the conclusion would be that finally the cytoplasm regulates quantitatively the formation of secondary nuclei and consequently the replication of nuclear DNA (Bonotto *et al.*, 1971a).

IV. Culture of *Acetabularia*

A. *Natural medium*

The "Erdschreiberlösung" (ESL) (Hämmerling, 1931) is the classical medium for laboratory culture of *Acetabularia* and of other marine algae (Föyn, 1934a, b; Provasoli *et al.*, 1957; Bonotto, 1976). Table II gives the composition used by Lateur and Bonotto (1973) for the culture of *Acetabularia mediterranea*.

Table II. Composition of the Natural Culture Medium for *Acetabularia mediterranea* (After Lateur and Bonotto, 1973)

Compounds	*Amount/l*
$NaNO_3$	100 mg
Na_2HPO_4	20 mg
Earth extract	2·5 ml
Natural sea water	1 l

Although earth extract had been utilized for a long time in several laboratories for the culture of marine algae, its chemical composition was entirely unknown. Analysis of the earth extract with the amino acid analyser has shown that all the usual amino acids are present together with ammonia and several unidentified minor compounds (Lateur and Bonotto, 1973). Moreover, it was found that the chemical composition of the extract varies with the type of the earth (Table III).

Table III shows also that differences in amino acid composition may exist even between two extracts from the same earth (compare Experiment I with Experiment II).

TABLE III. AMINO ACID COMPOSITION OF THE EXTRACT OBTAINED FROM TWO DIFFERENT TYPES OF EARTH (After Lateur and Bonotto, 1973)

Amino acid	*Earth of Brussels*		*Earth of Naples*	
(*% of micromoles*)	*Exp. I.*	*Exp. II*	*Exp. I*	*Exp. II*
Aspartic acid	12·12	14·76	11·51	12·99
Threonine	7·38	3·91	2·02	3·15
Serine	7·11	4·25	4·02	4·48
Glutamic acid	13·18	17·46	10·44	13·93
Proline	6·46	8·92	2·30	1·76
Glycine	16·55	11·01	6·35	8·05
Alanine	11·69	6·54	3·51	4·15
Valine	5·91	8·04	4·44	6·12
Methionine and derivatives	1·44	0·60	0·79	0·10
Isoleucine	3·23	6·59	6·61	6·24
Leucine	4·83	10·01	11·33	10·26
Tyrosine	3·08	2·62	2·98	2·81
Phenylalanine	7·01	5·30	33·67	25·94

The extract from the earth of Naples, which contains a high level of phenylalanine, enhances the growth of *Acetabularia*. It is not known, however, whether this result is due to the phenylalanine or to other unknown substances present in the earth extract. The study of the effect of phenylalanine on the growth of *Acetabularia* should give an answer to this question. According to recent investigations by Schweiger and Kretschmer (1975), unsupplemented sea water might replace the ESL medium if provided in a flow-through system. Finally, it has been discovered by Brändle *et al.* (1975) that proton dislocation results in a pH change in the culture medium of *Acetabularia*. These authors have observed also that under continuous light conditions the pH of the medium exhibits a rhythm with a period length of 23 hours.

B. *Artificial medium*

Since it is known that sea water may contain variable amounts of organic compounds (Bohling, 1970; Hoyt, 1970) and that the chemical composition of the earth extract is not constant (Table III), a suitable artificial medium for the culture of *Acetabularia* would overcome these drawbacks. Only a few attempts have been made to grow *Acetabularia* in artificial media (Keck, 1964; Shephard, 1970b).

Table IV gives the chemical composition of the artificial medium developed by Shephard (1970b). The medium contains a series of different salts, supplemented with vitamins, as recommended by Provasoli *et al.* (1957). According to Shephard (1970b) the use of this artificial medium permits a normal development of *Acetabularia.*

TABLE IV. COMPOSITION OF THE ARTIFICIAL MEDIUM DEVELOPED BY SHEPHARD FOR THE CULTURE OF *Acetabularia* (After Shephard, 1970b)

Compounds	*Amount/l*
NaCl	24 g
$MgSO_4 \cdot 7H_2O$	12 g
$CaCl_2 \cdot 2H_2O$	1 g
Tris	1 g
KCl	0·75 g
$NaNO_3$	40 mg
K_2HPO_4	1 mg
Na_2EDTA	12 mg
$ZnSO_4 \cdot 7H_2O$	2 mg
$Na_2MoO_4 \cdot 2H_2O$	1 mg
$FeCl_3 \cdot 6H_2O$	0·5 mg
$MnCl_2 \cdot 4H_2O$	0·2 mg
$CoCl_2 \cdot 6H_2O$	2μg
$CuSO_4 \cdot 5H_2O$	2 μg
$NaHCO_3$	100 mg
Thiamine-HCl	300 μg
p-Aminobenzoate	20 μg
Ca-Panthothenate	10 μg
Vitamin-B_{12}	4 μg
Distilled water	1 l

C. *Culture techniques*

Techniques for cultivating *Acetabularia* were first reported by Hämmerling (1931) and by Beth (1953a). Later on several authors tried to improve culture medium and techniques (Gibor and Izawa, 1963; Lateur, 1963; Puiseux-Dao, 1963; Keck, 1964; Schreiber *et al.*,

1964; Shephard, 1970b; Lateur and Bonotto, 1973; Schweiger and Kretschmer, 1975). Lateur and Bonotto (1973) have reported the time sequence for cultivating *Acetabularia mediterranea* in the laboratory (Table V).

TABLE V. TIME SEQUENCE FOR LABORATORY CULTURE OF *Acetabularia mediterranea* (After Lateur and Bonotto, 1973)

1.	Sterile collection of the cysts and exposure to the light	5 days
2.	Keeping of the cysts in the dark at 14°C	2 days
3.	Re-exposure of the cysts in the light	5 days
4.	Keeping again in the dark	2 days
5.	Re-exposure in the light and culture of zygotes with renewing of the medium	11 days
6.	Culture of the young plants (0·5 mm) removed from the glass walls of the vessels and grouped by 500	4 days
7.	Renewing of the medium and removing of the algae from the glass walls. The plants of 1 mm are grouped by 200	6 days
8.	Same treatment as in 7; the plants of 2 mm are grouped by 50	6 days
9.	Same treatment as in 7; the plants of 4 mm are grouped by 30 (the medium is renewed every 7 to 10 days)	42 days
10.	At the beginning of cap formation, the plants are grouped by 15	60 days
11.	Collection of caps with cysts and exposure to light	7 days
12.	Stocking of caps in the dark at 10°C	30 days
		180 days

Table V shows that, even if the growth of *Acetabularia* is relatively rapid, 180 days are necessary to pass from one generation to another. This rather long time constitutes a serious obstacle for genetical research. However, this obstacle could be overcome if species with a shorter life cycle were selected.

A major problem in cultivating *Acetabularia* is to obtain axenic plants. It has been shown that *Acetabularia* can support a treatment with various antibiotics and bacteriocides practically at all developmental stages (Gibor and Izawa, 1963; Lateur, 1963; Puiseux-Dao, 1963, 1970b; Shephard, 1970b; Lateur and Bonotto, 1973). However, although antibiotics and bacteriocides seem useful for cleaning *Acetabularia* from bacteria, they should be used very carefully and possibly for only a short time in order to prevent any damage to the cells. A frequent renewal of the sterile culture medium seems to avoid infections very effectively and provides at the same time new nutritive substances and regular aeration of the cultures (Puiseux-Dao, 1970b; Lateur and Bonotto, 1973).

Knowledge of germination of cysts and gamete release has remained till now rather indefinite (Lateur and Bonotto, 1973). The recent work of Koop (1975a, b, d) provides very useful information on experimental conditions and time sequence of cyst germination in *Acetabularia mediterranea*.

D. *Culture conditions*

Conditions influencing morphogenesis, chlorophyll content, photosynthetic activity and cell metabolism were summarized and discussed by Puiseux-Dao (1970b). Since *Acetabularia* species come from warm seas, they grow little, or not at all, at temperatures below 18°C. The optimum temperature for *A. mediterranea* is around 20°–22°C, whereas *A. crenulata* seems to grow better at 24°C. Possibly, differences exist between the other species of *Acetabularia*.

The effects of light conditions on *Acetabularia* have been studied by several authors (Arasaki, 1941; Beth, 1953a, 1955a and b; Dao, 1954a; Richter, 1962; Clauss, 1963, 1968, 1970a, 1973; Terborgh, 1963, 1965, 1966; Terborgh and Thimann, 1964, 1965; Richter and Kirschstein, 1966). Optimum growth for *Acetabularia mediterranea* was found for a light intensity of about 1300 lux (Dao, 1954a). According to Puiseux-Dao (1963, 1970b), the light intensities required for cell lengthening, cap formation and cap maturation in *Acetabularia mediterranea* are respectively 100, 500 and 800–1 000 lux. The studies so far accomplished on light effects allow the conclusion that morphogenesis, pigment and photosynthesis of the algae are controlled not only by the amount of light received but also by its quality (wavelength). Red light has inhibitory effects, which can be reversed by blue light (Clauss, 1970a, b). Moreover, the photoperiod seems to affect considerably the development of the algae. According to present knowledge on life histories of the marine algae *Bangia* and *Porphyra*, the formation of reproductive structures is induced by a photoperiodic, phytochrome-mediated system (Hoshaw and West, 1971). Unfortunately, hitherto no investigations have been made on the presence of phytochrome and on its eventual control of cap formation in *Acetabularia*. The use of the Phytotron (Blondon and Jacques, 1970) might make it possible to fill this gap in the future. The effects of temperature on the growth of *Acetabularia* have been studied by Beth (1956b). Further progress on the study of temperature and light effects on *Acetabularia* could be realized by a cultivation apparatus described by Edwards and van Baalen (1970). The apparatus can supply a temperature gradient at a right angle to an irradiance gradient and has been used successfully for studying effects of temperature and light intensity on growth and reproduction of several brown and red marine algae (Edwards and van Baalen, 1970).

V. Cellular Organization

A. *Polarity in* Acetabularia

Growth and development of *Acetabularia* have been studied by many authors (Hämmerling, 1931, 1932, 1953; Dao, 1954a; Beth, 1955a; Hämmerling and Werz, 1958; Terborgh and Thimann, 1964, 1965; Shephard, 1965a). It seems now well established that the growth of this alga occurs exponentially (Shephard, 1965a). The polarity of *Acetabularia* was first studied by Hämmerling (1931, 1936, 1955b), who discovered the now famous apico-basal and baso-apical morphogenetic gradients. Investigations by other authors have shown that the internal gradient concerns also the ultrastructure and the activity of the cell and of its organelles (Werz, 1959a, 1960b, c, d, 1961a, c, 1962; Issinger *et al.*, 1971; Boloukhère, 1972; Puiseux-Dao *et al.*, 1972a, b; Sironval *et al.*, 1972, 1973; Dujardin *et al.*, 1975).

B. *Sterile whorls: morphology and function*

During the morphological differentiation of *Acetabularia*, several regularly spaced whorls of polychotomic hairs (Fig. 7) appear on the apical region of the stalk (de Bary and Strasburger, 1877; Cramer, 1890; Dao, 1954a; Puiseux-Dao, 1962, 1963; Werz, 1965). In some species, the formation of the whorls is endogenous: they rise under a pectic membrane, which colours with ruthenium red (Valet, 1967a). This membrane, which was first observed by Cramer (1890), is absent in the most studied species, *A. mediterranea* (Valet, 1967a).

Biological observations have shown that, at least in *A. mediterranea*, the reproductive cap never develops without the previous formation of sterile whorls (Bonotto, unpublished; Lateur, personal communication). Anucleate fragments, on the other hand, are able to form one or several whorls before developing the reproductive cap (Puiseux-Dao, 1970b; Bonotto *et al.*, 1971d; Bonotto *et al.*, 1972d). Thus, the genetical information for whorls formation, as well as for cap formation (Hämmerling, 1963; Zetsche, 1966a; Brachet, 1968) must be present when the cell is enucleated. The utilization of these two types of genetical information (for whorls and cap formation respectively) is clearly subjected to a temporal regulation, as whorls and caps alternate at the tip of the stalk (Bonotto *et al.*, 1971d). These findings suggest that the whorls may play a role in the differentiation of the cell. It

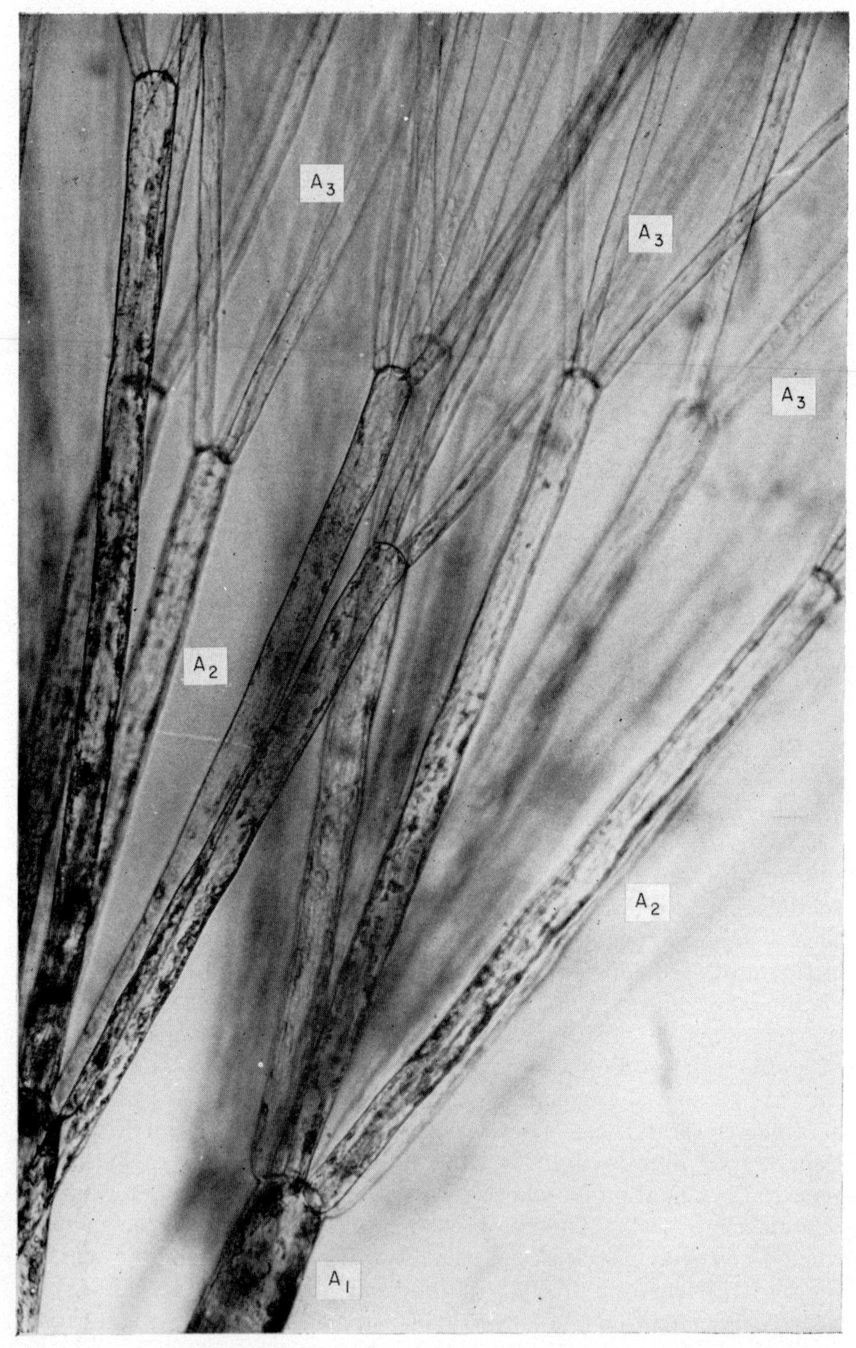

FIG. 7. *Acetabularia mediterranea*. Sterile whorls showing the articles of first (A1), second (A2) and third (A3) order. (From Declève *et al.*, 1972, with permission.)

was, in fact, found by Bonotto and Puiseux-Dao (1970) that some algae (about 5%) which lack the reproductive cap present several anomalies of the whorls. Such anomalies, mainly represented by an important enlargement of the first order articles of the hairs (Bonotto and Puiseux-Dao, 1970), appear also in algae irradiated with gamma radiations (Bonotto and Puiseux-Dao, 1970; Bonotto *et al.*, 1970b; Bonotto and Kirchmann, 1972a), cultivated in poor media (Dao, 1954a), treated with higher than optimal temperatures (Werz, 1970b) or treated with ethidium bromide (Dazy, personal communication). Since the teratological modifications of the whorls occur even in the absence of the nucleus, Bonotto and Puiseux-Dao (1970) suggested that they are probably due to a purely cytoplasmic mechanism acting on some long-lived messages originating from the nucleus. Whorls formation, like cap formation (Brachet, 1963b), probably requires a normal orderly synthesis of RNA and proteins: algae treated with actinomycin D (Boloukhère-Presburg, 1965) or with puromycin (Boloukhère-Presburg, 1966) lose their whorls and in the meantime the capacity of forming a reproductive cap.

We know from the early paper by Mangenot and Nardi (1931) and from more recent work (Boloukhère, 1969; Bonotto, 1969b; Declève *et al.*, 1972; Gibor, 1973a) that the whorls contain chloroplasts, mitochondria and all the other common constituents of the cell. Moreover, they are to be regarded as structures metabolically very active, since they are able to incorporate labelled precursors of DNA (^{3}H-thymidine), RNA (^{3}H-uridine) and proteins (^{3}H-leucine) (Bonotto, 1969b; Declève *et al.*, 1972). The whorls of *Acetabularia* increase considerably the surface area of the cell (Lateur, personal communication; Bonotto, 1969b; Gibor, 1973a), probably facilitating the uptake of nutritive substances present at low concentrations in the marine environment (Bonotto, 1969b; Gibor, 1973a). The efficacy of the whorls in the uptake of soluble substances has been demonstrated by Gibor (1973a) by immersing the entire cell in a dilute solution of neutral red or of methylene blue.

This author has also shown that the development of the whorls is directly controlled by light and by the availability of nutrients (Gibor, 1973a). It has been suggested by several authors (Bonotto, 1969b; Declève *et al.*, 1972; Gibor, 1973b) that the whorls may be the site of synthesis of cellular growth regulators, which could migrate into the stalk. In recent work, Gibor (1973b) has elegantly demonstrated, by cutting or by irradiating the whorls with UV, that these cellular structures influence the elongation of the stalk. The whorls, which possess numerous active chloroplasts, could also play an important role in photosynthesis by enhancing the cellular surface exposed to light.

Generally, the whorls are deciduous organelles. However, exceptionally, they can transform into lateral branches (Dao, 1954a; Bonotto, 1969b). In *A. mediterranea*, normal ovoidal cysts were found in the hairs of a whorl near the ripe cap (Bonotto, 1969b), a fact which supports the hypothesis that the reproductive cap is nothing more than a specialized whorl (Dao, 1954a; Valet, 1968b).

C. *Reproductive cap*

At maturity, *Acetabularia* cells develop one or more (according to the species) reproductive caps. Cap formation is certainly one of the most conspicuous morphogenetic processes of the cell and has been particularly studied by a number of authors (see literature in Puiseux-Dao, 1970). Most of the morphological features of the cap are species specific and are consequently particularly useful in interspecific graft experiments (Hämmerling, 1940, 1943b, 1946a, b; Beth, 1943b, c) and for taxonomical purposes (Valet, 1969a; Berger *et al.*, 1972, 1974; Schweiger *et al.*, 1974b). The morphology of the reproductive cap of *A. mediterranea*, cultivated in the laboratory or grown in the sea, has been the object of new observations in recent years (Bonotto, 1968, 1970; Bonotto and Janowski, 1969; Bonotto *et al.*, 1969a, 1971b; Janowski and Bonotto, 1969; Loni and Bonotto, 1971). Abnormal caps were found to occur in culture as well as in the sea, where some toxic or antibiotic substances may be present (Bonotto, 1970). In nature, the reproductive cap is partially or completely calcified (Fig. 3) and serves as a substratum for the growth of numerous epiphytes (Zavodnik, 1969a, b; van der Ben *et al.*, 1972). The morphology of the cysts contained in natural calcified caps has been studied by Bonotto *et al.* (1969b).

D. *Rhizoids: morphology and function*

Young *Acetabularia* cells at stages 2 and 3 (Bonotto and Kirchmann, 1970) are very sticky: in culture, they adhere tenaciously to the glass walls of the vessels by appendices of the basal stalk, which are called rhizoids. These latter develop very soon during the morphological differentiation of the cell. Thanks to the rhizoids, the young algae become fixed on solid substrata. It is not known whether the rhizoids produce adhesive substances.

In culture, the rhizoids are difficult to study, since their branches overlap and become entangled. However, in some cells the rhizoids

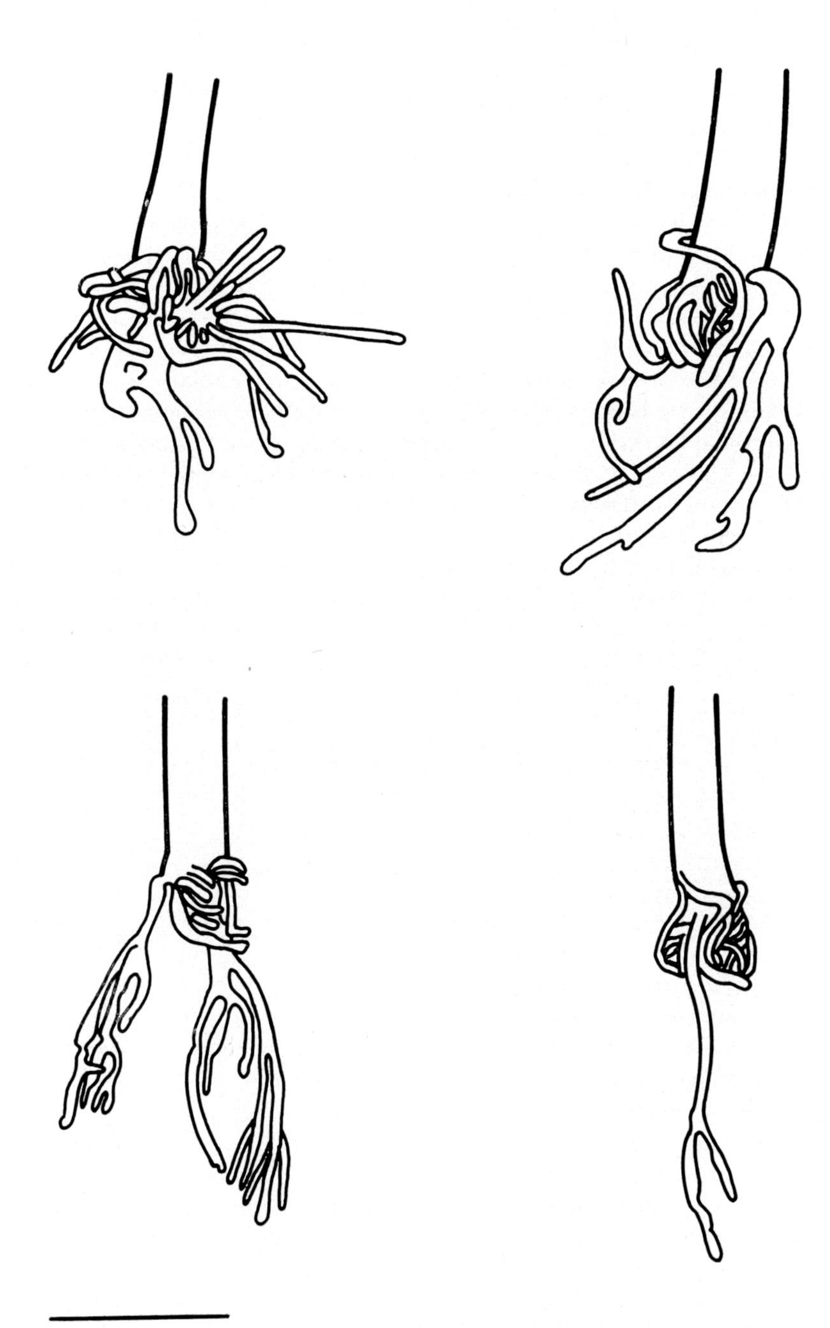

FIG. 8. *Acetabularia mediterranea*. Line drawings showing that rhizoids, like the sterile whorls, are branched. Scale = 1 mm.

are rather loose. A closer observation of loose rhizoids of *Acetabularia mediterranea* has shown that they are branched (Fig. 8). In a study on *Acetabularia* phototropism, rhizoids were found to grow away from the light (Bonotto and Puiseux-Dao, unpublished). It is well known that the nucleus rests in one of the branches of the rhizoids (Hämmerling, 1931). Most probably, in nature, this particular branch winters to produce a new stalk in the next year (de Bary and Strasburger, 1877; Schulze, 1939; Valet, 1969a; Berger *et al.*, 1974; Bonotto, 1975). Hämmerling (1936) has demonstrated that rhizoid formation is controlled by morphogenetic substances, which accumulate in the stalk according to a baso-apical gradient. Moreover, it has been found that after transplantation of a nucleus into an anucleate fragment, rhizoids arise in the region of the stalk where the nucleus rests (Hämmerling, 1955b; Zetsche, 1962).

Rhizoid formation may be induced in anucleate *A. mediterranea* cells by phenylethylalcohol (PEA) (de Vitry, 1965b). The author suggests that the genetical factors responsible for rhizoid formation are not exclusively present in the nucleus, but are also distributed in the cytoplasm, in a repressed form under normal cellular growth conditions. Recent studies on the morphological features of rhizoids in *A. mediterranea*, *A. major* and *A. cliftoni*, performed with the scanning electron microscope (Berger *et al.*, 1974; Schweiger *et al.*, 1974b) have shown that some differences exist between these three species.

E. *Nucleate and anucleate fragments*

The polar organization of *Acetabularia* allows one easily to obtain nucleate and anucleate fragments, simply by cutting off the basal part of the stalk, where the nucleus is located (Hämmerling, 1931, 1932). The basal nucleate fragment is able to regenerate a complete new alga and to form a reproductive cap, though on a shorter stalk (Beth, 1956a). The ultrastructure of regenerating basal nucleate fragments has been studied by Hoursiangou-Neubrun and Puiseux-Dao (1974) and by Tikhomirova *et al.* (1975). The anucleate fragment survives for several weeks and even months, thus being a very useful material for research on the biology and biochemistry of the anucleate systems (Keck, 1969; Woodcock and Miller, 1972a). The peculiar morphological organization of *Acetabularia* permits also the making of intra- and interspecific grafts (Beth, 1943b, c; Hämmerling, 1943b, 1946a, b; Maschlanka, 1943b, 1946; Werz and Zetsche, 1963; Puiseux-Dao *et al.*, 1970; Bonotto *et al.*, 1971a) and also the transplantation of nuclei or other cellular constituents (Hämmerling, 1955b; Werz, 1955; Richter,

1958b; Zetsche, 1963b; Keck, 1964; Pressman *et al.*, 1972). Transplantation of nuclei is currently utilized for studying nuclear dependency of chloroplast proteins (see Section X).

VI. Ultrastructure

A. *Cellular walls*

Most species of the *Dasycladaceae* growing in nature possess highly calcified cellular walls (Leitgeb, 1887). Many fossil species were preserved thanks to the calcification of their walls. In culture, *Acetabularia* cells are generally uncalcified. However, under unfavourable conditions, calcification may occur (Puiseux-Dao, 1970b; Lateur, personal communication). Recent investigations with the scanning electron microscope have shown that the cellular walls of *Acetabularia* are covered by a slime coating which is transparent when the plant is in a liquid medium (Berger *et al.*, 1974). As reported above (Section III, A) stalk and cap walls are constituted mainly by β-1,4-mannan, whereas cellulose is the predominant polysaccharide of the cyst wall (Iriki and Miwa, 1960; Frei and Preston, 1961, 1968; Miwa *et al.*, 1961; Mackie and Preston, 1968; Preston, 1968; Herth, 1975; Herth *et al.*, 1975). The sugars which constitute the cellular walls and their metabolism in the cell have been the object of many studies (Stich, 1951c; Clauss and Keck, 1959; Werz, 1960c, 1961a, 1963c, d, 1964a; Zetsche, 1965b, 1966a, e, 1967, 1968a, 1969a, 1972; Zetsche *et al.*, 1972). The viscoelastic properties of *A. crenulata* stalk wall were studied by Haughton *et al.* (1969). Such studies are necessary to understand how growth might depend on the mechanical properties of the cell wall.

According to Werz (1957d, 1965, 1966b), the wall of the apical region of the stalk increases both by apposition and by intussusception. At the base, it increases in thickness solely by apposition. The appearance of a whorl of sterile hairs or of the cap's primordia is preceded by a lysis of the existing wall (Werz, 1965, 1966b). In the apical region of the stalk, a network of special structures, characteristic of each species and probably constituted of anionic polysaccharides, was demonstrated by Werz (1960b, 1961a). Moreover, this author, working on *Acetabularia* protoplasts and cytoplasts, has elegantly demonstrated that cell wall formation is governed by the nucleus, which directs cytoplasm morphogenesis (Werz, 1968a, b, 1970a, b, 1972, 1974). When a foreign protein is introduced into the stalk, the cytoplasm synthesizes around it a new cellular wall (Werz, 1967). The ultrastructure of the wall of the whorls has been studied by Declève *et al.* (1972).

B. *The primary nucleus*

Early studies on the nucleus of *Acetabularia mediterranea*, *Batophora oerstedii* and *Cymopolia barbata* have been made with the light microscope (Hämmerling, 1931, 1955a, 1957, 1958, 1963; Schulze, 1939; Maschlanka, 1943a; Brachet, 1952; Werz, 1955; Puiseux-Dao, 1957, 1958a, b, 1959, 1960, 1962). It was found that in young germlings the nucleus possesses a network of chromatin, which stains with the Feulgen-technique (Feulgen and Rossenbeeck, 1924), and a small basophilic nucleolus. In growing plants, nuclear chromatin is less and less readily detectable and finally the giant resting nucleus becomes Feulgen-negative (Brachet, 1970). The nucleolus, however, increases considerably in size and shows a very strong basophilia, suggesting the presence of RNAs and basic proteins (Brachet, 1952; Puiseux-Dao, 1958a, b, 1960, 1962, 1963; Werz, 1959b). One of the most interesting findings was the demonstration of the so-called " nuclear emissions " around the nuclear membrane, in *Acetabularia mediterranea* as well as in *Batophora oerstedii* (Puiseux-Dao, 1958a, b, 1962, 1970b; Werz, 1961c, 1962). The " nuclear emissions " have been interpreted as perinuclear buds, containing substances of nuclear origin. A number of investigations on the action of darkness, various chemical inhibitors and UV radiation, led to the conclusion that the nucleus of the growing cell is very active in producing RNAs which, together with proteins, are transported into the cytoplasm (Brachet, 1951, 1952; Stich, 1951a, b, 1956b, 1959a; Hämmerling and Stich, 1954, 1956a, b; Brachet *et al.*, 1955, 1964; Werz, 1957a, b, c, 1959b, 1961b, c, 1962; Puiseux-Dao 1958b, 1960, 1962; Hämmerling and Hämmerling, 1959; Richter 1959a, b, c, Zetsche, 1964a, b, c, 1965a, 1966b, c; Spring *et al.*, 1974).

The use of the electron microscope and of the freeze-etching technique (Zerban *et al.*, 1973; Werz, 1974; Berger *et al.*, 1975a; Zerban and Werz, 1975a, b; Franke *et al.*, 1975b) made possible enormous progress in knowledge of the nucleus organization. The ultrastructure of the *Acetabularia* primary nucleus (called also " vegetative " or " resting " nucleus) has been studied by several authors (Crawley, 1963, 1964, 1965; Werz, 1964b, c, 1970a, b; Bouloukhère-Presburg, 1965, 1966, 1967; Van Gansen and Bouloukhère-Presburg, 1965; Bouloukhère, 1969, 1970; Burr and West, 1971; Woodcock and Miller, 1973b; Zerban *et al.*, 1973; Franke *et al.*, 1974; Schweiger *et al.*, 1974b; Spring *et al.*, 1974; Werz, 1974; Berger, 1975c; Berger *et al.*, 1975a; Franke *et al.*, 1975a, b).

Figure 9 shows the ultrastructure of the primary nucleus of *Acetabularia cliftoni* (= *Acetabularia peniculus* according to Valet, 1969a), as revealed by the freeze-etching technique, which seems suitable to prevent fixation artifacts (Zerban *et al.*, 1973). The nucleus is surrounded by a 100 nm thick layer, which communicates through junction channels with the perinuclear cytoplasm, where perinuclear bodies (" perinukleäre Körper ") are present together with other structures schematically represented in Fig. 10. The nucleus of *A. cliftoni* possesses about 3×10^6 nuclear pores to communicate with the surrounding cytoplasm (Zerban *et al.*, 1973). More recent investigations by Zerban and Werz (1975b) give the figure of 4·9–7·5 million pores per nucleus.

Fig. 9. *Acetabularia cliftoni* (= *A. peniculus* in the new classification of Valet, 1969a). Ultrastructure of the primary nucleus and its surroundings as revealed by the freeze-etching technique. Connections (→) between plasmalayer (pl) and perinuclear cytoplasm (CP). NP = nucleoplasm; M = mitochondrium; PK = perinuclear body (?). (From Zerban *et al.*, 1973, with permission.)

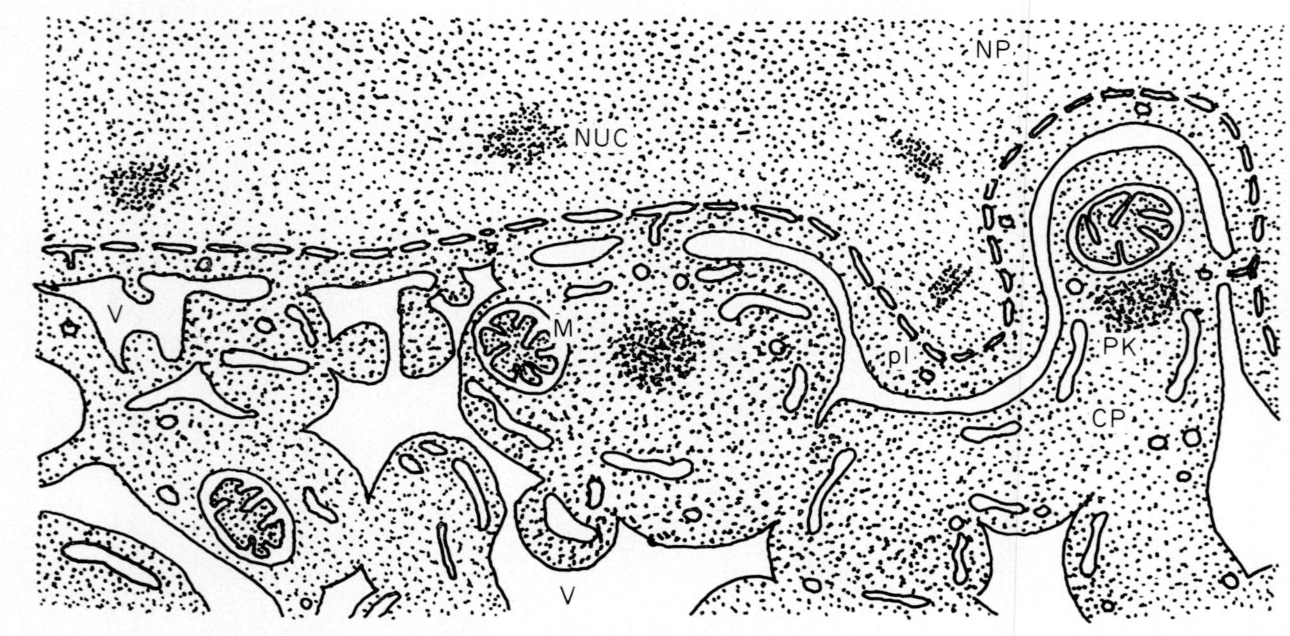

Fig. 10. *Acetabularia cliftoni* (= *A. peniculus* in the new classification of Valet, 1969a). Schematic drawing of the perinuclear cytoplasm. NUC = nucleolar components; NP = nucleoplasm; pl = plasmalayer; PK = perinuclear bodies; V = vacuole; M = mitochondria; CP = cytoplasm. (From Zerban *et al.*, 1973, with permission.)

Very recently the ultrastructure of the primary nucleus and of the perinuclear cytoplasm has been thoroughly reinvestigated by Franke *et al.* (1974) in the species *A. mediterranea* and *A. major*. Their very interesting observations are schematically represented in Fig. 11 which gives a detailed picture of the complex ultrastructural organization of the nucleus and its surroundings. According to these authors the total number of nuclear pores increases, in a two-month period, from 2 500 (in the zygote) to $2{\cdot}2 \times 10^6$ (in a full grown plant), corresponding to a net nuclear pore formation rate of about 23 pores per minute. The

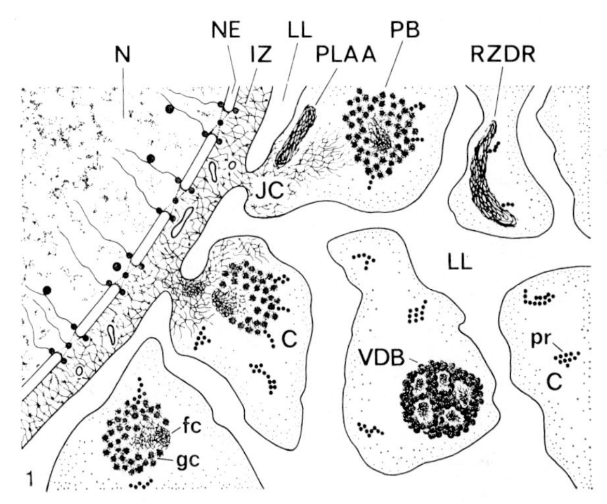

Fig. 11. Schematic drawing of the perinuclear zone of the giant primary nucleus of *Acetabularia*. N = nucleoplasm; NE = nuclear envelope; IZ = intermediate zone; LL = lacunar labyrinthum; JC = junction channel; PLAA = perinuclear lacuna associated aggregate; PB = perinuclear dense body; RZDR = reticulate zone dense rods; C = cytoplasm; VDB = vacuolated dense body; fc = fibrillar component; gc = granular component; pr = polyribosomes. (From Franke *et al.*, 1974, with permission.)

pore frequency reported for *Acetabularia* (Zerban *et al.*, 1973; Franke *et al.*, 1974) is one of the highest so far reported. However, the dramatic increase in the number of nuclear pores (about 900-fold) is less important than the simultaneous increase of the nuclear volume (about 40 000-fold). The most exciting finding by Franke *et al.* (1974) and by Berger and Schweiger (1975d) is the observation that the perinuclear dense bodies, which can rise to the figure of 20 000 around a full grown primary nucleus (Spring *et al.*, 1975a), react cytochemically in a way usually considered characteristic of chromatin. The authors suggest that these perinuclear bodies may include amplified DNA molecules, perhaps r-RNA cistrons, which would be transported into the cytoplasm. Nucleolar cistrons have been recently demonstrated in *A. mediterranea* and *A. major* (Spring *et al.*, 1974, 1976; Trendelenburg *et al.*, 1974a, b, 1975; Woodcock, 1975; Woodcock *et al.*, 1975).

Moreover, Spring *et al.* (1975b) have discovered, in the primary nucleus of *A. mediterranea*, lampbrush chromosomes (Fig. 12), suggesting the possibility that in *Acetabularia* meiosis begins very early in the young germling and is characterized by an extended meiotic prophase. Shortly afterwards, this hypothesis was corroborated by Koop (1975c), who found by microspectrophotometry that the primary nuclei of 24 hours old zygotes of *A. mediterranea* have 2·58 pg of DNA, while the secondary nuclei and those of cysts and gametes possess

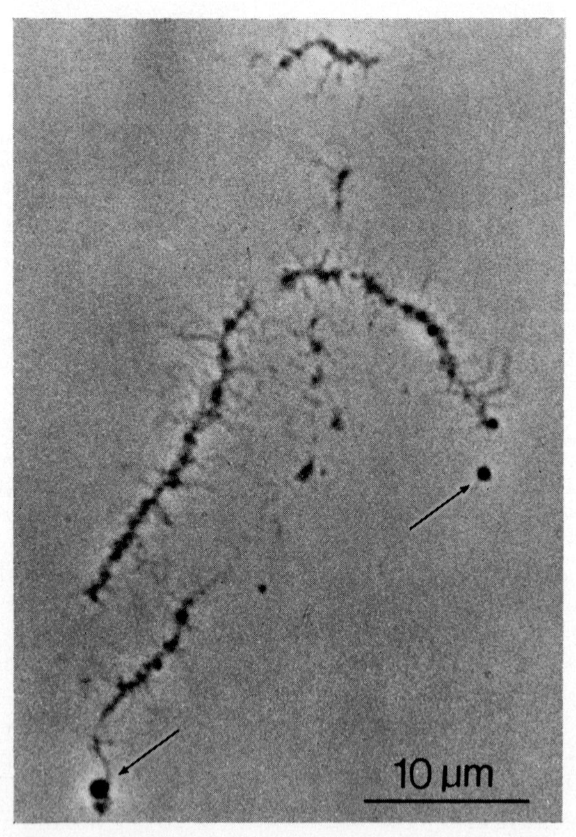

Fig. 12. *Acetabularia mediterranea.* One of the frequent " pair-associations " of lampbrush type chromosomes (not identifiable as bivalents), found in the primary nucleus. The arrows show (pre)terminal large globules on the chromosomes. (From Spring *et al.*, 1975b, with permission.)

only half this quantity. These data demonstrate that the primary nuclei of young zygotes are diploid, while the secondary nuclei and those of cysts and gametes are haploid. According to Koop (1975c), meiosis occurs either before or during fragmentation of the primary nucleus or immediately afterwards. If one combines the results obtained by Green (1973a), Spring *et al.* (1975b) and Koop (1975c), it appears clearly that *Acetabularia*, like other marine algae (Kornmann, 1938, 1970; Zetsche and Wutz, 1975), is characterized by a diplo-haplophasic cycle.

The primary nucleus of *Acetabularia* can be isolated, free from cytoplasmic contaminants and transplanted into a recipient cell (Hämmerling, 1963). However, until recently, its viability was limited to a few minutes (Sandakhchiev *et al.*, 1973; Brändle and Zetsche, 1973). Berger *et al.* (1975b) have observed that primary nuclei of *Acetabularia* may be stored at room temperature for 24 hours in an artificial medium without losing their viability. This improvement will certainly make possible *in vitro* experiments on the synthesis and the exchange of macromolecules between nucleus and cytoplasm (Brachet, 1967; Clérin *et al.*, 1975).

In vivo transport of macromolecules takes place not only from the nucleus to the cytoplasm but also from the cytoplasm to the nucleus (Dazy *et al.*, 1975a, b; Franke *et al.*, 1975a, b). Moreover, it has been demonstrated by Berger (1975c) that the cytoplasm causes a change in the morphology of the implanted primary nucleus in a way which depends upon its state (" young " or " old " cytoplasm).

C. *Secondary nuclei*

When the reproductive cap of *Acetabularia* is full grown, the giant (d = $\pm$ 100 μm in *A. mediterranea*) primary nucleus divides into a series of secondary nuclei. This process was first studied by Schulze (1939) and later on by many other authors (Puiseux-Dao, 1962a, 1966b; Valet, 1969a; Boloukhère, 1970; Woodcock and Miller, 1973b; Berger, 1975a; Berger *et al.*, 1975a; Koop, 1975c). However, the intimate mechanism of formation of the first secondary nuclei still remains enigmatic. The secondary nuclei formed in the rhizoids (Fig. 13) are small (d = $\pm$ 5 μm) and lack the buds as well as the cytoplasmic border found in the primary nucleus (Bouloukhère, 1970; Berger *et al.*, 1975a). They continue to divide for a limited period of time and are then transported towards the reproductive cap by cytoplasmic streaming (Puiseux-Dao, 1970b; Berger *et al.*, 1975a). Moreover, some mitotic divisions occur even while the nuclei are being

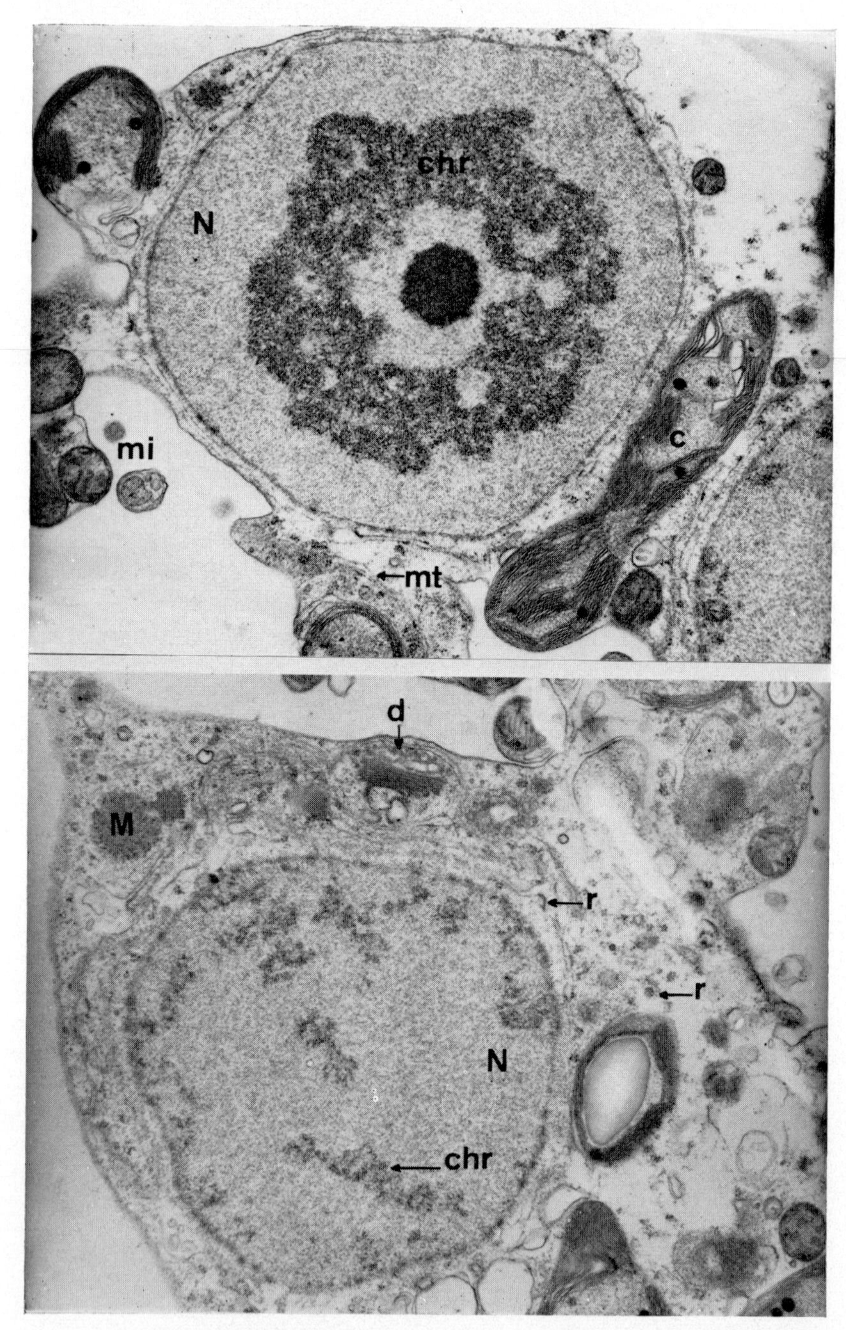

Fig. 13. *Acetabularia mediterranea.* General views of the secondary nuclei formed in the rhizoids and the cytoplasm surrounding them. In one nucleus (above) the chromatin fibres (chr) are condensed into a ring around the nucleolus, whereas in another nucleus (below), the slightly involuted chromatin fibres (chr) are aggregated in clumps. M = granular mass; N = nucleus; c = chloroplast; chr = chromatin; d = dictyosomes; mi = mitochondrion, mt = microtubules; r = ribosomes. × 15 000. (From Bouloukhère, 1970, with permission.)

transported towards the cap (Berger *et al.*, 1975a). The mechanism of migration of the secondary nuclei through the stalk remains to be elucidated. According to Woodcock (1971), after translocation in the cap rays, the secondary nuclei become anchored at fixed positions in the cytoplasm, by numerous microtubules. These latter could eventually contribute to increase the cytoskeletal rigidity of the region where the nucleus is located (Berger *et al.*, 1975a). In the cap, each secondary nucleus is associated with several perinuclear dense bodies, which can vary considerably in size (Woodcock, 1971; Woodcock and Miller, 1973a, b; Francke *et al.*, 1974; Berger *et al.*, 1975a). The function of these perinuclear bodies at present remains a mystery.

It is known from the extensive work performed by Werz (1968a, b, 1969a, 1970, 1974) that the secondary nuclei control cyst formation. Each cyst contains at first only one nucleus, from which, by successive mitoses, several nuclei arise. Finally, most frequently after a resting period, gametes are formed (Schulze, 1939; Dao, 1957; Puiseux-Dao, 1962; Woodcock and Miller, 1973a; Berger *et al.*, 1975a). The fusion of two gametes gives origin to a zygote (Crawley, 1966, 1970; Kellner and Werz, 1969; Berger *et al.*, 1975a). Recent work (Green, 1973a; Koop, 1975c) has demonstrated that in *Acetabularia* meiosis occurs before cyst formation, so that the secondary nuclei migrating into the cap are to be considered as haploid. By using several million swarmers, Spring *et al.* (1974) have found that the DNA content of the haploid nucleus is 1·29 picogram, a value which agrees with the microspectrophotometric DNA determinations performed by Koop (1975c) on the nuclei of zygotes, gametes, cysts and on the secondary nuclei from the rhizoid.

D. *Chloroplasts: morphological and physiological heterogeneity of the plastidal population*

A full grown *Acetabularia* cell may contain several million chloroplasts (Shephard, 1965b; Schweiger, 1969; Clauss *et al.*, 1970). Their structure has been described by several authors (Crawley, 1963, 1964; Bouck, 1964; Boloukhère-Presburg, 1965, 1966; Werz, 1965, 1966a; Van Gansen and Boloukhère-Presburg, 1965; Puiseux-Dao, 1966a, 1967, 1968; Boloukhère, 1969, 1970, 1972; Vanden Driessche and Hars, 1972a, b, c, 1973; Vanden Driessche *et al.* 1973a). These organelles are bounded by a double membrane and possess bundles of peripheral lamellae surrounding a stromatic zone, containing DNA, ribosomes and most frequently one or more carbohydrate granules (Puiseux-Dao, *et al.*, 1967). From a careful investigation of the relationship between

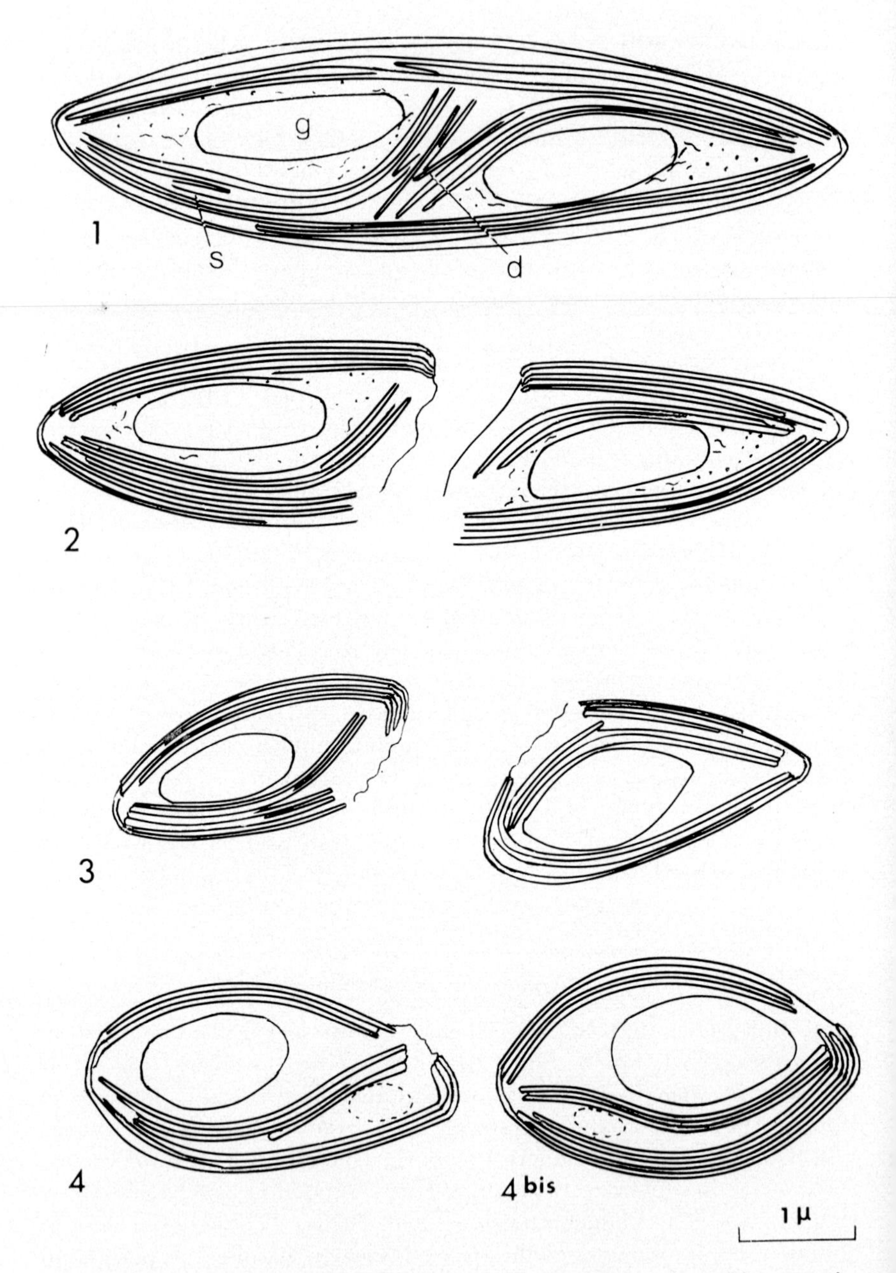

FIG. 14. Schematic representation of plastid division in *Acetabularia*. 1,2,3: successive stages of the division; 4, 4bis: formation of a new polysaccharide centre. d = diagonal lamellae; g = polysaccharide granule; s = saccules. (From Puiseux-Dao and Dazy, 1970, with permission.)

lamellae and carbohydrate granules, Puiseux-Dao and collaborators have concluded that the chloroplasts are constituted of one or more identical plastidal units (Puiseux-Dao, 1966a, 1967, 1968, 1970a, b; Puiseux-Dao and Gilbert, 1967; Puiseux-Dao, 1970a, b; Puiseux-Dao *et al.*, 1972a, b). Chloroplast division occurs by separation of the adjacent plastidal units, according to a process described in detail by Puiseux-Dao and Dazy (1970a, b) and illustrated in Figs 14 and 15.

The plastidal population of *Acetabularia* is heterogeneous (Lüttke *et al.*, 1975). This has been proved by different approaches. In the vegetative cell, the chloroplasts are distributed according to an apico-basal morphological (Puiseux-Dao, 1970a, b, 1972; Boloukhère, 1972; Puiseux-Dao *et al.*, 1972a, b; Hoursiangou-Neubrun and Puiseux-Dao, 1974, 1975; Hoursiangou-Neubrun *et al.*, 1975) and physiological gradient (Issinger *et al.*, 1971). In the apical region of the stalk, most of the chloroplasts contain well developed lamellae, divide normally and are relatively poor in carbohydrate granules. Their general aspect gradually changes from the apex to the base of the stalk, where many chloroplasts are large, have reduced lamellae, larger carbohydrate granules and seldom divide. This gradient is maintained under normal growth conditions. However, as illustrated in Fig. 16 (Puiseux-Dao *et al.*, 1972b) some morphological variations occur at two different times of the day, resulting from the circadian activity of the chloroplasts (Vanden Driessche and Hars, 1972a, b). Research on whole and on anucleate algae as well as on regenerating nucleate basal fragments, placed in darkness or treated by ribonuclease and by various inhibitors of nucleic acid and protein synthesis, has provided indirect evidence supporting the hypothesis that the chloroplast apico-basal gradient arises from a nuclear control on these organelles (Puiseux-Dao, 1970a, b; Puiseux-Dao and Dazy, 1970a, b; Hoursiangou-Neubrun *et al.*, 1792; Puiseux-Dao *et al.*, 1972a, b; Dubacq and Puiseux-Dao, 1974). According to Puiseux-Dao *et al.* (1972b), the nucleus controls several linked plastidal functions, namely the replication of the organelle itself, the synthesis of thylacoids and the utilization of reserve substances. This conclusion is supported by recent investigations on chloroplast multiplication in cytoplasts (Vanden Driessche *et al.*, 1972, 1973a, b). Recent work by Hoursiangou-Neubrun and Puiseux-Dao (1974) has shown that the spatial differentiation of the chloroplasts is a reversible process. In fact, at the young apex of regenerating basal fragments, where substances of nuclear origin accumulate, active apical plastids arise from the old basal ones. Moreover, in the absence of the nucleus, apical chloroplasts gradually transform into organelles of the basal type (Clauss and Keck, 1959; Shephard, 1965a; Hoursiangou-Neubrun and

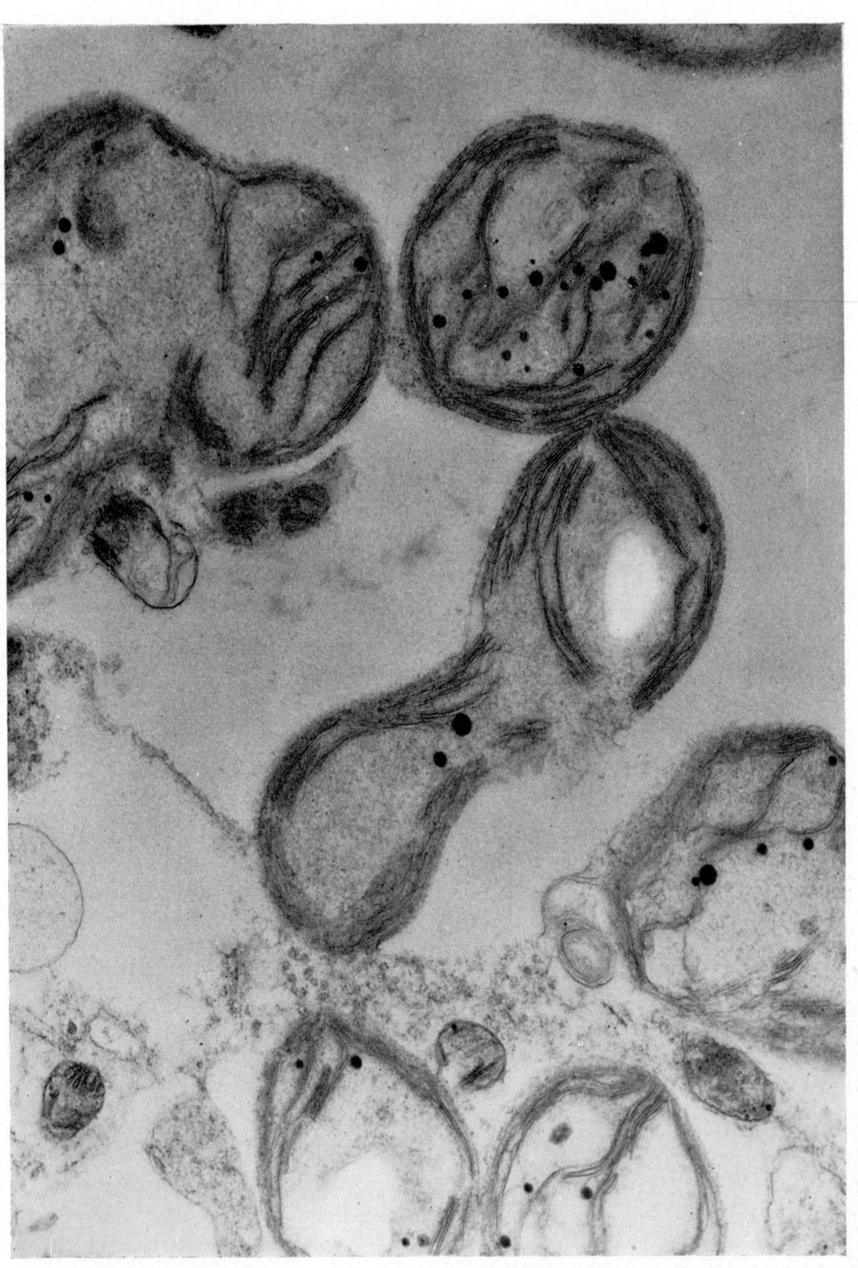

FIG. 15. Ultrastructural aspect of a dividing chloroplast. A few small mitochondria are also visible on this picture. (Courtesy of Simone Puiseux-Dao.)

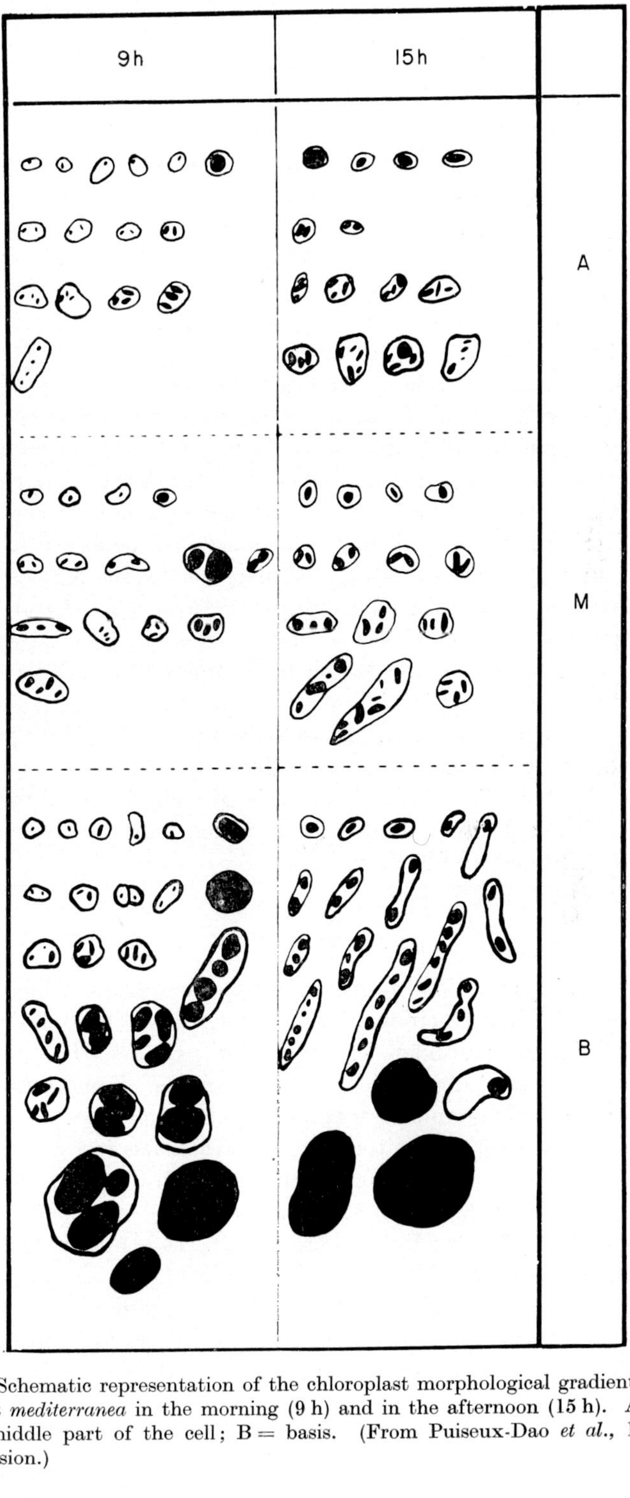

9h 15h

A

M

B

FIG. 16. Schematic representation of the chloroplast morphological gradient in *Acetabularia mediterranea* in the morning (9 h) and in the afternoon (15 h). A = apex; M = middle part of the cell; B = basis. (From Puiseux-Dao *et al.*, 1972, with permission.)

Puiseux-Dao, 1974). A rejuvenation of the basal chloroplasts has been observed by Bouloukhère (1970, 1972) during the formation of secondary nuclei. The same author has demonstrated that the ultrastructure of the chloroplasts depends, not only on the particular region of the cell where they are located (spatial differentiation), but also on the particular stage of the biological cycle of the alga (temporal differentiation) and on the experimental conditions (Bouloukhère, 1972). The chloroplasts present in the sterile whorls are sometimes particularly elongate (Mangenot and Nardi, 1931; Declève *et al.*, 1972). Unfortunately their function in the cell remains at present unknown.

The heterogeneity of *Acetabularia* chloroplasts has been recently studied by means of the fluorescence emission technique at the temperature of liquid nitrogen (Sironval *et al.*, 1972, 1973; Dujardin *et al.*, 1975). Examination of the 77°K fluorescence emission spectra provide a good criterion for distinguishing chloroplasts of the apical from those of the basal region of the stalk and may be useful for research on chloroplast morphogenesis. The ultrastructural apico-basal gradient of the chloroplasts is correlated with a gradient in their physiological activity. It has been demonstrated by Issinger *et al.* (1971), that the photosynthetic capacity is highest in the apical region and lowest in the basal region of the stalk. It is to be noted that the apico-basal gradient of the chloroplasts is maintained in the cell in spite of the protoplasmic streamings (Kamiya, 1960; Bouck, 1964; Williams and Terborgh, 1970; Puiseux-Dao, 1972), which are sensitive to cytochalasin B (Brachet and Tencer, 1973).

E. *Mitochondria*

Acetabularia contains numerous mitochondria, which appear heterogenous both in their form and in their size (Bouloukhère, 1970; Declève *et al.*, 1972; Vanden Driessche *et al.*, 1973a). They are recognizable by their cristae and their dense granules. Some mitochondria are bigger than others and possess many parallel cristae. It was found by Bouloukhère (1972) that mitochondria ultrastructure varies according to the biological cycle and the metabolic activity of the cell. Alterations of their morphology has been also observed after enucleation or dark treatment (Werz, 1966a; Bouloukhère, 1972). It is not known whether *Acetabularia* mitochondria, like the chloroplasts, are distributed in the cell according to a morphological apico-basal gradient (spatial differentiation). Mitochondria clustered in strings of beads or forming buds were observed by Bouloukhère (1970) in the course of formation of secondary nuclei.

Frequently the small mitochondria are located close to the chloroplast membrane (Declève *et al.*, 1972). This observation raises the question whether a relationship exists between chloroplasts and mitochondria in *Acetabularia*.

F. *Microtubules*

Microtubules have been found in *Acetabularia* by several authors (Boloukhère, 1970; Woodcock, 1971; Woodcock and Miller, 1973b; Berger *et al.*, 1975; Zerban and Werz, 1975a). Bundles of microtubules were observed in the cytoplasm surrounding the fragmenting primary nucleus as well as the migrating secondary nuclei (Boloukhère, 1970). According to Woodcock (1971), microtubules play a role in anchoring the secondary nuclei in the cap rays, whereas Berger *et al.* (1975a) think that they might contribute to an increase in the cytoskeletal rigidity. Typical spindle and flagella microtubules were described by Berger *et al.* (1975a) in *A. mediterranea*. The flagella microtubules have been recently studied by Zerban and Werz (1975a) by the freeze-etching technique.

G. *Cytoplasts*

When *Acetabularia* cells are cut, droplets of cytoplasm are spontaneously released from the stalk in the culture medium. They are called cytoplasts (Gibor, 1965). Cytoplasts represent an anucleate system, which can be used for studying the influence of the nucleus on cytoplasm differentiation (Werz, 1968b, 1970a, b), on chloroplast multiplication (Vanden Driessche *et al.*, 1972, 1973a, b) and on Golgi apparatus (Werz, 1964d). The ultrastructure of cytoplasts has been studied in detail by Vanden Driessche *et al.* (1973a).

VII. Morphogenesis

A. *The main morphogenetic processes in* Acetabularia

The term morphogenesis has been used for *Acetabularia* to designate the morphological differentiation of the cell as a whole. Three main morphogenetic processes can be distinguished during the biological cycle of the alga: (1) the formation of the different organs of the cell, namely the stalk, the rhizoid, the sterile whorls and the reproductive cap; (2) the closing of each secondary nucleus into a specialized survival organ, the cyst; (3) the formation of gametes into the cyst. This latter is thus to be regarded as a gametangium (Werz, 1974). Each one of these morphogenetic processes is very complex and is accompanied by important changes at the ultrastructural and the biochemical levels. The fine structural changes which occur during morphogenesis in *Acetabularia* have been recently discussed in an excellent review by Werz (1974).

B. *The " morphogenetic substances " of Hämmerling*

Hämmerling first demonstrated that *Acetabularia* has a single nucleus, located in the rhizoid, and that this giant cell is therefore a unicellular uninucleate organism (1931). This important discovery allowed him to perform a series of experiments, which have remained of prime importance for research on *Acetabularia*. Hämmerling first showed that the algae can easily be cut into two halves, nucleate and anucleate, and that the anucleate fragment is capable not only of surviving for weeks and even months, but also of regenerating such complex structures as rhizoid, sterile whorls and the reproductive cap (Hämmerling, 1932, 1934a, b, c, d). Further work by Hämmerling (1935, 1936, 1940, 1943a, b, 1944, 1946a, b, 1953, 1955a, b, 1957, 1958) and his collaborators (Beth, 1943a, b, c, 1953b, 1955a, b; Maschlanka, 1943a, b; Werz, 1955; Hämmerling *et al.*, 1958), mainly based on merotomy experiments, interspecific grafts and transplantation of isolated nuclei into anucleate fragments, led to the following main conclusions: (1) the nucleus produces " morphogenetic substances ", which are species-specific; (2) these substances are transferred to the cytoplasm, where they become distributed according to an apico-basal (cap forming substances) or to a baso-apical (rhizoid forming substances) gradient; (3) the species-specific morphogenetic substances produced by the nucleus accumulate and are stored in the cytoplasm for a long period, until they express morphogenesis; (4) the expression of a species-specific morphogenesis is induced by non species-specific substances.

According to Werz (1974), the non species-specific substances discovered by Beth (1943b), are probably special proteins which act as triggers in *Acetabularia* morphogenesis.

C. *The informational RNA concept of Brachet*

Some twenty years after the first observations by Hämmerling (1931, 1932), Brachet and collaborators undertook a series of investigations on the biochemistry of *Acetabularia* (Brachet, 1951, 1952a, b, 1957, 1958a, b, 1959, 1960a, b; Brachet and Chantrenne, 1951, 1952,

1953, 1956; Chantrenne *et al.*, 1952, 1953; Vanderhaeghe, 1952, 1954, 1963; Brachet and Brygier, 1953; Brachet and Szafarz, 1953; Baltus, 1955, 1959; Brachet *et al.*, 1955; Vanderhaeghe and Szafarz, 1955; Vanderhaeghe-Hougardy and Baltus, 1962).

It was found, around 1955, that anucleate *Acetabularia* can synthesize proteins, including functional enzymes. Since at that time it was already known that the synthesis of specific proteins is under genetic control, Brachet proposed, shortly before the discovery of the messenger ribonucleic acids (mRNAs), that the information present in the nuclear genes must be transported to the cytoplasm in the form of stable molecules of RNA (Brachet, 1960a, b, 1961a, b, 1962a, b, 1963a, c, d). The informational RNA concept of Brachet (1960a, b) is of fundamental importance for research on *Acetabularia* as well as on other living organisms. According to this concept, the Hämmerling's species-specific morphogenetic substances are identical with stable mRNAs. This assumption, which is shared also by Hämmerling (1963a, b) and by other authors (Zetsche, 1964c, 1965a, 1966b, c, d, 1968b; Hämmerling and Zetsche, 1966; Schweiger, 1964, 1967, 1968, 1969, 1970a, b; Werz, 1969a, 1970a, b, 1974; Puiseux-Dao, 1970b; Bonotto *et al.*, 1971d), has received further support from recent investigations on the effects of various chemical and physical factors on *Acetabularia* morphogenesis and from the study by Alexeev *et al.* (1974, 1975) on the influence of 40S ribonucleoprotein particles (RNP) on non-growing anucleate fragments of *Acetabularia crenulata* (see section VII, E).

D. *Effects of chemical and physical factors on growth and morphogenesis in* Acetabularia

A number of chemical substances (metabolites, antimetabolites, antibiotics, various inhibitors of oxidative phosphorylation, of sulphur metabolism, of nucleic acid and protein synthesis, hormones, etc.) and several physical factors (temperature, light, Ultraviolet-, X-, and Gamma-rays) have been used as tools for investigations on *Acetabularia* growth and morphogenesis. Table VI gives a list of the chemical substances used in research on *Acetabularia*, together with the references. As pointed out by Werz (1969a, b, 1974), several inhibitors are particularly useful tools for research on *Acetabularia* morphogenesis. However, their use has allowed only indirect evidence to be obtained on the mechanisms which control the main morphognetic events in *Acetabularia*. Altogether, the results obtained by submitting the algae to various treatments by chemical or by physical factors have added

TABLE VI. CHEMICAL SUBSTANCES USED AS TOOLS IN RESEARCH ON *Acetabularia* GROWTH AND MORPHOGENESIS

Substance	*References*
Actinomycin	Schweiger and Schweiger (1963), Brachet and Denis (1963), Brachet *et al.* (1974), de Vitry (1964a), Zetsche (1964b, c), Bouloukhère-Presburg (1965), de Vitry (1965b), Werz (1968a, 1969a, 1970a)
Amino acids	Dao (1956), Brachet (1958a), Olszewska and Brachet (1961), de Vitry (1962b), Puiseux-Dao (1957, 1962), Werz (1963a, b)
Aminopterin	de Vitry (1962b)
Antimetabolites	Brachet (1958a), de Vitry (1962b)
Beta-mercaptoethanol	Brachet (1958a, 1959, 1962a, b), Brachet and Delange-Cornil (1959), Brachet *et al.* (1963)
Bromodeoxyuridine	de Vitry (1962a)
Camptothecin	Brachet (1974)
Cicloheximide	Brachet (1967a), Bonotto *et al.* (1969c)
Cloramphenicol	Brachet *et al.* (1964), Bonotto *et al.* (1969c)
Colchicin	Werz (1969a, b)
Concanavalin	Brachet (1974)
Cordycepin	Brachet (1974), Kloppstech (1975a)
Cytochalasin	Herth *et al.* (1972), Brachet and Tencer (1973)
Dithiodiglycol	Brachet (1962a), de Vitry (1962b)
Deoxyribonuclease	de Vitry (1962b)
Ethidium bromide	Brachet (1968), Dazy (1969), Heilporn and Limbosch (1969, 1970, 1971b), Dazy *et al.* (1970), Puiseux-Dao and Dazy (1970)
Ethionine	Brachet (1958a, b)
Fluorodeoxyuridine	de Vitry (1962a, 1965b)
Fluorouridine	de Vitry (1962a)
Glutathione	de Vitry (1962b)
Heparin	De Carli and Brachet (1968)

Hydroxyurea	Brachet (1967b, 1968), Heilporn-Pohl and Limbosch-Rolin (1969), Heilporn and Limbosch (1970)
Isoproterenol	Vanden Driessche (1976)
Lead	Kirchmann and Bonotto (1972), Bonotto *et al.* (1972b), Bonotto and Kirchmann (1973)
Lindane	Borghi *et al.* (1972, 1973), Puiseux-Dao *et al.* (1972d)
Lipoic acid	Brachet (1961a, b, 1962a, b)
Metabolites	Brachet (1958a), de Vitry (1962b)
Morphactins	Vanden Driessche (1974a)
Papaverin	Vanden Driessche (1976)
Para-fluorophenylalanine	Zetsche (1966b), Zetsche *et al.* (1970)
Phenylethylalcohol	de Vitry (1965b)
Puromycin	Brachet (1963b), Brachet *et al.* (1964), Zetsche (1965a, 1966c), de Vitry (1965b), Bouloukhère-Presburg (1966, 1967), Werz (1968a, 1969a, 1970a), Bonotto *et al.* (1969c)
Ribofuranosylbenzimidazole	de Vitry (1962b)
Ribonuclease	Stich and Plaut (1958), Puiseux-Dao (1958a, b, 1962), de Vitry (1962b), Brachet and Six (1966), Puiseux-Dao and Dazy (1970)
Rifampicin	Brachet (1970), Brändle and Zetsche (1971), Puiseux-Dao *et al.* (1972c), Bonotto *et al.* (1972c)
Selenium	Werz (1961b)
Tetracycline	Bonotto *et al.* (1969c)
Theophyllin	Vanden Driessche (1976)
Hormones	
Giberellin	Zetsche (1963a), Bonotto and Kirchmann (1974)
Indoleacetic acid	Dao (1964b)
Indolacetonitrile	Thimann and Beth (1959)
Kinetin	de Vitry (1962b), Zetsche (1963a), Spencer (1968)
Methylindoleacetate	Thimann and Beth (1959)
Naphthalene acetic acid	Thimann and Beth (1959)
2-3-5-Triiodobenzoic acid	Thimann and Beth (1959), Spencer (1968)

credibility to the conclusions drawn by Hämmerling and collaborators (Hämmerling, 1934a, 1953, 1963) on the basis of merotomy and graft experiments.

E. *New evidence in favour of the informational RNA concept*

It has been emphasized by Brachet (1970) that the full demonstration that the " morphogenetic substances " of Hämmerling (1934a) are really mRNAs, will not be given until it is possible to isolate highly purified RNA preparations which, after injection into a non-growing anucleate fragment of the stalk, will induce a species-specific morphogenesis.

This demonstration has been partially achieved by Alexeev *et al.* (1974, 1975). These authors have shown, in fact, that the injection, into non-growing anucleate fragments of *Acetabularia crenulata*, of native 40 S RNP ($\rho = 1 \cdot 4$ g/cm^3) purified by ultracentrifugation on a sucrose gradient, induces the initiation of a limited morphogenesis (structures resembling sterile whorls and rudiments of caps). Interestingly enough, no effects were observed after injection of ribosomes or of heated (denatured) 40 S RNP particles. The rather limited regeneration induced by the injected RNP particles is probably ascribable to a lack of optimal conditions, as pointed out by the authors themselves. The results obtained by Alexeev *et al.* (1974, 1975) are very promising for further work on *Acetabularia* morphogenesis. RNP particles sedimenting with 120S ($\rho = 1 \cdot 395$ g/cm^3) have been also detected in *Acetabularia mediterranea* by Kloppstech (1972). Moreover, 20–100S RNP particles ($\rho = 1 \cdot 4$ g/cm^3) were found in the same species by Sandakhchiev (1973). It is known that RNP particles of similar density ($\rho = 1 \cdot 4$ g/cm^3) to those found in *Acetabularia* and containing what is usually considered mRNA, were also detected in animal cells (Spirin and Nemer, 1965; Spirin *et al.*, 1965; Spirin, 1969; Lucanidin *et al.*, 1972a, b; Baeyens *et al.*, 1974; Georgiev, 1974).

Further accurate work is required in order to characterize the RNA present in the RNP particles detected in *Acetabularia*. It is to be hoped that this RNA can be checked for its activity in an appropriate assay system. For instance, the activity of the 9S haemoglobin messenger RNA from reticulocytes has been assayed by injection into *Xenopus* oocytes (Lane *et al.*, 1972).

F. *New trends in* Acetabularia *morphogenesis*

According to current ideas, it is admitted that the " morphogenetic substances " of Hämmerling (1934a) are identical with mRNAs

(Brachet, 1960a, b), which specify the synthesis of the enzymes necessary for the morphogenetic events. The extensive work of Zetsche and collaborators (Zetsche, 1965a, b, c, 1966a, b, c, d, e, 1968a, b, 1969a, 1972; Zetsche *et al.*, 1970, 1972) showed, in fact, that morphogenesis in *Acetabularia* is realized under a regulatory mechanism which takes place at the level of translation. However, as emphasized by Werz (1974), post-translational processes are also important for the realization of morphogenesis. This author has recalled the importance not only of synthetic but also of lytic events, which occur in the already existing cell wall (Werz, 1965, 1966b, 1969a, b, 1970a, b, 1974). Moreover, dictyosomes (Werz, 1964d) and certain polypeptides associated with the polysaccharides of the cellular wall would play an important role in morphogenesis (Werz, 1974).

The importance of the cytoplasmic membrane in the initiation of morphogenesis has been recently discussed by several authors (Werz, 1970a, b, 1974; Bonotto *et al.*, 1971d; Puiseux-Dao, 1972). According to Bonotto *et al.* (1971d) and to Puiseux-Dao (1972), the cytoplasmic membrane of the apical region of the stalk, would contain specific receptors, capable of interacting with the morphogenetic substances. Only certain configurations of complex molecular structures of the membrane would assure its "competence", that is the capability to react to an inducing stimulus and thus initiate morphogenesis (Bonotto *et al.*, 1971d). The accumulation of the morphogenetic substances at the apex of the cell would be facilitated by hydrodynamic as well as by electrical forces (Puiseux-Dao, 1972; Novák, 1975). However, the trigger mechanism for the activation of the accumulated substances remains unknown. Since the light dependency of morphogenesis is well established (Beth, 1953a, 1955a, b; Richter, 1962; Clauss, 1963, 1968, 1970a, b, 1973; Terborgh and Thimann, 1964, 1965), the activation of these substances could be controlled by a light-dependent mechanism. Recent graft experiments by Puiseux-Dao *et al.* (1970) between *Acetabularia mediterranea* and *A. peniculus* have shown that the morphogenetic substances are not irreversibly bound to the apical membrane. The same conclusion has been drawn by Chirkova *et al.* (1970) and Sandackchiev *et al.* (1972) by submitting anucleate fragments to centrifugation in a basal direction, and by Bonotto *et al.* (1972d) by investigating basal cap formation after enucleation. Moreover, several investigations on abnormal morphogenesis (Bonotto, 1968, 1969a, 1971, 1973; Bonotto and Puiseux-Dao, 1970; Bonotto and Kirchmann, 1971b), and on cap formation in branched cells (Bonotto and Janowski, 1968; Bonotto and Bonnijns-Van Gelder, 1969; Bonotto *et al.*, 1970a, 1972a; Bonotto, 1971, 1975) or in anucleate fragments showing reversion of polarity (Bonotto *et al.*, 1972d) have permitted the following conclusions: (1) The morphogenetic substances are able to move from a storage region of the apical stalk to a new growth region of the basal stalk. (2) Identical morphogenetic substances (presumably mRNAs) must be used (translated) at different times, since cap formation is asynchronous at the two ends of anucleate fragments or at the apices of branched cells; consequently, the lifetime of a species-specific morphogenetic substance may vary in different regions of the same cell. (3) Since anucleate fragments may supply sufficient morphogenetic substances to two growth regions of the stalk (apical and basal ends, respectively) instead of one (apical region only), the total amount of morphogenetic substances stored in the cytoplasm must exceed the real needs of the cell; however, one cannot exclude the possibility that these substances may be used repeatedly or that they may even be replicated (Bonotto *et al.*, 1971d). Finally, although morphogenesis is controlled only indirectly by the nuclear genes, several authors have reported that the nucleus has some kind of direct inhibitory effects on the cytoplasm (Beth, 1953b; Brachet *et al.*, 1955; Zetsche *et al.*, 1970; Bonotto *et al.*, 1972d). In fact, after the removal of the nucleus, cap formation may even be accelerated (Beth, 1953b; Bonotto *et al.*, 1972d) and protein synthesis may undergo a temporary stimulation (Brachet *et al.*, 1955). Altogether, early and more recent investigations show that morphogenesis in *Acetabularia* constitutes a series of very complex events. It appears, therefore, that real progress can be made in future research on morphogenesis only if biological and biochemical experiments are coupled with ultrastructural, physicochemical and electrophysiological studies. In this context, time lapse filming methods applied to the study of *Acetabularia* morphogenesis (Martinov *et al.*, 1974; Puiseux-Dao and Hoursiangou-Neubrun, personal communication; Yazykov *et al.*, 1975) may prove useful.

VIII. DNA

A. *DNA of the primary nucleus and of the secondary nuclei*

In young *Acetabularia* cells, the nucleus possesses Feulgen stainable chromatin and a small basophilic nucleolus. In adult cells, the giant nucleus becomes generally Feulgen-negative (Brachet, 1970). Probably, the DNA is too diluted in the enlarged nucleus to give a positive reaction. However, a very faint staining of the nucleoplasm has been obtained by Spring *et al.* (1974). These authors have also observed that nucleolar aggregates are positively stained, especially in their cortical regions, and that sometimes a faintly positive reaction seems to occur in the perinuclear dense bodies. These latter may include amplified DNA molecules (rDNA), which are transported into the cytoplasm (Spring *et al.*, 1974). rRNA cistrons have been recently demonstrated in *Acetabularia mediterranea*, *A. major* and *Batophora oerstedii* (Spring *et al.*, 1974; Trendelenburg *et al.*, 1974a, b, 1975; Berger and Schweiger, 1975; Liddle *et al.*, 1975; Woodcock, 1975; Woodcock *et al.*, 1975). Moreover, lampbrush chromosomes have been discovered in the nucleoplasm of the primary nucleus of *Acetabularia mediterranea* and *Batophora oerstedii* (Spring *et al.*, 1975b; Liddle *et al.*, 1975). Spring *et al.* (1975b) have investigated the ultrastructural organization of the lampbrush chromosomes by the electron microscope. However, the DNA of these chromosomes has not yet been studied. It has been found that the primary nucleus of young zygotes is diploid, while the secondary nuclei are haploid, since meiosis occurs before, during or immediately after the fragmentation of the primary nucleus (Koop, 1975c). The DNA content of the haploid nucleus of *A. mediterranea* was determined as 1·29 picogram (Spring *et al.*, 1974).

B. *DNA of chloroplasts and mitochondria*

Before the discovery by Ris and Plaut (1962) that chloroplasts and mitochondria possess DNA, Brachet, in collaboration with A. Ficq (see: Brachet, 1958b) showed that nucleate and anucleate fragments of *Acetabularia* were able to incorporate ^{3}H-thymidine into a compound resistant to ribonuclease digestion or to Feulgen hydrolysis (Feulgen and Rossenbeck, 1924). Later on, the presence of DNA in the chloroplasts of *Acetabularia* has been definitely demonstrated by a fluorometric procedure (Baltus and Brachet, 1963; Gibor and Izawa, 1963). Moreover, Shephard (1965a) showed that chloroplasts are able to multiply in the absence of the nucleus. Although on a number of occasions indirect evidence has implied that chloroplastic DNA was capable of self replicating in the absence of the nucleus (Brachet, 1958b; Baltus and Brachet, 1963; Gibor and Izawa, 1963; Brachet *et al.*, 1964; de Vitry, 1964a, b, 1965a, b; Zetsche, 1964b, c; Schweiger and Berger, 1964; Goffeau and Brachet, 1965; Janowski, 1965; Shephard, 1965a, b; Chapman *et al.*, 1966), final proof was obtained in 1966 by Heilporn-Pohl and Brachet. They showed, by the fluorometric method of Kissane and Robbins (1958), that anucleate fragments of *Acetabularia* are capable of a net synthesis of DNA (Heilporn-Pohl and Brachet, 1966). Electron microscope examination of thin sections of fixed plastids has revealed the presence of typical fibrilla DNA-containing areas (Puiseux-Dao, 1966a; Puiseux-Dao *et al.*, 1967

Werz and Kellner, 1968b; Boloukhère, 1970; Woodcock and Bogorad, 1970b). Moreover DNA strands were observed in spread preparations of chloroplasts obtained from anucleate fragments (Werz and Kellner, 1968a, b; Green and Burton, 1970; Green *et al.*, 1970; Woodcock and Bogorad, 1970b). The DNA of *Acetabularia* mitochondria has not yet been visualized, although several studies have suggested its presence in these organelles (Green *et al.*, 1967, 1970; Heilporn and Limbosch, 1969, 1970, 1971a, b.).

C. *Amount of DNA per chloroplast*

Early reports give a value of 10^{-16} g of DNA per chloroplast, which corresponds to 6×10^7 daltons (Baltus and Brachet, 1963; Gibor and Izawa, 1963). More recent investigations (Green, 1972, 1973b; Green and Padmanabhan, 1975; Green *et al.*, 1975) have shown that chloroplasts may contain large amounts of DNA: up to 1·5 and $4{\cdot}7 \times 10^9$ daltons in *Acetabularia mediterranea* and *Acetabularia cliftoni* respectively (Green, 1973b). Moreover, a value of $1{\cdot}9 \times 10^9$ daltons has been found by Mazza and Bonotto (unpublished) for a single chloroplast of *Acetabularia mediterranea.* However, the DNA content of individual chloroplasts may vary considerably and DNA could not be detected in 65–80% of the plastidal population (Woodcock and Bogorad, 1968, 1970b). Variability in DNA content has been found also in chloroplasts from other plant cells (Herrmann, 1970; Hermann and Kowallik, 1970). According to Green (1973b) *Acetabularia* chloroplasts could have a small number of copies of their genome.

D. *Visualization of* Acetabularia *chloroplast DNA*

Acetabularia chloroplast DNA seems to assume different configurations according to the electron microscope techniques employed for its visualization. In fact, DNA from osmotically shocked chloroplasts (Green and Burton, 1970; Green *et al.*, 1970; Woodcock and Bogorad, 1970b) is preferentially, although not exclusively, released as large patches of entangled molecules (Fig. 17) often radiating from a centre or " displays " (Green and Burton, 1970), while DNA extracted by concentrated salt solutions (Werz and Kellner, 1968a, b) or by Sarkosyl (Green, 1972, 1973b) shows essentially a linear configuration with molecules of various sizes, very few of them reaching a total length of about 200 μm (Green, 1973b). DNA prepared by molecular sieving on sepharose (Lurquin *et al.*, 1974; Heyn *et al.*, 1974; Bonotto *et al.*, 1975) is represented by aggregates ranging from 40 to 300 μm and by linear molecules of various sizes. Very few minicircles were

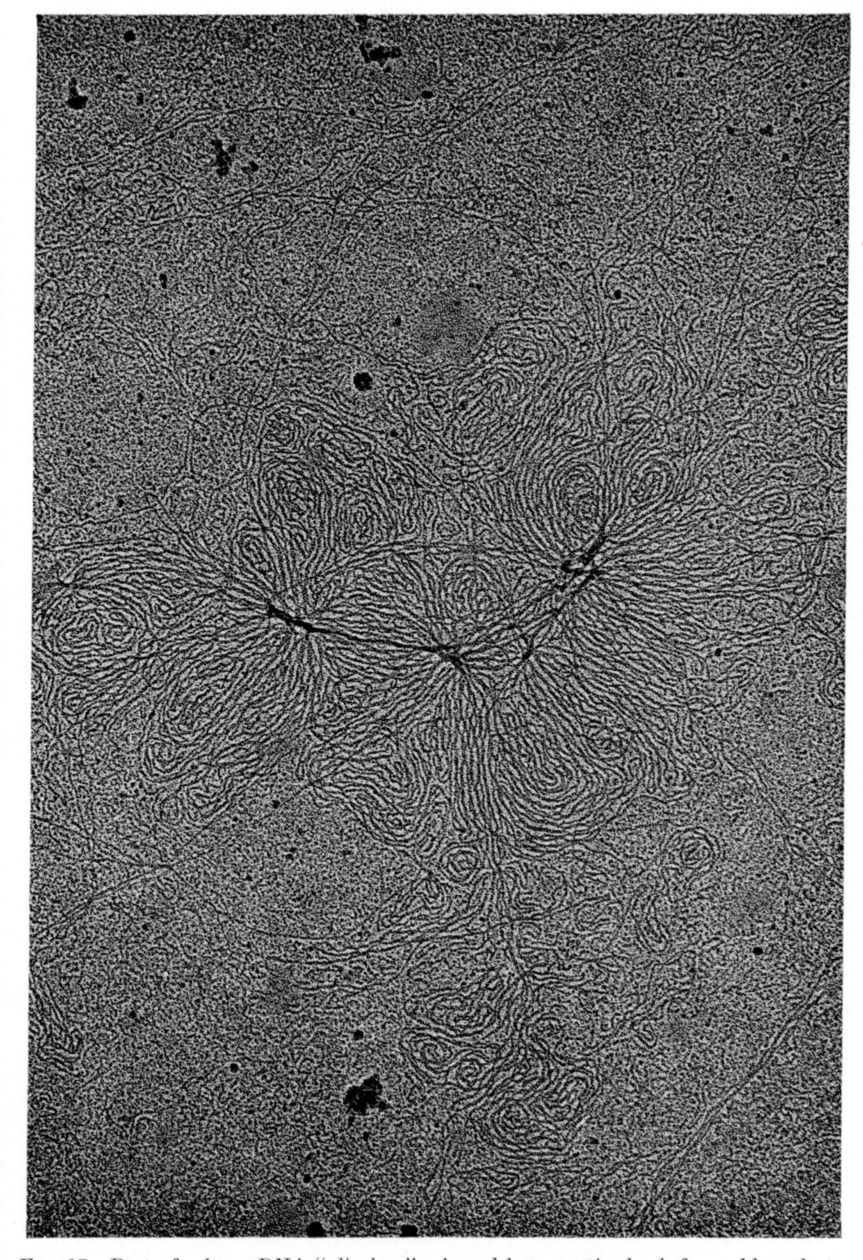

FIG. 17. Part of a large DNA " display " released by osmotic shock from chloroplasts of the basal part of the stalk of *Acetabularia mediterranea*. × 17 300.

also found in this DNA preparation (Mazza and Bonotto unpublished). These authors have also observed Cairns forks, both in circular and linear molecules, and have suggested that *Acetabularia* chloroplast DNA may replicate according to the Cairns model (Cairns, 1963), as in bacteria. Recently, small circular DNA molecules were found in chloroplasts of *Acetabularia cliftoni*, lysed by Sarkosyl (Green, 1973c; Green and Padmanabhan, 1975) and in osmotically shocked chloroplasts of *Acetabularis mediterranea* (Fig. 18). These molecules could represent a minor component of the chloroplast genome, as suggested

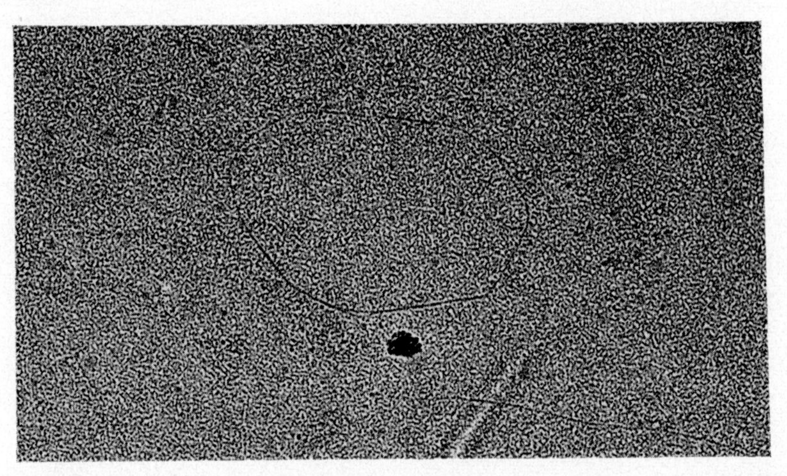

FIG. 18. Circular DNA molecule released by osmotic shock from chloroplasts of the apex of *Acetabularia mediterranea*. × 31 100.

by Green (1973c). However, their origin deserves further accurate investigation, since chloroplast preparations may be contaminated by mitochondria.

E. *Buoyant density of the* Acetabularia *DNAs*

The buoyant density of the DNAs present in whole and in anucleate *Acetabularia* has been determined by CsCl gradient centrifugation. The values found by various authors for DNAs extracted from the plant or from contaminant micro-organisms are given in Table VII, together with the references. The table shows that different buoyant densities have been reported for nuclear (1·696, 1·702, 1·712 g/cm^3), chloroplastic (1·695, 1·702, 1·703, 1·704, 1·706, 1·712, 1·717 g/cm^3) and mitochondrial (1·709, 1·714, 1·715 g/cm^3) DNA. The reasons for such

TABLE VII. BUOYANT DENSITY OF DNAs FROM *Acetabularia* AND FROM ITS BACTERIAL CONTAMINANTS

Buoyant density g/cm³	*Origin of the DNA*	*References*
1·695	Chloroplasts	Gibor (1967)
1·696	Micro-organisms	Green *et al.* (1967)
1·696	Supposed cytoplasmic	Lurquin *et al.* (1972a, b), Bonotto *et al.* (1973)
1·696	Supposed nuclear	Heilporn and Limbosch (1971a)
1·698	Micro-organisms	Green *et al.* (1967), Green (1972)
1·702	Supposed nuclear	Green *et al.* (1967)
1·702	Chloroplasts	Green (1972, 1973b)
1·703	Chloroplasts	Green *et al.* (1967)
1·704	Chloroplasts	Green *et al.* (1967, 1970), Heilporn and Limbosch (1971a, b) Lurquin *et al.* (1974), Bonotto *et al.* (1975)
1·704	Micro-organisms	Green *et al.* (1967)
1·706	Chloroplasts	Green (1973b, c)
1·709	Mitochondria	Berger and Schweiger (1969)
1·712	Supposed nuclear	Berger and Schweiger (1969)
1·712	Chloroplasts (minor component)	Green (1973c)
1·713	Micro-organisms	Green *et al.* (1967)
1·714	Mitochondria	Green *et al.* (1967, 1970)
1·715	Mitochondria	Heilporn and Limbosch (1971a, b)
1·716	Micro-organisms	Green (1972)
1·717	Chloroplasts	Berger and Schweiger (1969)
1·718	Supposed cytoplasmic	Lurquin *et al.* (1972a, b), Bonotto *et al.* (1973)
1·718	Micro-organisms	Green (1972)
1·722	Unknown	Green *et al.* (1967)
1·723	Micro-organisms	Green *et al.* (1967)
1·731	Micro-organisms	Green *et al.* (1967)

a disparity of values have not yet been explained. This discrepancy could be due, at least in part, to inaccuracy in buoyant density determination or to eventual bacterial contaminations. However, one cannot exclude the possibility that in *Acetabularia* the nucleus and the organelles contain more than one species of DNA. For instance, the chloroplasts of *Acetabularia cliftoni* possess a major component ($\rho = 1{\cdot}706$ g/cm³) together with a minor one ($\rho = 1{\cdot}712$ g/cm³), constituted by small circular molecules (Green, 1973c). The buoyant density of the amplified rDNA (Spring *et al.*, 1974; Trendelenburg *et al.*, 1974a, b, 1975; Berger and Schweiger, 1975; Liddle *et al.*, 1975; Woodcock,

1975; Woodcock *et al.*, 1975) and of the lampbrush chromosomes DNA (Liddle *et al.*, 1975; Spring *et al.*, 1975b) has not yet been determined.

IX. RNA

A. *RNA and ribosomes*

Autoradiographic and biochemical experiments have shown that whole as well as anucleate *Acetabularia* are capable of synthesizing RNA (Brachet and Szafarz, 1953; Brachet *et al.*, 1955; Vanderhaeghe and Szafarz, 1955; Vanderhaeghe, 1963; Richter, 1957, 1958b, 1959a, b, c, 1966; Naora *et al.*, 1959, 1960; Schweiger and Bremer, 1960a, b, 1961; Sutter *et al.*, 1961; Ceska, 1962; de Vitry, 1963, 1964b, 1965a, b, c; Janowski, 1963, 1965, 1969; Baltus and Quertier, 1966; Brachet and Six, 1966; Berger, 1967a, b; Schweiger *et al.*, 1967a, c; Bonotto *et al.*, 1968; Janowski *et al.*, 1968; Dillard and Schweiger, 1968, 1969; Farber *et al.*, 1968; Farber, 1969a, b; Bonotto, 1969c; Bonotto *et al.*, 1969c; Bonotto and Kirchmann, 1969a, b; Vanden Driessche and Bonotto, 1969a; Janowski and Bonotto, 1970; Dillard, 1970; Vanden Driessche *et al.*, 1970; Schweiger, 1970b; Woodcock and Bogorad, 1970a; Brändle and Zetsche, 1971, 1973; Fox and Shephard, 1972; Spring *et al.*, 1974; Klopstech, 1975a Kloppstech and Schweiger, 1975b) and ribosomes (Janowski, 1966, 1976, 1969; Baltus *et al.*, 1968; Bonotto 1969c; Janowski *et al.*, 1969; Janowski and Bonotto, 1970; Lurquin *et al.*, 1972a; Kloppstech and Schweiger, 1973a, b, 1975a; Berger and Schweiger, 1975a; Dazy *et al.*, 1975a, b; Kloppstech, 1975b). According to Werz (1960b, d) the RNA is distributed in the cell along an apico-basal gradient. Electron microscopical investigations have shown that ribosomes and polyribosomes are distributed in the same way (Werz, 1965; Van Gansen and Boloukhère-Presburg, 1965). However, biochemical investigations have shown that in *Acetabularia major* cytosol ribosomes are evenly distributed throughout the cell (Kloppstech and Schweiger, 1975a).

It is now well established that *Acetabularia* chloroplast and cytosol ribosomes have a sedimentation coefficient of 70S and 80S respectively (Janowski, 1966, 1967, 1969; Baltus *et al.*, 1968; Janowski *et al.*, 1969; Bonotto, 1969c; Janowski and Bonotto, 1970; Kloppstech and Schweiger, 1973a, b, 1975a; Berger and Schweiger, 1975a; Kloppstech, 1975b). Moreover, 80S ribosomes from *Acetabularia major* have been visualized by Schweiger *et al.* (1974b). Ribosomes from *Acetabularia* mitochondria have not yet been studied, due to the difficulty in obtaining a pure fraction of these organelles. However, labelled ribosomal and soluble RNA have been extracted from a crude mitochondrial fraction (Schweiger *et al.*, 1967; Bonotto, 1969c; Janowski and Bonotto, 1970).

Total RNA from *Acetabularia* sediments in a sucrose gradient in two peaks of 25S and 16S respectively. Polyacrylamide gel electrophoresis has shown that total *Acetabularia* rRNA can be separated into four species: 26S and 18S, corresponding to the cytosol ribosomes and 23S and 16S, corresponding to the chloroplast ribosomes (Schweiger, 1970b; Woodcock and Bogorad, 1970a; Fox and Shephard, 1972; Kloppstech and Schweiger, 1973a). The existence of a stable 15S RNA, which might be contained in a stable 30S particle, has been demonstrated by Janowski and Bonotto (1970). Further investigations are needed in order to know the nature of this particularly stable RNA species. Most of the cell RNA has been attributed to the chloroplasts (Naora *et al.*, 1960; Fox and Shephard, 1972). This assumption has been questioned by Schweiger (1970b), who suggests that the greater part of the faster sedimenting rRNA (25S) is of nuclear origin.

Indirect evidence has been obtained that isolated chloroplasts are able to synthesize mRNA (Goffeau and Brachet, 1965). Moreover, Berger (1967a, b) has demonstrated that synthesis of rRNA and tRNA occurs in chloroplasts after their isolation. Similar results have been obtained by Fox and Shephard (1972) by supplying $^{14}CO_2$ to isolated chloroplasts. RNA synthesis in the chloroplasts of whole as well as anucleate cells shows a daily variation (Vanden Driessche and Bonotto, 1968a, 1969a, b; Bonotto *et al.*, 1972d).

The nature of the RNA synthesized in the primary nucleus has been recently studied by Spring *et al.* (1974). These authors have proposed a scheme for nuclear-cytoplasmic pre-rRNA processing in *Acetabularia mediterranea*. In *Acetabularia* only very little, if any, pre-rRNA material seems to be lost during the processing (Spring *et al.*, 1974, 1976; Trendelenburg *et al.*, 1974a; Berger and Schweiger, 1975a; Kloppstech and Schweiger, 1975a; Woodcock *et al.*, 1975). Recently, polyadenylated RNA, presumably mRNA, has been isolated from whole and from anucleate *Acetabularia mediterranea* (Kloppstech and Schweiger, 1975b).

B. *Redundancy of the rRNA cistrons*

Early light microscopical investigations have revealed a large amount of nucleolar material in the primary nucleus of *Acetabularia* (Hämmerling, 1931; Schulze, 1939; Puiseux-Dao, 1962, 1966b). The nucleolus may occupy up to 50% of the nuclear volume. Moreover,

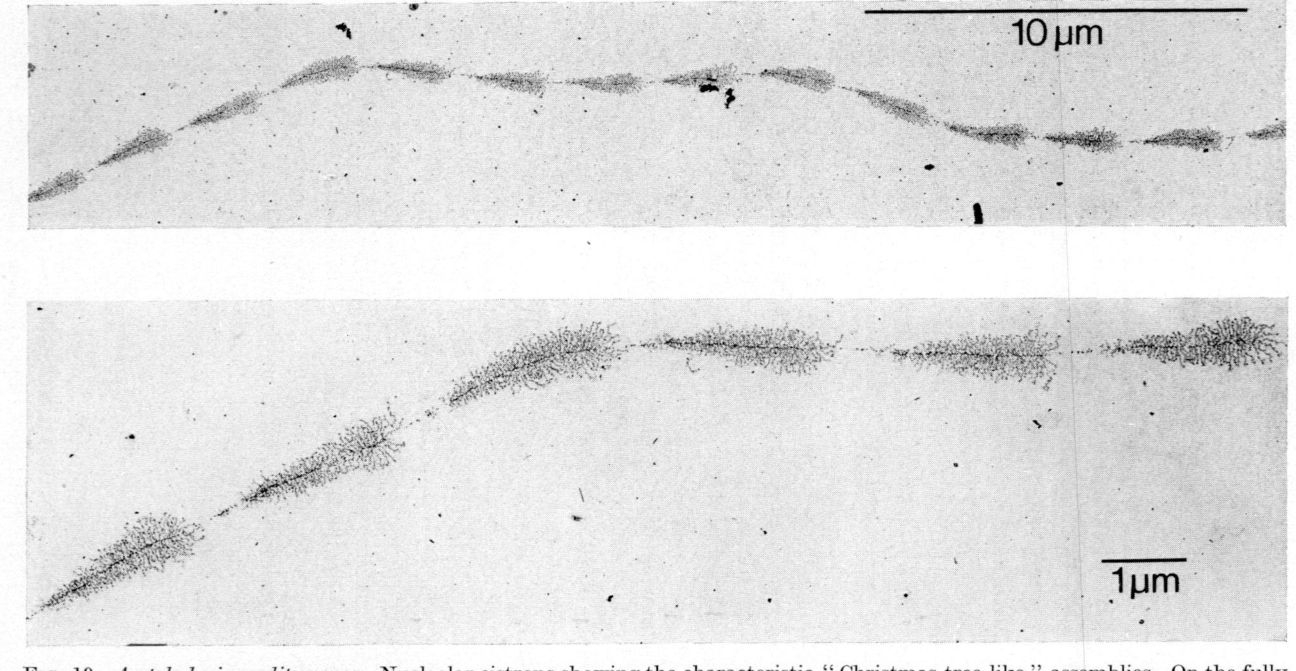

Fig. 19. *Acetabularia mediterranea.* Nucleolar cistrons showing the characteristic "Christmas-tree-like" assemblies. On the fully extended strand, fibril-covered regions are separated from each other by "spacers". (From Spring *et al.*, 1976, with permission.)

after disruption, it gives rise to distinct nucleolar subunits or droplets (Trendelenburg *et al.*, 1974a; Berger and Schweiger, 1975). Electron microscopical investigations of spread nucleolar material have demonstrated the existence of the characteristic "Christmas-tree-like" assemblies (Fig. 19), which correspond to transcriptionally active cistrons (Spring *et al.*, 1974, 1976; Trendelenburg *et al.*, 1974a, b, 1975; Berger and Schweiger, 1975 a b, c; Liddle *et al.*, 1975; Woodcock, 1975 Woodcock *et al.*, 1975a, b). The active cistrons are identified by their adherent nascent ribonucleoprotein fibrils, which show a typical length gradient. Generally, the fibril-covered regions are separated from each other by naked regions or "spacers". Some spacers, however, were covered by short fibrils (Trendelenburg *et al.*, 1974a). Moreover, "tail-to-tail", and also "head-to-head" cistron arrangements were observed indicating that the DNA matrix may be transcribed in both directions (Trendelenburg *et al.*, 1974a; Berger and Schweiger, 1975a, c). The total number of pre-rRNA cistrons has been estimated as about $1{\cdot}3 \times 10^4$ for *A. mediterranea* (Trendelenburg *et al.*, 1974a) and as 3–$3{\cdot}8 \times 10^4$ for *A. major* (Berger and Schweiger, 1975a; Kloppstech and Schweiger, 1975). These values are an estimate of the degree of redundancy for rRNA genes in *Acetabularia.* Probably the amplification of rDNA occurs through a greater part of the vegetative phase. It has been suggested that extrachromosomal copies of rDNA might be contained in the perinuclear bodies (Franke *et al.*, 1974; Spring *et al.*, 1974).

C. *Informative ribonucleoprotein particles*

Recent investigations (Kloppstech, 1972; Sandakhchiev, 1973; Alexeev *et al.*, 1974, 1975) have revealed the existence in *Acetabularia* of ribonucleoprotein particles having the same buoyant density ($\rho = 1{\cdot}4$ g/cm^3) as the "informosomes" of animal cells (Spirin, 1969). Alexeev *et al.* (1974, 1975) have obtained strong evidence supporting the hypothesis that these particles are identifiable with the "morphogenetic substances" of Hämmerling (1934a). If the results of Alexeev *et al.* (1974, 1975) should be confirmed, *Acetabularia* might be revealed as a very useful object for studying long-lived plant informosomes.

X. Protein

A. *Protein synthesis in the presence and in the absence of the nucleus*

Studies directly or indirectly concerned with protein synthesis in whole and in anucleate *Acetabularia* have been performed by many authors (Brachet and Chantrenne, 1951, 1952, 1953, 1956; Brachet and Brygier, 1953; Chantrenne *et al.*, 1953; Vanderhaeghe, 1954, 1963; Giardina, 1954; Brachet *et al.*, 1955; Stich and Kitiyakara, 1957; Werz, 1957a, 1960a, d, 1962; Clauss, 1958, 1959, 1961, 1962a, b, c; Keck and Clauss, 1958; Baltus, 1959; Richter, 1959a, b, c; Werz and Hämmerling, 1959; Bremer and Schweiger, 1960; Keck, 1960, 1961, 1963, 1969; Olszewska and Brachet, 1960, 1961; Clauss and Werz, 1961; Olszewska *et al.*, 1961; Bremer *et al.*, 1962; Keck and Choules, 1963; Chauduri and Spencer, 1968; Craig and Gibor, 1970; Ceron and Johnson, 1971). It has been found that after the removal of the nucleus, protein synthesis not only continues at an appreciable rate, but it may even undergo a temporary stimulation (Brachet *et al.*, 1955). However, two to three weeks after the enucleation total protein synthesis progressively diminishes in anucleate fragments (Brachet *et al.*, 1955; Hämmerling *et al.*, 1958; Richter, 1959a; Werz and Hämmerling, 1959; Vanderhaeghe-Hougardy and Baltus, 1962). At this time photosynthesis or respiration are hardly affected (Chantrenne *et al.*, 1952; Hämmerling *et al.*, 1958).

TABLE VIII. *Acetabularia* ENZYMES SYNTHESIZED IN THE ABSENCE OF THE NUCLEUS

Enzyme	*References*
Acid phosphatase	Keck and Clauss (1958), Keck (1960, 1961), Vanderhaeghe-Hougardy and Baltus (1962), Keck and Choules (1963), Spencer and Harris (1964), Triplett *et al.* (1965), Keck (1969), Brachet (1970)
Aldolase	Baltus (1955, 1959), Brachet *et al.* (1955)
Catalase	Brachet and Chantrenne (1953)
Invertase	Keck and Clauss (1958)
Lactic dehydrogenase	Reuter and Schweiger (1969), Schweiger (1970a, b)
Malic dehydrogenase	Schweiger *et al.* (1967b, 1969, 1972), Schweiger (1970a, b), Sandakhchiev and Niemann (1972) Sandakhchiev *et al.* (1973)
Phosphoglucomutase	Grieninger and Zetsche (1972)
Phosphoglucose-isomerase	Zetsche *et al.* (1970), Grieninger and Zetsche (1972)
Phosphorylase	Clauss (1959)
Ribonuclease	Schweiger (1966)
Thymidine kinase	Bannwarth (1969, 1972a, b, 1975a, b) Bannwarth and Schweiger (1975)
UDPG-4-epimerase	Zetsche (1966e, 1972)
UDPG-pyrophosphorylase	Zetsche (1968a, b, 1969a, 1972), Zetsche *et al.* (1970)

A number of enzymes have been found to increase in whole cells as well as in anucleate fragments (Brachet and Lang, 1965). Table VIII gives a list of those enzymes which were found to increase even in the absence of the nucleus.

Several other *Acetabularia* specific proteins were investigated: enzymes involved in general sugar metabolism (Zetsche 1965b, 1966a, b, d, e, 1967, 1968a, b, 1969a, 1972; Zetsche *et al.*, 1970; Bachmann and Zetsche, 1975), or concerned with the pentose pathway (Hellebust *et al.*, 1967; Boege *et al.*, 1975), NAD-kinase and NAD-pyrophosphorylase (Bannwarth *et al.*, 1969; Bannwarth, 1972c), thymidine kinase (Bannwarth, 1969, 1972a, b, 1975a, b; Bannwarth and Schweiger, 1975) and protein kinase (Pai *et al.*, 1975a, b). This last enzyme could be one of the trigger proteins in the induction process for differentiation (Pai *et al.*, 1975b). Studies on specific enzymes are of great interest for a better understanding of the fine mechanism which regulates cellular differentiation in *Acetabularia*.

B. *Nuclear dependency of chloroplast proteins*

Experiments by Bachmayer (1970) have shown that N-formylmethionyl-tRNA (fMet-tRNA) is most probably involved in the initiation of protein synthesis in *Acetabularia* chloroplasts. Chloroplasts isolated from anucleate fragments are capable of incorporating radioactive amino acids into proteins (Brachet and Goffeau, 1964; Goffeau and Brachet, 1965; Goffeau, 1969a, b, 1970). It has been suggested that chloroplast DNA codes not only for structural proteins but also for other water-insoluble as well as water-soluble proteins (Goffeau, 1970). However, direct evidence that *Acetabularia* chloroplast DNA is able to code for a specific chloroplast protein is still lacking, even though one of the acid phosphatase isozymes found by Triplett *et al.* (1965) could be under chloroplastic control (Keck, 1969). On the contrary, it has been demonstrated, by nuclear transplantation experiments, that several chloroplast proteins are coded for by the nuclear genes (Schweiger, 1970a, b; Apel and Schweiger, 1972). Chloroplast nuclear dependency has so far been reported for: malic dehydrogenase (Schweiger *et al.*, 1967b, 1969, 1972a; Sandackchiev *et al.*, 1973; Berger *et al.*, 1974), lactic dehydrogenase (Reuter and Schweiger, 1969), membrane proteins (Apel, 1970, 1972a, b, 1975 a, b; Apel and Schweiger, 1972, 1973; Schweiger *et al.*, 1972a; Apel *et al.*, 1975) and ribosomal proteins (Kloppstech and Schweiger, 1971, 1972, 1973a, b, 1974). According to Kloppstech and Schweiger (1974), at least most of the proteins of the 44S subunit of chloroplast ribosomes are synthesized on 80S cytosol ribosomes. Although the existent experimental data suggest that several chloroplast proteins are coded for by nuclear genes, the possibility remains that nuclear control is restricted to secondary structural modifications of the proteins.

From the reported studies it appears that in *Acetabularia* the nucleus probably regulates chloroplast functions by coding for enzymes or other proteins, which are involved in the activity of this organelle. Nuclear dependence of some chloroplast proteins has been reported in the literature also for other plant material (Anderson and Levin, 1970; Margulies, 1971; Mets and Bogorad, 1971). Moreover, it has been demonstrated that in *Chlamydomonas* two genetic systems (nuclear and plastid genes) cooperate in the synthesis of chloroplast ribosomes (Bogorad *et al.*, 1975). The existence of an intergenomic cooperation in *Acetabularia* would explain the apparently contradictory results reported in the literature on dependency of chloroplast proteins (Goffeau, 1970; Apel and Schweiger, 1972).

XI. Lipids

Several studies have been made on the composition and synthesis of *Acetabularia* lipids (Dubacq, 1970, 1971, 1972, 1973; Brush and Percival, 1972; Dubacq and Puiseux-Dao, 1974; Dubacq and Hoursiangou-Neubrun, 1975). It has been found that in algae kept for two weeks in darkness the total fatty acids content decreased to 50% of that observed in light. Experiments with ^{14}C-acetate or with ^{14}C-carbonate have shown that the synthesis of palmitic acid sharply decreases in the dark. The synthesis of oleic acid is less affected by darkness. Polyunsaturated fatty acids are also reduced in dark grown algae (Dubacq, 1971). In further work, Dubacq (1972, 1973) has demonstrated that lipid composition in *Acetabularia* changes during its morphological differentiation. Young cells (stage 4, according to Bonotto and Kirchmann, 1970) contain a high level of palmitic and oleic acids and are relatively poor in polyunsaturated acids, such as linoleic and linolenic acid. These latter are abundant in higher plants and in many green algae. During the growth of the cap (stage 7), the amount of linoleic and linolenic acid increases considerably (Dubacq, 1972). When cysts are formed (stage 10), lipid content of the cell decreases. Dubacq (1973) has noticed that, in contrast to young cells, full grown *Acetabularia* have a fatty acid composition which is more typical for marine algae. The same author has reported that monogalactosyldiglyceride (MGDG) is present essentially in the membranes of active chloroplasts. The basal region of the stalk, which possesses old plastids, is indeed characterized by having a low content of membrane lipids and relatively saturated fatty acids (Dubacq, 1973). By using galactolipids as markers of the chloroplast lamellae, Dubacq and Puiseux-Dao (1974) have concluded that probably some thylacoid transformations which occur in the plastids in the dark are not linked to active membrane biosynthetic processes. Dubacq and Puiseux-Dao (1974) have also found that isolated chloroplasts, incubated in the light, synthesized new fatty acids very actively.

Investigations by Dubacq (1973) and by Brush and Percival (1972) have demonstrated that differences in lipid composition may exist between species of *Acetabularia:* one particular fatty acid, the C20: 4, is present in *A. crenulata*, but it is absent in *A. mediterranea.* If these results were to be confirmed, the C20: 4 fatty acid could be used as a marker in interspecific graft experiments for more sophisticated studies on lipid synthesis in *Acetabularia.*

XII. Polyphosphates

Phosphorus-containing compounds play an important role in numerous metabolic processes. Of particular interest are the polyphosphates, which serve as energy as well as phosphate donors in various organisms (Iwamura and Kuwashima, 1964; Conesa, 1969; Niemeyer and Richter, 1969; Rensen, 1969; Sauer *et al.*, 1969). They are found in bacteria, blue green algae, fungi, lower and higher plants and also in animal cells (references in Niemeyer and Richter, 1972). Polyphosphates condensed in granules were first detected in *Acetabularia* by Stich (1953) and by Thilo *et al.* (1956). Several cytochemical and biochemical investigations made by Stich and Hämmerling (Stich, 1953, 1955, 1956a, b, 1959; Stich and Hämmerling, 1953; Hämmerling and Stich, 1954) have emphasized the importance of polyphosphates in the metabolism of *Acetabularia.* Polyphosphate content in *Acetabularia* is connected with photosynthesis and is affected by any treatment which reduces the energy level of the cell (Stich, 1953, 1956a; Scherbaum, 1973). Stich has observed also that chemical substances which inhibit polyphosphate synthesis, cause a size reduction of nucleus and nucleolus (Stich, 1956a). More recently, Niemeyer and Richter (1972) succeeded in isolating rapidly labelled condensed phosphates from whole as well as from anucleate *A. mediterranea* and *A. crenulata.* The ^{32}P-labelled polyphosphates were submitted to column chromatography on methylated serum albumin and Kieselgur (MAK), and eluted together with DNA (Richter, 1966). The complete separation of labelled polyphosphates from the nucleic acids was achieved by column chromatography on QAE Sephadex A50. Finally, their identification was obtained by two dimensional thin layer chromatography on a mixture of Cellex MX and Cellulose, followed by autoradiography on X-ray film.

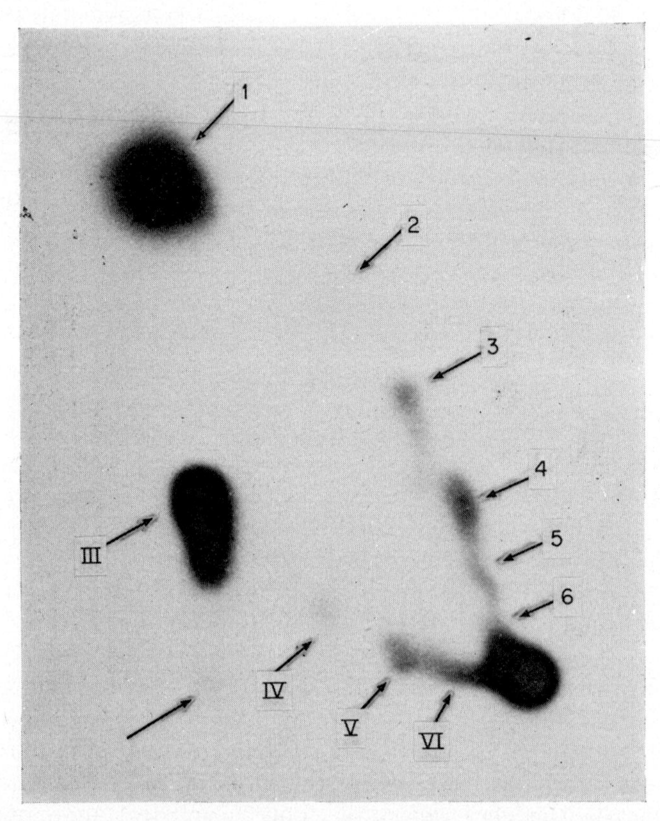

FIG. 20. Two-dimensional thin layer chromatography and autoradiography of ^{32}P-labelled inorganic phosphates from *Acetabularia*. 1, orthophosphate; 2–6, linear oligophosphates with two to six residues; III–VI: cyclic metaphosphates with three to six residues. (From Niemeyer and Richter, 1972, with permission.)

By means of these different techniques, Niemeyer and Richter (1972) have found that the rapidly labelled condensed inorganic phosphates of *Acetabularia* consisted respectively of compounds having more than 10 phosphate residues, of linear oligophosphates and of cyclic metaphosphates (Fig. 20).

Polyphosphates may appear as contaminants in RNA preparations of *Acetabularia*. However, in this case the polyphosphates can be separated from the RNA by washing the ethanol precipitate with 60% ethanol at −20°C, in which they are soluble (Woodcock and Bogorad, 1970a). This procedure, initially devised for the purification of RNA, may well be useful for the preparation of polyphosphates for subsequent purification.

XIII. Photosynthesis

A. Acetabularia *pigments*

Acetabularia possesses the chlorophylls a and b, the carotenoids β-carotene, lutein, luteinepoxid, violaxanthin, neoxanthin and the ketohydroxycarotenoids astaxanthin, 3-hydroxyechinenone, 3,3′-dihydroxyechinenone, 3-hydroxycanthaxanthin and an unknown yellow pigment (Clauss, 1958; Richter, 1958a; Kleinig, 1967; Egger and Kleinig, 1967; Kleinig and Egger, 1967; Shephard *et al.*, 1968; Moore, 1971a and b, 1972). Most of the *Acetabularia* pigments increase in the absence of the nucleus (Clauss, 1958; Hämmerling *et al.*, 1958; Richter, 1958a; Zetsche, 1969b; Brändle and Zetsche, 1971; Moore, 1972). Experiments with isolated chloroplasts (Shephard *et al.*, 1968; Moore, 1971a, b, 1972) have shown that the pigment synthetic pathways are plastidal and that they remain functional up to two months after the enucleation. Therefore, it has been suggested that the *Acetabularia* chloroplast genome contains the information needed for pigment synthesis (Moore, 1972). However, the study of the effects of chloramphenicol, cycloheximide, rifampicin and tetracycline on the cell and on its chlorophyll content (Bonotto *et al.*, 1969c; Zetsche, 1969b; Brändle and Zetsche, 1971) suggests that the synthesis of the photosynthetic pigments could depend both on chloroplastic and on nuclear information (Vanden Driessche, 1973a). In this context the paper by Trench and Smith (1970) is to be cited. Very recent experiments by Tschismadia (1975) have shown that chloroplasts isolated from *Acetabularia mediterranea* are able to synthesize plastoquinones.

B. *Photosynthesis in whole and anucleate* Acetabularia

The rate of photosynthesis in whole or in anucleate *Acetabularia* can be estimated by measuring O_2-evolution or by counting incorporated ^{14}C-radioactivity after administration of $^{14}CO_2$ (Brachet and Chantrenne, 1951, 1952; Brachet and Brygier, 1953; Terborgh, 1966; Clauss, 1972, 1975). It is now well established that the photosynthetic activity of *Acetabularia* can be influenced by light of different wavelengths as well as by supplied exogenous substances such as nicotinamide adenine dinucleotide (Vanden Driessche and Bonotto, 1971).

Photosynthesis is inhibited by red light (Terborgh, 1966; Clauss, 1968; Schael and Clauss, 1968; Clauss, 1970a, b, 1972; Vettermann, 1972, 1973) and can be restimulated by very small doses ($<10^{-8}$ Einstein/cm^2) of violet or blue radiation (Terborgh, 1966). This author has also observed that flashes of blue light as brief as 2·5 seconds are effective in potentiating photosynthetic O_2-evolution. The beneficial effect of blue light on photosynthesis has been confirmed by Clauss (1970a, 1972). White light assures a normal photosynthesis but, when it is given in excess of the growth requirements, *Acetabularia* cells excrete organic matter into the culture medium (Terborgh, 1963).

In white as well as in blue light, chloroplasts divide normally, while their division is reduced in red light (Schmid and Clauss, 1974, 1975). In chloroplasts, red light reduces also chlorophyll and protein synthesis (Clauss, 1968; Schmid and Clauss, 1974, 1975), possibly by decreasing the RNA-content of the cell (Richter and Kirschstein, 1966). Moreover, it affects the general metabolism and the morphogenesis of the cell (Richter, 1962; Clauss, 1963, 1968, 1970a, 1973; Terborgh, 1965). Blue light, on the contrary, has opposite effects both on metabolism and morphogenesis (Clauss, 1970a).

It has been found by Clauss (1975) that the rate of photosynthesis, calculated on the basis of protein content and chlorophyll content, decreases with increasing cell size. He suggested that this fact may be an expression of the apico-basal gradient of photosynthetic capacity existing within the cell (Issinger *et al.*, 1971).

Anucleate algae are able to photosynthesize for several weeks after the removal of the nucleus, but some metabolic changes occur (Clauss, 1958, 1968, 1970a, 1972; Clauss and Keck, 1959; Werz and Hämmerling, 1959; Clauss and Werz, 1961; Zetsche *et al.*, 1972; Vanden Driessche, 1969a, 1972a, 1973a; Vanden Driessche and Bonotto, 1972) and abundant storage granules appear in the plastids (Puiseux-Dao, 1970b; Puiseux-Dao *et al.*, 1972b). According to Vettermann (1972, 1973) the nucleus exercises a control on the chloroplastic polysaccharids, which may be constituted of starch (Werz and Clauss, 1970) and probably not of other additional sugars such as inulin or polyfructosans (Vanden Driessche and Bonotto, 1967, 1968b; Vanden Driessche, 1969b, 1970a). Inulin, whose content is highest in cells with caps (Clauss, 1975), is a reserve material in *Acetabularia* and other related Dasycladaceae (Nägeli, 1863; Leitgeb, 1897; Chadefaud, 1938; du Merac, 1953, 1955, 1956; Puiseux-Dao, 1962). Chloroplast preparations can be contaminated by a fructosan, as it has been found by Winkenbach *et al.* (1972a). These authors suggest that fructosan synthesis and storage are associated with or take place in the cell vacuole.

C. *Photosynthesis in isolated chloroplasts*

Isolated chloroplast preparations are particularly advantageous for the study of carbon pathways of photosynthesis (Shephard *et al.*, 1968; Moore, 1971a, b, 1972). Undamaged chloroplasts can be easily prepared from *Acetabularia* simply by cutting into small pieces the long cylindrical stalk in an appropriate buffer solution. However, after preparation, most *Acetabularia* chloroplasts are found in small droplets containing some cytoplasm and surrounded by a membrane (Gibor, 1965; Shephard *et al.*, 1968; Bidwell *et al.*, 1969; Bidwell, 1972). Most of these cytoplasmic droplets may be disrupted by passage through a nucleopore filter (Shephard and Levin, 1972). Isolated *Acetabularia* chloroplasts show a photosynthetic rate which approaches that observed *in vivo* (Shephard *et al.*, 1968; Shephard, 1970a; Dodd and Bidwell, 1971a; Shephard and Bidwell, 1972). Photosynthesis, measured as CO_2 uptake, was essentially similar for chloroplasts *in vivo*, in a liquid medium, or in an " artificial leaf " constituted by a double layer of 74 mesh nylon bolting cloth stretched on a plastic frame (Dodd and Bidwell, 1971a). *Acetabularia* chloroplasts evolve CO_2 in light (Bidwell *et al.*, 1969; Dodd and Bidwell, 1971a). Moreover, the pH of the medium and the presence of various substances (nitrate, nitrite, ammonia) influence the flow of carbon into photosynthetic intermediates (Bidwell *et al.*, 1970; Bidwell, 1971; Dodd and Bidwell, 1971b; Winkenbach *et al.*, 1972a, b). Sucrose, which is the most obvious end product of carbon fixation, is probably a precursor of fructosans, which could be formed in the cytoplasm (Smestad *et al.*, 1972; Shephard and Bidwell, 1972, 1973).

XIV. Phototropism

In nature, *Acetabularia* grows towards the light. The length of the stalk and the size of the reproductive cap depend mostly on light intensity. The plants growing in shallow waters have short stalks and large caps; those growing in deep waters possess long stalks and small caps. Very long (10 cm) *Acetabularia* having a small cap were observed by Brachet on a Greek amphora found at a depth of 20 m (Brachet, 1965).

The first observations on the phototropism of *Acetabularia* were made in 1937 on the species *Acetabularia calyculus* growing in the Red Sea (Nasr, 1939). By placing the collected plants in a glass basin next to a window, this author observed that the old plants exposed their caps perpendicularly towards the direction of light. Young plants without cap (stage 4) showed also a clear positive phototropism.

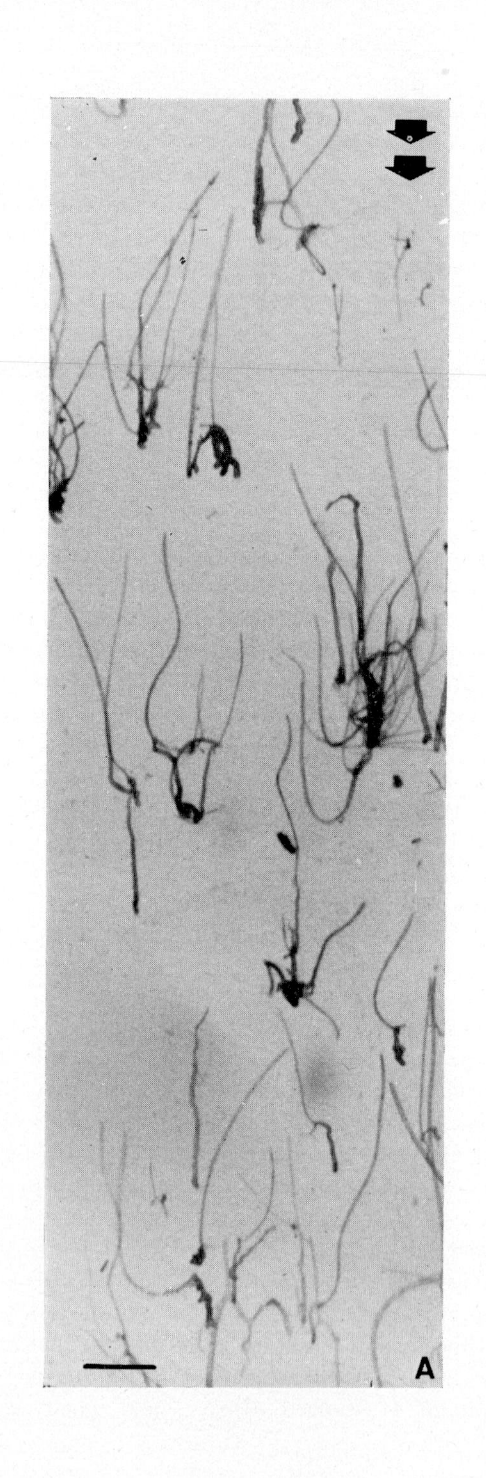
A

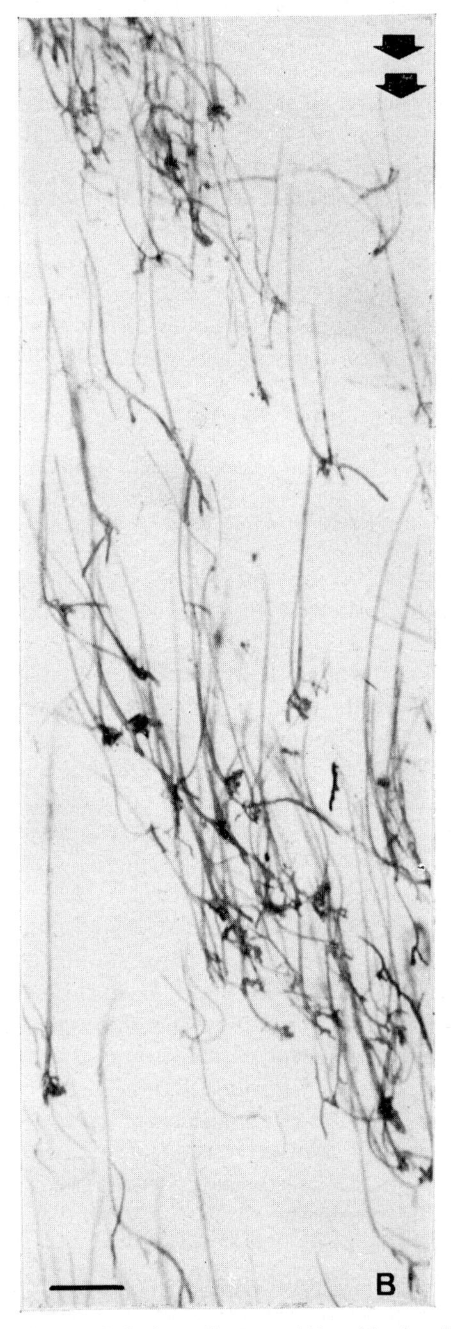

Fig. 21. Phototropism in *Acetabularia mediterranea* (A) and in *Acetabularia crenulata* (B). The arrows indicate light direction. Scale = 5 mm.

However, the same author was unable to observe any phototropism in the species *A. moebii* and *A. mediterranea* (Nasr, 1939). In contrast with this early report, a clear positive phototropism of the stalk has been occasionally observed in *A. mediterranea* (Puiseux-Dao, 1962; Lateur and Bonotto, unpublished). New investigations by Bonotto and Puiseux-Dao (unpublished) performed under controlled conditions, have shown that in *A. mediterranea* as well as in *A. crenulata*, the stalk displays a positive phototropism (Fig. 21). The rhizoids, however, grow opposite to light. These observations do not permit any hypothesis on the origin of phototropism in *Acetabularia*. Nevertheless, they emphasize the usefulness of this giant cell as a tool for research on phototropism.

XV. Rhythmic Activity in *Acetabularia*

A. *Rhythmic activities in whole and in anucleate* Acetabularia

It is known that living organisms possess biological rhythms, which oscillate with different frequencies (Bünning, 1967; Sweeney, 1969). *Acetabularia* cells display rhythmic activities of both short (2–360 min) and long (about 24 h) duration. Von Klitzing (1969) has reported that the activity of lactic dehydrogenase and malic dehydrogenase oscillates with a frequency of about 50 min. The same author has found that the concentration of an unknown " Substance E ", which absorbs in UV light, varies with a frequency comprised between 150 and 360 min. However, the concentration of ATP varies with a much shorter frequency (2–4 min only) in the presence as well as in the absence of the nucleus (von Klitzing, 1969). Probably a rapid exchange of ATP occurs between chloroplasts and surrounding cytoplasm, as suggested by the findings of several authors (Strotmann and Berger, 1968, 1969; Strotmann and Heldt, 1969). However, the circadian rhythm in ATP content of the chloroplasts (see below) may be taken as an indication that ATP produced in the chloroplasts is only partially exchangeable (Vanden Driessche, 1973a).

These findings are not contradictory, since the existence of a circadian rhythm in ATP content in the chloroplasts (Vanden Driessche, 1970a, d, 1973a) does not necessarily exclude the possibility of short duration oscillations inside the rhythm. A number of daily variations were found in *Acetabularia*. They concern various structural, physiological and biochemical activities of the cell. Table IX gives the daily variations so far reported in the literature together with the respective references. Most of them are associated with the chloroplasts. The

TABLE IX. DAILY VARIATIONS IN STRUCTURAL, PHYSIOLOGICAL AND BIOLOGICAL ACTIVITIES OF *Acetabularia*

Daily variation	*References*
ATP content of chloroplasts	Vanden Driessche (1970a, d, 1973a)
DNA synthesis	Bonotto *et al.* (1972d)
Fructose (inulin) content of chloroplasts	Vanden Driessche and Bonotto (1967, 1968b), Vanden Driessche (1969b)
Hill activity of chloroplasts	Vanden Driessche (1974b)
Number of carbohydrate granules in chloroplasts	Puiseux-Dao and Gilbert (1967), Puiseux-Dao (1968, 1970b)
Photosynthesis	Sweeney and Haxo (1961), Richter (1963), Schweiger *et al.* (1964a, b), Schweiger and Schweiger (1965), Vanden Driessche (1966a, b, 1967a, b, c, 1970a, b, c, 1971a, b, c, d, 1972a, b, 1973a, b), Hellebust *et al.* (1967), Sweeney *et al.* (1967), Terborgh and McLeod (1967), von Klitzing and Schweiger (1969), Vanden Driessche *et al.* (1970), Mergenhagen and Schweiger (1971, 1973, 1975a, b), Schweiger (1971), Mergenhagen (1972a, b), Sweeney (1972), Zetsche *et al.* (1972), Vanden Driessche and Hars (1973), Karahashian and Schweiger (1975, 1976), Vanden Driessche and Hayet (1975)
Plastid division	Vanden Driessche and Hellin (1972)
Polysaccharide content	Vanden Driessche (1970a, 1972g, 1973a), Vanden Driessche and Hars (1973)
Polysaccharide synthesis	Glory and Vanden Driessche (1976)
Protein synthesis	Bonotto *et al.* (1972d)
Proton flux	Brändle *et al.* (1975)
RNA synthesis	Vanden Driessche and Bonotto (1968a, 1969a, b), Bonotto *et al.* (1972d)
Shape of chloroplasts	Vanden Driessche (1966a, 1967c)
Transcellular current	Novák and Sironval (1976)
Ultrastructure of chloroplasts	Vanden Driessche and Hars (1972a, b, c), Vanden Driessche *et al.* (1973b)

rhythm in photosynthesis is one of the most studied (Fig. 22). Its sensitivity to various antibiotics (actinomycin D, chloramphenicol, cycloheximide, puromycin, rifampicin), to RNase and to morphactins (anti-auxin compounds) has been checked (Vanden Driessche, 1966b, 1967c; Sweeney *et al.*, 1967; Mergenhagen and Schweiger, 1971, 1975; Vanden Driessche and Hayet, 1975). As RNase paradoxically increases RNA content in *Acetabularia* (Brachet and Six, 1966), results obtained with this enzyme should be considered with caution. The daily variation in DNA synthesis has been followed by incubating anucleate fragments of *A. mediterranea* in the presence of ^{3}H-thymidine during 60 min (Bonotto *et al.*, 1972d). It deserves further accurate investigation in order to check if ^{3}H-thymidine is incorporated only into DNA. The daily variation in protein synthesis has been shown by labelling experiments with ^{3}H-leucine (Bonotto *et al.*, 1972d). Further work is needed in order to know more precisely this important metabolic activity. The rhythm in fructose (inulin) content of the chloroplasts has to be reinvestigated, as recent reports suggest that only starch is synthesized in these organelles (Werz and Clauss, 1970; Vettermann, 1972, 1973).

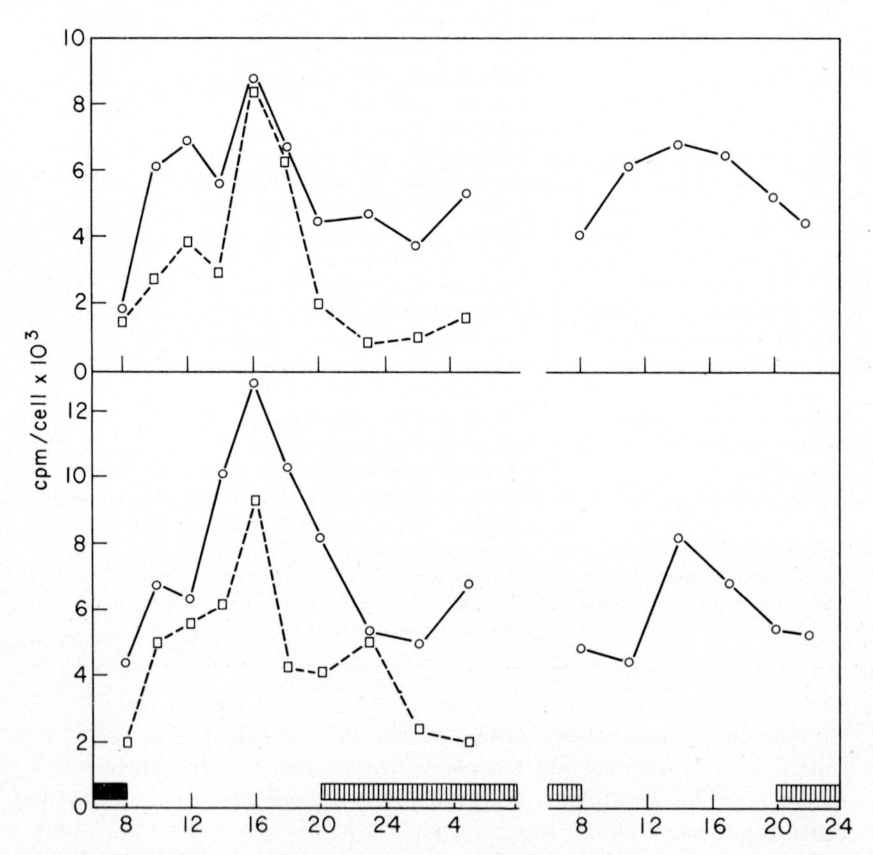

FIG. 22. *Acetabularia mediterranea.* Circadian rhythm in photosynthesis, shown by changes in the capacity of ^{14}C-incorporation in continuous light after light-dark cycles of 12:12. Upper part, anucleate cells; lower part, nucleate cells. ○—○ total ^{14}C fixation; □ -- □ ^{14}C incorporation into soluble fraction. (From Zetsche *et al.*, 1972, with permission.)

B. *Nuclear control of chloroplast circadian rhythms*

The possible molecular mechanisms underlying circadian rhythms in *Acetabularia* have been discussed by various authors (Vanden Driessche, 1970a, b, c, 1973a, b; Schweiger, 1970a, 1971; Sweeney, 1972).

The problem of the origin of circadian rhythms remains at present unsolved. The study of the effects of various antibiotics (actinomycin D, chloramphenicol, cycloheximide, puromycin) led to the conclusion that in *Acetabularia* the nucleus controls chloroplast rhythms, possibly by means of long-lived mRNAs (Vanden Driessche, 1967c, 1973a; Schweiger, 1971). Furthermore, experiments with rifampicin have shown that a daily transcription of chloroplast genome is not implied in the circadian rhythm of photosynthesis (Vanden Driessche *et al.*, 1970). However, a daily translation on 80S ribosomes seems required for a normal expression of the rhythm (Karakashian and Schweiger, 1975, 1976; Mergenhagen and Schweiger, 1975b). The major role the nucleus plays in the control of the chloroplast photosynthetic rhythm has been emphasized by elegant graft or transplantation experiments (Schweiger *et al.*, 1964a; Vanden Driessche, 1967b, 1973a; Schweiger, 1971). These experiments have shown that the nucleus is able to phase the rhythm of photosynthesis (Schweiger *et al.*, 1964a; Schweiger, 1971) or to restore it in algae without (R^-) rhythmicity (Vanden Driessche, 1967b, 1973a). In recent work, Mergenhagen and Schweiger (1974) have investigated whether intercellular synchronization of circadian rhythmicity occurs in *Acetabularia*. They have found that intercellular synchronization does not take place, at least when the phases of individual cells differ by 180°. These authors have also found that during cyst formation the photosynthetic rhythm is lost, probably on account of the disruption of intracellular organization which takes place at this stage (Mergenhagen and Schweiger, 1975a).

C. *Significance of circadian rhythms in* Acetabularia

It is not known whether circadian rhythms have a direct influence on *Acetabularia* morphogenesis, as " arythmic " algae may form a cap (Brachet, 1970).

However, the existence of circadian rhythms in *Acetabularia* shows that this unicellular organism can " measure time ". Time measurement seems not the sole prerogative of rhythmic phenomena.

The fact that *Acetabularia* shows a precise temporal organization (Schweiger, 1970a, 1971; Vanden Driessche, 1970a, 1973a; Zetsche,

1972) suggests that time measurement could be of great importance in cellular differentiation.

XVI. Electrophysiology of *Acetabularia*

Electrophysiological studies have been made on the species *A. crenulata* or *A. mediterranea* by many authors (Schilde, 1966, 1968; Gradmann, 1970a, b; Saddler, 1970a, b, c, 1971; Gradmann and Bentrup, 1970; Hansen and Gradmann, 1971; Novák, 1972, 1975; Novak and Bentrup, 1972a, b; Brändle *et al.*, 1975; Gläsel and Zetsche, 1975; Gradmann and Bokelok, 1975; Novák and Sironval, 1975; Zubarev *et al.*, 1975). Research in this field is very important not only for studying electrical behaviour and ionic regulation in *Acetabularia*, but also for elucidating how genetic information in a cell is realized only locally, that is, how the cell gains the " positional information " (Wolpert, 1969). It has been found by Novak and Bentrup (1972a, b) that anucleate stalk fragments of *A. mediterranea* generate a transcytoplasmic electric current, which may change its sign for some hours after illumination has started. However, after 29 hours this transcytoplasmic electric current flows steadily towards the end of the fragment where outgrowth takes place. Growth can be detected microscopically only after 45 hours. More recent experiments by Novák (1975) and by Novak and Sironval (1975) on anucleate posterior stalk segments have shown that when the apical and the basal ends of the fragments are isolated electrically, no regeneration occurs. When electrical connection between the two ends of the anucleate fragments is established by a bridge (agar-artificial sea water), normal regeneration takes place. The authors have concluded that a transcellular electrical current must flow through the fragments if they are to regenerate. Possibly, transcellular current and current pulses cause polar transport of the morphogenetic substances towards the tip of the fragment, thus starting the regeneration process (Novák, 1975; Zubarev *et al.*, 1975). The results obtained by Novák (1975), Novák and Sironval (1975), Zubarev and Rogatykh (1975), and by Zubarev *et al.* (1975) underline the importance of electrical currents in the realization of spatial differentiation in *Acetabularia*.

XVII. Hormones

A. *Endogenous hormones in* Acetabularia

Tandler first reported that *Acetabularia* contains indole compounds in the vacuole, which is particularly rich in oxalic acid (Tandler,

1962a, b, c). More direct evidence for the presence of indol-3-acetic acid (IAA) in *Acetabularia* was recently obtained by Vanden Driessche and Delegher-Langohr (1975). They have detected in extracts of *A. mediterranea* an auxin-like substance, which is biologically active and which moves slightly faster than commercial IAA during chromatography. These authors have also reported that exogenous ^{14}C-IAA is bound to the membrane of *Acetabularia*.

The presence of IAA in marine algae has been questioned by Buggeln and Craigie (1971), who were unable to detect endogenous IAA in 10 species of marine algae, including *A. mediterranea*. According to the authors, if auxins exist in marine algae, their levels are below those detectable by present chemical methods.

A significant amount of 3′:5′-cyclic AMP (c-AMP) has been found in *A. mediterranea* by Vanden Driessche *et al.* (1975). According to Legros *et al.* (1973) and to Vanden Driessche (1976), c-AMP would play a physiological role in *Acetabularia*. They have found that exogenous c-AMP enhances respiration rate in the cells. Finally, on the basis of biological investigations, it has been suggested that the whorls of *Acetabularia* may be the site of synthesis of cellular growth regulators (see Section V, B).

B. *Effects of various hormones on* Acetabularia

Acetabularia has been tested for its responsiveness to various plant hormones or related substances: indole acetic acid (IAA), naphtalene acetic acid (NAA), 2-3-5-triiodobenzoic acid, giberellin and kinetin (Dao, 1954b, 1956; Thimann and Beth, 1959; Zetsche, 1963a; Spencer, 1968). The significance of these hormones for the morphogenesis of *Acetabularia* has been discussed by Puiseux-Dao (1970b). In general, plant hormones have limited effects on *Acetabularia*. Recent work by Bonotto and Kirchmann (1974) has shown that irradiated *Acetabularia*, treated with gibberellic acid, partially recover from the radiation damage. The action of animal hormones on *Acetabularia* has been also checked. Insulin enhances oxygen consumption of nucleate and anucleate algae grown in darkness (Legros *et al.*, 1967, 1971, 1972, 1974; Legros and Conard, 1973), while aldosterone provokes a transient inhibition (Legros *et al.*, 1972, 1974).

Oxidized insulin has no more metabolic effect (Legros *et al.*, 1971; Legros and Conard, 1973).

Moreover, a competitive relationship has been observed between insulin and ouabain, a substance which blocks the active transport of ions (Legros *et al.*, 1973). Studies with ^{125}-I-insulin have indicated that receptor sites for insulin exist on the membrane of *Acetabularia* (Legros and Conard, 1974; Hanson *et al.*, 1975; Legros *et al.*, 1975a, b, c) and on isolated chloroplasts (Dumont *et al.*, 1975). These findings raise the question of the nature and specificity of the insulin receptors on membrane and chloroplasts. Exogenous c-AMP and theophyllin have been found to stimulate oxygen consumption (Legros *et al.*, 1973). The hypothesis by Legros *et al.* (1973) that c-AMP plays a physiological role is supported by the work of Vanden Driessche *et al.* (1975) and of Vanden Driessche (1976,) who have demonstrated its presence in *Acetabularia* (see Section XVII, A).

XVIII. Radiobiology of *Acetabularia*

A. *Biological and biochemical effects of radiations*

The morphological and physiological features of *Acetabularia* make this organism particularly useful for radiobiological experiments. Since the first investigations on the effects of X- and UV-rays (Bacq *et al.*, 1955; Hämmerling, 1956; Six, 1956a, b), a number of papers on this subject have been published. Table X gives the type of radiations used, together with the respective references.

Table X. Radiations Used in Research on *Acetabularia*

Type of radiation	*References*
X-rays	Bacq *et al.* (1955, 1957), Hämmerling (1956), Six (1956b, 1958), Six and Puiseux-Dao (1961)
Ultraviolet rays	Errera and Vanderhaeghe (1957), Six (1956a, 1958), Brachet and Olszewska (1960), Olszewska *et al.* (1961), Werz and Hämmerling (1961), Sandakhchiev *et al.* (1972)
Gamma rays (from ^{60}Co)	Bonotto and Kirchmann (1969a, 1971a, 1972a, b, 1974), Bonotto *et al.* (1970b, 1971c, 1972a, b, d, 1973), Bonotto and Puiseux-Dao (1970), Kirchmann and Bonotto (1970, 1973a), Netrawali (1970), Sandakhchiev *et al.* (1972), Bonotto (1975)

Acetabularia cells are particularly radioresistant to X-rays (Bacq *et al.*, 1955). Anucleate fragments, however, are more easily damaged by X-rays than whole cells. This is also the case for UV-rays (Errera and Vanderhaeghe, 1957). Graft experiments between an irradiated fragment and an unirradiated one of *A. mediterranea* and of *A. crenulata* have suggested that the species specific morphogenetic substances stored within the cytoplasm and their synthesis are inhibited by X-rays (Six and Puiseux-Dao, 1961) as well as by UV-rays (Werz and Hämmerling, 1961). Six (1956a, 1958) has found that the most important growth reduction in anucleate fragments of *A. mediterranea* occurs for UV-rays having a wavelength of 254 nm, which corresponds to the zone of maximum absorption of nucleic acids. Biochemical studies by Brachet and Olszewska (1960) and by Olszewska *et al.* (1961) have confirmed the sensitivity of nucleic acid synthesis to UV-rays. Experiments with ^{35}S-methionine (Brachet and Olszewska, 1960; Olszewska and Brachet, 1960, 1961; Olszewska *et al.*, 1961) have shown that UV irradiation interferes with the synthesis of sulphur-containing proteins, which play an important part in the morphogenesis of *Acetabularia* (Brachet, 1959).

Gamma rays have similar effects on whole and on anucleate cells, a dose of about 125 krad being necessary to stop almost completely cap initiation (Fig. 23). The synthesis of DNA, RNA and ribosomes is strongly inhibited by gamma rays (Bonotto *et al.*, 1970b, 1971c, 1972d, 1973; Bonotto and Kirchmann, 1969a, 1971a; Kirchmann and Bonotto, 1970, 1973a). However, the synthesis of protein was found to continue at an appreciable rate in the irradiated cells (Bonotto *et al.*, 1970, 1971c), at least for some hours after the irradiation. Probably, protein synthesis in gamma irradiated cells takes place, at least in part, on pre-existing molecules of mRNA and on pre-existing ribosomes (Bonotto *et al.*, 1972d). It is not known if gamma rays inactivate the morphogenetic substances stored in the cytoplasm, as is the case for X- and UV-rays (Six and Puiseux-Dao, 1961; Werz and Hämmerling, 1961). Pre-treatment with 2-β-aminoethylisothiourea (AET) protects the cells from radiation damage (Netrawali, 1970).

B. *Heteromorphoses induced by radiations*

Whole or anucleate *Acetabularia*, irradiated with X-rays (Bacq *et al.*, 1955) or gamma rays (Bonotto *et al.*, 1970b) present several types of morphological anomalies: loss of the sterile whorls, abnormal enlargement of the whorls, branching of the stalk (Fig. 24), enlargement of the apex of the cell, abnormal cap formation and reversion from the reproductive to the vegetative phase (Bonotto and Puiseux-Dao, 1970; Bonotto and Kirchmann, 1972a, b; Kirchmann and Bonotto 1973a). Apical stalk enlargement increases with the radiation dose, whereas branching of the stalk shows a maximum in the dose range comprised between 100 and 150 krad (Fig. 23). In branched cells,

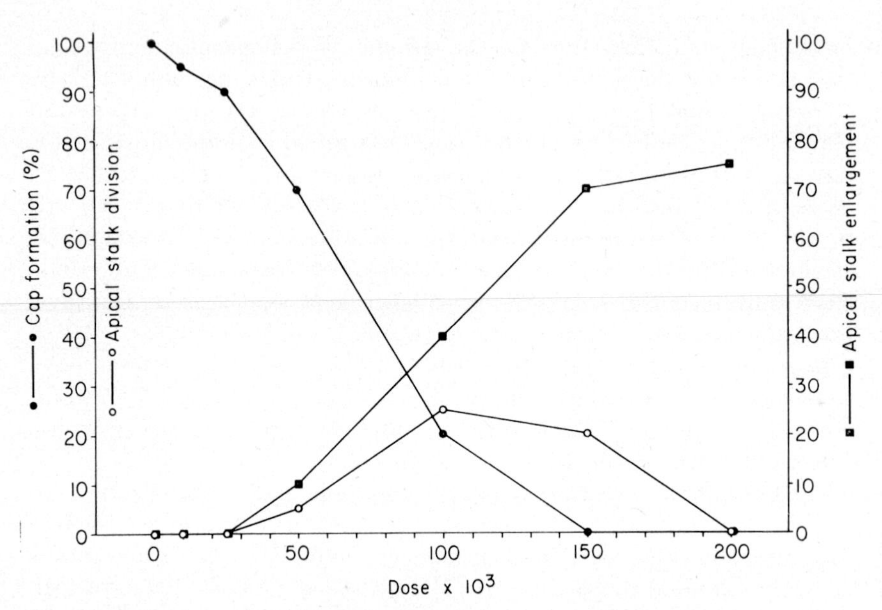

FIG. 23. Biological effects of gamma irradiation on *Acetabularia mediterranea*. (From Bonotto *et al.*, 1970, with permission.)

provided the dose of radiations is not higher than 100 krad, every branch may form a reproductive cap (Fig. 24). Branched cells are very useful for investigations on spatial differentiation in *Acetabularia* (Bonotto, 1975).

XIX. POLLUTION

Acetabularia is a useful tool not only for research on cellular and molecular biology, but also on pollution. *A. mediterranea* and *A. crenulata* have been used to study the incorporation of ^3H released from the nuclear reactors into the environment (Kirchmann and Bonotto, 1971, 1972, 1973b). It has been shown, by biochemical methods, that an appreciable amount of ^3H is incorporated into proteins and nucleic acids (Kirchmann and Bonotto, 1971, 1973b; Bonotto and Kirchmann, 1975; Bonotto *et al.*, 1976). *Acetabularia mediterranea* and *Dasycladus clavaeformis* have been tested for their resistance against heavy metals (Bonotto *et al.*, 1972b; Kirchmann and Bonotto, 1972; Tsekos *et al.*, 1972; Bonotto and Kirchmann, 1973). Lead, supplied as $Pb(NO_3)_2$, inhibits morphogenesis in whole as well as in anucleate fragments of *A. mediterranea*. Moreover, lead also induces some teratological effects (Bonotto and Kirchmann, 1973). The biochemical effects of lead in *Acetabularia* remain to be investigated.

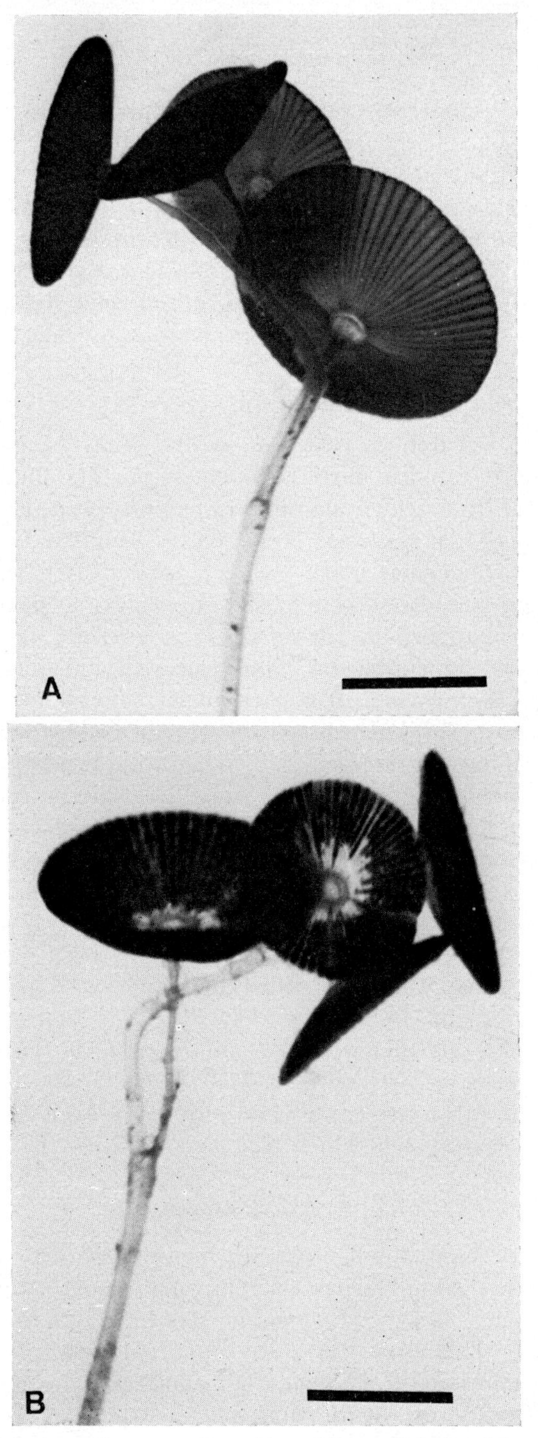

FIG. 24. *Acetabularia mediterranea*. Branching of the stalk induced by gamma irradiation. Cell with 4 reproductive caps before (A) and after (B) cyst formation. Scale = 5 mm. These cells contains more cysts than the normal ones.

Lindane, a chlorinated insecticide, inhibits growth and morphogenesis of whole *Acetabularia* and regeneration of nucleate basal fragments (Borghi *et al.*, 1972, 1973; Puiseux-Dao *et al.*, 1972d). Moreover, this substance induces a reduction of the thylacoids and of the storage material in chloroplasts. It has also an inhibitory effect on the synthesis of nucleic acids and proteins (Borghi *et al.*, 1973). Apparently lindane lowers the metabolic activity of the cell. Its main sites of action may be the metabolically active membranes (Borghi *et al.*, 1973).

XX. Concluding Remarks

Progress in research on *Acetabularia* and related Dasycladales has been particularly rapid in these last five years. The foundation of an " International Research Group on *Acetabularia* " and the organization of three international symposia have greatly stimulated investigations on these giant unicellular marine algae. New ideas have arisen from these meetings and what is equally important, a sincere wish for international collaboration. *Acetabularia* is certainly one of the most useful organisms for studies on cellular growth and differentiation in the absence of the nucleus. It is particularly suitable for investigating the subtle relationships which exist between nucleus and cytoplasm and for examining the intergenomic cooperation inside the eucaryotic cell. The possibility of obtaining anucleate cells or cell fragments, regenerating nucleate basal parts, cytoplasts, isolated nuclei and organelles permits various types of *in vivo* and *in vitro* biological and biochemical experiments. Moreover, elegant intra- and interspecific grafts are possible with different species of *Acetabularia.* The easy realization of nuclear transplantations increases even more the value of this alga, which was once called by Brachet " one of the most fascinating living objects in the world."

We hope that this review, which summarizes the most significant recent advances in research on *Acetabularia*, will induce more and more workers, in the future, to investigate the many exciting problems of the life of this marine alga which still remain unsolved.

XXI. Acknowledgements

The authors thank Mrs J. Romeyer for the typewriting of the text. They thank also the kind help of the following persons: Mrs A. Bonotto, Mrs C. Huysmans, Mrs L. Steylemans, Mrs L. Weyts, Mr G. Bas and Mr G. Nuyts. This work has been supported by the " Fonds de la Recherche Scientifique Fondamentale Collective ", by EMBO and by the NATO Research Grant no. 1027.

XXII. References

Agardh, C. A. (1824). Systema Algarum. p. 192. Lund.

Agardh, C. A. (1827). Auszählung einiger in den österreichischen Ländern gefundenen neuen Gattungen und Arten von Algen, nebst ihrer Diagnostik und beigefügten Bemerkungen. *Flora*, **10**, 625–640.

Agardh, C. A. (1828). Species Algarum. Vol. 2, pp. 15–17. Gryphiae.

Agardh, J. G. (1854). Nya algformer. *Öfversight af Kongl. Vet. Akad. Förhandlingar*, **11**, 107–111.

Agardh, J. G. (1863). Species genera et ordines algarum. Vol. 1 Lund.

Agardh, J. G. (1887). Till Algernes Systematik, VIII. Siphoneae. *Lunds Universitets Årsskrift*, **23**, 1–174.

Alexeev, A. B., Betina, M. I., Stvolinsky, S. L., Yazykov, A. A. and Zubarev, T. N. (1974). Evidence for a relationship between some ribonucleoprotein-complexes and the " morphogenetic substances " of *Acetabularia*. *Plant Science Letters*, **3**, 297–302.

Alexeev, A. B., Stvolinsky, S. L., Tikhomirova, L. A., Yazykov, A. A. and Zubarev, T. N. (1975). Messenger ribonucleoproteins of *Acetabularia*. *Protoplasma*, **83**, 168.

Anderson, L. E. and Levin, D. A. (1970). Chloroplast aldolase is controlled by a nuclear gene. *Plant Physiology*, **46**, 819–820.

Apel, K. (1970). Kerndeterminierte Synthese von " unlöslichen " Chloroplastenproteinen in *Acetabularia*. pp. 1–72. Dissertation, University of Hamburg.

Apel, K. (1972a). Origin of genetic information and biosynthesis of chloroplast membrane proteins. *Protoplasma*, **75**, 472.

Apel, K. (1972b). Isolation and characterization of chloroplast membrane proteins. *Protoplasma*, **75**, 472–473.

Apel, K. (1975a). Fractionation and characterization of chloroplast membranes of *Acetabularia mediterranea*. *Protoplasma*, **83**, 168–169.

Apel, K. (1975b). Fractionation of the chloroplast membrane of *Acetabularia mediterranea*. *Plant Physiology* (Suppl.) **56**, 32.

Apel, K. and Schweiger, H. G. (1972). Nuclear dependency of chloroplast proteins in *Acetabularia*. *European Journal of Biochemistry*, **25**, 229–238.

Apel, K. and Schweiger, H. G. (1973). Sites of synthesis of chloroplast-membrane proteins. Evidence for three types of ribosomes engaged in chloroplast-protein synthesis. *European Journal of Biochemistry*, **38**, 373–383.

Apel, K., Miller, K. R. and Bogorad, L. (1975). Topography of the chloroplast membrane of *Acetabularia mediterranea*. *Plant Physiology*, (Suppl.) **56**, 32.

Arasaki, S. (1941). Some experiments on the influence of light conditions on the development of *Acetabularia calyculus*. *Suisan gakkwaiho*, **8**, 290–297.

Arasaki, S. (1942). On the life history of *Acetabularia calyculus* Quoy et Gaimard. *Botany Magazine, Tokyo*, **56**, 383–391.

Arnoldi, W. (1912). Algologische Studien. Zur Morphologie einiger Dasycladaceen (*Bornetella*, *Acetabularia*). *Flora*, **104**, 86–100.

Bachmann, P. and Zetsche, K. (1975). Activities of a mannose-polymerizing enzyme and GDP-mannose pyrophosphorylase during the morphogenesis of *Acetabularia mediterranea*. *Protoplasma*, **83**, 169.

Bachmayer, H. (1970). Initiation of protein synthesis in intact cells and in isolated chloroplasts of *Acetabularia mediterranea*. *Biochimica et Biophysica Acta*, **209**, 584–586.

Bacq, Z. M., Damblon, J. and Herve, A. (1955). Radiorésistance d'une algue *Acetabularia mediterranea* Lamour., *Comptes rendus des Séances de la Societé de Biologie, Paris*, **149,** 1512–1515.

Bacq, Z. M., Vanderhaeghe, F., Damblon, J., Errera, M. and Herve, A. (1957). Effets des rayons X sur *Acetabularia mediterranea. Experimental Cell Research*, **12,** 639–648.

Baeyens, W., Goutier, R. and Vangheel, V. (1974). Radiation-induced alterations in the synthesis of ribosomal and informosomal particles in rat liver. *Strahlentherapie*, **147,** 660–669.

Baltus, E. (1955). Influence du noyau sur le maintien de l'aldolase dans le cytoplasme. *Congrès International de Biochimie. Résumés des Communications*, 3e *Congr. Brussels, p.* 76.

Baltus, E. (1959). Evolution de l'aldolase dans les fragments anucléés d'*Acetabularia. Biochimica et Biophysica Acta*, **33,** 337–339.

Baltus, E. and Brachet, J. (1963). Presence of deoxyribonucleic acid in the chloroplast of *Acetabularia mediterranea. Biochimica et Biophysica Acta*, **76,** 490–492.

Baltus, E. and Quertier, J. (1966). A method for the extraction and characterization of RNA from subcellular fractions of *Acetabularia. Biochimica et Biophysica Acta*, **119,** 192–194.

Baltus, E., Edström, J. E., Janowski, M., Hanocq-Quertier, J., Tencer, R. and Brachet, J. (1968). Base composition and metabolism of various RNA fractions in *Acetabularia mediterranea. Proceedings of the National Academy of Science*, **59,** 406–413.

Bannwarth, H. (1969). Untersuchungen über eine mögliche Kontrolle des Zellkerns über den Pyridinnukleotidhaushalt bei *Acetabularia mediterranea*, pp. 1–51. Diplomarbeit, Stuttgart-Hohenheim.

Bannwarth, H. (1972a). Measurement of *in vivo* incorporation of ^{3}H-thymidine into the acid soluble thymidine phosphates of *Acetabularia mediterranea. Protoplasma*, **75,** 473.

Bannwarth, H. (1972b). Regulation of thymidine phosphorylating enzymes in *Acetabularia. Protoplasma*, **75,** 473–474.

Bannwarth, H. (1972c). Localization of NAD-pyrophosphorylase in *Acetabularia mediterranea. Protoplasma*, **75,** 474.

Bannwarth, H. (1975a). Some characteristics of the thymidine phosphorylating enzyme (TPE) in *Acetabularia. Protoplasma*, **83,** 169–170.

Bannwarth, H. (1975b). Effects of transcription and translation inhibitors on the regulation of the thymidine phosphorylating enzyme (TPE) in *Acetabularia. Protoplasma*, **83,** 170.

Bannwarth, H. and Schweiger, H. G. (1975). Regulation of thymidine phosphorylation in nucleate and anucleate cells of *Acetabularia. Proceedings of the Royal Society of London*, B, **188,** 203–219.

Bannwarth, H., Siebert, G. and Schweiger, H. G. (1969). Pyridinnucleotide in *Acetabularia mediterranea* vor und nach Zellkernentfernung. Hoppe-Seyler's *Zeitschrift für Physiologische Chemie*, **350,** 1475–1476.

Berger, S. (1967a). RNA-Synthese in isolierten Chloroplasten von *Acetabularia*, pp. 1–71. Dissertation, University of Cologne.

Berger, S. (1967b). RNA-Synthesis in *Acetabularia*. II. RNA-Synthesis in isolated chloroplasts. *Protoplasma*, **64,** 13–25.

Berger, S. (1975a). Formation and ultrastructure of secondary nuclei in the rhizoid of *Acetabularia mediterranea*. *Protoplasma*, **83,** 170.

Berger, S. (1975b). *Acetabularia haemmerlingi*, a new species. *Protoplasma*, **83,** 171.

Berger, S. (1975c). Ultrastructure of nuclei after implantation. *Protoplasma*, **83,** 171.

Berger, S. and Schweiger, H. G. (1969). Synthesis of chloroplast DNA in *Acetabularia*. *Physiological Chemistry and Physics*, **1,** 280–292.

Berger, S. and Schweiger, H G. (1975a). 80S ribosomes in *Acetabularia major*. Redundancy of rRNA cistrons. *Protoplasma*, **83,** 41–50.

Berger, S. and Schweiger, H. G. (1975b). An apparent lack of nontranscribed spacers in rDNA of a green alga. *Molecular & General Genetics* **139**, 269–275.

Berger, S. and Schweiger, H. G. (1975c). Ribosomal DNA in different members of a family of Green Algae (*Chlorophyta*, *Dasycladaceae*): an electron microscopical study. *Planta*, **127,** 49–62.

Berger, S. and Schweiger, H. G. (1975d). The ultrastructure of the nucleocytoplasmic interface in *Acetabularia*. *In* "Molecular Biology of Nucleocytoplasmic Relationships" (S. Puiseux-Dao, ed.), pp. 243–250. Elsevier, Amsterdam and New York.

Berger, S., Apel, K., Schweiger, M. and Schweiger, H. G. (1972). Some morphological features of *Dasycladaceae*. *Protoplasma*, **75,** 475.

Berger, S., Sandakhchiev, L. S. and Schweiger, H. G. (1974). Fine structural and biochemical markers of *Dasycladaceae*. *Journal de Microscopie*, **19,** 89–104.

Berger, S., Herth, W., Franke, W. W., Falk, H., Spring, H. and Schweiger, H. G. (1975a). Morphology of the nucleo-cytoplasmic interactions during the development of *Acetabularia* cells. II. The generative phase. *Protoplasma*, **84,** 223–256.

Berger, S., Niemann, R. and Schweiger, H. G. (1975b). Viability of *Acetabularia* nucleus after 24 hours in an artificial medium. *Protoplasma*, **85,** 115–118.

Berthold, G. (1880). Die geschlechtliche Fortpflanzung von *Dascyladus clavaeformis* Ag. *Botanische Zeitung*, **38,** 648–651.

Bertoloni, A. (1819). "Amoenitates italicae"; p. 231; pp. 274–280. Bologne.

Beth, K. (1943a). Entwicklung und Regeneration von *Acicularia schenckii*. *Zeitschrift für induktive Abstammungs-und Vererbungslehre*, **81,** 252–270.

Beth, K. (1943b). Ein- undzweikernige Transplantate zwischen *Acetabularia mediterranea* und *Acicularia schenckii*. *Zeitschrift für induktive Abstammungs- und Vererbungslehre*, **81,** 271–312.

Beth, K. (1943c). Ein- und zweikernige Transplantate verschiedener Acetabulariaceen. *Naturwissenschaften*, **31,** 206–207.

Beth, K. (1953a). Experimentelle Untersuchungen über die Wirkung des Lichtes auf die Formbildung von kernhaltigen und kernlosen *Acetabularia* Zellen. *Zeitschrift für Naturforschung*, **8b,** 334–342.

Beth, K. (1953b). Über den Einfluss des Kernes auf die Formbildung von *Acetabularia* in verschiedenen Entwicklungsstadien. *Zeitschrift für Naturforschung*, **8b,** 771–775.

Beth, K. (1955a). Beziehungen zwischen Wachstum und Formbildung in Abhängigkeit von Licht und Temperatur bei *Acetabularia*. *Zeitschrift für Naturforschung*, **10b,** 267–276.

Beth, K. (1955b). Unterschiedliche Beeinflussung von Wachstum und Teilung durch Veränderung von Licht und Temperatur. *Zeitschrift für Naturforschung*, **10b,** 276–281.

Beth, K. (1956a). Amputation and Stiellänge bei *Acetabularia*. *Naturwissenschaftten*, **13,** 307–308.

Beth, K. (1956b). Verfrühte Hutbildung bei *Acetabularia* als Folge nächtlicher Kältebehandlung. *Naturwissenschaften*, **13,** 308–309.

Bidwell, R. G. S. (1971). Products of photosynthesis by *Acetabularia* chloroplasts: possible control mechanisms. *Proceedings of the 2nd International Congress on Photosynthesis Research, Stresa*. pp. 1927–1934.

Bidwell, R. G. S. (1972). Impurities in preparations of *Acetabularia* chloroplasts. *Nature, London*, **237,** 169.

Bidwell, R. G. S., Levin, W. B. and Shephard, D. C. (1969). Photosynthesis, photorespiration and respiration of chloroplasts from *Acetabularia mediterranea*. *Plant Physiology*, **44,** 946–954.

Bidwell, R., Levin, W. B. and Shephard, D. C. (1970). Intermediates of photosynthesis in *Acetabularia mediterranea* chloroplasts. *Plant Physiology*, **45,** 70–75.

Blondon, F. and Jacques, R. (1970). Action de la lumière sur l'initiation florale du *Lolium temulentum L.*: spectre d'action et rôle du phytochrome. *Comptes rendus hebdomadaires des Séances de l'Académie des Sciences, Paris*, **270,** 947–950.

Boege, U., Boege, F. and Zetsche, K. (1975). On the mechanism of activation of glyceraldehyde 3-phosphate dehydrogenase of *Acetabularia*. *Protoplasma*, 83, 171–172.

Boergesen, F. (1925). Note on the development of the young thallus of *Cymopolia barbata* (*L.*) Lamour. *Nuova Notarisia*, **36,** 211–214.

Boergesen, F. (1940). Some marine algae from Mauritius. 1. *Chlorophyceae. Kongelige Danske Videnskabernes Selskabs, Biol. Meddel.*, **15,** 1–81.

Boergesen, F. (1946). Some marine algae from Mauritius. An additional list of species to part 1. *Chlorophyceae. Kongelige Danske Videnskabernes Selskabs, Biol. Meddel.* **20,** 5–60.

Bogorad, L., Davidson, J. N., Hanson, M. R. and Mets, L. (1975). Intergenomic co-operation in the synthesis of chloroplast ribosomes of *Chlamydomonas*. *XII International Botanical Congress, Leningrad, Abstracts*, **2,** 398.

Bohling, H. (1970). Untersuchungen über freie gelöste Aminosäuren in Meerwasser. *Marine Biology*, **6,** 213–225.

Boloukhère, M. (1969). Ultrastructure de l'algue *Acetabularia mediterranea* au cours du cycle biologique et dans différentes conditions expérimentales. pp. 1–117. Thèse, Université Libre de Bruxelles.

Boloukhère, M. (1970). Ultrastructure of *Acetabularia mediterranea* in the course of the formation of secondary nuclei. *In* "Biology of *Acetabularia*" (J. Brachet and S. Bonotto, eds.) pp. 145–175. Academic Press, New York and London.

Boloukhère, M. (1972). Différenciation spatiale et temporelle des chloroplastes d'*Acetabularia mediterranea*. *Journal de Microscopie*, **13,** 401–416.

Boloukhère-Presburg, M. (1965). Effet de l'actinomycine D sur l'ultrastructure des chloroplastes et du noyau d'*Acetabularia mediterranea*. *Journal de Microscopie*, **4,** 363–372.

Boloukhère-Presburg, M. (1966). Effets de la puromycine sur l'ultrastructure d'*Acetabularia mediterranea*. *Journal de Microscopie*, **5,** 619–628.

Boloukhère-Presburg, M. (1967). Effets de la puromycine sur l'ultrastructure de l'algue *Acetabularia mediterranea*. Congress of Microscopy, Brussels.

Bonotto, S. (1968). Sur la formation des chapeaux anormaux chez *Acetabularia mediterranea*. *Protoplasma*, **66,** 55–61.

Bonotto, S. (1969a). Un tipo particolare di morfogenesi anormale in *Acetabularia mediterranea*. *Giornale Botanico Italiano*, **103,** 153–161.

Bonotto, S. (1969b). Quelques observations sur les verticilles d'*Acetabularia mediterranea*. *Bulletin de la Société royale de Botanique de Belgique*, **102,** 165–179.

Bonotto, S. (1969c). Synthèse du RNA et morphogenèse chez *Acetabularia mediterranea*, pp. 1–46. Thèse, Université Libre de Bruxelles.

Bonotto, S. (1970). Morphogenèse d'*Acetabularia mediterranea* dans la mer et en laboratoire. *Bulletin de la Société royale Botanique de Belgique*, **103,** 213–223.

Bonotto, S. (1971). Différents types de morphogenèse élaborés par une même cellule chez *Acetabularia mediterranea*. *Bulletin de la Société royale Botanique de Belgique*, **104,** 47–55.

Bonotto, S. (1973). Some teratological types of morphological differentiation in *Acetabularia mediterranea*. *Atti Associazione Genetica Italiana*, **18,** 104–108.

Bonotto, S. (1975). Morphogenesis in normal, branched and irradiated *Acetabularia mediterranea*. *Pubblicazioni della Stazione Zoologica di Napoli*, **39,** Suppl. 1, 96–107.

Bonotto, S. (1976). Cultivation of plants: Multicellular plants. *In* " Marine Ecology " (O. Kinne, ed.), Vol. 3. Part 1, pp. 467–501. Wiley-Interscience, London, New York, Sydney, Toronto.

Bonotto, S. and Bonnijns-van Gelder, E. (1969). A simple method to obtain high frequency of stalk division in *Acetabularia mediterranea*. *Plant Physiology*, **44,** 1738–1741.

Bonotto, S. and Janowski, M. (1968). Dichotomie du siphon chez *Acetabularia mediterranea*. *Bulletin de l'Academie royale de Belgique*, (Classe des Sciences), 54, 1369–1377.

Bonotto, S., and Janowski, M. (1969). Quelques observations sur la forme du chapeau d'*Acetabularia mediterranea*. *Bulletin de la Société royale de Botanique de Belgique*, **102,** 257–265.

Bonotto, S. and Kirchmann, R. (1969a). Preferential inhibition of RNA synthesis induced by gamma radiations in the unicellular alga *Acetabularia mediterranea*. *Giornale Botanico Italiano*, **103,** 601–602.

Bonotto, S. and Kirchmann, R. (1969b). Comparative study of the effects of rifampicin and actinomycin D on the synthesis of DNA, RNA and proteins in *Acetabularia meditarreana*. *Giornale Botanico Italiano*, **103,** 601.

Bonotto, S., and Kirchmann, R. (1970). Sur les processus morphogénétiques d'*Acetabularia mediterranea*. *Bulletin de la Société royale de Botanique de Belgique*, **103,** 255–272.

Bonotto, S. and Kirchmann, R. (1971a). Effets des radiations gamma sur la synthèse des acides nucléiques et des protéines chez *Acetabularia mediterranea*. *Bulletin de la Société royale de Botanique de Belgique*, **104,** 125–135.

Bonotto, S. and Kirchmann, R. (1971b). Sur un type particulier de morphogenèse anormale chez *Acetabularia crenulata*. *Bulletin de la Société royale de Botanique de Belgique*, **104,** 5–12.

Bonotto, S. and Kirchmann, R. (1972a). Sur un type particulier de morphogenèse anormale provoquée par les radiations gamma chez l'*Acetabularia mediterranea*. *Giornale Botanico Italiano*, **106,** 21–27.

Bonotto, S. and Kirchmann, R. (1972b). Dose dependent stalk division in gamma irradiated whole *Acetabularia* cells. *Archives de Biologie*, **83**, 207–214.

Bonotto, S. and Kirchmann, R. (1973). Lead effects on the morphogenesis of the unicellular alga *Acetabularia*. *Annales de Gembloux*, **79**, 121–130.

Bonotto, S. and Kirchmann, R. (1974). Radiobiology of normal and branched *Acetabularia mediterranea*. *VIIIth International Seaweed Symposium, Bangor*, U.K., 17—24 August, 1974. Abstracts of Papers, B 50.

Bonotto, S. and Kirchmann, R. (1975). Incorporation of ^{3}H from tritiated water in the unicellular algae *Acetabularia mediterranea* and *Chlamydomonas reinhardi*. XII *International Botanical Congress, Leningrad, Abstracts*, **1**, 35,

Bonotto, S., and Puiseux-Dao, S. (1970). Modification de la morphologie des verticilles et differenciation chez l'*Acetabularia mediterranea*. *Comptes rendus hebdomadaires des Séances de l'Académie des Sciences, Paris*, **270**, 1100–1103.

Bonotto, S., Janowski, M., Vanden Driessche, T. and Brachet, J. (1968). Effet spécifique de la rifampicine sur la synthèse du RNA chez *Acetabularia mediterranea*. *Archives Internationales de Physiologie et de Biochimie*, **76**, 919–920.

Bonotto, S., Bonnijns-van Gelder, E., Felluga, B. and Netrawali, M. S. (1969a). Shape of the outer end of the cap rays and cyst formation in *Acetabularia mediterranea*. *Giornale Botanico Italiano*, **103**, 385–393.

Bonotto, S., Felluga, B. and Akşiyote, J. (1969b). Quelques observations sur la morphologie des cystes d'*Acetabularia mediterranea*. *Protoplasma*, **67**, 407–412.

Bonotto, S., Goffeau, A., Janowski, M., Vanden Driessche, T. and Brachet, J. (1969c). Effects of various inhibitors of protein synthesis on *Acetabularia mediterranea*. *Biochimica et Biophysica Acta*, **174**, 704–712.

Bonotto, S., Bonnijns-van Gelder, E., Fagniart, E. and Kirchmann, R. (1970a). Polycap formation in *Acetabularia mediterranea*. I. Spontaneous stalk division. *Giornale Botanico Italiano*, **104**, 117–130.

Bonotto, S., Kirchmann, R., Janowski, M. and Netrawali, M. S. (1970b). Effects of gamma-radiation on *Acetabularia mediterranea*. *In* " Biology of *Acetabularia* ". (J. Brachet and S. Bonotto, eds.) pp. 255–271. Academic Press, New York and London.

Bonotto, S., Kirchmann, R. and Manil, P. (1971a). Cell engineering in *Acetabularia*: a graft method for obtaining large cells with two or more reproductive caps. *Giornale Botanico Italiano*, **105**, 1–9.

Bonotto, S., Kirchmann, R., Puiseux-Dao, S. and Valet, G. (1971b). Chapeaux en forme de cloche et relations nucléo-cytoplasmiques chez l'*Acetabularia mediterranea*. *Comptes rendus hebdomadaires des Séances de l'Académie des Sciences, Paris*, **272**, 545–548.

Bonotto, S., Lurquin, P., and Kirchmann, R. (1971c). Radiation effects on transcription and translation in *Acetabularia mediterranea*. *Studia Biophysica (Berlin)*, 29, 23–28.

Bonotto, S., Puiseux-Dao, S., Kirchmann, R. and Brachet, J. (1971d). Faits et hypothèses sur le contrôle de l'alternance morphogénétique: croissance végétative—différenciation de l'appareil reproducteur chez les *Acetabularia mediterranea* Lamouroux, *A. crenulata* Lamouroux et *Halicoryne spicata* (Kützing) Solms-Laubach. *Comptes rendus hebdomadaires des Séances de l'Académie des Sciences, Paris*, **272**, 392–395.

Bonotto, S., Lurquin, P., Goutier, R., Kirchmann, R., and Maisin, J. R. (1972). Radiobiology of *Acetabularia*. *Protoplasma*, **75,** 476.

Bonotto, S., Bonnijns-van Gelder, E., Fagniart, E., Kirchmann, R. (1972a). Polycap formation in *Acetabularia mediterranea*. III. Superposed caps on the stalk. *Bulletin de la Société royale de Botanique de Belgique*, **105,** 45–55.

Bonotto, S., Fagniart, E., Kirchmann, R. (1972b). Biological effects of lead in the unicellular marine alga *Acetabularia mediterranea*. Book of Abstracts, pp. 54–64. ESNA Meeting, Ispra, Italy, 21–23 March, 1972.

Bonotto, S., Kirchmann, R., Dazy, A. C., Puiseux-Dao, S. (1972c). Réversibilité des effets de la rifampicine sur la différenciation morphologique d'*Acetabularia mediterranea*. *Bulletin de la Société royale de Botanique de Belgique*, **105,** 35–43.

Bonotto, S., Lurquin, P., Baugnet-Mahieu, L., Goutier, R., Kirchmann, R. and Maisin, J. R. (1972d). Biology and Radiobiology of anucleate *Acetabularia mediterranea*. In " Biology and Radiobiology of Anucleate Systems ", Vol. 2: Plant Cells (S. Bonotto, R. Goutier, R. Kirchmann and J. R. Maisin, eds.), pp. 339–368, Academic Press, New York and London.

Bonotto, S., Lurquin, P., and Kirchmann, R. (1973). Effect of gamma-radiation on the synthesis of cytoplasmic DNA's in *Acetabularia*. *International Journal of Radiation Biology*, **23,** 184.

Bonotto, S., Lurquin, P., Baeyens, W., Charles, P., Hoursiangou-Neubrun, D., Mazza, A., Tramontano, G. and Felluga, B. (1975). Plastid DNA in *Acetabularia mediterranea*. *Protoplasma*, **83,** 172–173.

Bonotto, S., Ndoite, I. O., Nuyts, G., Fagniart, E. and Kirchmann, R. (1976). Study of the distribution and biological effects of ^3H on the algae *Acetabularia, Chlamydomonas and Porphyra*. *International Conference on Molecular and Microdistribution of Radioisotopes and Biological Consequences*, Jülich, 2–4 October 1975. In press.

Borghi, H., Bonotto, S., Puiseux-Dao, S. and Hoursiangou-Neubrun, D. (1972). Effects of lindane on *Acetabularia mediterranea*. *Protoplasma*, **75,** 476.

Borghi, H., Puiseux-Dao, S., Bonotto, S. and Hoursiangou-Neubrun, D. (1973). The effects of lindane on *Acetabularia mediterranea*. *Protoplasma*, **78.** 99–112.

Bory de St Vincent, J. B. (1832). " Expédition scientifique de Morée," Vol. 3, pp. 329–330. Paris.

Bouck, G. B. (1964). Fine structure in *Acetabularia* and its relation to protoplasmic streaming. *In* " Primitive Motile Systems in Cell Biology ". (Allen, R. D. and Kamiya, N., eds.), pp. 7–18. Academic Press, New York and London.

Brachet, J. (1951). Quelques effets cytologiques et cytochimiques des inhibiteurs des phosphorylations oxydatives. *Experientia*, **7,** 344–345.

Brachet, J. (1952a). Quelques effets des inhibiteurs des phosphorylations oxydatives sur les fragments nucléés et anucléés d'organismes unicellulaires. *Experientia*, **8,** 347–351.

Brachet, J. (1952b). The role of the nucleus and the cytoplasm in synthesis and morphogenesis. *Symposium of the Society of Experimental Biology*, **6,** 173–199.

Brachet, J. (1957). " Biochemical Cytology ", pp. 1–535. Academic Press, New York and London.

Brachet, J. (1958a). The effects of various metabolites and antimetabolites on the regeneration of fragments of *Acetabularia mediterranea*. *Experimental Cell Research*, **14,** 650–651.

Brachet, J. (1958b). New observations on biochemical interactions between nucleus and cytoplasm in *Amoeba* and *Acetabularia*. *Experimental Cell Research* (Suppl), **6,** 78–96.

Brachet, J. (1959). The role of sulfhydryl groups in morphogenesis. *Journal of Experimental Zoology*, **142,** 115–139.

Brachet, J. (1960a). Le contrôle de la synthèse des protéines. *Scientia*, **54,** 119–124.

Brachet, J. (1960b). Ribonucleic acids and the synthesis of cellular proteins. *Nature, London*, **186,** 194–199.

Brachet, J. (1961a). Nucleocytoplasmic interactions in unicellular organisms. *In* " The Cell " (Brachet, J. and Mirsky, A. E., eds.), pp. 771–841, Academic Press, New York and London.

Brachet, J. (1961b). Morphogenetic effects of lipoic acid on amphibian embryos. *Nature, London*, **189,** 156–157.

Brachet, J. (1962a). Effects of β-mercaptoethanol and lipoic acid on morphogenesis. *Nature, London*, 193, 87–88.

Brachet, J. (1962b). Nucleic acids in development. *Journal of Cellular and Comparative Physiology*, (Suppl.) **1,** 60, 1–18.

Brachet, J. (1963a). The role of the nucleic acids in the process of induction, regulation and differentiation in the amphibian embryo and the unicellular alga *Acetabularia mediterranea*. *In* " Symposium on Biological Organization ", Varenna, 1962 (R. J. C. Harris, ed.) pp. 167–182. Academic Press, New York and London.

Brachet, J. (1963b). The effects of puromycin on morphogenesis in amphibian eggs and *Acetabularia mediterranea*. *Nature, London*, **199,** 714–715.

Brachet, J. (1963c). Acides ribonucléiques " messagers " et morphogenèse. *Bulletin de l'Académie Royale de Belgique* (*classe des Sciences*), **49,** 862–887.

Brachet, J. (1963d). Le rôle du noyau cellulaire dans les synthèses et la morphogenèse. *Nova Acta Leopoldina*, **26,** 17–27.

Brachet, J. (1964). Nouvelles observations sur le rôle des acides nucléiques dans la morphogenèse. *In* "Acidi Nucleici e loro funzione biologica ", pp. 288–315. Convegno Antonio Baselli, Milano, 16–18 settembre 1963.

Brachet, J. (1965a). Le rôle des acides nucléiques dans la morphogenèse. *L'Année Biologique*, **4,** 21–48.

Brachet, J. (1965b). The role of nucleic acids in morphogenesis. *Progress in Biophysics and Molecular Biology*, **15,** 99–127.

Brachet, J. (1965c). Le contrôle de la synthèse des protéines en l'absence du noyau cellulaire. Faits et hypothèses. *Bulletin de l'Académie Royale de Belgique* (Classe des Sciences), 51, 257–276.

Brachet, J. (1965d). L'*Acetabularia*. *Endeavour*, **24,** 155–161.

Brachet, J. (1966). L'énigme des acides désoxyribonucléiques (DNA) cytoplasmiques. *Scientia*, **101,** 1–6.

Brachet, J. (1967a). Protein synthesis in the absence of the nucleus. *Nature, London*, **219,** 650–655.

Brachet, J. (1967b). Effects of hydroxyurea on development and regeneration. *Nature, London*, **214,** 1132–1133.

Brachet, J. (1967c). Morphogenèse et synthèse des protéines en l'absence du noyau cellulaire. *In* " De l'Embryologie Expérimentale à la Biologie Moléculaire " (E. Wolff, ed.) pp. 5–42. Dunod, Paris.

Brachet, J. (1967d). Exchange of macromolecules between nucleus and cytoplasm. *Protoplasma*, **63,** 86–87.

Brachet, J. (1968). Synthesis of macromolecules and morphogenesis in *Acetabularia.* In " Current Topic in Developmental Biology " (A. Monroy and A. Moscona, eds.) Vol. 3, pp. 1–36. Academic Press, New York and London.

Brachet, J. (1970). Concluding remarks. In " Biology of *Acetabularia* ". (J. Brachet and S. Bonotto, eds.) pp. 273–291, Academic Press, New York and London.

Brachet, J. (1971). Nucleocytoplasmic interactions in morphogenesis. *Proceedings of the Royal Society*, B., **178,** 227–243.

Brachet, J. (1972a). Morphogenesis and synthesis of macromolecules in the absence of the nucleus. *In* " Biology and Radiobiology of Anucleate Systems ", (Bonotto, S., Goutier, R., Kirchmann, R. and Maisin, J. R., eds.) Vol. 2, pp. 1–26. Academic Press, New York and London.

Brachet, J. (1972b). Morfogenese. *Natuur en Techniek*, **40,** 144–159.

Brachet, J. (1974). The effects of a few inhibitors on morphogenesis in *Acetabularia*, University of Paris VII, 10–11 July 1974. Book of Abstracts, pp. 1–2.

Brachet, J. (1975). Nucleocytoplasmic interactions in cell differentiation. *In* " Molecular Biology of Nucleocytoplasmic Relationships " (Puiseux-Dao, S. ed.) pp. 187–201, Elsevier, Amsterdam and New York.

Brachet, J. and Bonotto, S. (1970). " Biology of *Acetabularia* ". Proceedings First International Symposium on *Acetabularia*, p. 300. Academic Press, New York and London.

Brachet, J. and Brygier, J. (1953). Le rôle de la lumière dans la régénération et l'incorporation de CO_2 radioactif chez *Acetabularia mediterranea. Archives Internationales de Physiologie*, **61,** 246–247.

Brachet, J. and Chantrenne, H. (1951). Protein synthesis in nucleated and non-nucleated halves of *Acetabularia mediterranea* studied with carbon-14 dioxyde. *Nature, London*, **168,** 950.

Brachet, J. and Chantrenne, H. (1952). Incorporation de $^{14}CO_2$ dans les protéines des chloroplastes et des microsomes de fragments nucléés et anucléés d'*Acetabularia mediterranea. Archives Internationales de Physiologie*, **60,** 547–549.

Brachet, J. and Chantrenne, H. (1953). La synthèse adaptative de la catalase dans des fragments nucléés d'*Acetabularia mediterranea. Archives Internationales de Physiologie*, **61,** 248–249.

Brachet, J., and Chantrenne, H. (1956). The function of nucleus in the synthesis of cytoplasmic proteins. *Cold Spring Harbor Symposium on Quantitative Biology*, **21,** 329–337.

Brachet, J., and Delange-Cornil, M. (1959). Recherches sur le rôle des groupes sulphydriles dans la morphogenèse. *Developmental Biology*, **1,** 79–100.

Brachet, J. and Denis, H. (1963). Effects of actinomycin D on morphogenesis. *Nature, London*, **198,** 205–206.

Brachet, J. and Goffeau, A. (1964). Le rôle des acides désoxyribonucléiques (DNA) dans la synthèse des protéines chloroplastiques chez *Acetabularia mediterranea. Comptes rendus hebdomadaires des Séances de l'Académie des Sciences, Paris*, **259,** 2899–2901.

Brachet, J. and Lang, A. (1965). The role of the nucleus and the nucleocytoplasmic interactions in morphogenesis. *In* " Handbuch der Pflanzenphysiologie " (W. Ruhland, ed.). Vol. 15, pp. 1–40. Springer-Verlag, Berlin.

Brachet, J. and Olszewska, M. J. (1960). Influence of localized ultraviolet irradiation on the incorporation of adenine-8-^{14}C and DL-methionine-^{35}S in *Acetabularia mediterranea*. *Nature, London*, **187,** 954–955.

Brachet, J. and Six, N. (1966). Quelques observations nouvelles sur les relations entre la synthèse des acides ribonucléiques et la morphogenèse chez *Acetabularia*. *Planta*, **68,** 225–239.

Brachet, J. and Szafarz, D. (1953). L'incorporation d'acide orotique radioactif dan les fragments nucléés et anucléés d'*Acetabularia mediterranea*. *Biochimica et Biophysica Acta*, **12,** 588–589.

Brachet, J. and Tencer, R. (1973). Effects of cytochalasin B on morphogenesis in tunicate and amphibian eggs and in the unicellular alga *Acetabularia*. *Acta Embryologiae Experimentalis*, 83–104.

Brachet, J., Chantrenne, H. and Vanderhaeghe, F. (1955). Recherches sur les interactions biochimiques entre le noyau et le cytoplasme chez les organismes unicellulaires. II. *Acetabularia mediterranea*. *Biochimica et Biophysica Acta*, **18,** 544–563.

Brachet, J., Decroly, M. and Quertier, J. (1963). Groupes sulfhydrilés et morphogénèse. III. Etude biochimique des effets du mercaptoéthanol sur les embryons de Batraciens et l'algue *Acetabularia mediterranea*. *Developmental Biology*, **6,** 113–131.

Brachet, J., Denis, H. and de Vitry, F. (1964). The effects of actinomycin D and puromycin on morphogenesis in amphibian eggs and *Acetabularia mediterranea*. *Developmental Biology*, **9,** 398–434.

Brändle, E. and Zetsche, K. (1971). Die Wirkung von Rifampicin auf die RNA- und Proteinsynthese sowie die Morphogenese und den Chlorophyllgehalt kernhaltiger und kernloser *Acetabularia*-Zellen. *Planta*, **99,** 46–55.

Brändle, E. and Zetsche, K. (1973). Zur Lokalisation der α-Amanitin sensitiven RNA-Polymerase in Zellkernen von *Acetabularia*. *Planta*, **111,** 209–217.

Brändle, E. P. O., Kötter, R. and Zetsche, J. (1975). Changes in pH in culture solution of *Acetabularia* caused by proton flux. *Protoplasma*, **83,** 173.

Bremer, H. J. and Schweiger, H. G. (1960). Der NH_3-Gehalt kernhaltiger und kernloser Acetabularien. *Planta*, **55,** 13–21.

Bremer, H. J., Schweiger, H. G. and Schweiger, E. (1962). Das Verhalten der freien Aminosäuren in kernhaltigen und kernlosen Acetabularien. *Biochimica et biophysica Acta*, **56,** 380–382.

Brown, R. (1819). Cited in Valet, G. (1969a).

Buggeln, R. G. and Craigie, J. S. (1971). Evaluation of evidence for the presence of indole-3-acetic acid in marine algae. *Planta*, **97,** 173-178.

Bünning, E. (1967). " The Physiological Clock " (2nd Edition). Springer Verlag, Berlin.

Cairns, J. (1963). The bacterial chromosome and its manner of replication as seen by autoradiography. *Journal of Molecular Biology*, **6,** 208–213.

Ceron, G. and Johnson, E. M. (1971). Control of protein synthesis during the development of *Acetabularia*. *Journal of Embryology and Experimental Morphology*, **26,** 323–338.

Ceska, M. (1962). Biosynthesis of ribonucleic acids in *Acetabularia mediterranea*. *Archives Internationales de Physiologie et Biochimie*, **70,** 566.

Chadefaud, M. (1938). Les plastes et l'amylogénèse chez les Dasycladacées. (Algues vertes, Siphonales). *Comptes rendus hebdomadaires des Séances de l'Académie des Sciences, Paris*, **206,** 362–364.

Chamberlain, Y. M. (1958). *Dasycladus ramosus*, a new species of *Dasycladus* from Inhaca Island and Peninsula, Portuguese East Africa. *South African Botany*, **24,** 119–121.

Chantrenne, H., Van Halteren, M. and Brachet, J. (1952). La respiration de fragments nucléés d'*Acetabularia mediterranea*. *Archives Internationales de Physiologie*, **60,** 187.

Chantrenne, H., Brachet, J. and Brygier, J. (1953). Quelques données nouvelles sur le rôle du noyau cellulaire dans le métabolisme des protéines chez *Acetabularia mediterranea*. *Archives Internationales de Physiologie*, **61,** 419–420.

Chapman, C. J., Nugent, N. A. and Schreiber, R. W. (1966). Nucleic acid synthesis in the chloroplasts of *Acetabularia mediterranea*. *Plant Physiology*, **41,** 589–592.

Chaudhuri, T. K. and Spencer, P. (1968). Amino acid uptake in *Acetabularia*. *Protoplasma*, **66,** 255–259.

Chirkova, L. I., Pikalov, A. V., Kisseleva, E. V., Betina, M. I., Khristolubova, N. B., Nikoro, Z. S. and Sandakhchiev, L. S. (1970). Subcellular localization of the morphogenetic factors in *Acetabularia mediterranea*. *Ontogenez.*, **1,** 42–54. (In Russian.)

Church, A. H. (1895). The structure of the thallus of *Neomeris dumetosa* Lamour. *Annals of Botany*, **9,** 581–608.

Cinelli, F. (1969). Primo contributo alla conoscenza della vegetazione algale bentonica del litorale di Livorno. *Pubblicazioni della Stazione zoologica di Napoli*, **37,** 545–566.

Cinelli, F. (1971). Alghe bentoniche di profondità raccolte alla Punta S. Pancrazio nell' Isola d'Ischia (Golfo di Napoli). *Giornale Botanico Italiano*, **105,** 207–236.

Clauss, H. (1958). Über quantitative Veränderungen der Chloroplasten und cytoplasmatischen Proteine in kernlosen Teilen von *Acetabularia mediterranea*. *Planta*, **52,** 334–339.

Clauss, H. (1959). Das Verhalten der Phosphorylase in kernhaltigen und kernlosen Teilen von *Acetabularia mediterranea*. *Planta*, **52,** 534–542.

Clauss, H. (1961). Über den Einbau von ^{35}S in *Acetabularia mediterranea*. *Zeitschrift für Naturforschung*, **16b,** 770–771.

Clauss, H. (1962a). Zur Frage der Beteiligung intrazellulärer Hemmfaktoren bei der Proteinsynthese von *Acetabularia*. *Zeitschrift für Naturforschung*, **17b,** 339–341.

Clauss, H. (1962b). Über die Intensität der Proteinsynthese von *Acetabularia* vor und nach der Hutbildung. *Zeitschrift für Naturforschung*, **17b,** 342–344.

Clauss, H. (1962c). Über die Synthese von *Acicularia* spezifischen Esterasen in *Acicularia-Acetabularia* Transplantaten. *Naturwissenschaften*, **49,** 523–524.

Clauss, H. (1963). Über den Einfluss von Rot- und Blaulicht auf das Wachstum kernhaltiger Teile von *Acetabularia mediterranea*. *Zeitschrift für Naturforschung*, **50,** 719.

Clauss, H. (1968). Beeinflussung der Morphogenese, Substanzproduktion und Proteinzunahme von *Acetabularia mediterranea* durch sichtbare Strahlung. *Protoplasma*, **65,** 49–80.

Clauss, H. (1970a). Effect of red and blue light on morphogenesis and metabolism of *Acetabularia mediterranea*. *In* " Biology of *Acetabularia* ". (J. Brachet and S. Bonotto, eds) pp. 177–191. Academic Press, New York and London.

Clauss, H. (1970b). Der Einfluss von Rot- und Blaulicht auf die Hillaktivität von *Acetabularia*-Chloroplasten. *Planta*, **91,** 32–37.

Clauss, H. (1972). Der Einfluss von Rot- und Blaulicht auf die Photosynthese von *Acetabularia mediterranea* und auf die Verteilung des assimilierten Kohlenstoffs. *Protoplasma*, **74,** 357–379.

Clauss, H. (1973). Regulation der Morphogenese von *Acetabularia* durch Licht. *In* " Vom Gen zum Genus ", Pressedienst Wissenschaft der FU Berlin, No. 3, pp. 45–59.

Clauss, H. (1975). Photosynthese -Intensität und Stoffwechsel der *Acetabularia*-Zelle in Abhängigkeit von ihrem Entwicklungszustand. *Protoplasma*, **83,** 147–166.

Clauss, H. and Keck, K. (1959). Über die löslichen Kohlenhydrate der Grünalge *Acetabularia mediterranea* und deren quantitative Veränderungen in kernhaltigen und kernlosen Teilen. *Planta*, **52,** 543–553.

Clauss, H. and Werz, G. (1961). Über die Geschwindigkeit der Proteinsynthese in kernlosen und kernhaltigen Zellen von *Acetabularia. Zeitschrift für Naturforschung*, **16b,** 161–165.

Clauss, H., Lüttke, A., Hellman, F. and Reinert, J. (1970). Chloroplastenvermehrung in kernlosen Teilstücken von *Acetabularia mediterranea* und *Acetabularia cliftonii* und ihre Abhängigkeit von inneren Faktoren. *Protoplasma*, **69,** 313–329.

Clérin, P., Vanden Driessche, T. and Brachet, J. (1975). Migration of cytoproteins into the nucleus of *Acetabularia*. *In* " Molecular Biology of Nucleocytoplasmic Relationships " (S. Puiseux-Dao, ed.) pp. 305–311, Elsevier, Amsterdam and New York.

Collins, F. S. (1909). The green algae of North America. *Tufts College Studies*, Science Series, **2,** 79–480.

Conesa, A. P. (1969. Possibilités d'utilisation des polyphosphates par la plante. *Comptes rendus hebdomadaires des Séances de l'Academies des Sciences, Paris*, **268,** 2063–2066.

Craig, I. W., Gibor, A. (1970). Biosynthesis of proteins involved with photosynthetic activity in enucleated *Acetabularia* sp. *Biochimica et Biophysica Acta*, **217,** 488–495.

Cramer, C. (1888). Über die verticillierten Siphoneen, besonders *Neomeris* und *Cymopolia. Neue Denkschriften der Allgemeiner Schweizerischen Gesellschaft für die gesamten Naturwissenschaften*, **30,** 3–50.

Cramer, C. (1890). Über die verticillierten Siphoneen, besonders *Neomeris* und *Bornetella. Neue Denkschriften der Allgemeinen Schweizerischen für die Gesellschaft für die gesamten Naturwissenschaften*, **32,** 4–48.

Cramer, C. (1895). Über *Halicoryne wrightii* Harvey. *Vierteljahrsschrift der Naturforschenden Gesellschaft in Zürich*, **40,** 265–277.

Crawley, J. C. W. (1963). The fine structure of *Acetabularia. Experimental Cell Research*, **32,** 368–378.

Crawley, J. C. W. (1964). Cytoplasmic fine structure in *Acetabularia. Experimental Cell Research*, **35,** 497–506.

Crawley, J. C. W. (1965). The fine structure of isolated *Acetabularia* nuclei. *Planta*, **65,** 205–217.

Crawley, J. C. W. (1966). Some observations on the fine structure of the gametes and zygotes of *Acetabularia*. *Planta*, **69,** 365–376.

Crawley, J. C. W. (1970). The fine structure of the gametes and zygotes of *Acetabularia*. In " Biology of *Acetabularia* ", (J. Brachet and S. Bonotto, eds) pp. 73–83. Academic Press, New York and London.

Dao, S. (1954a). Comportement de l'*Acetabularia mediterranea* Lam. en culture. Etude de sa croissance. *Revue génerale de Botanique*, **61,** 573–606.

Dao, S. (1954b). Action de l'acide indol acétique sur l'*Acetabularia mediterranea* Lamour. en culture. *Comptes rendus hebdomadaires des Séances de l'Académie des Sciences, Paris*, **338,** 2340–2341.

Dao, S. (1956). A propos de l'action du tryptophane sur l'*Acetabularia mediterranea* Lamour., *Comptes rendus hebdomadaires des Séances de l'Academie des Sciences, Paris*, **243,** 1552–1554.

Dao, S. (1957). La gamétogénèse chez l'*Acetabularia mediterranea* Lamouroux. *Comptes rendus hebdomadaires des Séances des l'Académie des Sciences,* **243,** 1552–1554.

Dao, S. (1958). Recherches caryologiques chez le *Neomeris annulata Dickie*. *Revue Algologique*, n.s., **3,** 192–201.

Dawson, E. Y. (1956). Algae of Southern Marshalls. *Pacific Science*, **10,** 25–66.

Dawson, E. Y. (1966). " Marine botany. An Introduction." Holt, Rinehart and Winston, New York.

Dazy, A. C. (1969). Action du bromure d'éthidium sur la morphogénèse d'*Acetabularia mediterranea* et la structure de ses plastes. Stage de troisième cycle, Paris.

Dazy, A. C. and Woodcock, C. L. F. (1972). An auto-radiographic assay for protein synthesis in *Acetabularia*. *Protoplasma*, **75,** 476–477.

Dazy, A. C., Borghi, H., Gauchery, J. and Puiseux-Dao, S. (1975a). RNA transport from the nucleus to the cytoplasm. *Protoplasma*, **83,** 173–174.

Dazy, A. C., Borghi, H., Gauchery, J. and Puiseux-Dao, S. (1975b). RNA transport from the nucleus to the cytoplasm in *Acetabularia mediterranea*. In "Molecular Biology of Nucleocytoplasmic Relationships " (S. Puiseux-Dao, ed.) pp. 251–257, Elsevier, Amsterdam and New York.

Dazy, A. C., Matthys, E and Puiseux-Dao, S. (1970). Action du bromure d'éthidium sur les organites à DNA satellite de cellules végétales (*Amphidinium carteri, Dinophyceae et Acetabularia mediterranea, Dasycladaceae*). *Bulletin Société botanique de France*, **117,** 311–320.

de Bary, A. and Strasburger, E. (1877). *Acetabularia mediterranea*. *Botanische Zeitung*, **35,** 713–758.

Decaisne, J. (1842). Mémoires sur les corallines ou polypiers calcifères. *Ann. Sci. Nat. Bot. Serie.*, **2,** 18, 96–128.

De Carli, H. and Brachet, J. (1968). Action de l'héparine sur la morphogenèse. *Bulletin de l'Académie royale de Belgique* (*Classe des Sciences*), **54,** 1158–1164.

Declève, A., van Gorp, U., Bouloukhère, M. and Bonotto, S. (1972). Biochemical and ultrastructural investigations on the whorls of *Acetabularia mediterranea*. *In* " Biology and Radiobiology of Anucleate Systems ", (Bonotto, S., Goutier, R., Kirchmann, R. and Maisin, J. R., eds.), Vol. 2, pp. 259–293. Academic Press, New York and London.

Derbès, A. and Solier, A. (1856). Mémoire sur quelques points de la physiologie des algues. Suppl. *Comptes rendus hebdomadaires des Séances de l'Académie des Sciences, Paris*, **1**, 1–120.

De Toni, G. B. (1889). " Sylloge Algarum ", Vol. I, pp. 409–423. Padoue.

de Vitry, F. (1962a). Etude de l'action de la 5-fluorodéoxyuridine sur la croissance et la morphogenèse d'*Acetabularia mediterranea. Experimental Cell Research*, **25**, 697–699.

de Vitry, F. (1962b). Action des métabolites et antimétabolites sur la croissance et la morphogenèse d'*Acetabularia mediterranea. Protoplasma*, **55**, 313–319.

de Vitry, F. (1963). Etude autoradiographique de l'incorporation de la ^{3}H-5-méthylcytosine chez *Acetabularia mediterranea. Experimental Cell Research*, **31**, 376–381.

de Vitry, F. (1964a). Etude autoradiographique de l'incorporation de l'actinomycine ^{14}C chez *Acetabularia mediterranea. Comptes rendus hebdomadaires des Séances de l'Academie des Sciences, Paris*, **258**, 4829–4831.

de Vitry, F. (1964b). Etude autoradiographique des effets de la 5-fluorodéoxyuridine, de l'actinomycine et de la puromycine chez *Acetabularia mediterranea. Developmental Biology*, **9**, 484–504.

de Vitry, F. (1965a). Etude du métabolisme des acides nucléiques chez *Acetabularia mediterranea*. I. Incorporation de précurseurs de DNA, de RNA et des protéines chez *Acetabularia mediterranea. Bulletin de la Société de Chimie Biologique*, **47**, 1325–1351.

de Vitry, F. (1965b). Etude du métabolisme des acides nucléiques chez *Acetabularia mediterranea*. II. Mise en évidence du rôle du DNA nucléaire à l'aide du phényléthylalcool et de la 5-fluorodéoxyuridine. Effets biologiques de l'actinomycine D et de la puromycine. *Bulletin de la Société de Chimie Biologique*, **47**, 1352–1372.

de Vitry, F. (1965c). Etude du métabolisme des acides nucléiques chez *Acetabularia mediterranea*. III. Etude autoradiographique des effets de l'actinomycine D et de la puromycine. *Bulletin de la Société de Chimie Biologique*, **47**, 1372–1394.

Dickie, G. (1874). On the algae of Mauritius. *Journal of the Linnean Society*, **14**, 190–202.

Dickie, G. (1876). Contributions to the Botany of the Expeditions of H.M.S. " Challenger." Algae, Chief Polynesian. *Journal of the Linnean Society*, **15**, 235–246.

Dillard, W. L. (1970). RNA synthesis in *Acetabularia. In* " Biology of *Acetabularia* " (Brachet, J. and Bonotto, S., eds.), pp. 13–14. Academic Press, New York and London.

Dillard, W. L. and Schweiger, H. G. (1968). Kinetics of RNA synthesis in *Acetabularia. Biochimica et Biophysica Acta*, **169**, 561–563.

Dillard, W. L. and Schweiger, H. G. (1969). RNA synthesis in *Acetabularia*. III. The kinetics of RNA synthesis in nucleate and enucleated cells. *Protoplasma*, **67**, 87–100.

Dodd, W. A. and Bidwell, R. G. S. (1971a). Photosynthesis and gas exchange of *Acetabularia* chloroplasts in an artificial leaf. *Nature, London*, **234**, 45–47.

Dodd, W. and Bidwell, R. (1971b). The effect of pH on the products of photosynthesis in $^{14}CO_2$ by chloroplast preparations from *Acetabularia mediterranea. Plant Physiology*, **47**, 779–783.

Donati, V. (1750). "Saggio della storia marina dell'Adriatico". Venezia. Translated into French: "Essai sur l'histoire naturelle de la mer Adriatique". La Haye 1758.

Dubacq, J. P. (1970). Mise au point d'une méthode d'isolement des chloroplastes d'Acétabulaires. Premières analyses des lipides des algues entières et des plastes isolés. Rapport de stage, Université de Paris.

Dubacq, J. P. (1971). Effet de l'éclairement sur la composition en acides gras de l'algue *Acetabularia mediterranea* Lam. *Comptes rendus hebdomadaires des Séances de l'Académie des Sciences, Paris*, **273,** 1941–1944.

Dubacq, J. P. (1972). Fatty acid composition of *Acetabularia mediterranea* (Lamour) during the morphogenesis of the cap. *Protoplasma*, **75,** 477.

Dubacq, J. P. (1973). Lipid composition of *Acetabularia mediterranea* (Lamour). Variations during cap morphogenesis. *Protoplasma*, **76,** 373–385.

Dubacq, J. P. and Hoursiangou-Neubrun, D. (1975). Fatty acids biosynthesis in *Acetabularia mediterranea*. *Protoplasma*, **83,** 174.

Dubacq, J. P. and Puiseux-Dao, S. (1974). Morphological transformations and lipid synthesis in chloroplasts of dark-treated *Acetabularia* (Lamouroux). *Plant Science Letters*, **3,** 241–250.

Dujardin, E., Bonotto, S., Sironval, C. and Kirchmann, R. (1975). La différenciation plastidale chez l'Acétabulaire étudiée par l'émission de flourescence à 77°K. *Plant Science Letters*, **5,** 209–216.

Dumont, I., Kruyen, F. and Legros, I. (1975). Fixation spécifique d'insuline à des chloroplastes isolés d'*Acetabularia mediterranea*. *Comptes rendus des Séances de la Société de Biologie, Paris*, **169,** 250–253.

du Merac, M. L. (1953). A propos de l'inuline des Acétabularies. *Revue genérale de botanique* **60,** 689–706.

du Merac, M. L. (1955). Sur la présence de fructosanes chez *Dasycladus vermicularis* (Scopoli) Krasser. *Comptes rendus hebdomadaire des Séances de l'Académie des Sciences, Paris*, **241,** 88–90.

du Merac, M. L. (1956). Une Néoméridée fructosanifère: *Cympolia barbata* (L.) Harv. *Comptes rendus hebdomadaire des Séances de l'Académie des Sciences, Paris*, **243,** 714–717.

Edwards, P. and van Baalen, C. (1970). An apparatus for the culture of benthic marine algae under varying regimes of temperature and light-intensity. *Botanica marina*, **13,** 42–43.

Egerod, E. L. (1952). An analysis of the siphonous Chlorophycophyta with special reference to the Siphonocladales, Siphonales and Dasycladales of Hawaï. *University of California Publication, Botany*, **25,** 325–453.

Egger, K. and Kleinig, H. (1967). Die Ketocarotinoide in *Adonis annua* L. III. Vergleich mit synthetischen Substanzen. *Phytochemistry*, **6,** 903–905.

Ellis, J. (1755). "An essay towards a natural history of the Corallines". p. 103. London.

Ellis, J. and Solander, D. (1786). The natural history of many curious and uncommon zoophytes collected from various parts of the globe, 208 pp. London.

Errera, M. and Vanderhaeghe, F. (1957). Effets des rayons UV sur *Acetabularia mediterranea*. *Experimental Cell Research*, **13,** 1–10.

Farber, F. E. (1969a). Studies on RNA metabolism in *Acetabularia mediterranea*. I. The isolation of RNA and labelling studies on RNA of whole plants and plant fragments. *Biochimica et Biophysica Acta*, **174,** 1–11.

Farber, F. E. (1969b). Studies on RNA metabolism in *Acetabularia mediterranea.* II. The localization and stability of RNA species; the effects on RNA metabolism of dark period and actinomycin D. *Biochimica et Biophysica Acta*, **174,** 12–22.

Farber, F. E., Cape, M., Decroly, M. and Brachet, J. (1968). The *in vitro* translation of *Acetabularia mediterranea* RNA. *Proceedings of the National Academy of Science*, **61,** 843–846.

Feldmann, J. and Feldmann, G. (1947). Additions à la flore des algues marines de l'Algérie. *Bulletin de la Société d'Histoire Naturelle, Afrique du Nord*, **38,** 80–91.

Feulgen, R. and Rossenbeck, H. (1924). Mikroskopisch-chemischer Nachweis einer Nukleinsäure vom Typus der Thymonukleinsäure und die darauf beruhende elektive Färbung von Zellkernen in mikroskopischen Präparaten. Hoppe-Seyler's *Zeitschrift für physiologische Chemie*, **135,** 203–248.

Fox, D. K. and Shephard, D. C. (1972). Nucleic acid species and their synthesis in chloroplasts of *Acetabularia. Journal of Cell Biology*, **55,** 77a.

Føyn, B. (1934a). Lebenszyklus, Cytologie und Sexualität der Chlorophycee *Cladophora suhriana* Kützing. *Archiv für Protistenkunde*, **83,** 1–56.

Føyn, B. (1934b). Lebenszyklus und Sexualität der Chlorophycee *Ulva lactuca* L. *Archiv für Protistenkunde*, **83,** 154–177.

Franke, W. W., Berger, S., Falk, H., Spring, H., Scheer, U., Herth, W. Trendelenburg, M. F. and Schweiger, H. G. (1974). Morphology of the nucleo-cytoplasmic interactions during the development of *Acetabularia* cells. I. The vegetative phase. *Protoplasma*, **82,** 249–282.

Franke, W. W., Berger, S., Scheer, U. and Trendelenburg, M. F. (1975a). The nuclear envelope and the perinuclear lacunar labyrinthum. *Protoplasma*, **83,** 174–175.

Franke, W. W., Spring, H., Scheer, U. and Zerban, H. (1975b). Growth of the nuclear envelope in the vegetative phase of the green alga *Acetabularia. Journal of Cell Biology*, **66,** 681–689.

Frei, E. and Preston, R. D. (1961). Variants in the structural polysaccharides of algal cell walls. *Nature, London*, 939–943.

Frei, E. and Preston, R. D. (1968). Non-cellulosic structural polysaccharides in algal cell walls. III. Mannan in siphonous green algae. *Proceedings of the Royal Society*, B. **169,** 127–145.

Gall, J. G. (1966). Techniques for the study of lampbrush chromosomes. *In* "Methods in Cell Physiology", Vol. II, pp. 37–60 (D. M. Prescott, ed.), Academic Press, New York and London.

Georgiev, G. P. (1974). Precursor of mRNA (Pre-mRNA) and ribonucleoprotein particles containing Pre-mRNA. *In* "The Cell Nucleus", pp. 67–108 (H. Busch, ed.). Academic Press, New York and London.

Giaccone, G. (1968a). Raccolte di fitobentos nel Mediterraneo Orientale. *Giornale Botanico Italiano*, **102,** 217–228.

Giaccone, G. (1968b). Contributo allo studio fitosociologico dei popolamenti algali nel Mediterraneo Orientale. *Giornale Botanico Italiano*, **102,** 485–506.

Giaccone, G. (1969a). Associazioni algali e fenomeni secondari di vulcanismo nelle acque marine di Vulcano (Mar Tirreno). *Giornale Botanico Italiano*, **103,** 353–366.

Giaccone, G. (1969b). Raccolte di fitobentos sulla banchina continentale italiana. *Giornale Botanico Italiano*, **103,** 485–514.

Giaccone, G. (1970). The climax problem in the deep regions of the Mediterranean Sea. *Thalassia Jugoslavica*, **6,** 195–199.

Giaccone, G. (1971). Contributo allo studio dei popolamenti algali del Basso Tirreno. *Annali dell 'Università di Ferrara* (N. S.), Sez. IV, Bot., **4,** 17–43.

Giaccone, G. (1972). Struttura, ecologia e corologia dei popolamenti a Laminarie dello Stretto di Messina e del Mare di Alboran. *Memorie di Biologia Marina e di Oceanografia*, **2,** 37–59.

Giaccone, G. and Pignatti, S. (1967). Studi sulla produttività primaria del fitobentos nel Golfo di Trieste. II. *La Nova Thalassia*, **3,** (2), 28 pp.

Giaccone, G., Scammacca, B., Cinelli, F., Sartoni, G. and Furnari, G. (1972). Studio preliminare sulla tipologia della vegetazione sommersa del Canale di Sicilia e isole vicine. *Giornale Botanico Italiano*. **106,** 211–229.

Giardina, G. (1954). Role of the nucleus in the maintenance of the protein level in the alga *Acetabularia mediterranea*. *Experientia*, **10,** 215.

Gibor, A. (1965). Surviving cytoplasts *in vitro*. *Proceedings of the National Academy of Science*, **54,** 1527–1531.

Gibor, A. (1966). *Acetabularia:* a useful giant cell. *Scientific American*, **215,** 118–124.

Gibor, A. (1967). DNA synthesis in chloroplasts. *In* " Biochemistry of Chloroplasts ". Vol. 2, pp. 321–328. (T. W. Goodwin, ed.). Academic Press, New York and London.

Gibor, A. (1972). Semiautonomy in chloroplasts. *Protoplasma*, **75,** 477.

Gibor, A. (1973a). Observations on the sterile whorls of *Acetabularia*. *Protoplasma*, **78,** 195–202.

Gibor, A. (1973b). *Acetabularia:* Physiological role of their deciduous organelles. *Protoplasma*, **78,** 461–465.

Gibor, A. and Izawa, M. (1963). The DNA content of the chloroplasts of *Acetabularia*. *Proceedings of the National Academy of Science*, **50,** 1164–1169.

Gilbert, W. J. (1943). Studies on Philippine *Chlorophyceae*. I. The *Dasycladaceae*. *Publication of the Michigan Academy of Science, Arts and Letters*, **28,** 15–35.

Gläsel, R. M. and Zetsche, K. (1975). ^{36}Chloride fluxes of *Acetabularia*. *Protoplasma*, **83,** 175.

Glory, M. and Vanden Driessche, T. (1976). Circadian rhythm of polysaccharide synthesis in *Acetabularia*. *Archives Internationales de Physiologie et de Biochimie*, **84,** 48–50.

Goffeau, A. (1969a). Incorporation of amino acids into the soluble and membrane bound proteins of chloroplasts isolated from enucleated *Acetabularia*. *Biochimica et Biophysica Acta*, **174,** 340–350.

Goffeau, A. (1969b). Régulation de la synthèse et de l'activité de protéines appartenant aux structures membranaires subcellulaires. Dissertation, Bruxelles.

Goffeau, A. (1970). Amino acids incorporation by chloroplasts isolated from anucleate *Acetabularia*. *In* " Biology of *Acetabularia* " (J. Brachet and S. Bonotto, eds.), pp. 239–254. Academic Press, New York and London.

Goffeau, A. and Brachet, J. (1965). Deoxyribonucleic acid-dependent incorporation of amino-acids in the proteins of the chloroplasts isolated from anucleate *Acetabularia* fragments. *Biochimica et Biophysica Acta*, **95,** 302–313.

Gradmann, D. (1970a). Einfluss von Licht, Temperatur und Aussenmedium auf das elektrische Verhalten von *Acetabularia crenulata*. Dissertation, Tübingen.

Gradmann, D. (1970b). Einfluss von Licht, Temperatur und Aussenmedium auf des elektrische Verhalten von *Acetabularia crenulata*. *Planta*, **93**, 323–353.

Gradmann, D. and Bentrup, F. W. (1970). Light-induced membrane potential charges and rectification in *Acetabularia*. *Naturwissenschaften*, **57**, 46.

Gradmann, D. and Bokeloh, G. (1975). Energy consumption of the electrogenic pump in *Acetabularia mediterranea*. *Protoplasma*, **83**, 172.

Green, B. R. (1972). Studies on *Acetabularia* chloroplast DNA. *Protoplasma*, **75**, 478.

Green, B. R. (1973a). Evidence for the ocurrence of meiosis before cyst formation in *Acetabularia mediterranea* (*Chlorophyceae*, Siphonales). *Phycologia*, **12**, 233–235.

Green, B. R. (1973b). The genetic potential of *Acetabularia* chloroplasts. *Journal of Cell Biology*, **59**, 123a.

Green, B. R. (1973c). Small circular DNA molecules associated with *Acetabularia* chloroplasts. *Proceedings of the Canadian Federation of Biological Societies*, **17**, 20.

Green, B. R. and Burton, H. (1970). *Acetabularia* chloroplast DNA: electron microscopic visualization. *Science, New York*, **188**, 981–982.

Green, B. R., Burton, H., Heilporn, V. and Limbosch, S. (1970). The cytoplasmic DNA's of *Acetabularia mediterranea*: their structure and biological properties. *In* "Biology of *Acetabularia*" (J. Brachet and S. Bonotto, eds.), pp. 35–59. Academic Press, New York and London.

Green, B. R., Heilporn, V., Limbosch, S., Bolouhère, M. and Brachet, J. (1967). The cytoplasmic DNAs of *Acetabularia mediterranea*. *Proceedings of the National Academy of Science*, **58**, 1351–1358.

Green, B. R. and Padmanabhan, U. (1975). Kinetic complexity of *Acetabularia* chloroplast DNA. *Journal of Cell Biology*, **67**, 144a.

Green, B. R., Padmanabhan, U. and Muir, B. L. (1975). The kinetic complexity of *Acetabularia* chloroplast DNA. *XII International Botanical Congress, Leningrad. Abstracts*, **2**, 403.

Gušić, I. (1970). The algal genera *Macroporella*, *Salpingoporella* and *Pianella* (*Dasycladaceae*). *Taxon*, **19**, 257–261.

Hämmerling, J. (1931). Entwicklung und Formbildungsvermögen von *Acetabularia mediterranea*. I. Die normale Entwicklung. *Biologisches Zentralblatt*, **51**, 633–647.

Hämmerling, J. (1932). Entwicklung und Formbildungsvermögen von *Acetabularia mediterranea*. II. Das Formbildungsvermögen kernhaltiger und kernloser Teilstücke. *Biologisches Zentralblatt*, **52**, 42–61.

Hämmerling, J. (1934a). Über formbildende Substanzen bei *Acetabularia mediterranea*, ihre räumliche und zeitliche Verteilung und ihre Herkunft. *Archiv für Entwicklungsmechanik der Organismen*, **131**, 1–81.

Hämmerling, J. (1934b). Entwicklungsphysiologische und genetische Grundlagen der Formbildung bei der Schirmalge *Acetabularia*. *Naturwissenschaften*, **22**, 829–836.

Hämmerling, J. (1934c). Regenerationsversuche an kernhaltigen und kernlosen Zellteilen von *Acetabularia Wettsteinii*. *Biologisches Zentralblatt*, **54**, 650–665.

Hämmerling, J. (1934d). Über die Geschlechtsverhältnisse von *Acetabularia mediterranea* und *Acetabularia Wettsteinii*. *Archiv für Protistenkunde*, **83**, 57–97.

Hämmerling, J. (1935). Über Genomwirkungen und Formbildungsfähigkeit bei *Acetabularia*. *Archiv für Entwicklungsmechanik der Organismen*, **132,** 424–462.

Hämmerling, J. (1936). Studien zum Polaritätsproblem. *Zoologische Jahrbücher*, **56,** 441–483.

Hämmerling, J. (1939). Über die Bedingungen der Kernteilung und der Zystenbildung bei *Acetabularia mediterranea*. *Biologisches Zentralblatt*, **59,** 158–193.

Hämmerling, J. (1940). Transplantationsversuche zwischen *Acetabularia mediterranea und Acetabularia crenulata*. *Note dell' Istituto Italo-Germanico di Biologia Marina di Rovigno d' Istria* **2,** 3–7.

Hämmerling, J. (1943a). Entwicklung und Regeneration von *Acetabularia crenulata*. *Zeitschrift für induktive Abstammangs- und Vererbungslehre* **81,** 84–113.

Hämmerling, J. (1943b). Ein- und zweikernige Transplantate zwischen *Acetabularia mediterranea und Acetabularia crenulata*. *Zeitschrift für induktive Abstammangs- und Vererbungslehre*, **81,** 114–180.

Hämmerling, J. (1944). Zur Lebenweise, Fortpflanzung und Entwicklung verschiedener *Dasycladaceae*. *Archiv für Protistenkunde*, **97,** 7–56.

Hämmerling, J. (1946a). Neue Untersuchungen über die physiologischen und genetischen Grundlagen der Formbildung. *Naturwissenschaften*, **11,** 361–365.

Hämmerling, J. (1946b). Dreikernige Transplantate zwischen *Acetabularia crenulata* und *mediterranea*. *Zeitschrift für Naturforschung*, **1,** 337–342.

Hämmerling, J. (1952). Über die Fortpflanzung von *Cymopolia*, nebst Bemerkungen über *Neomenis* und *Dasycladus*. *Biologisches Zentralblatt*, **71,** 1–10.

Hämmerling, J. (1953). Nucleocytoplasmic relationships in the development of *Acetabularia*. *International Revue of Cytology*, **2,** 475–498.

Hämmerling, J. (1955a). Über mehrkernige Acetabularien und ihre Entstehung. *Biologisches Zentralblatt*, **74,** 420–427.

Hämmerling, J. (1955b). Neuere Versuche über Polarität und Differenzierung bei *Acetabularia*. *Biologisches Zentralblatt*, **74,** 545–554.

Hämmerling, J. (1956). Wirkungen von UV- und Röntgenstrahlen auf kernlose und kernhaltige Teile von *Acetabularia*. *Zeitschrift für Naturforschung*, **11b,** 217–221.

Hämmerling, J. (1957). Nucleus and Cytoplasm in *Acetabularia*. *8th International Congress on Botany, Paris, Comm. Sect.*, **10,** 87–103.

Hämmerling, J. (1958). Über die wechselseitige Abhängigkeit von Zelle und Kern. *Zeitschrift für Naturforschung*, **13b,** 440–448.

Hämmerling, J. (1963a). Nucleocytoplasmic interactions in *Acetabularia* and other cells. *Annual Review of Plant Physiology*, **14,** 65–92.

Hämmerling, J. (1963b). The role of the nucleus in differentiation in *Acetabularia*. *Symposia of the Society of Experimental Biology*, **17,** 127–137.

Hämmerling, J. (1964). Gibt es bei Dasycladaceen Zoosporen? *Ann. Biol.*, **3,** 33–35.

Hämmerling, J. and Hämmerling, Ch. (1959). Kernaktivität bei aufgehobener Photosynthese. *Planta*, **52,** 516–527.

Hämmerling, J. and Stich, H. (1954). Über die Aufnahme von ^{32}P in kernhaltige und kernlose Acetabularien. *Zeitschrift für Naturforschung*, **9b,** 149–155.

Hämmerling, J., and Stich, H. (1956a). Einbau und Ausbau von ^{32}P in Nukleolus, nebst Bemerkungen über intra- und extra-nukleäre Proteinsynthese. *Zeitschrift für Naturforschung*, **11b,** 158–161.

Hämmerling, J. and Stich, H. (1956b). Abhängigkeit des ^{32}P Einbaues in den Nukleolus von Energiezustand der Cytoplasmas, sowie vorläufige Versuche über Kernwirkungen während der Abbauphase des Kernes. *Zeitschrift für Naturforschung*, **11b**, 162–165.

Hämmerling, J. and Werz, G. (1958). Über den Wuchsmodus von *Acetabularia*. *Zeitschrift für Naturforschung*, **13b**, 449–454.

Hämmerling, J. and Zetsche, K. (1966). Zeitliche Steuerung der Formbildung von *Acetabularia*. Cytoplasmatische Regulation versus Genaktivierung. *Umschau*, **15**, 489–492.

Hämmerling, J., Clauss, H., Keck, K., Richter, G. and Werz, G. (1958). Growth and protein synthesis in nucleated and anucleated cells. *Experimental Cell Research*, Suppl., **6**, 210–226.

Hansen, U. P. and Gradmann, D. (1971). The action of sinusoidally modulated light on the membrane potential of *Acetabularia*. *Plant and Cell Physiology*, **12**, 335–348.

Hanson, B., Legros, F. and Conard, V. (1975). Influence de l'obscurité sur la réceptivité à l'insuline de l'algue unicellulaire *Acetabularia mediterranea*. *Comptes rendus des Séances de la Société de Biologie, Paris*, **169**, 254–256.

Harvey, W. H. (1857). Cited in: Valet, G. (1969a).

Harvey, W. H. (1858). Nereis Boreali-Americana. Part III. Chlorospermae. *Smithsonian Contributions to Knowledge*, **10**, 1–51.

Harvey, W. H. (1859). Characters of new algae, chiefly from Japan and adjacent regions, collected by Charles Wright in the North Pacific Exploring Expedition under Captain John Rodgers. *Proceedings of the American Academy of Arts and Sciences*, **4**, 327–335.

Haughton, P. M., Sellen, D. B. and Preston, R. D. (1969). The viscoelastic properties of algal cell walls. *Proceedings of the International Seaweed Symposium*, **6**, 453–462.

Heilporn-Pohl, V. and Brachet, J. (1966). Net DNA synthesis in anucleate fragments of *Acetabularia mediterranea*. *Biochimica et Biophysica Acta*, **119**, 429–431.

Heilporn, V. and Limbosch, S. (1969). Etuds des effets du bromure d'éthidium sur *Acetabularia mediterranea*. *Archives Internationales de Physiologie et Biochimie*, **77**, 383–384.

Heilporn, V. and Limbosch, S. (1970). Effects of hydroxyurea and ethidium bromide on *Acetabularia mediterranea*. In "Biology of *Acetabularia*" (J. Brachet and S. Bonotto, eds.) pp. 67–72. Academic Press, New York and London.

Heilporn, V. and Limbosch, S. (1971a). Recherches sur les acides désoxyribonucléiques *d'Acetabularia mediterranea*. *European Journal of Biochemistry*, **22**, 573–579.

Heilporn, V. and Limbosch, S. (1971b). Les effets du bromure d'éthidium sur *Acetabularia mediterranea*. *Biochimica et Biophysica Acta*, **240**, 94–108.

Heilporn-Pohl, V. and Limbosch-Rolin, S. (1969). Effets de l'hydroxyurée sur *Acetabularia mediterranea*. *Biochimica et Biophysica Acta*, **174**, 220–229.

Hellebust, J. A., Terborgh, J. and McLeod, G. C. (1967). The photosynthetic rhythm of *Acetabularia crenulata*. II. Measurements of photoassimilation of carbon dioxyd and the activities of enzymes of the reductive pentose cycle. *Biological Bulletin*, **133**, 670–678.

Herrmann, R. G. (1970). Multiple amounts of DNA related to the size of chloroplasts. I. An autoradiographic study. *Planta*, **90**, 80–96.

Herrmann, R. G. and Kowallik, K. V. (1970). Multiple amounts of DNA related to the size of chloroplasts. I. Comparison of electron-microscopic and auto-radiographic data. *Protoplasma*, **69,** 365–372.

Herth, W. (1975). Cellulose in *Acetabularia* cyst walls. *Protoplasma*, **83,** 175.

Herth, W., Franke, W. W. and van der Woude, W. J. (1972). Cytochalasin stops tip growth in plants. *Naturwissenschaften*, **59,** 38–39.

Herth, W., Kuppel, A. and Franke, W. W. (1975). Cellulose in *Acetabularia* cyst walls. *Journal of Ultrastructure Research*, **50,** 289–292.

Heyn, R. F., Hermans, A. K. and Schilperoort, R. A. (1974). Rapid and efficient isolation of highly polymerized plant DNA. *Plant Science Letters*, **2,** 73–78.

Hoshaw, R. W. and West, J. A. (1971). Morphology and life histories. *In* " Selected Papers in Phycology " (J. R. Rosowski and B. C. Parker, eds.), pp. 153–158. University of Nebraska, Lincoln.

Hoursiangou-Neubrun, D. and Puiseux-Dao, S. (1974). Modifications du gradient apicobasal de la population plastidale chez l'*Acetabularia mediterranea*. *Plant Science Letters*, **2,** 209–219.

Hoursiangou-Neubrun, D. and Puiseux-Dao, S. (1975). Studies on the hetero-geneity of the plastid population in one *Acetabularia* cell. *XII. International Botanical Congress*, Leningrad, *Abstracts*, **2,** 293.

Hoursiangou-Neubrun, D., Puiseux-Dao, S., Dubacq, J.-P. and Borghi, H. (1972). Comparative studies on lamellar morphogenesis in nucleated and anucleated *Acetabularia mediterranea*. *Protoplasma*, **75,** 478–479.

Howe, M. A. (1901). Observations on the algal genera *Acicularia* and *Acetabulum*. *Bulletin of the Torrey Botany Club*, **28,** 321–334.

Howe, M. A. (1904). Notes on Bahamas Algae. *Bulletin of the Torrey Botany Club*, **31,** 93–100.

Howe, M. A. (1905a). Phycological studies. I. New *Chlorophyceae* from Florida and the Bahamas. *Bulletin of the Torrey Botany Club*, **32,** 241–252.

Howe, M. A. (1905b). Phycological studies. II. New *Chlorophyceae*, new *Rhodo-phyceae*, and miscellaneous notes. *Bulletin of the Torrey Botany Club*, **32,** 563–586.

Howe, M. A. (1909). Phycological studies. IV. The genus *Neomeris* and notes on the other Siphonales. *Bulletin of the Torrey Botany Club*, **36,** 75–104.

Howe, M. A. (1932). Marine algae from the Islands of Panay and Negros (Philip-pines) and Niuafoou (between Samoa and Fiji). *Journal of the Washington Academy of Science*, **22,** 167–170.

Hoyt, J. W. (1970). High molecular weight algal substances in the sea. *Marine Biology*, **7,** 93–99.

Iriki, Y. and Miwa, T. (1960). Chemical nature of the cell wall of the green algae, *Codium*, *Acetabularia* and *Halicoryne*. *Nature, London*, **185,** 178–179.

Issinger, O., Maass, I. and Clauss, H. (1971). Photosyntheseintensität der Stielregionen von *Acetabularia mediterranea*. *Planta*, **101,** 360–364.

Iwamura, T. and Kuwashima, S. (1964). Formation of adenosine 5′- triphosphate from polyphosphate by a cell-free extract from *Chlorella*. *Journal of General and Applied Microbiology*, **10,** 83–86.

Janowski, M. (1963). Incorporation de phosphore radioactif dans les acides ribonucléiques de fragments nucléés et anucléés d'*Acetabularia mediterranea*. *Archives Internationales de Physiologie et de Biochimie*, **71,** 819–820.

Janowski, M. (1965). Synthèse chloroplastique d'acide nucléiques chez *Acetabu-laria mediterranea*. *Biochimica et Biophysica Acta*, **103,** 399–408.

Janowski, M. (1966). Detection of ribosomes and polysomes in *Acetabularia mediterranea*. *Life Sciences*, **5,** 2113–2116.

Janowski, M. (1967). Incorporation d'uridine-^{3}H dans les ribosomes et dans les polyribosomes d'*Acetabularia mediterranea*. *Archives Internationales de Physiologie et de Biochime*, **75,** 172.

Janowski, M. (1969). Le métabolisme du RNA et des ribosomes chez *Acetabularia mediterranea*. Thèse, Université Libre de Bruxelles.

Janowski, M. and Bonotto, S. (1969). Les loges du chapeau d'*Acetabularia mediterranea*. *Bulletin de la Société royale de Botanique de Belgique*, **102,** 267–276.

Janowski, M. and Bonotto, S. (1970). A stable RNA species in *Acetabularia mediterranea*. *In* " Biology of *Acetabularia* ", (J. Brachet and S. Bonotto, eds.), pp. 1–34. Academic Press, New York and London.

Janowski, M., Bonotto, S. and Brachet, J. (1968). Cinétique de l'incorporation d'uridine-^{3}H dans les RNA d'*Acetabularia mediterranea*. *Archives Internationales de Physiologie et de Biochimie*, **76,** 934–935.

Janowski, M., Bonotto, S. and Bouloukhère, M. (1969). Ribosomes of *Acetabularia mediterranea*. *Biochimica et Biophysica Acta*, **174,** 525–535.

Joly, A. B., Cordeiro-Marino, N., Ugadim, Y., Yamaguishi-Tomita, N. and Pinheiro, F. (1965). New marine algae from Brazil. *Arq. Est. Biol. Univ. Cearà*, **5,** 79–92.

Kamiya, N. (1960). Physics and chemistry of protoplasmic streaming. *Annual Review of Plant Physiology*, **11,** 323–340.

Karakashian, M. W. and Schweiger, H. G. (1975). 80S protein synthesis provides a component of the *Acetabularia* circadian clock. *Journal of Cell Biology*, **67,** 200a.

Karakashian, M. W. and Schweiger, H. G. (1976). Evidence for a cycloheximide-sensitive component in the biological clock of *Acetabularia*. *Experimental Cell Research*, **98,** 303–312.

Keck, K. (1960). Nucleo-cytoplasmic interactions in the synthesis of species-specific proteins in *Acetabularia*. *Biochemical and Biophysical Research Communications*, **3,** 56–61.

Keck, K. (1961). Nuclear and cytoplasmic factors determining the species specificity of enzyme proteins in *Acetabularia*. *Annals of the New York Academy of Science*, **94,** 741–752.

Keck, K. (1963). The nuclear control of synthetic activities in *Acetabularia*. *Proceeding of the XVI International Congress of Zoology*, **3,** 203–207.

Keck, K. (1964). Culturing and experimental manipulation of *Acetabularia*. *In* " Methods in Cell Physiology " (D. M. Prescott, ed.). Vol **1,** pp. 189–213. Academic Press, New York and London.

Keck, K. (1969). Metabolism of enucleated cells. *International Review of Cytology*, **26,** 191–233.

Keck, K. and Choules, E. A. (1963). An analysis of cellular and subcellular systems which transform the species character of acid phosphatase in *Acetabularia*. *Journal of Cell Biology*, **18,** 459–469.

Keck, K. and Clauss, H. (1958). Nuclear control of enzyme synthesis in *Acetabularia*. *Botanical Gazette*, **120,** 43–49.

Kellner, G. and Werz, G. (1969). Die Feinstruktur des Augenfleckes bei *Acetabularia* Gameten und sein Verhalten nach der Gametenfusion. *Protoplasma*, **67,** 117–120.

Kirchmann, R. J. and Bonotto, S. (1970). Protein synthesis in irradiated *Acetabularia* and *Phaseolus*. *In* " Improving Plant Protein by Nuclear Techniques ", pp. 411–418. International Atomic Energy Agency, Vienna.

Kirchmann, R. and Bonotto, S. (1971). Pénétration et distribution du tritium dans l'algue marine *Acetabularia. Revue Internationale d'Oceanographic Médicale*, **24,** 138–139.

Kirchmann, R. and Bonotto, S. (1972). *Acetabularia*, a useful tool for research on pollution, *Protoplasma*, **75,** 479.

Kirchmann, R. J. and Bonotto, S. (1973a). Effets biologiques et biochimiques des rayonnements gamma sur l'algue marine *Acetabularia mediterranea.* In " Radioactive Contamination of the Marine Environment ", pp. 527–541. International Atomic Energy Agency, Vienna.

Kirchmann, R. and Bonotto, S. (1973b). Pénétration et distribution du tritium dans l'algue marine *Acetabularia mediterranea. Atti del 5° Colloquio Internazionale di Oceanografia medica, Messina*, pp. 325–333.

Kleinig, H. (1967). Die Bildung von Sekundärcarotinoiden in *Acetabularia mediterranea. Berichte der Deutschen botanischen Gesellschaft*, **79,** 126–130.

Kleinig, H. and Egger, K. (1967). Ketocarotinoidester in *Acetabularia mediterranea* Lam. *Phytochemistry*, **6,** 611–619.

Kloppstech, K. (1972). Ribonucleoprotein particles in *Acetabularia. Protoplasma*, **75,** 479–480.

Kloppstech, K. (1975a). Poly-A containing RNA in *Acetabularia. Protoplasma*, **83,** 177.

Kloppstech, K. (1975b). Cytosol-ribosomes in *Acetabularia*: Distribution and migration within the cell. *Protoplasma*, **83,** 176–177.

Kloppstech, K. and Schweiger, H. G. (1971). Nuclear coded ribosomal proteins in chloroplasts of *Acetabularia. Abstract Communication of the 7th Meeting of the FEBS, Varna*, p. 155.

Kloppstech, K. and Schweiger, H. G. (1972). Nuclear dependence of chloroplast ribosomal proteins in *Acetabularia. In* " Biology and Radiobiology of Anucleate Systems ", (S. Bonotto, R. Goutier, R. Kirchmann and J. R. Maisin, eds.), Vol. 2: Plant Cells, pp. 127–133. Academic Press, New York and London.

Kloppstech, K. and Schweiger, H. G. (1973a). Nuclear genome codes for chloroplast ribosomal proteins in *Acetabularia.* I. Isolation and characterization of chloroplast ribosomal particles. *Experimental Cell Research*, **80,** 63–68.

Kloppstech, K. and Schweiger, H. G. (1973b). Nuclear genome codes for chloroplast ribosomal proteins in *Acetabularia.* II. Nuclear transplantation experiments. *Experimental Cell Research*, **80,** 69–78.

Kloppstech, K. and Schweiger, H. G. (1974). The site of synthesis of chloroplast ribosomal proteins. *Plant Science Letters*, **2,** 101–105.

Kloppstech, K. and Schweiger, H. G. (1975a). 80 S ribosomes in *Acetabularia major.* Distribution and transportation within the cell. *Protoplasma*, **83,** 27–40.

Kloppstech, K. and Schweiger, H. G. (1975b). Polyadenylated RNA from *Acetabularia. Differentiation*, **4,** 115–123.

Köhler, K. (1957). Neue Untersuchungen über die Sexualität bei *Dasycladus* und *Chaetomorpha. Archiv für Protistenkunde*, **102,** 209–217.

Koop, H. U. (1975a). Germination of cysts in *Acetabularia mediterranea. Protoplasma*, **84,** 137–146.

Koop, H. U. (1975b). Influence of light on the germination of cysts of *Acetabularia mediterranea*. *Protoplasma*, **83,** 177–178.

Koop, H. U. (1975c). Über den Ort der Meiose bei *Acetabularia mediterranea*. *Protoplasma*, **85,** 109–114.

Koop, H. U. (1975d). Multinuclear stages of the life cycle of *Acetabularia mediterranea*. *Protoplasma*, **86,** 351–362.

Kornmann, P. (1938). Zur Entwicklungsgeschichte von *Derbesia* und *Halicystis*. *Planta*, **28,** 464–470.

Kornmann, P. (1970). Advances in marine phycology on the basis of cultivation. *Helgoländer Wissenschaftliche Meeresuntersuchungen*, **20,** 39–61.

Koster, J. (1937). Algues marines des Ilots Itu-Aba, Sand Caye et Nam-Yit, situés à l'ouest de l'ile Palawan. *Blumea*, Suppl. **1** (J. J. Smith Jubilee Volume), 219–228.

Kuntze, O. (1891). "Revisio Generum Plantarum", pars 2, pp. 877–881. Leipzig.

Kützing, F. T. (1843). "Phycologia Generalis oder Anatomie, Physiologie und Systemkunde der Tange", pp. 311–318. Leipzig.

Kützing, F. T. (1849). "Species Algarum", pp. 508–511. Leipzig.

Kützing, F. T. (1856). "Tabulae Phycologicae", Vol. 6, pp. 32–33. Nordhausen.

Kützing, F. T. (1863). "Diagnosen und Bemerkungen zu dreiundsiebenzig neuen Algenspecies", 19 pp. Nordhausen.

Kützing, F. T. (1866). "Tabulae Phycologicae", Vol. 16, plate 1. Nordhausen.

Lamouroux, J. V. (1812). Extrait d'un mémoire sur la classification des polypiers coralligènes non entièrement pierreux. *Nouv. Bull. Soc. Philom., Paris*, **3,** 181–188.

Lamouroux, J. (1816). "Histoire de polypiers coralligènes flexibles, vulgairement nommés Zoophytes", pp. 216–317. Caen.

Lamouroux, J. (1821). "Exposition méthodique des genres de l'ordre des polypiers", pp. 14–20. Paris.

Lateur, L. (1963). Une technique de culture pour l'*Acetabularia meditarranea*. *Rev. Algol.* (n.s.), **1,** 26–37.

Lateur, L. and Bonotto, S. (1973). Culture of *Acetabularia mediterranea* in the laboratory. *Bulletin de la Société royale de Botanique de Belgique*, **106,** 17–38.

Lane, C. D., Marbaix, G. and Gurdon, J. B. (1972). 9S haemoglobin messenger RNA from reticulocytes and its assay in living frog cells. *In* "Biology and Radiobiology of Anucleate Systems" (S. Bonotto, R. Goutier, R. Kirchmann and J. R. Maisin, eds.). Vol. 1. Bacteria and Animal cells, pp. 101–113. Academic Press, New York and London.

Legros, F. and Conard, V. (1973). Effects of insulin on oxygen uptake of *Acetabularia mediterranea*. *Hormone Research*, **4,** 107–113.

Legros, F. and Conard, V. (1974). Radioiodinated insulin fixation to a marine alga: *Acetabularia mediterranea*. *Archives Internationales de Physiologie et de Biochimie*, **82,** 359–361.

Legros, F., Saines, M. and Conard, V. (1967). Competitive effects of insulin and ouabain on metabolism of *Acetabularia mediterranea*. *Archives Internationales de Physiologie et de Biochimie*, **81,** 745–754.

Legros, F., Rogister, C. and Conard, V. (1971). Modifications of oxygen uptake of *Acetabularia mediterranea* by cristalline and oxidized insulins. *In* "Structure-activity Relationships of Protein and Polypeptide Hormones" (M. Margoulies and F. C. Greenwood, eds.), *Excerpta Medica, International Congress Series*, **241,** 542–543.

Legros, F., Saines, M., Renard, M. and Conard, V. (1972). Influence of insulin and aldosterone and the metabolism of *Acetabularia mediterranea. Protoplasma*, **75,** 480–481.

Legros, F., Saines, M. and Conard, V. (1973). Action du 3′:5′-AMP cyclique et de la théophylline sur la respiration d'*Acetabularia mediterranea. Archives Internationales de Physiologie et de Biochimie*, **81,** 164–166.

Legros, F., Saines, M., Renard, M. and Conard, V. (1974). Nuclear influence on oxygen uptake modifications induced by insulin and aldosterone in *Acetabularia mediterranea. Plant Science Letters*, **2,** 339–345.

Legros, F., Dumont, I., Hanson, B., Jeanmart, J., and Conard, V. (1975a). A metabolic influence on ^{125}I-insulin fixation to *Acetabularia mediterranea. Protoplasma*, **83,** 178.

Legros, F., Hanson, B., Dumont, I., Jeanmart, J. and Conard, V. (1975b). Nuclear influence on the membrane fixation of insulin. Studies on *Acetabularia mediterranea. In* " Molecular Biology of Nucleocytoplasmic Relationships " (S. Puiseux-Dao, ed.) pp. 299—304. Elsevier, Amsterdam and New York.

Legros, F., Uytdenhoef, P., Dumont, I., Hanson, B., Jeanmart, J., Massant, B. and Conard, V. (1975c). Specific binding of insulin to the unicellular alga *Acetabularia mediterranea. Protoplasma*, **86,** 119–134.

Leitbeg, H. (1887). Die Inkrustation des Membran von *Acetabularia. Sitzungberichte der matematischnaturwissenschaftlichen Classe der Kaiserlichen Akademie der Wissenschaften*, **96,** 13–37.

Levring, T., Hoppe, H. A., and Schmid, O. J. (1969). " Marine algae. A Survey of Research and Utilization." Cram, De Gruyter and Co., Hamburg.

Lewin, R. A. (1970). Marine algae: biology of *Acetabularia. Science, New York*, **170,** 725–726.

Liddle, L., Berger, S. and Schweiger, H. G. (1975). Ultrastructure of the nucleus of *Batophora oerstedii* (*Chlorophyta, Dasycladaceae*). *Journal of Cell Biology*, **67,** 242a.

Linnaeus, C. (1758). " Systema Naturae," Vol. 1, pp. 789–798. 10th Ed. Holmiae.

Linnaeus, C. (1767). " Systema Naturae ", Vol. 1, 324 pp., 12th Ed. Holmiae.

Loni, M. C. and Bonotto, S. (1971). Statistical studies on cap morphology in normal and branched *Acetabularia mediterranea. Archives de Biologie, Liège*, **82,** 225–244.

Lukanidin, E. M., Olsnes, S. and Pihl, A. (1972a). Antigenic difference between informofers and protein bound to polyribosomal RNA from rat liver. *Nature New Biology*, **240,** 90–92.

Lukanidin, E. M., Zalmanzon, E. S., Komaromi, L., Samarina, O. P. and Georgiev, G. P. (1972b). Structure and function of informofers. *Nature New Biology*, **238,** 193–197.

Lurquin, P., Baeyens, W. and Bonotto, S. (1972a). Effects of acriflavine, ethidium bromide and rifampicin on nucleic acids and protein synthesis in anucleate fragments of *Acetabularia mediterranea. In* " Biology and Radiobiology of Anucleate Systems ", (S. Bonotto, R. Goutier, R. Kirchmann and J. R. Maisin, eds.), Vol. 2: Plant Cells, pp. 321–327. Academic Press, New York and London.

Lurquin, P., Baeyens, W. and Bonotto, S. (1972b). DNA synthesis in *Acetabularia mediterranea. Protoplasma*, **75,** 481.

Lurquin, P., Baeyens, W. and Bonotto, S. (1974). Rapid preparation of algal DNA. *VIIIth International Seaweed Symposium, Bangor, U.K.*, 17–24 August, 1974. Abstracts of Papers, B35.

Lüttke, A., Rahmsdorf, U. and Schmid, R. (1976). Heterogeneity in chloroplasts of Siphonacious Algae as compared with higher plant chloroplasts. *Zeitschrift fur Naturforschung*, **31c**, 108–110.

Mackie, W. and Preston, R. D. (1868). The occurence of mannan microfibrils in the green algae *Codium fragile* and *Acetabularia crenulata*. *Planta*, **79,** 249–253.

Mangenot, G. and Nardi, R. (1931). Les plastes de l'*Acetabularia mediterranea* Lamour. *Travaux cryptogamiques*, *Paris*, **11,** 459–463.

Margulies, M. M. (1971). Concerning the sites of synthesis of proteins of chloroplast ribosomes and of fraction I protein (Ribulose, 1,5–diphosphate carboxylase). *Biochemical and Biophysical Research Communications*, **44,** 539–545.

Maschlanka, H. (1943a). Zytologische Untersuchungen an Algen aus der Familie der Dasycladaceen. *Naturwissenschaften*, **31,** 548–549.

Maschlanka, H. (1943b). Zweikernige Transplantate zwischen *Acetabularia crenulata* und *Acicularia Schenkii*. *Naturwissenschaften*, **31,** 549.

Maschlanka, H. (1946). Kernwirkungen in artgleichen und artverschiedenen *Acetabularia*-Transplantaten. *Biologisches Zentralblatt*, **65,** 167–176.

Mazé, H. and Schram, A. (1870.) " Essai de classification des algues de la Guadeloupe ", 2nd Ed., pp. 83–84. Basse-Terre.

Mergenhagen, D. (1972a). A method for recording a circadian rhythm in a single cell. *Protoplasma*, **75,** 481–482.

Mergenhagen, D. (1972b). Effects of inhibitors on a circadian rhythm in a single cell. *Protoplasma*, **75,** 482.

Mergenhagen, D. and Schweiger, H. G. (1971). A method for recording a circadian rhythm in a single cell and in cell fragments. *In* " Proceedings of the First European Biophysics Congress ", (E. Broda, A. Locker, H. Springer-Lederer, eds.), pp. 497–501. Verlag der Wiener Medzinischen Akademie, Baden.

Mergenhagen, D. and Schweiger, H. G. (1973). Recording the oxygen production of a single *Acetabularia* cell for a prolonged period. *Experimental Cell Research*, **81,** 360–364.

Mergenhagen, D. and Schweiger, H. G. (1974). Circadian rhythmicity: Does intercellular synchronization occur in *Acetabularia? Plant Science Letters*, **3,** 387–389.

Mergenhagen, D. and Schweiger, H. G. (1975a). Circadian rhythm of oxygen evolution in cell fragments of *Acetabularia mediterranea*. *Experimental Cell Research*, **92,** 127–130.

Mergenhagen, D. and Schweiger, H. G. (1975b). The effect of different inhibitors of transcription and translation on the expression and control of circadian rhythm in individual cells of *Acetabularia*. *Experimental Cell Research*, **94,** 321–326.

Mets, L. J. and Bogorad, L. (1971). Mendelian and uniparental alterations in erythromycin binding by plastid ribosomes. *Science, New York*, **174,** 707–709.

Miwa, I., Iriki, Y. and Suzuki, T. (1961). Mannan and xylan as essential cell wall constituents of some siphoneous green algae. *Chimie et physico-chimie des principes immédiats tirés des algues. Coll. Int. Centre Nat. Rech. Sci.*, **103,** 135–144.

Möbius, M. (1889). Bearbeitung der von H. Schenck in Brasilien gesammelten Algen. *Hedwigia*, **28,** 309–347.

Moore, D. F. (1971a). *In vivo* and *in vitro* pigment synthesis in normal and enucleate algae. pp. 1–136. Thesis, Case Western Reserve University, U.S.A.

Moore, F. D. (1971b). Pigment synthesis by chloroplasts isolated from *Acetabularia. 11th Annual Meeting of the American Society of Cell Biology, Abstract.*

Moore, F. D. (1972). Pigment synthesis in nucleate and enucleate *Acetabularia mediterranea. Protoplasma*, **75,** 482–483.

Moorjani, S. A. (1970). Notes on Kenya *Acetabularia* Lamouroux, (Chlorophyta). *Journal of the East Africa Natural History Society and National Museum*, **28,** 47–52.

Munier-Chalmas, M. (1877). Observations sur les algues calcaires appartenant au groupe des Siphonées verticillées (Dasycladées Harv.) et confondues avec les Foraminifères. *Comptes rendus hebdomadaires des Séances de l'Académie des Sciences, Paris*, **85,** 814–818.

Munier-Chalmas, M. (1879). Observations sur les algues calcaires confondues avec les foraminifères et appartenant au groupe des Siphonées dichotomes. *Bulletin de la Société Geologique*, Series III, **7,** 661–670.

Nägeli, C. (1863). Sphaerocrystalle in *Acetabularia. Botanische Mitteilungen aus den Tropen*, **1,** 206–216.

Naora, H., Richter, G. and Naora, H. (1959). Further studies on the synthesis of RNA in enucleate *Acetabularia mediterranea. Experimental Cell Research*, **16,** 434–436.

Naora, H., Naora, H. and Brachet, J. (1960). Studies on independent synthesis of cytoplasmic ribonucleic acids in *Acetabularia mediterranea. Journal of General Physiology*, **43,** 1083–1102.

Nasr, A. H. (1939). On the phototropism of *Acetabularia calyculus* Quoy et Gaimard. *Revue algologique*, **11,** 347–350.

Netrawali, M. S. (1970). Chemical radioprotection in the absence of the nucleus in *Acetabularia mediterranea. Radiation Botany*, **10,** 365–370.

Niemeyer, R. and Richter, G. (1969). Schnellmarkierte Polyphosphate und Metaphosphate bei der Blaualge *Anacystis nidulans. Archiv für Mikrobiologie*, **69,** 54–59.

Niemeyer, R., and Richter, G. (1972). Rapidly labelled polyphosphates in *Acetabularia. In* " Biology and Radiobiology of Anucleate Systems ". (S. Bonotto, R. Goutier, R. Kirchmann and J. R. Maisin, eds.) Vol. 2, Plant cells, pp. 225–236. Academic Press, New York and London.

Nizzamuddin, M. (1964). The life history of *Acetabularia möbii* Solm-Laubach. *Annals of Botany*, N. S. **28,** 77–81.

Novák, B. (1972). An electrophysiological study on the spatial differentiation of plant cells. pp. 1–81. Doctoral thesis, University of Tübingen.

Novák, B. (1975). Sustained current pulses and transcellular current during the regeneration of *Acetabularia mediterranea. Protoplasma*, **83,** 178.

Novák, B. and Bentrup, W. (1972a). An electrophysiological study of regeneration in *Acetabularia mediterranea. Planta*, **108,** 227–244.

Novák, B. and Bentrup, F. W. (1972b). Transcytoplasmic electric currents through regenerating stalk segments of *Acetabularia mediterranea. Protoplasma*, **75,** 483.

Novák, B. and Sironval, C. (1975). Inhibition of regeneration of *Acetabularia mediterranea* enucleated posterior stalk segments by electrical isolation. *Plant Science Letters*, **5**, 183–188.

Novák, B. and Sironval, C. (1976). Circadian rhythm of the transcellular current in regenerating enucleated posterior stalk segments of *Acetabularia mediterranea*. *Plant Science Letters*, **6**, 273–283.

Okamura, K. (1907). Icones of Japanese Algae. Vol. 1, pp. 217–228. Tokyo.

Okamura, K. (1912). Icones of Japanese Algae. Vol. 2, pp. 177–179. Tokyo.

Okamura, K. (1913). Icones of Japanese Algae. Vol. 3, pp. 21–22. Tokyo.

Okamura, K. (1916). Nippon Sorui Meii. cited in Valet, G. (1969a).

Okamura, K. (1932). Icones of Japanese Algae. Vol. 6, pp. 68–74. Tokyo.

Olszewska, M. J. and Brachet, J. (1960). Incorporation de la DL-méthionine-^{35}S dans l'algue *Acetabularia mediterranea*. *Archives Internationales de Physiologie et de Biochimie*, **68**, 693–694.

Olszewska, M. J. and Brachet, J. (1961). Incorporation de la DL-méthionine-^{35}S dans les fragments nucléés et anucléés d'*Acetabularia mediterranea*. *Experimental Cell Research*, **22**, 370–380.

Olszewska, M., de Vitry, F. and Brachet, J. (1961). Influence d'irradiations UV localisées sur l'incorporation de l'adénine-8-^{14}C, de l'uridine-^{3}H et de la DL-méthione-^{35}S dans l'algue *Acetabularia mediterranea*. *Experimental Cell Research*, **24**, 58–63.

Pai, H. S., Dehm, D., Schweiger, M., Rahmsdorf, H. J., Ponta, H., Hirsch-Kauffmann, M. and Schweiger, H. G. (1975a) Protein Kinase of *Acetabularia*. *Protoplasma*, **83**, 179.

Pai, H. S., Dehm, D., Schweiger, M., Rahmsdorf, H. J., Ponta, H., Hirsch-Kauffmann, M. and Schweiger, H. G. (1975b). Protein Kinase of *Acetabularia*. *Protoplasma*, **85**, 209–218.

Pallas, P. S. (1766). Elenchus Zoophytorum, pp. 430–432. Frankfurt am Main.

Parkinson, J. (1640). Theatrum botanicum, p. 1303. Thos. Cotes, London.

Pia, J. (1920). Die Siphoneae verticillatae vom Karbon bis zur Kreide. *Abhandlungen der Zoologisch-botanischen Gesellschaft in Wien*, **11**, 1–263.

Pia, J. (1926). Pflanzen als Gesteinsbildner. pp. 105–149. Berlin.

Pia, J. (1927). Dasycladaceae (Siphoneae verticillatae). *In* " Handbuch der Paläo-Botanik " (M. Hirmer, ed.), pp. 61–87. München und Berlin.

Pignatti, S. and Giaccone, G. (1967). Studi sulla produttività primaria del fitobentos nel Golfo di Trieste. I. Flora sommersa del Golfo di Trieste. *Nova Thalassia*, **3**, 1, 17 pp.

Pignatti, S., de Cristini, P. and Rizzi, L. (1967). Le associazioni algali della Grotta delle Viole nell' Isola di S. Domino (Is. Tremiti). *Giornale Botanico Italiano*, **101**, 117–126.

Popa, E. and Dragastan, D. (1973). Alge si foraminifere Triasice (Anisian-ladinian) din estul Padurii Craiului (Muntii Apuseni). *St. cerc. geol. geofiz. geogr., Seria geologie*, **18**, 425–442.

Pressman, E. K., Levin, I. M. and Sandhakchiev, L. S. (1972). Assembly of *Acetabularia mediterranea* cell from nucleus, cytoplasm, and cell wall. *Protoplasma*, **75**, 484.

Preston, R. D. (1968). Plants without cellulose. *Scientific American*, **218**, 6, 102–108.

Provasoli, L., McLaughlin, J. J. A. and Droop, M. R. (1957). The development of artificial media for marine algae. *Archiv für Mikrobiologie*, **25**, 392–428.

Puiseux-Dao, S. (1957). Comportement de fragments anucléés de *Batophora Oerstedi* J. Ag. (Dasycladacée) dans l'eau de mer contenant soit un acide aminé, soit une auxine, *Comptes rendus hebdomadaires des Séances de l'Académie des Sciences, Paris*, **245**, 2371–2374.

Puiseux-Dao, S. (1958a). Action de la ribonucléase sur le noyau de *Batophora Oerstedii* J. Ag. (Dasycladacées). As above, **246**, 1076–1079.

Puiseux-Dao, S. (1958b). A propos du comportement du noyau chez le *Batophora Oerstedii* J. Ag. (Dasycladacées) cultivé, soit à l'obscurité, soit en présence de ribonucléase. As above, **246**, 2286–2288.

Puiseux-Dao, S. (1959). Endomitoses dans le noyau primaire du *Batophora oerstedii* Ag. (Dasycladacées). As above, **249**, 1139–1141.

Puiseux-Dao, S. (1960). Le comportement du noyau chez le *Batophora oerstedii* Ag. (Dasycladacées) privé de lumière; ce que l'on peut en déduire sur la structure des nucléoles. As above, **250**, 176–178.

Puiseux-Dao, S. (1962). Recherches biologiques et physiologiques sur quelques Dasycladacées. *Rev. Gén. Bot.*, **819**, 409–503.

Puiseux-Dao, S. (1963). Les Acétabulaires, matériel de la laboratoire. *Année biologique*, **2**, 99–154.

Puiseux-Dao (1965). Morphologie et morphogenèse chez les Dasycladacées. Travaux dédiés à L. Plantefol, pp. 147–170. Masson, Paris.

Puiseux-Dao, S. (1966a). L'ultrastructure et la division des plastes chez l'*Acetabularia mediterranea*, Dasycladacées. *Sixth International Congress on Electron Microscopy, Kyoto*, 377–378.

Puiseux-Dao, S. (1966b). Siphonales and Siphonocladales. *In* " The chromosomes of the algae ". (M. B. E. Godwin, ed.), Ch. 1, pp. 52–77, Edward Arnold, London.

Puiseux-Dao, S. (1967). L'unité plastidiale et sa réplication chez l'*Acetabularia mediterranea* placée dans des conditions expérimentales diverses. *Journal de Microscopie*, **6**, 78a.

Puiseux-Dao, S. (1968). Evolution de la population des plastes chez l'*Acétabularia mediterranea*, Dasycladacées. *Comptes rendus hebdomadaires des Séances de l'Académie des Sciences, Paris*, **266**, 1382–1384.

Puiseux-Dao, S. (1972). Problèmes de morphogenèse et biologie moléculaire chez les Acétabulaires. *Bulletin Société botanique de France*, Mémoires (Coll. Morphologie), 71–88.

Puiseux-Dao, S. (1970a). Le problème du contrôle nucléaire du fonctionnement des plastes chez l'*Acetabularia*. *Comptes rendus hebdomadaires des Séances de l'Académie des Sciences, Paris*, **270**, 358–361.

Puisuex-Dao, S. (1970b). *Acetabularia* and cell biology, 162 pp. Logos Press, London.

Puiseux-Dao, S. (1975a). Third Symposium on *Acetabularia*. *Protoplasma*, **83**, 167–183.

Puiseux-Dao, S. (1975b). Molecular Biology of Nucleocytoplasmic Relationships. 328p. Elsevier, Amsterdam and New York.

Puiseux-Dao, S., and Dazy, A. C. (1970a). La population plastidiale chez l'*Acetabularia mediterranea*: Ultrastructure et réplication de l'unité plastidiale, croissance de la population et contrôle nucléaire de cette croissance. *Bulletin Société phycologique de France*, no. 15, 80.

Puiseux-Dao, S. and Dazy, A. C. (1970b). Plastid structure and the evolution of plastids in *Acetabularia*. *In* " Biology of *Acetabularia* " (J. Brachet and S. Bonotto, eds.), pp. 111–122. Academic Press, New York and London.

Puiseux-Dao, S. and Gilbert, A. M. (1967). Rythme de réplication de l'unité plastidiale chez l'*Acetabularia mediterranea* placée dans diverses conditions d'éclairement. As above, **265,** 870–873.

Puiseux-Dao, S., Gibello, D. and Hoursiangou-Neubrun, D. (1967). Techniques de mise en évidence du DNA dans les plastes. *Comptes rendus hebdomadaires des Séances de l'Académie de Sciences, Paris,* **265,** 406–408.

Puiseux-Dao, S., Valet, G. and Bonotto, S. (1970). Greffes interspécifiques uninucléées, *Acetabularia mediterranea* et *A. peniculus* et mobilité des substances morphogénétiques dans le cytoplasme. *Comptes rendus hebdomadaires des Séances de l'Académie des sciences, Paris,* **271,** 1354–1357.

Puiseux-Dao, S., Dazy, A. C., Hoursiangou-Neubrun, D. and Borghi, H. (1972a). Observations on the apicobasal gradient of the plastidal population in *Acetabularia*. *Protoplasma,* **75,** 484–485.

Puiseux-Dao, S., Dazy, A. C., Hoursiangou-Neubrun, D. and Matthys, E. (1972b). First observations on nuclear control on plastidal membrane morphogenesis. *In* " Biology and Radiobiology of Anucleate Systems ", (S. Bonotto, R. Goutier, R. Kirchmann and J. R. Maisin, eds.). Vol. 2. pp. 101–125, Academic Press, New York and London.

Puiseux-Dao, S., Aksiyote-Benbasat, J. and Bonotto, S. (1972c). Effets biologiques de la rifampicine chez l'*Acetabularia mediterranea*. *Comptes rendus hebdomadaires des Séances de l'Académie des Sciences, Paris,* **274,** 1678–1681.

Puiseux-Dao, S., Marano, F., Levain, N., Borghi, H. and Bonotto, S. (1972d). The action of lindane on cellular morphogenesis and multiplication. ESNA Meeting, Budapest, Hungary, 26–29 September, 1972. Book of Abstracts, p. 42.

Quoy and Gaimard (1824). Genre Acétabulaire—*Acetabularia* Lamx., Acétabulaire à petit godet, *Acetabularia caliculus*. *In* " Zoologie du voyage de l'Uranie et la Physicienne " (Feycinet, ed.), p. 621. Paris.

Rafinesque-Schmaltz, C. S. (1810). " Caratteri di alcuni nuovi generi e nuove specie di animali e piante della Sicilia ", pp. 88–89. Palermo.

Raviv, V. and Lorch, J. (1970). *Verticilloporella*, a new mesozoic genus of *Dasycladaceae*, with discussion on *Munieria* and *Actinoporella*. *Israel Journal of Botany,* **19,** 225–235.

Reinbold, Th. (1902). Flora of Koh Chang. Part IV. Chlorophyceae. *Botanisk Tidsskrift,* **24,** 187–201.

Reuss, A. E. (1861). Über die fossiele Gattung *Acicularia* d'Arch. *Sitzung-berichte der Deutschen Akademie der Wissenschaften zu Wien,* Cl. 43, Abt. 1, 7–10.

Reuter, W. and Schweiger, H. G. (1969). Kernkontrollierte Lactatdehydrogenase in *Acetabularia*. *Protoplasma,* **68,** 357–368.

Richter, G. (1957). Zur Frage der RNS-Synthese in kernlosen Teilen von *Acetabularia*. *Naturwissenschaften,* **44,** 520–521.

Richter, G. (1958a). Das Verhalten der Plastidepigmente in kernlosen Zellen und Teilstücken von *Acetabularia mediterranea*. *Planta,* **52,** 259–275.

Richter, G. (1958b). Regeneration und RNS-Synthese bei der Einwirkung eines artfremden Zellkernes auf gealterte kernlose Zellteile von *Acetabularia mediterranea*. *Naturwissenschaften,* **45,** 629–630.

Richter, G. (1959a). Die Auswirkungen der Zellkern-Entfernung auf die Synthese von Ribonucleinsäure und Cytoplasmaproteinen bei *Acetabularia mediterranea*. *Biochimica et Biophysica Acta*, **34,** 407–419.

Richter, G. (1959b). Die Auslösung kerninduzierter Regeneration bei gealterten kernlosen Zellteilen von *Acetabularia* und ihre Auswirkungen auf die Synthese von Ribonucleinsäure und Cytoplasmaproteinen. *Planta*, **52,** 554–564.

Richter, G. (1959c). Das Verhalten von Ribonucleinsäure und löslichen Cytoplasma–Proteinen in UV-bestrahlten kernhaltigen und kernlosen Zellen von *Acetabularia*. *Zeitschrift für Naturforschung*, **14b,** 100–104.

Richter, G. (1962). Die Wirkung von blauer und roter Strahlung auf die Morphogenese von *Acetabularia*. *Naturwissenschaften*, **49,** 238.

Richter, G. (1963). Die Tagespriodik der Photosynthese bei *Acetabularia* und ihre Abhängigkeit von Kernaktivität, RNS- und Proteinsynthese. *Zeitschrift für Naturforschung*, **18,** 1085–1089.

Richter, G. (1966). Pulse-labelling of nucleic acids and polyphosphates in normal and anucleate cells of *Acetabularia*. *Nature, London*, **212,** 1363.

Richter, G. and Kirschtein, M. J. (1966). Regeneration und Photosynthese-Leistung kernhaltiger Zell-Teilstücke von *Acetabularia* in blauer und roter Strahlung. *Zeitschrift für Pflanzenphysiologie*, **54,** 106–117.

Ris, H. and Plaut, W. (1962). Ultrastructure of DNA-containing areas in the chloroplast of *Chlamydomonas*. *Journal of Cell Biology*, **13,** 383–391.

Rizzi, L., Pignatti, S. and de Cristini, P. (1967a). Contribuzione alla flora algologica del litorale garganico meridionale fra Manfredonia e Mattinata. *Giornale Botanico Italiano*, **101,** 131–132.

Rizzi, L., Pignatti, S. and Froglia, C. (1967b). Flora delle acque circostanti I'Isola di Pianosa (Isole Tremiti). *Giornale Botanico Italiano*, **101,** 237–239.

Roth, A. G. (1806). " Catalecta Botanica ", Vol. 3, pp. 315–317. Leipzig.

Saddler, H. D. W. (1970a). The ionic relations of *Acetabularia mediterranea*. *Journal of Experimental Botany*, **21,** 345–359.

Saddler, H. D. W. (1970b). The membrane potential of *Acetabularia mediterranea*. *Journal of General Physiology*, **55,** 802–821.

Saddler, H. D. W. (1970c). Fluxes of sodium and potassium in *Acetabularia mediterranea*. *Journal of Experimental Botany*, **21,** 605–616.

Saddler, H. D. W. (1971). Spontaneous and induced changes in the membrane potential and resistance of *Acetabularia mediterranea*. *Journal of Membrane Biology*, **5,** 250–260.

Sandakhchiev, L. S. (1973) *Ontogenez*, **4,** 323 (in Russian), cited by Alexeev *et al.* (1974).

Sandakhchiev, L. S. and Niemann, R. (1972). Changes of the MDH isozyme pattern after heterologous nucleus transplantation. *Protoplasma*, **75,** 485.

Sandakhchiev, L. S., Puchkova, L. I., Pikalov, A. V., Khristolubova, N. B. and Kiseleva, E. V. (1972). Subcellular localization of morphogenetic factors in anucleate *Acetabularia* at the stages of genetic information transfer and expression. *In* " Biology and Radiobiology of Anucleate Systems ", (S. Bonotto, R. Goutier, R. Kirchmann, J. R. Maisin, eds.) Vol. 2, pp. 297–320. Academic Press, New York and London.

Sandakhchiev, L. S., Niemann, R., and Schweiger, H. G. (1973). Kinetics of changes of malic dehydrogenase isozyme pattern in different regions of *Acetabularia* hybrids. *Protoplasma*, **76,** 403–415 (1973).

Sartoni, G. (1974). Contributo alla conoscenza della flora algale bentonica di Sar Uanle (Somalia meridionale). *Giornale Botanico Italiano*, **108,** 281–303.

Sauer, H. W., Goodman, E. M., Babcock, K. L. and Rusch, H. P. (1969). Polyphosphate in the life cycle of *Physarum polycephalum* and its relation to RNA synthesis. *Biochimica et Biophysica Acta*, **195,** 401–409.

Schael, U. and Clauss, H. (1968). Die Wirkung von Rotlicht und Blaulicht auf die Photosynthese von *Acetabularia mediterranea*. *Planta*, **78,** 98–114.

Scherbaum, O. H. (1963). Acid-soluble phosphates in nucleate and enucleate *Acetabularia*. I. Paper-chromatographic patterns. *Biochimica et Biophysica Acta*, **72,** 509–515.

Schilde, C. (1966). Zur Wirkung des Lichtes auf das Ruhepotential der grünen Pflanzenzelle. *Planta*, **71,** 184–188.

Schilde, C. (1968). Schnelle photoelektrische Effekte der Alge *Acetabularia*. *Zeitschrift für Naturforschung*, **23b,** 1369–1376.

Schmid, R. and Clauss, H. (1974). Die Vermehrung der Chloroplasten von *Acetabularia* im Rot- und Blaulicht. *Protoplasma*, **82,** 283–287.

Schmid, R. and Clauss, H. (1975). Multiplication and protein content of chloroplasts of *Acetabularia mediterranea* in blue light after prolonged irradiation with red light. *Protoplasma*, **85,** 315–325.

Schreiber, R. W., Nugent, N. A. and Chapman, C. J. (1964). An improved technique for the culture of *Acetabularia*. *Experimental Cell Research*, **36,** 421–422.

Schulze, K. L. (1939). Cytologische Untersuchungen an *Acetabularia mediterranea* und *Acetabularia Wettsteinii*. *Archiv für Protistenkunde*, **92,** 179–225.

Schussnig, B. (1930). Phytologische Beiträge. III. *Acetabularia Wettsteinii* n. sp. im Mittelmeer. *Österreich Botanische Zeitschrift*, **79,** 333–339.

Schweiger, E., Walraff, H. G. and Schweiger, H. G. (1964a). Endogenous circadian rhythm in cytoplasm of *Acetabularia*: Influence of the nucleus. *Science, New York*, **146,** 658–659.

Schweiger, E., Walraff, H. G. and Schweiger, H. G. (1964b). Über tagesperiodische Schwankungen der Sauerstoffbilanz kernhaltiger und kernloser *Acetabularia mediterranea*. *Zeitschrift für Naturforschung*, **19b,** 499–505.

Schweiger, H. G. (1964). Nuclear functions in erythroid cells and in *Acetabularia*. *Excerpta Medica*, **77,** 40.

Schweiger, H. G. (1966). Ribonuclease-Aktivität in *Acetabularia*. *Planta*, **68,** 247–255.

Schweiger, H. G. (1967). Regulations-probleme in der einzelligen Alge *Acetabularia*. *Arzneimittelforschung*, **17,** 1433–1438.

Schweiger, H. G. (1968). *Acetabularia* als zellbiologisches Objekt. *Mitteilungen aus der Max-Planck-Gesellschaft*, **1,** 3–24.

Schweiger, H. G. (1969). Cell Biology of *Acetabularia*. *Current Topics in Microbiology and Immunology*, **50,** 1–36.

Schweiger, H. G. (1970a). Regulatory problems in *Acetabularia*. *In* "Biology of *Acetabularia*" (J. Brachet, S. Bonotto, eds.), pp. 3–12. Academic Press, New York and London.

Schweiger, H. G. (1970b). Synthesis of RNA in *Acetabularia*. *Symposium of the Society of Experimental Biology*, **24,** 327–344.

Schweiger, H. G. (1971). Circadian rhythms: subcellular and biochemical aspects. *Proceedings of the International Symposium on Circadian Rhythmicity*, Wageningen, 157–174.

Schweiger, H. G. and Berger, S. (1964). DNA-dependent RNA synthesis in chloroplasts of *Acetabularia*. *Biochimica et Biophica Acta*, **77,** 533–535.

Schweiger, H. G. and Berger, S. (1972). Second symposium on *Acetabularia*. *Protoplasma*, **75,** 471–492.

Schweiger, H. G. and Berger, S. (1975). *Acetabularia haemmerlingii*, a new species. *Nova Hedwigia*, **26,** 33–43.

Schweiger, H. G. and Bremer, H. J. (1960a). Das Verhalten verschiedener P-Fraktionen in kernhaltigen und kernlosen *Acetabularia mediterranea*. *Zeitschrift für Naturforschung*, **15b,** 395–400.

Schweiger, H. G. and Bremer, H. J. (1960b). Nachweis cytoplasmatischer Ribonucleinsäuresynthese in kernlosen Acetabularien. *Experimental Cell Research*, **20,** 617–618.

Schweiger, H. G. and Bremer, H. J. (1961). Cytoplasmatische RNA-Synthese in kernlosen Acetabularien. *Biochimica et Biophysica Acta*, **51,** 50–59.

Schweiger, H. G. and Kretschmer, H. (1975). Unsupplemented seawater as culture medium for *Acetabularia*. *Protoplasma*, **83,** 179.

Schweiger, H. G. and Schweiger, E. (1963). Zur Wirkung von Actinomycin C auf *Acetabularia*. *Naturwissenschaften*, **50,** 620–621.

Schweiger, H. G. and Schweiger, E. (1965). The role of the nucleus in a cytoplasmic diurnal rhythm. *In* " Circadian Clocks ". Proceeding of the Feldafing Summer School, 7–18 September 1964 " (J. Aschoff, ed.), pp. 195–197. North-Holland Publishing, Amsterdam.

Schweiger, H. G., Berger, S., Dillard, W. L. and Gibor, A. (1967a). RNA synthesis in *Acetabularia*. *7th International Congress of Biochemistry, Tokyo*, August 19–25, Symp. VII, pp. 4–5.

Schweiger, H. G., Master, R. W. P. and Werz, G. (1967b). Nuclear control of a cytoplasmic enzyme in *Acetabularia*. *Nature, London*, **216,** 554–557.

Schweiger, H. G., Dillard, W. L., Gibor, A. and Berger, S. (1967c). RNA-synthesis in *Acetabularia*. I. RNA-synthesis in enucleated cells. *Protoplasma*, **64,** 1–12.

Schweiger, H. G., Werz, G. and Reuter, W. (1969). Tochtergenerationen von heterologen Implantaten bei *Acetabularia*. *Protoplasma*, **68,** 354–356.

Schweiger, H. G., Apel, K. and Kloppstech, K. (1972a). Source of genetic information of chloroplast proteins in *Acetabularia*. *Advances in Biosciences*, **8,** 249–262.

Schweiger, H. G., Berger, S., Apel, K. and Schweiger, M. (1972b). *Acetabularia major*, a useful tool in cell biology. *Protoplasma*, **75,** 485–486.

Schweiger, H. G., Berger, S., Bonotto, S. and Focken, H. (1974a). *Acetabularia* and other *Dasycladaceae*: Bibliography of the reprint collection of the Max-Planck-Institut für Zellbiologie. Reports from the Max-Planck-Institut für Zellbiologie, Wilhelmshaven, n° 1.

Schweiger, H. G., Berger, S., Kloppstek, K., Apel, K. and Schweiger, M. (1974b). Some fine structural and biochemical features of *Acetabularia major* (*Chloroplyta*, *Dasycladaceae*) grown in the laboratory. *Phycologia*, **13,** 11–20.

Schweiger, H. G. Bannwarth, H., Berger, S. and Kloppstech, K. (1975). *Acetabularia*, a cellular model for the study of nucleocytoplasmic interactions. In " Molecular Biology of Nucleocytoplasmic Relationships " (S. Puiseux-Dao, ed.) pp. 203—215. Elsevier, Amsterdan and New York.

Scopoli, J. A. (1772). " Flora carniolica ", Vol. 2, pp. 411–413. Wien.

Segonzac, G. (1970a). Essai de classement de quelques Acétabulariacées tertiaires (algues calcaires). *Bulletin de la Societé d'Histoire naturelle, Toulouse*, **106,** 333–340.

Segonzac, G. (1970b). Dasycladales nouvelles du Sparnacien des Pyrénées ariégeoises. *Comptes rendus hebdomadaires des Séances de l'Académie des Sciences, Paris*, **270,** 1881–1884.

Setschell, W. A. (1926). Tahitian Algae collected by W. A. Setchell, C. B. Setchell and H. E. Parks. *University of California Publications in Botany*, **12,** 61–142.

Shephard, D. (1965a). Chloroplast multiplication and growth in the unicellular alga *Acetabularia mediterranea. Experimental Cell Research*, **37,** 93–110.

Shephard, D. (1965b). An autoradiographic comparison of the effects of enucleation and actinomycin D on the incorporation of nucleic acids and protein precursors by *Acetabularia mediterranea. Biochimica et Biophysica Acta*, **108,** 635–643.

Shephard, D. C. (1970a). Photosynthesis in chloroplasts isolated from *Acetabularia mediterranea. In* " Biology of *Acetabularia* " (J. Brachet, and S. Bonotto, eds.), pp. 195–212. Academic Press, New York and London.

Shephard, D. C. (1970b). Axenic culture of *Acetabularia* in a synthetic medium. *In* " Methods in Cell Physiology " (D. Prescott, ed.), Vol. 4, pp. 49–69. Academic Press, New York and London.

Shephard, D. C. (1970c). Protein synthesis in a chloroplast isolate from *Acetabularia. Journal of Cell Biology*, **47,** 188a–189a.

Shephard, D. C. (1971). Algal model. Book review of *Acetabularia* and Cell Biology by Puiseux-Dao. *Science, New York*, **172,** 834.

Shephard, D. C. and Bidwell, R. G. S. (1972). Photosynthesis and carbon metabolism in a chloroplast preparation from *Acetabularia. Protoplasma*, **75,** 486–487.

Shephard, D. C. and Bidwell, R. G. S. (1973). Photosynthesis and carbon metabolism in a chloroplast preparation from *Acetabularia. Protoplasma*, **76,** 289–307.

Shephard, D. C. and Levin, W. B. (1972). Biosynthesis in isolated *Acetabularia* chloroplasts. I. Protein amino acids. *Journal of Cell Biology*, **54,** 279–294.

Shephard, D. C., Levin, W. B., and Bidwell, R. G. S. (1968). Normal photosynthesis by isolated chloroplasts. *Biochemical and Biophysical Research Communications*, **32,** 413–420.

Silva, P. C. (1952). A review of nomenclatural conservation in the algae from the point of view of the type method. *University of California Publications in Botany*, **25,** 241–323.

Sironval, C., Bonotto, S., Kirchmann, R., Hoursiangou-Neubrun, D. and Puiseux-Dao, S. (1972). Chloroplast heterogeneity in *Acetabularia mediterranea* studied by means of low temperature fluorescence techniques and electron microscopy. *Protoplasma*, **75,** 487.

Sironval, C., Bonotto, S. and Kirchmann, R. (1973). Sur l'hétérogénéité plastidiale chez l'acétabulaire. Etude de la fluorescence émise à la température de l'azote liquide. *Plant Science Letters*, **1,** 47–52.

Six, E. (1956a). Die Wirkung von Strahlen auf *Acetabularia*. I. Die Wirkung von ultravioletten Strahlen auf kernlose Teile von *Acetabularia mediterranea*. *Zeitschrift für Naturforschung*, **11b,** 463–470.

Six, E. (1965b). Die Wirkung von Strahlen auf *Acetabularia*. II. Die Wirkung von Röntgenstrahlen auf kernlose Teile von *Acetabularia mediterranea*. *Zeitschrift für Naturforschung*, **11b,** 598–603.

Six, E. (1958). Die Wirkung von Strahlen auf *Acetabularia*. III. Die Wirkung von Röntgenstrahlen und ultravioletten Strahlen auf kernhaltige Teile von *Acetabularia mediterranea*. *Zeitschrift für Naturforschung*, **13b,** 6–14.

Six, E. and Puiseux-Dao, S. (1961). Die Wirkung von Strahlen auf Acetabularien. IV. Röntgenstrahlenwirkungen in zweikernigen Transplantaten. *Zeitschrift für Naturforschung*, **16b,** 832–835.

Smestad, B., Percival, E. and Bidwell, R. G. S. (1972). Metabolism of soluble carbohydrates in *Acetabularia mediterranea* cells. *Canadian Journal of Botany*, **50,** 1357–1361.

Solms-Laubach, H. Graf zu. (1893). Über die Algengenera *Cymopolia*, *Neomeris* und *Bornetella*. *Annales du Jardin botanique de Buitenzorg*, **11,** 61–97.

Solms-Laubach, H., Graf zu. (1895). Monograph of the *Acetabulariae*. *Transactions of the Linnean Society, London*, 2nd Series, Botany, **5,** part 1, 1–39.

Sonder, W. (1871). Die Algen des tropischen Australiens. Abhandlungen aus dem Gebiete der Naturwissenschaften (Hamburg), **4,** 35–74.

Spencer, T. (1968). Effects of Kinetin on the phosphates enzymes of *Acetabularia*. *Nature, London*, **217,** 62–64.

Spencer, T. and Harris, H. (1964). Regulation of enzyme synthesis in an enucleate cell. *The Biochemical Journal*, **91,** 282–286.

Spirin, A. S. (1969). Informosomes. *European Journal of Biochemistry*, **10,** 20–35.

Spirin, A. S. and Nemer, M. (1965). Messenger RNA in early sea-urchin embryos: cytoplasmic particles. *Science, New York*, **150,** 214–217.

Spirin, A. S., Belitsina, N. V. and Lerman, M. I. (1965). Use of formaldehyde fixation for studies on ribonucleoprotein particles by caesium chloride density-gradient centrifugation. *Journal of Molecular Biology*, **14,** 611–615.

Spring, H., Trendelenburg, M. F., Scheer, U., Franke, W. W. and Herth, W. (1974). Structural and biochemical studies of the primary nucleus of two algal species, *Acetabularia mediterranea* and *Acetabularia major*. *Cytobiologie*, **10,** 1–65.

Spring, H., Franke, W. W., Falk, H. and Berger, S. (1975a). Perinuclear dense bodies in *Acetabularia*: Some ultrastructural and cytochemical aspects. *Protoplasma*, **83,** 180.

Spring, H. Scheer, U., Franke, W. W. and Trendelenburg, M. F. (1975b). Lampbrush-type chromosomes in the primary nucleus of the green alga *Acetabularia mediterranea*. *Chromosoma*, **50,** 25–43.

Spring, H., Krohne, G., Franke, W. W., Scheer, U. and Trendelenburg, M. F. (1976). Homogeneity and heterogeneity of sizes of transcriptional units and spacer regions in nucleolar genes of *Acetabularia*. *Journal de Microscopie et de Biologie Cellulaire*, **25,** 107—116.

Stich, H. (1951a). Experimentelle karyologische und cytochemische Untersuchungen an *Acetabularia mediterranea*. Ein Beitrag zur Beziehung zwischen Kerngrösse und Eiweissynthese. *Zeitschrift für Naturforschung*, **6b,** 319–326.

Stich, H. (1951b). Trypaflavin und Ribonucleinsäure. Untersucht an Mäusegeweben, *Condylostoma sp.* und *Acetabularia mediterranea. Naturwissenschaften*, **38,** 435–436.

Stich, H. (1951c). Das Vorkommen von Kohlenhydraten im Ruhekern und während der Mitose. *Chromosoma*, **4,** 429–438.

Stich, H. (1953). Der Nachweis und das Verhalten von Metaphosphaten in normalen, verdunkelten und Trypaflavinbehandelten Acetabularien. *Zeitschrift für Naturforschung*, **8b,** 36–44.

Stich, H. (1955). Synthese und Abbau der Polyphosphaten von *Acetabularia* nach autoradiographischen Untersuchungen des ^{32}P-Stoffwechsels. *Zeitschrift für Naturforschung*, **10b,** 281–284.

Stich, H. (1956a). Änderungen von Kern und Polyphosphaten in Abhängigkeit von dem Energiegehalt des Cytoplasmas bei *Acetabularia. Chromosoma*, **7,** 693–707.

Stich, H. (1956b). Bau und Funktion der Nukleolen. *Experientia*, **12,** 7–14.

Stich, H. (1959). Changes in nucleoli related to alteration in cellular metabolism. *In* " Developmental Cytology " (D. Rudnick, ed.), pp. 105–122. Ronald Press, New York.

Stich, H. and Hämmerling, J. (1953). Der Einbau von ^{32}P in die Nukleolarsubstanz der Zellkernes von *Acetabularia mediterranea. Zeitschrift für Naturforschung*, **8b,** 329–333.

Stich, H. and Kitiyakara, A. (1957). Self-regulation of protein synthesis in *Acetabularia. Science, New York*, **126,** 1019–1020.

Stich, H. and Plaut, W. (1958). The effects of ribonuclease on protein synthesis in nucleated and enucleated fragments of *Acetabularia. Journal of Biophysical and Biochemical Cytology*, **4,** 119–121.

Strotmann, H. and Berger, S. (1968). Austausch von Adeninnukleotiden durch die Chloroplasten Membran. *Berichte der deutschen botanischen Gesellschaft*, **81,** 306.

Strotmann, H. and Berger, S. (1969). Adenine nucleotide translocation across the membrane of isolated *Acetabularia* chloroplasts. *Biochemical and Biophysical Research Communications*, **35,** 20–26.

Strotmann, H. and Heldt, H. W. (1969). Phosphate containing metabolites participating in photosynthetic reactions of *Chlorella pyrenoidosa*. *In* " Progress in Photosynthesis Research " (H. Metzner, ed.), Vol. 3, 1131–1140. International Union of Biological Sciences, Tübingen.

Sutter, R. P., Whitman, S. L. and Webster, G. (1961). Cytoplasmic formation of the ribonucleic acid of ribosomes. *Biochimica et Biophysica Acta*, **49,** 233–235.

Svedelius, N. (1923). Zur Kenntnis der Gattung *Neomeris. Svensk. Bot. Tidskrift*, **17,** 449–471.

Sweeney, B. (1969). " Rhythmic Phenomena in Plants ". Academic Press, London.

Sweeney, B. (1972). The *Acetabularia* rhythm paradoxes. *Protoplasma*, **75,** 488.

Sweeney, B. M. and Haxo, F. T. (1961). Persistence of a photosynthetic rhythm in enucleated *Acetabularia. Science, New York*, **134,** 1361–1363.

Sweeney, B. M., Tuffli, C. F. and Rubin, R. H. (1967). The circadian rhythm in photosynthesis in *Acetabularia* in the presence of actinomycin D, puromycin and chloramphenicol. *Journal of General Physiology*, **50,** 647–659.

Tandler, C. J. (1962a). Oxalic acid and Potassium in *Acetabularia. Naturwissenschaften*, **49,** 112.

Tandler, C. J. (1962b). A naturally occuring crystalline indolyl derivative in *Acetabularia. Naturwissenschaften*, **49,** 213–214.

Tandler, C. J. (1962c). Bound indoles in *Acetabularia. Planta*, **59,** 91–107.

Taylor, W. R. (1928) Marine algae of Florida with special reference to the Dry Tortugas. *Carnegie Institution of Washington*, **25,** 66–70.

Taylor, W. R. (1945). Pacific marine algae of the Allan Hancock expeditions to the Galapagos Islands. *Allan Hancock Pacific Expeditions, First Series*, **12,** 58–61.

Taylor, W. R. (1950). Plants of Bikini and other northern Marshall Islands. *University of Michigan Studies, Scientific Series*, **18,** 49–67.

Taylor, W. R. (1960). "Marine algae of the eastern tropical and subtropical coasts of the Americas". pp. 97–107. University of Michigan, Ann Arbor.

Terborgh, J. (1963). Studies on the growth and morphogenesis of *Acetabularia crenulata*. Thesis, Harvard University.

Terborgh, J. W. (1965). Effects of red and blue light on the growth and morphogenesis of *Acetabularia crenulata. Nature, London*, **40,** 1360–1363.

Terborgh, J. W. (1966). Potentiation of photosynthetic oxygen evolution in red light by small quantities of monochromatic blue light. *Plant Physiology*, **41,** 1401–1410.

Terborgh, J. W. and McLeod, G. C. (1967). The photosynthetic Rhythm of *Acetabularia crenulata*. I. Continuous measurements of oxygen exchange in alternating light-dark regimes and in constant light of different intensities. *Biological Bulletin*, **133,** 659–669.

Terborgh, J. W. and Thimann, K. V. (1964). Interactions between daylength and light intensity in the growth and chlorophyll content of *Acetabularia crenulata. Planta*, **63,** 83–98.

Terborgh, J. W. and Thimann, K. V. (1965). The control of development in *Acetabularia crenulata* by light. *Planta*, **64,** 241–253.

Thilo, E., Grunze, H., Hämmerling, J. and Werz, G. (1956). Über Isolierung und Identifizierung der Polyphosphate aus *Acetabularia mediterranea. Zeitschrift für Naturforschung*, **11b,** 266–270.

Thimann, K. V. and K. Beth (1959). Action of auxins on *Acetabularia* and the effect of enucleation. *Nature, London*, **183,** 946–948.

Tikhomirova, L. A., Betina, M. I., Fomina, O. V., Yazykov, A. A. and Zubarev, T. N. (1975). Ultrastructure of regenerating cells of *Acetabularia mediterranea*. XII *International Botanical Congress, Leningrad, Abstracts*, Vol. **1,** 46.

Tolomio, C. (1973). Fitoplancton e fitobentos lungo le coste calabro-campane Mar Terreno). *Primo Contributo. Giornale Botanico Italiano*, **107,** 87–100.

Tournefort, J. P. (1700). Observations sur les plantes qui naissent au fond de la mer. *Mémoire de l'Académie Royale des Sciences, Paris*.

Tournefort, J. P. (1719). Institutiones Rei Herbariae. T. III, Paris.

Trench, R. K. and Smith, D. C. (1970). Synthesis of pigment in symbiotic chloroplasts. *Nature, London*, **227,** 196–197.

Trendelenburg, M. F., Spring, H., Scheer, U. and Franke, W. W. (1974a). Morphology of nucleolar cistrons in a plant cell, *Acetabularia mediterranea. Proceedings of the National Academy of Science*, **71,** 3626–3630.

Trendelenburg, M. F., Franke, W.W., Spring, H. and Scheer, U. (1974b). Ultrastructure of transcription in the nucleoli of the green algae *Acetabularia major* and *A. mediterranea*. *In* "Proceedings of the 9th FEBS Meeting" (E. J. Hidvegi, J. Sümegi and P. Venetianer, eds.), 33, 159–168, Akademiai Kiado, North Holland, Budapest, Amsterdam.

Trendelenburg, M. F., Spring, H., Scheer, U. and Franke, W. W. (1975). Demonstration of active nucleolar cistrons of the *Acetabularia* nucleus with the spreading technique. *Protoplasma*, **83,** 180.

Triplett, E. L., Steens-Lievens, A. and Baltus, E. (1965). Rates of synthesis of acid phosphatases in nucleate and enucleate *Acetabularia* fragments. *Experimental Cell Research*, **38,** 366–378.

Tsekos, I., Haritonidis, S. and Diannelidis, Th. (1972). Protoplasmaresistenz von Meeresalgen und Meeresanthophyten gegen Schwermetallsalze. *Protoplasma*, **75,** 45–65.

Valet, G. (1967a). Sur l'origine endogène des rameaux verticillés chez certaines Dasycladales. *Comptes rendus hebdomadaires des Séances de l'Académie des Sciences, Paris*, 265, 1175–1178.

Valet, G. (1967b). Algues marines de la Nouvelle-Calédonie. I. Chlorophycées. *Nova Hedwigia*, **15,** 29–63.

Valet, G. (1968a). Contribution à l'étude des Dasycladales, pp. 1–216. Thesis, University of Paris.

Valet, G. (1968b). Contribution à l'étude des Dasycladales. 1. Morphogenèse. *Nova Hedwigia*, **16,** 21–82.

Valet, G. (1969a). Contribution à l'étude des Dasycladales. 2. Cytologie et reproduction. 3. Révision systématique. *Nova Hedwigia*, **17,** 551–644.

Valet, G. (1969b). Le monde étrange des Dasycladales. *Bulletin Société Phycologique de France*, Nos. 13–14.

Valet, G. (1972). Revised systematics of Dasycladales. *Protoplasma*, **75,** 488–489.

Valet, G. and Segonzac, G. (1969). Les genres *Chalmasia* et *Halycoryne* (Algues Acétabulariacées). *Bulletin Société géologique de France*, **11,** 124–127.

Vanden Driessche, T. (1966a). Circadian rhythms in *Acetabularia*: Photosynthetic capacity and chloroplast shape. *Experimental Cell Research*, **42,** 18–30.

Vanden Driessche, T. (1966b). The role of the nucleus in the circadian rhythms of *Acetabularia mediterranea*. *Biochimica et Biophysica Acta*, **126,** 456–470.

Vanden Driessche, T. (1967a). Les horloges biologiques. *Revue générale de Science*, **74,** 85–95.

Vanden Driessche, T. (1967b). Experiments and hypothesis on the role of RNA in the circadian rhythm of photosynthetic capacity in *Acetabularia mediterranea*. *Nachrichten Akademie der Wissenschaften in Göttingen*. II. Math.-Phys. Klasse, **10,** 108–109.

Vanden Driessche, T. (1967c). The nuclear control of the chloroplasts circadian rhythms. *Science Progress, Oxford*, **55,** 293–303.

Vanden Driessche, T. (1969a). Light-induced metabolic variations in *Acetabularia mediterranea*. Third International Biophysics Congress of the International Union for Pure and Applied Biophysics. Cambridge, Massachusetts, U.S.A., August 29–September 3, 1969, Book of Abstracts, p. 127.

Vanden Driessche, T. (1969b). The influence of constant light on the inulin content of the chloroplasts in *Acetabularia mediterranea*. *In* " Progress in Photosynthesis Research " (H. Metzner, ed.), Vol. 1, 450–457. International Union of Biological Sciences, Tübingen.

Vanden Driessche, T. (1970a). Temporal regulation in *Acetabularia*. *In* " Biology of *Acetabularia* " (J. Brachet and S. Bonotto, eds.), pp. 213–236, Academic Press, New York and London.

Vanden Driessche, T. (1970b). Rhythms in unicellular organisms: their biochemical regulation. Interdisciplinary conference on the experimental and clinical study of HNF, Mariánské Lázně, *Activitas Nervosa Superior*, **12,** 140–141.

Vanden Driessche, T. (1970c). Les rythmes circadiens chez les unicellularires. *Journal of Interdisciplinary Cycle Research*, **1,** 21–42.

Vanden Driessche, T. (1970d). Circadian variation in ATP content in the chloroplasts of *Acetabularia mediterranea*. *Biochimica et Biophysica Acta*, **205,** 526–528.

Vanden Driessche, T. (1971a). Possible diversity in basic mechanisms of biological oscillations. *Journal of Interdisciplinary Cycle Research*, **2,** 133–145.

Vanden Driessche, T. (1971b). Circadian rhythms in whole and in anucleate *Acetabularia*. *Gegenbaurs morphologisches Jahrbuch*, (*Leipzig*). **117,** 81–83.

Vanden Driessche, T. (1971c). Les rhythmes circadiens, mécanismes de régulation cellulaire. *La Recherche*, **2,** 255–261.

Vanden Driessche, T. (1971d). Structural and functional rhythms in the chloroplasts of *Acetabularia*: molecular aspects of the circadian system. *In* " Biochronometry " (M. Menaker, ed.), pp. 612–622. National Academy of Sciences, Washington, D.C.

Vanden Driessche, T. (1972a). Some photosynthetic activities of the chloroplasts of *Acetabularia* in the presence and in the absence of the nucleus. *In* " Biology and Radiobiology of Anucleate Systems ", (S. Bonotto, R. Goutier, R. Kirchmann and J. R. Maisin, eds.), Vol. 2, pp. 53–73. Academic Press, New York and London.

Vanden Driessche, T. (1972b). Biological rhythms in plants under various light and temperature conditions. *International Journal of Biometereology*, **16,** Suppl., 127–151.

Vanden Driessche, T. (1973a). The chloroplasts of *Acetabularia*. The control of their multiplication and activities. *Sub-cellular Biochemistry*, **2,** 33–67.

Vanden Driessche, T. (1973b). A population of oscillators: A working hypothesis and its compatibility with the experimental evidence. *International Journal of Chronobiology*, **1,** 253–258.

Vanden Driessche, T. (1974a). Effects of morphactins on growth and differentiation of *Acetabularia*. *Protoplasma*, **81,** 323–334.

Vanden Driessche, T. (1974b). Circadian rhythm in the Hill reaction of *Acetabularia*. Proceedings of the Third International Congress on Photosynthesis, Rehovot (M. Avron, ed.), **1,** 745–751, Elsevier, Amsterdam.

Vanden Driessche, T. (1976). c-AMP and morphogenesis in *Acetabularia*. *Differentiation*, **5,** 119–126.

Vanden Driessche, T. and Bonotto, S. (1967). Nature du matériel accumulé par les chloroplasts d'*Acetabularia mediterranea*. *Archives Internationales de Physiologie et de Biochimie*, **75,** 186–187.

Vanden Driessche, T. and Bonotto, S. (1968a). Variations journalières de l'incorporation d'uridine dans le RNA d'*Acetabularia*. *Archives Internationales de Physiologie et de Biochimie*, **76,** 959–960.

Vanden Driessche, T. and Bonotto, S. (1968b). Le rythme circadien de la teneur en inuline chloroplastique d'*Acetabularia mediterranea*. *Archives Internationales de Physiology et de Biochimie*, **76,** 205–206.

Vanden Driessche, T. and Bonotto, S. (1969a). Rhythms in *Acetabularia*. *Rassegna di Neurologia Vegetativa*, **23,** 113–128.

Vanden Driessche, T. and Bonotto, S. (1969b). The circadian rhythm in RNA synthesis in *Acetabularia mediterranea*. *Biochimica et Biophysica Acta*, **179,** 58–66.

Vanden Driessche, T. and Bonotto, S. (1971). Effect of NAD on photosynthesis and carbohydrate synthesis in *Acetabularia*. *Proceedings of the 2nd International Congress on Photosynthesis, Stresa*, pp. 2059–2070.

Vanden Driessche, T. and Bonotto, S. (1972). *In vivo* activity of the chloroplasts of *Acetabularia* in continuous light and in light-dark cycles. *Archives de Biologie, Liège*, **83,** 89–104.

Vanden Driessche, T. and Delegher-Langohr, V. (1975). Presence of an auxin-like substance in *Acetabularia*. *Protoplasma*, **83,** 181.

Vanden Driessche, T. and Hars, R. (1972a). Variations circadiennes de l'ultrastructure des chloroplastes d'*Acetabularia*. I. Algues entières. *Journal de Microscopie*, **15,** 85–90.

Vanden Driessche, T. and Hars, R. (1972b). Variations circadiennes de l'ultrastructure des chloroplastes d'*Acetabularia*. II. Algues anucléées. *Journal de Microscopie*, **15,** 91–98.

Vanden Driessche, T. and Hars, R. (1972c). Substructure of the chloroplasts of *Acetabularia* and rate of physiological activities. *International Union for Pure and Applied Biophysics*. Moscow, August, 7–14, 1972. Book of Abstracts, pp. 63–64.

Vanden Driessche, T. and Hars, R. (1973). Ultrastructure of the chloroplasts of *Acetabularia mediterranea* and rate of physiological activities. *Archives de Biologie, Bruxelles*, **84,** 539–551.

Vanden Driessche, T. and Hayet, M. (1975). Circadian rhythm of photosynthesis as influenced by anti-auxin. *Protoplusma*, **83,** 181.

Vanden Driessche, T. and Hellin, J. (1972). Le rythme circadien de division des chloroplastes d'*Acetabularia mediterranea*. *Archives Internationales de Physiologie et de Biochimie*, **80,** 626–627.

Vanden Driessche, T., Bonotto, S. and Brachet, J. (1970). Inability of rifampicin to inhibit circadian rhythmicity in *Acetabularia* despite inhibition of RNA synthesis. *Biochimica et Biophysica Acta*, **224,** 631–634.

Vanden Driessche, T. Hellin, J. and Hars, R. (1972). Limitations in chloroplast multiplication in *Acetabularia mediterranea*. *Protoplasma*, **75,** 489–490.

Vanden Driessche, T., Hars, R., Hellin, J. and Bouloukhère, M. (1973a). The substructure of cytoplasts obtained from *Acetabularia mediterranea*. *Journal of Ultrastructure Research*, **42,** 479–490.

Vanden Driessche, T., Hellin, J. and Hars, R. (1973b). Limitations in chloroplast multiplication in *Acetabularia mediterranea*. *Protoplasma*, **76,** 465–472.

Vanden Driessche, T. Moens, U. and Kram, R. (1975). Presence of c-AMP in *Acetabularia*. *Protoplasma*, **83,** 181–182.

vand der Ben, D., Felluga, B. and Bonotto, S. (1972). Epiphytes on *Acetabularia mediterranea* from the Isle of Ischia (Italy). *Protoplasma*, **75**, 475.

Vanderhaeghe, F. (1952). Mesures de croissance de fragments nucléés et énucléés d'*Acetabularia mediterranea*. *Archives Internationales de Physiologie et de Biochimie*, **60**, 190–191.

Vanderhaeghe, F. (1954). Les effets de l'énucléation sur la synthèse des protéines chez *Acetabularia mediterranea*. *Biochimica et Biophysica Acta*, **15**, 281–287.

Vanderhaeghe, F. (1963). Rôle des acides nucléiques dans la synthèse des protéines chez *Acetabularia mediterranea*. *Bulletin de la Société française de Physiologie Vegétale*, **9**, 67–77.

Vanderhaeghe, F. and Szafarz, D. (1955). Enucléation et synthèse d'acideribonucléique chez *Acetabularia mediterranea*. *Archives Internationales de Physiologie et de Biochimie*, **63**, 267–268.

Vanderhaeghe-Hougardy, F. and Baltus, E. (1962). Effets de l'énucléation sur le maintien de la phosphatase acide dans le cytoplasme d'*Acetabularia mediterranea*. *Archives Internationales de Physiologie et de Biochimie*, **70**, 414–415.

Van Gansen, P. and Bolouk hère-Presburg, M. (1965). Ultrastructure de l'algue unicellulaire *Acetabularia mediterranea*. *Journal de Microscopie*, **4**, 347–362.

van Rensen, J. J. S. (1969). Polyphosphate formation in *Scenedesmus* in relation to photosynthesis. *In* " Progress in Photosynthesis Research " (H. Metzner, ed.), Vol. 3, 1769–1776. International Union of Biological Sciences, Tübingen.

Vettermann, W. (1972). Mechanism of the light-dependent accumulation of starch in chloroplasts of *Acetabularia* and its regulation. *Protoplasma*, **75**, 490–491.

Vettermann, W. (1973). Mechanism of the light-dependent accumulation of starch in chloroplasts of *Acetabularia*, and its regulation. *Protoplasma*, **76**, 261–278.

von Klitzing, L. (1969). Oszillatorische Regulationserscheinungen in der einzelligen Grünalge *Acetabularia*. *Protoplasma*, **68**, 341–350.

von Klitzing, L. and Schweiger, H. G. (1969). A method for recording the circadian rhythm of the oxygen balance in a single cell of *Acetabularia mediterrannea*. *Protoplasma*, **67**, 327–332.

von Martens, G. (1866). Die preussische Expedition nach Ost-Asien. *Botanischer Teil. Die Tange. Berlin*, pp. 24–25; 62–68; 126–127.

Weber-Van Bosse, A. (1913). Liste des algues du Siboga. I. Myxophyceae, Chlorophyceae, Phaeophyceae. *Siboga Expeditie* (1913–1928), **59**, 88–114.

Werz, G. (1953). Über die Kernverhältnisse der Dasycladaceen, besonders von *Cympolia barbata* (L.) Harv. *Archiv für Protistenkunde*, **99**, 148–155.

Werz, G. (1955). Kernphysiologische Untersuchungen an *Acetabularia*. *Planta*, **46**, 113–153.

Werz, G. (1957a). Eiweissvermehrung in ein- und zweikerningen Systemen von *Acetabularia*. *Experientia*, **13**, 79.

Werz, G. (1957b). Über die Wirkung von Cobalt-II-nitrat auf Kern un Cytoplasma von *Acetabularia mediterranea*. *Experientia*, **13**, 279.

Werz, G. (1957c). Die Wirkung von Trypaflavin auf Kern und Cytoplasma von *Acetabularia mediterranea*. *Zeitschrift für Naturforschung*, **12b**, 559–563.

Werz, G. (1957d). Membranbildung bei kernlosen wachsenden und nicht wachsenden Teilen von *Acetabularia mediterranea*. *Zeitschrift für Naturforschung*, **12b**, 739–740.

Werz, G. (1959a). Über polare Plasmaunterschiede bei *Acetabularia*. *Planta*, **53,** 502–521.

Werz, G. (1959b). Weitere Untersuchungen zum Problem der Kernaktivität bei gesenktem Zellstoffwechsel. *Planta*, **53,** 528–533.

Werz, G. (1960a). Über die Wirkung von Dunkelheit auf den Proteingehalt kernhaltiger und kernloser Zellteile von *Acetabularia mediterranea*. *Zeitschrift für Naturforschung*, **15b,** 85–90.

Werz, G. (1960b). Anreicherung von Ribonucleinsäure in der Wuchszone von *Acetabularia mediterranea*. *Planta*, **55,** 22–37.

Werz, G. (1960c). Über Struckturierungen der Wuchszone von *Acetabularia mediterranea*. *Planta*, **55,** 38–56.

Werz, G. (1960d). Distribution and origin of cytoplasmic RNA and basic protein and their relations to morphogenetic substances in *Acetabularia*. 10ème *Congrès International de Biologie Cellulaire, Paris*, pp. 16–17,

Werz, G. (1961a). Unterschiede der Protoplasten-Strukturierungen bei verschiedenen Dasycladaceen und ihre Abhängigkeit von artspezifischen Kernwirkungen. *Planta*, **56,** 490–498.

Werz, G. (1961b). Uber die Beeinflussung der Formbildung von *Acetabularia* durch Selenat. *Planta*, **57,** 250–257.

Werz, G. (1961c). Zur Frage der Herkunft und Verteilung cytoplasmatischer Ribonucleinsäure und ihre Beziekungen zu " morphogenetischen Substanzen" bei *Acetabularia mediterranea*. *Zeitschrift für Naturforschung*, **16b,** 126–129.

Werz, G. (1962). Zur Frage der Elimination von Ribonucleinsäure und Protein aus dem Zellkern von *Acetabularia mediterranea*. *Planta*, **57,** 636–655.

Werz, G. (1963a). Morphogenetische Wirkungen von Aminosäuren bei Acetabularien. I. Wirkungen von 3-C-Aminosäuren und verwandten Derivaten. *Planta*, **60,** 205–210.

Werz, G. (1963b). Morphogenetische Wirkungen von Aminosäuren bei Acetabularien. II. Die Beeinflussung der Morphogenese durch einige 4-C-Aminosäuren. *Planta*, **60,** 211–215.

Werz, G. (1963c). Vergleichende Zellmembraneanalysen bei verschiedenen Dasycladaceen. *Planta*, **60,** 322-330.

Werz, G. (1963d). Das Fernsehmikroskop und seine Anwendung für qualitative und quantitative Untersuchungen in der Zellforschung. *Zeitschrift für wissenschaftliche Mikroskopie und für mikroskopische Technik*, **65,** 265 278.

Werz, G. (1964a). Unterschiede in der Zusammensetzung von Mannanen aus verschiedenen Dasycladaceen-Arten. *Planta*, **60,** 540–542.

Werz, G. (1964b). Feinstrukturuntersuchungen an Ribosomen des perinukleären Plasmas von *Acetabularia*. *Planta*, **62,** 191–193.

Werz, G. (1964c). Untersuchungen zur Feinstruktur des Zellkernes und des perinukleären Plasmas von *Acetabularia*. *Planta*, **62,** 255–271.

Werz, G. (1964d). Electronenmikroskopische Untersuchungen zur Genese des Golgi-Apparates (Dictyosomen) und ihrer Kernabhängigkeit bei *Acetabularia*. *Planta*, **63,** 366–381.

Werz, G. (1965). Determination and realization of morphogenesis in *Acetabularia*. *Brookhaven Symposium in Biology*, **18,** 185–203.

Werz, G. (1966a). Morphologische Veränderungen in Chloroplasten und Mitochondrien von verdunkelten *Acetabularia*-Zellen. *Planta*, **68,** 256–268.

Werz, G. (1966b). Primärvorgänge bei der Realisation der Morphogenese von *Acetabularia*. *Planta*, **69,** 53–57.

Werz, G. (1967). Induktion von Zellwandbildung durch Fremdprotein bei *Acetabularia*. *Naturwissenschaften*, **54,** 374.

Werz, G. (1968a). Plasmatische Formbildung als Voraussetzung für die Zellwandbildung bei der Morphogenese von *Acetabularia*. *Protoplasma*, **65,** 81–96.

Werz, G. (1968b). Differenzierung und Zellwandbildung in isoliertem Cytoplasma aus *Acetabularia*. *Protoplasma*, **65,** 349–357.

Werz, G. (1969a). Morphogenetic processes in *Acetabularia*. *In* " Inhibitors tools in cell research " (Th. Bücher, and H. Sies, eds.), pp. 167–186. Springer-Verlag, Mosbach.

Werz, G. (1969b). Wirkungen von Colchicin auf die Morphogenese von *Acetabularia*. *Protoplasma*, **67,** 67–78.

Werz, G. (1970a). Mechanisms in cell wall formation in *Acetabularia*. In " Biology of *Acetabularia* " (J. Brachet and S. Bonotto, eds.), pp. 125–144. Academic Press, New York and London.

Werz, G. (1970b). Cytoplasmic control of cell wall formation in *Acetabularia*. *Current Topics in Microbiology and Immunology*, **51,** 27–62.

Werz, G. (1972). Ultrastructure of *Acetabularia*. *Protoplasma*, **75,** 491.

Werz, G. (1974). Fine structural aspects of morphogenesis in *Acetabularia*. *International Revue of Cytology*, **38,** 319–367.

Werz, G. and Clauss, H. (1970). Über die chemische Natur der Reserve-Polysaccharide in *Acetabularia*-chloroplasten. *Planta*, **91,** 165–168.

Werz, G. and Hämmerling, J. (1959). Proteinsynthese in wachsenden und nicht wachsenden kernlosen Zellteilen von *Acetabularia*. *Planta*, **53,** 145–161.

Werz, G. and Hämmerling, J. (1961). Über die Beeinflussung artspezifischer Formbildungsprozesse von *Acetabularia* durch UV-Bestrahlung. *Zeitschrift für Naturforschung*, **16b,** 829–832.

Werz, G. and Kellner, G. (1968a). Isolierung und elektroninmikroskopische Charakterisierung von Desoxyribonucleinsäure aus Chloroplasten kernloser *Acetabularia*-Zellen. *Zeitschrift für Naturforschung*, **23b,** 1018.

Werz, G. and Kellner, G. (1968b). Molecular characteristics of chloroplast DNA of *Acetabularia* cells. *Journal of Ultrastructure Research*, **24,** 109–115.

Werz, G. and Zetsche, K. (1963). Autoradiographische Untersuchungen an verdunkelten einkernigen Transplantaten von *Acetabularia mediterranea*. *Planta*, **59,** 563–568.

Williams, N. S. and Terborgh, J. (1970). Protoplasmic streaming in red light potentiated by small quantities of blue light in *Acetabularia crenulata*. *Plant Physiology*, **46,** 1.

Winkenbach, F., Parthasarathy, M. V. and Bidwell, R. G. S. (1972a). Sites of photosynthetic metabolism in cells and chloroplast preparations of *Acetabularia mediterranea*. *Canadian Journal of Botany*, **50,** 1367–1375.

Winkenbach, F., Grant, B. R. and Bidwell, R. G. S. (1972b). The effects of nitrate, nitrite, and ammonia on photosynthetic carbon metabolism of *Acetabularia* chloroplast preparations compared with spinach chloroplasts and whole cells of *Acetabularia* and *Dunaliella*. *Canadian Journal of Botany*, **50,** 2545–2551.

Wolpert, L. (1969). Positional information and the spatial pattern of cellular differentiation. *Journal of Theoretical Biology*, **25,** 1–47.

Womersley, H. B. S. (1955). New Marine *Chlorophyta* from southern Australia. *Pacific Science*, **9,** 387–395.

Woodcock, C. L. F. (1970). The anchoring of nuclei by cytoplasmic microtubules in *Acetabularia mediterranea*. *Journal of Cell Biology*, **47,** 231a.

Woodcock, C. L. F. (1971). The anchoring of nuclei by cytoplasmic microtubules in *Acetabularia*. *Journal of Cell Science*, **8,** 611–621.

Woodcock, C. L. F. (1975). Nucleolar fine structure in *Acetabularia mediterranea*. *Protoplasma*, **83,** 182.

Woodcock, C. L. F. and Bogorad, L. (1968). Evidence for wide disparity in the amount of DNA per plastid in *Acetabularia mediterranea*. 8th Ann. Meeting of the Am. Soc. for Cell Biol., *Journal of Cell Biology*, **39,** 144a–145a.

Woodcock, C. L. F. and Bogorad, L. (1970a). On the extraction and characterization of ribosomal RNA from *Acetabularia*. *Biochimica et Biophysica Acta*, **224,** 639–643.

Woodcock, C. L. F. and Bogorad, L. (1970b). Evidence for variation in the quantity of DNA among plastids of *Acetabularia*. *Journal of Cell Biology*, **44,** 361–375.

Woodcock, C. L. F. and Miller, G. J. (1972a). The anucleate cell as a system for studying chloroplast functions. *In* " Biology and Radiobiology of Anucleate Systems " (S. Bonotto, R. Goutier, R. Kirchmann and J. R. Maisin, eds.), Vol. 2, pp. 75–98. Academic Press, New York and London.

Woodcock, C. L. F. and Miller, G. J. (1972b). Ultrastructural features of gametogenesis in *Acetabularia mediterranea*. *Protoplasma*, **75,** 491–492.

Woodcock, C. L. F. and Miller, G. J. (1973a). Ultrastructural features of the life cycle of *Acetabularia mediterranea*. I. Gametogenesis. *Protoplasma*, **77,** 313–329.

Woodcock, C. L. F. and Miller, G. J. (1973b). Ultrastructural features of the life cycle of *Acetabularia mediterranea*. II. Events associated with the division of the primary nucleus and the formation of cysts. *Protoplasma*, **77,** 331–341.

Woodcock, C. L. F., Stanchfield, J. E. and Gould, R. R. (1975a). Morphology and size of ribosomal cistrons in two plant species: *Acetabularia mediterranea* and *Chlamydomonas reinhardi*. *Plant Science Letters*, **4,** 17–23.

Woodcock, C. L. F., Stanchfield, J. E. and Gould, R. R. (1975b). Direct visualization of information transfer events. In " Molecular Biology of Nucleocytoplasmic Relationships " (S. Puiseux-Dao, ed.) pp. 235–241, Elsevier, Amsterdam and New York.

Woronine, H. (1861). Recherches sur les algues marines *Acetabularia* Lamx et *Espera* Dcne. *Annales des Sciences Naturelles*, **16,** 200–214.

Yamada, Y. (1925). Studien über die Meeresalgen von der Insel Formosa. I. *Chlorophyceae*. *Botany Magazine*, *Tokyo*, **39,** 77–95.

Yamada, Y. (1933). Notes on some Japanese algae. *Journal of the Faculty of Science, Hokkaido Imperial University*, Ser. **5, 2,** 277–285.

Yamada, Y. (1934). The marine *Chlorophyceae* from Ryukyu, especially from the vicinity of Nawa. *Journal of the Faculty of Science, Hokkaido Imperial University*, Ser. **5, 3,** 33–88.

Yamada, Y. and Tanaka, T. (1938). The marine algae from the island of Yonakuni. *Institute of Algological Research Scientific Papers, Hokkaido Imperial University*, **2,** 53–86.

Yazykov, A. A., Zubarev, T. N. and Batkilin, E. M. (1975). Morphogenesis and cytoplasmic streaming. Microfilm, time 15 min., XII *International Botanical Congress, Leningrad, Abstracts*, **1**, 48.

Zanardini, G. (1878). Phyceae Papuanae Novae. *Nuovo Giornale Botanico Italiano*, **10,** 34–40.

Zavodnik, D. (1969a). La communauté à *Acetabularia mediterranea* Lamour., dans l'Adriatique du Nord. *Int. Revue Ges. Hydrobiol.*, **54,** 543–551.

Zavodnik, D. (1969b). Some problems concerning the study of phytal communities. *Thalassia Jugoslavica*, **5,** 451–456.

Zerban, H. and Werz, G. (1975a). Ultrastructure of flagellar microtubules in the green algae *Acetabularia mediterranea* and *Dunaliella salina* as revealed in freeze-etch preparations. *Cytobiologie*, **11,** 314–320.

Zerban, H. and Werz, G. (1975b). Changes in frequency and total number of nuclear pores in the life cycle of *Acetabularia. Experimental Cell Research*, **93,** 472–477.

Zerban, H., Wehner, M. and Werz, G. (1973). Über die Feinstruktur des Zellkerns von *Acetabularia* nach Gefrierätzung. *Planta*, **114,** 239–250.

Zetsche, K. (1962). Die Aktivität implantierter Zellkerne von *Acetabularia* bei aufgehobener Photosynthese. *Naturwissenschaften*, **49,** 404–405.

Zetsche, K. (1963a). Der Einfluss von Kinetin und Giberellin auf die Morphogenese kernhaltiger und kernloser Acetabularien. *Planta*, **59,** 624–634.

Zetsche, K. (1963b). Das morphologische und physiologische Verhalten implantierter Zellkerne bei *Acetabularia mediterranea. Planta*, **60,** 331–338.

Zetsche, K. (1964a). Polyploide-Effekte und ihr Zustandekommen bei mehrkernigen Transplantaten von *Acetabularia mediterranea. Planta*, **61,** 142–152.

Zetsche, K. (1964b). Der Einfluss von Actinomycin D auf die Abgabe morphogenetischer Substanzen aus dem Zellkern von *Acetabularia mediterranea. Naturwissenschaften*, **51,** 18–19.

Zetsche, K. (1964c). Hemmung der Synthese morphogenetischer Substanzen in Zellkern von *Acetabularia mediterranea* durch Actinomycin D. *Zeitschrift für Naturforschung*, **19b,** 751–759.

Zetsche, K. (1965a). Anreicherung von morphogenetischen Substanzen in Lichtpflanzen von *Acetabularia mediterranea* unter dem Einfluss von Puromycin. *Planta*, **64,** 119–128.

Zetsche, K. (1965b). Nachweiss von Guanosindiphosphat-Mannose- und Uridindiphosphat-Glucose-Pyrophosphorylase in *Acetabularia. Planta*, **64,** 129–137.

Zetsche, K. (1965c). Übertragung und Realisierung genetischer Information bei der Morphogenese von *Acetabularia. Berichte der deutschen botanischen Gesellschaft*, **78,** 87–88.

Zetsche, K. (1966a). Nachweiss von Enzymen des Galactosestoffwechsels in der Grünalge *Acetabularia mediterranea. Planta*, **68,** 240–246.

Zetsche, K. (1966b). Entkopplung morphogenetischer Prozesse in *Acetabularia mediterranea* durch P-fluorphenylalanin. *Planta*, **68,** 360–370.

Zetsche, K. (1966c). Anreicherung von Proteinsynthese induzierenden Substanzen in *Acetabularia mediterranea* unter dem Einfluss von Puromycin. *Zeitschrift für Naturforschung*, **21b,** 88–90.

Zetsche, K. (1966d). Regulation der zeitlichen Aufeinanderfolge von Differenzierungsvorgängen bei *Acetabularia. Zeitschrift für Naturforschung*, **21b,** 376–379.

Zetsche, K. (1966e). Regulation der UDP-Glucose 4-Epimerase-Synthese in kernhaltigen und kernlosen Acetabularien. *Biochimica et Biophysica Acta*, **124,** 332–338.

Zetsche, K. (1967). Unterschiedliche Zusammensetzung von Stiel- und Hutzellwand bei *Acetabularia mediterranea*. *Planta*, **76,** 326–334.

Zetsche, K. (1968a). Regulation der UDPG-Pyrophosphorylaseaktivität in *Acetabularia*. I. Morphogenese und UDPG-Pyrophosphorylase-Synthese in kernhaltigen und kernlosen Zellen. *Zeitschrift für Naturforschung*, **23b,** 369–376.

Zetsche, K. (1968b). Steuerung der Zelldifferenzierung bei der Grünalge *Acetabularia*. *Biologische Rundschau*, **6,** 97–112.

Zetsche, K. (1969a). Regulation der UDPG.-Pyrophosphorylaseaktivität in *Acetabularia*. II. Unterschiedliche Synthese des Enzymes in verschiedenen Zellregionen. *Planta*, **89,** 244–253.

Zetsche, K. (1969b). Die Wirkung von RNA- und Proteinsynthese-inhibitoren auf den Chlorophyllgehalt kernhaltiger und kernloser Acetabularien. *Planta*, **89,** 284–298.

Zetsche, K. (1972). The temporal and spatial control of enzyme synthesis and activity in *Acetabularia*. *Protoplasma*, **75,** 492.

Zetsche, K. and Wutz, M. (1975). Regulation of the heteromorphic life cycle of the siphonale green alga *Derbesia-Halicystis*. *Protoplasma*, **83,** 182–183.

Zetsche, K., Grieninger, G. E. and Anders, J. (1970). Regulation of enzyme activity during morphogenesis of nucleate and anucleate cells of *Acetabularia*. *In* " Biology of *Acetabularia* " (J. Brachet and S. Bonotto, eds.), pp. 87–110. Academic Press, New York and London.

Zetsche, K., Brändle, P. O. and Streichner, K. (1972). Photosynthesis in nucleate and anucleate cells of *Acetabularia*: The pathway of carbon. *In* " Biology and Radiobiology of Anucleate Systems ", (S. Bonotto, R. Goutier, R. Kirchmann and J. R. Maisin, eds.) Vol. 2, pp. 239–258. Academic Press, New York and London.

Zubarev, T. N., Melkumian, V. G., Rogatykh, N. P., Yazykov, A. A. and Yassinovsky, V. G. (1975). On a possible signal role of the action biopotential in *Acetabularia mediterranea*. XII *International Botanical Congress, Leningrad, Abstracts*, **1**, 48.

Zubarev, T. N. and Rogatykh, N. P. (1975). On the nature of the signal inducing the synthesis of " morphogenetic substances " in the *Acetabularia* nucleus in cap regeneration. In " Molecular Biology of Nucleocytoplasmic Relationships " (S. Puiseux-Dao, ed.) pp. 259–262, Elsevier, Amsterdam and New York.

Note

Professor Simone Puiseux-Dao (University of Paris VII) has recently edited the book " Molecular Biology of Nucleocytoplasmic Relationships, " 328 pages, 1975, Elsevier, Amsterdam/New York. It contains together with many interesting papers on nucleocytoplasmic relationships, 8 reports specifically on *Acetabularia*.

Adv. mar. Biol., Vol. 14, 1976, pp. 251–284

A STUDY IN ERRATIC DISTRIBUTION: THE OCCURRENCE OF THE MEDUSA *GONIONEMUS* IN RELATION TO THE DISTRIBUTION OF OYSTERS

C. Edwards

Dunstaffnage Marine Research Laboratory
Oban, Argyll, Scotland

I. Introduction

The sporadic occurrence of medusae of the genus *Gonionemus* in European coastal waters and in aquaria has excited interest because of the enigmatic distribution of the species. Medusae were first found in 1859 in the Gulf of Georgia on the Pacific coast of the United States of America by Alexander Agassiz (L. Agassiz, 1862, p. 350). Later they were reported from the American Atlantic coast at Woods Hole, Massachusetts and a few other localities in southern New England (Murbach, 1895; Mayer, 1910, pp. 343–8). The medusae have since been found to be widespread on North Pacific coasts from Puget Sound and British Columbia to the Aleutian Islands, Kurile Islands, Japan, Sakhalin, the Russian Far East Maritime Territory, China and Vietnam. They do not, however, occur in arctic waters, and the genus is of temperate and warm-temperate distribution on North Pacific coasts. In Europe medusae have been found on the Mediterranean coast of France, and at various localities on the western and north-western coasts from Brittany to Norway and the British Isles. The polyp, originally described as *Haleremita cumulans* (Schaudinn, 1894), has also been found in a few places in Europe.

It has been supposed that the medusa was transported from America to the Oslofjord in the tank of a ship (Kramp, 1922; Broch, 1924, p. 473; 1929, p. 488), that the medusa was carried across the Atlantic to the French coast entangled in seaweed attached to a ship's hull (Teissier, 1932), and that the medusa or the hydroid was carried from the Pacific to the Irish Sea on the bottom of a ship (Russell, 1953, p. 399). Werner (1950a, p. 494) considered it more likely that the species was introduced into Europe in the polyp stage rather than as medusae. To explain its appearance in the Rantumbecken on Sylt in the North Friesian Wattenmeer in 1947–49 he thought the species had probably been introduced in the polyp stage by shipping or flying boats during the 1939–45 war from some other European locality where it was already established, perhaps from the Oslofjord, where the species had, as above mentioned, been found in 1921. Leloup (1948) attributed the presence of medusae at Ostend in 1946–47 to the introduction of the hydroid on oysters, *Crassostrea virginica*, imported from the east coast of the United States in 1939–40. Tambs-Lyche (1964) reported having found medusae in 1958 in western Norway and discussed the problem of the distribution of the species. He considered that it had been spread by man, probably in the polyp stage in fouling on ships' hulls; but he was not convinced that the species was originally American and thought that the evidence equally supported a European origin. Kramp (1968, p. 137) concluded that the species is carried in the polyp stage attached to algae growing on ships' bottoms, whence medusae may be liberated, establishing the species in favourable localities.

The finding of medusae and polyps in aquarium jars at the Dunstaffnage Laboratory in 1974 and 1975 and the occurrence of a medusa in 1966 in the Marine Station, Millport, have led me to analyse the records of *Gonionemus* (Table II, p. 272) and to attempt to explain the distribution. An understanding of the life-history, habits and habitat of the medusa and its hydroid is essential to discussion of the distribution and the possible modes of transport of the species, and a summary is given from the literature (A. Agassiz, 1865; Goto, 1903; Hargitt, 1905; Joseph, 1925; Mayer, 1910, pp. 344–5; Perkins, 1903; Picard, 1951; Schaudinn, 1894; Tambs-Lyche, 1964; Teissier, 1932; Uchida, 1929; Werner, 1950a, b). The medusae are strictly littoral, occurring in shallow water close inshore and in coastal lagoons, where algae (such as laminarians, *Himanthalia*, *Sargassum*, and fucoids) or sea-grasses (*Zostera*, *Posidonia*, *Ruppia*) are present. They have adhesive tentacles, by which they cling to weeds or other objects. At times, especially at nightfall or on dull days, they swim to the surface, turn over, and sink with tentacles spread out to catch organisms, the process being repeated with little respite. Occasionally a medusa may be seen swimming in clear water away from weeds. Eggs shed by

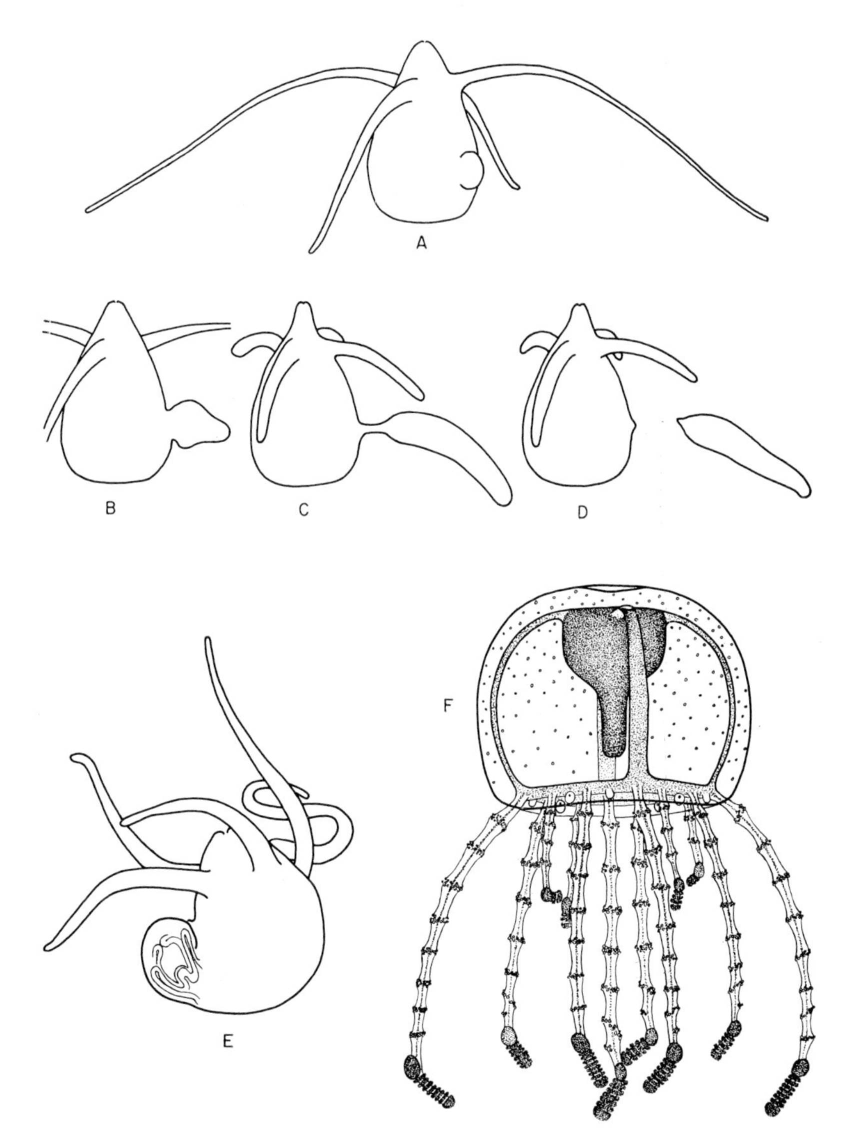

Fig. 1. Stages in the life-history of *Gonionemus*. A–D, successive stages in the production by a polyp of an asexual bud and its severance as a frustule (redrawn from Perkins, 1903, Figs. 2, 5, 7, 8); E, polyp bearing a gonophore-bud (redrawn from Joseph, 1925, Fig. 40); F, medusa, diameter 1·51 mm, height 1·15 mm, shortly after release from a polyp in the Dunstaffnage Marine Research Laboratory 22 April 1974.

medusae settle on objects on the bottom or on weeds, and they develop into polyps, which are small, solitary and enclosed in a slimy sheath masked with algal or other detritus and are difficult to find. The species reproduces in two ways: by eggs produced sexually by free-swimming medusae budded from the polyps, and by the asexual production from the polyps of outgrowths ("frustules") which become vermiform and larva-like, leading a free-creeping life (days to a few weeks) before settling to transform into polyps (Fig. 1). The species has the potentiality for continued and long-term asexual multiplication of polyps without sexual reproduction.

II. *Gonionemus* at Dunstaffnage and Millport

Gonionemus medusae were found at the Dunstaffnage Laboratory in April to June 1974 in aquarium jars in which I was keeping colonies of *Sarsia* hydroids growing on stones, collected in February and April 1973 in various low-water-mark localities in the neighbourhood. The medusae were removed to other jars and reared on *Artemia* nauplii. Examination of the stones revealed several *Gonionemus* polyps, some bearing single medusa-buds. These stones were a few of many collected and kept in the laboratory under similar conditions. None of the other stones bore *Gonionemus*. I conclude from my knowledge of the collecting of the stones and of their treatment in the laboratory that it is very unlikely that they bore *Gonionemus* when collected and that fairly certainly the *Gonionemus* were introduced into my jars with sea-water drawn from the open circulating supply of the laboratory, probably as frustules produced by polyps in the sea outside. Medusae have not been seen in the sea here. The stones had been kept at various temperatures, some at 6°C, some at 10°C, and others at the temperature of the circulating sea-water. Notwithstanding these differences in temperature the *Gonionemus* medusae appeared in various jars at about the same time, indicating that some factor other than temperature had brought the polyps simultaneously to reproductive condition—fairly certainly the regular feeding with *Artemia* nauplii. Medusae liberated and kept at 6°C made very slow growth and died while small and immature, showing that this is too low a temperature for development and supporting my belief that feeding, not temperature, had caused the medusae to be produced. Medusae released and kept at the temperature of the circulating sea-water (9–11°C) attained maturity at about 7 mm diameter with about 56 tentacles (Table I). This is considerably smaller than the size (15–20 mm diameter; 60–80 tentacles) attained in nature (Kramp, 1961, p. 223; Mayer, 1910,

TABLE I. *Gonionemus* MEDUSAE REARED AT 9–11°C APRIL–MAY 1974

Diameter (mm)	*Height (mm)*	*Number of tentacles and rudiments*	*Number of vesicles*	*Remarks*
1·62	1·35	17	4	
1·98	—	20	4	
2·12	1·44	21	4	
2·20	1·76	30	11	
2·55	1·67	28	4	
2·64	1·76	32	8	
4·6	3·5	35	12	
4·8	3·7	40	12	
6·0	4·0	49	14	Mature male
7·0	4·0	56	8	Mature male

p. 342). The medusae were male and the sinusoidal testes occupied the proximal three-quarters of the radial canals. Russell (1953, p. 402) reared some medusae and found that his largest specimens attained only 10 mm diameter. My specimens developed relatively few vesicles. The eastern American *Gonionemus murbachi* was described as having at least half as many vesicles as tentacles, the Pacific American *Gonionemus vertens* as having about the same number of vesicles as tentacles, and the Japanese *Gonionemus depressum* as having twice as many vesicles as tentacles. I consider that my specimens did not attain their natural size and complement of vesicles and tentacles under the abnormal conditions of laboratory rearing, particularly temperature. In nature medusae appear in summer and autumn, in warmer conditions than those in which mine were reared.

Further medusae have been liberated from polyps in my aquarium jars in the Dunstaffnage Laboratory in March, May and September 1975 at temperatures ranging from 6° to 10°C.

At the Marine Station, Millport, a very young medusa was found on 16 March 1966 in a jar containing a clinker which had been dredged nine days before near the laboratory in 12 m depth. Evidently the clinker bore one or more polyps (these were not searched for).

III. OCCURRENCE OF *Gonionemus* IN THE BRITISH ISLES

In the British Isles *Gonionemus* has been found (Table II, p. 272) at the Dove Marine Laboratory, the Marine Biological Station, Port Erin, the Marine Station, Millport, the Dunstaffnage Laboratory, in a private aquarium at Aberdour, Fife, and in a small temporary coastal lagoon at Cuckmere Haven, Sussex. This led me at first to suppose that the species is widespread, but that, because of its cryptic habits, it is not likely to be detected except where laboratories are established. However, there are no records of the species from other coastal laboratories and, further, only one specimen was found at Millport, where I had made an intensive study of the medusan fauna for several years.

Since the species lives in very shallow water with weeds, I consider that it is unlikely to be transported on ships' hulls. I agree with Werner (1950a, p. 494) that it is more likely to have been introduced as polyps than as medusae. Because of its habits long-distance dispersal across oceans and seas by natural means would be extremely difficult and most improbable: only slow and coastwise dispersal is likely in nature.

For a number of years oyster investigations were conducted at the Marine Station, Millport (Millar, 1961), and I have been led to suspect the implication of oysters in the transport of *Gonionemus*. Oysters live in shallow waters, their shells support a rich epifauna (Korringa, 1951; Möbius, 1877), and they have been very extensively transported and transplanted. They have been the agencies of introduction of several foreign species of molluscs: the slipper-limpet *Crepidula fornicata* (L.) was introduced from North America into England from about 1880 onwards with American east-coast oysters *Crassostrea virginica* (Orton, 1912); the oyster-drill *Urosalpinx cinerea* (Say) was similarly brought from North America to England, probably at about the same time (Orton and Winckworth, 1928), though Cole (1942) was inclined to think it was later; and the rock-boring bivalve *Petricola pholadiformis* Lamarck was first recorded on the east coast of England in 1890, probably introduced from North America in the same way (Haas, 1938, p. 442; Purchon, 1968, p. 388).

Oyster investigations at Millport were carried out from 1946 to 1968, and the following summary is based on Dr R. H. Millar's 1961 report and on information received personally from him. From time to time consignments of 2–3 years-old imported oysters were received at Millport and used for experimental layings in a number of localities on the west coast of Scotland in Argyll, Mull, Soay, Wester Ross and Lewis. The consignments were held at Millport in large concrete tanks filled from the sea and discharging to the sea, and some of the oysters were retained in smaller tanks for observation and experiment. The majority were *Ostrea edulis* imported from Brittany. One consignment of *O. edulis* was received from Bergen, Norway. Some Portuguese oysters, *Crassostrea angulata*, were also received both from Brittany and (via an English grower) from Portugal. The sole specimen of *Gonionemus* found at Millport was on a clinker dredged near the Marine Station, and the source may well have been imported oysters. Concerning the occurrence of *Gonionemus* at Dunstaffnage, the nearest localities in which oysters were laid were Loch Melfort, Balvicar Bay, Clachan, Loch Feochan, Loch Creran and Loch Don.

In 1969 a consignment of *Ostrea edulis* from western Ireland was laid in Loch Creran by Scottish Sea Farms Ltd, followed by small quantities from Gairloch (Wester Ross) and Loch Scridain (Mull). These were " native " oysters, but it is impossible to say whether the beds from which they came had always been free of imported oysters. Stocks of *Crassostrea gigas* have since been maintained in Loch Creran by this company, but these have been derived from laboratory-bred stocks reared at the Fisheries Experiment Station at Conwy, North Wales, and they cannot have been the source of introduction of *Gonionemus*.

At the Marine Biological Station, Port Erin, two medusae, reported by Russell (1953, p. 399) and identified by him as *Gonionemus vertens*, were found in a storage tank on 2 December 1924. Examination of the annual reports of the station has led me to the work done there in 1896–99 by Professor W. A. Herdman and co-workers (Herdman and Boyce, 1899) on oysters and their diseases. They used *Ostrea edulis* from various localities in England, Wales, Isle of Man, Channel Islands, Netherlands, Belgium, west coast of France, and Italy; *Crassostrea virginica*, some imported direct from America to Liverpool and some that had been relaid at various localities in England, Wales and Ireland for longer or shorter periods; and *Crassostrea angulata* from England and south-west France. There is the probability that *Gonionemus* in the polyp stage was brought to Port Erin on oysters and that the species persisted in the laboratory tanks or in the sea close by.

In the Dove Marine Laboratory tanks Miss Robson (1913, p. 27; 1914, p. 90) found five very young medusae, which she provisionally named *Cladonema* sp. Joseph (1924, p. 129; 1925, p. 377) identified them as *Gonionemus* from Robson's figure. In May 1951 medusae were again found in the tanks there, together with polyps, and some medusae were sent to Plymouth, where they were reared by Russell (1953, pp. 402–3, pl. 35), who identified them as *Gonionemus murbachi*. It is likely that the species had persisted there. I have been unable to find any reference to oysters having been kept in the Dove Marine Laboratory, but have noticed references to oysters in reports on work done on the Northumberland mussel beds by staff of the laboratory. In the Northumberland Sea Fisheries' Committee's Report for 1899, pp. 49–50, it is stated that oysters were brought for fattening to beds at Warham and Fenham in the northern part of the county. Again, in the Committee's Report for 1906, p. 38, it is stated: " A few oysters occur here [Holy Island], but they are not native; the true old natives are extinct ". Bulstrode (1896, p. 50) reported on oyster ponds in Budle Bay in the same area, " wherein oysters of various ages and from divers sources are, or have been, placed for the purposes of storage, growth and fattening ". Interest in oysters was then widespread and specimens may well have been kept in the Dove Laboratory for observation.

Gonionemus medusae were found at Aberdour, Fife, by Howe (1959) in his private aquarium, which he had stocked with stones from the shore there. Aberdour was formerly noted for its native oysters, which, as elsewhere in the Firth of Forth, have long been extinct through overfishing. It is likely that imported oysters have at some time been laid there, but I have found no record. At Cuckmere Haven, Sussex, twelve medusae were collected by the late Dr W. J. Rees and Mr R. C. Vernon on 26 July 1963 in " brackish water in some abandoned workings behind the sea-wall, obviously open to the sea at high tides " (personal communication from Mr Vernon). Mr D. T. Streeter tells me that small temporary lagoons are occasionally formed behind the shingle bank, but that there are no permanent pools. There is no history of this place having been used for laying oysters. Probably *Gonionemus* polyps are established in places along the Sussex coast and asexually produced frustules can enter lagoons, where, if conditions are suitable, medusae can be produced in summer. Alternatively, stones or shells bearing polyps may be thrown by high seas over the shingle banks into lagoons. There has of course been a long history of oyster cultivation in various parts of the Sussex and Hampshire coasts.

After drafting this paper I heard from Dr P. F. S. Cornelius of the Coelenterate Section, British Museum (Natural History) that *Gonionemus* medusae had just been found, on 19 September 1975, in an aquarium at the museum, maintained at 15°C by the Bryozoa Section (Miss P. L. Cook). They were 1 cm in diameter and evidently several weeks old. Search had revealed the presence of reproductive polyps on a dead shell of *Laevicardium crassum* (Gmelin) and of some small recently-released medusae. The shell was part of a consignment of shell-gravel received by Miss Cook from Plymouth, dredged on 24 June 1975 on the offshore Eddystone grounds. No other polyps could be found. From information kindly supplied by Miss Cook about the aquarium and its contents it is almost certain that the source of the *Gonionemus* was the shell gravel. The occurrence of the species offshore is most unusual. (The medusae and the polyps have been preserved in the British Museum and given registered numbers 1975.9.29.1a, 1975.9.29.1b.)

IV. Occurrence of *Gonionemus* in North America

On the Pacific coast of North America the medusa *Gonionemus vertens* has been reported from the Gulf of Georgia (A. Agassiz, in: L. Agassiz, 1862, p. 350; A. Agassiz, 1865, pp. 128–30), Victoria Harbour, Vancouver Island and Puget Sound (Murbach and Shearer, 1903, p. 184), and several localities in the Strait of Georgia (Foerster, 1923, p. 264). A. Agassiz found it quite commonly in July 1859 swimming in patches of kelp, and Foerster states that it is common. Ricketts and Calvin (1952) say that the medusa is locally abundant in sheltered places from Sitka southwards to the region of Puget Sound, mainly inhabiting eelgrass beds. As mentioned in the introduction, *Gonionemus* is widespread around the North Pacific temperate coasts (see Table II, p. 272). In Japanese waters medusae are common. Recently (Todd *et al.*, 1966) the medusa has been found in a lagoon on the Santa Barbara campus of the University of California, almost 1 000 miles south of Puget Sound, and it is supposed that the species has been accidentally introduced.

In contrast with the widespread distribution and abundance of the species on the Pacific coast of North America, the occurrence of *Gonionemus* on the Atlantic coast is remarkable. Perkins (1903, pp. 750–1) says: " In spite of the fact that the ' eel-pond ' at the centre of the village of Woods Hole, a small body of water connected with the outer harbor by a narrow inlet, is easy of access to collectors, and that numerous students of jelly-fishes had investigated the waters around Woods Hole summer after summer for a number of years, *Gonionema* was never found in the Atlantic Ocean until 1894. During that summer a number of specimens were taken from the eel-pond, the creature having made an astonishingly sudden appearance upon the scene. It seems incredible that *Gonionema* could have been living in this small body of water for any time previously, or at any rate that any number of individuls had been there. But the jelly-fish at once secured a good foothold, and since the first summer it has been very plentiful." Mayer (1910, p. 344) says: " It has continued to reappear each summer in large numbers in this small pond since its discovery in 1894. The medusa has been found occasionally in Woods Hole Harbor outside the Eel Pond, and it is reported also from Noank, Connecticut, and from Hadley Harbor, Muskegat Island (Murbach)." The medusa continued to be plentiful in the Eel Pond until 1930 (Rugh, 1930), but with the disappearance of the eelgrass *Zostera marina* there in 1931 (when the plant died widely on the Atlantic coast) the medusa disappeared also (Teissier, 1950, pp. 6–7; Werner, 1950a, p. 498). The species has not been reported from any other part of the North American Atlantic coast, despite the very considerable work done on the fauna at many places along this coast.

The sudden appearance of *Gonionemus* at Woods Hole in 1894 and its extremely restricted distribution in southern New England strongly point to its introduction there. That it came with oysters from the Pacific in the course of trade seems unlikely, because the trade in live oysters in the second half of the nineteenth century was in the reverse direction, the east-coast oyster *Crassostrea virginica* being transported to the Pacific coast, especially the San Francisco region, from at least as early as 1849 (Barrett, 1963, p. 21) to meet the demand by the many easterners who had poured into California. Some live oysters, but chiefly canned ones, had been shipped from east-coast ports by sea and transported across the Isthmus to the west coast for onward shipment to California. With the completion of the transcontinental Central Pacific Railroad in 1869 and of other transcontinental railroads in later years a big trade developed in the despatch of fresh east-coast oysters to California for immediate consumption, for planting and as seed (Barrett, 1963, p. 26; Ingersoll, 1887, p. 538). Although scientific and commercial interest in the introduction of the Japanese oyster *Crassostrea gigas* (now called the Pacific oyster) into the United States Pacific coast began in the 1870s, leading this century to the development of a Pacific oyster industry based on Japanese seed imports (Barrett, 1963, pp. 48–9), the Pacific oyster was not commercially planted on the east coast.

There were tentative efforts in the 1880s to transport various clams from the Pacific coast to the Atlantic (Stearns, 1883), and in the Report for 1885 (pp. L, XCIX-C) of the United States Commission of Fish and Fisheries it is stated that large numbers of the clam *Tapes staminea* were brought from Puget Sound to Woods Hole and planted in various localities in the vicinity of the station, and that there were other species which it was proposed to transport in a similar manner. Accordingly it is possible that *Gonionemus* was introduced into Woods Hole with these Pacific clams. The buried mode of life of clams, however, renders them unlikely hosts of the hydroid. If seaweed or eelgrass had been used for packing the clams, the hydroid may have been introduced in this material, but there is no information on this point.

In the 1880s, at the U.S. Fish Commission's Laboratories, John A. Ryder was engaged in extensive investigations into the biology of oysters. Two artificial ponds were constructed at Woods Hole in 1884 for the propagation of oysters, and use of the adjacent portion of the Eel Pond was granted to the Commission (U.S. Commission of Fish and Fisheries, Report for 1884, pp. XXXVIII, LXV). In this report it states: "It was shown in the course of the experiments that oysters might be successfully grown at Woods Holl, a locality in which that mollusk had never, so far as we know, been indigenous." Ryder had earlier been investigating the comparative anatomy of the American east-coast oyster and the several European species at the Commission's laboratory at Saint Jerome, Maryland (U.S. Commission of Fish and Fisheries, Report for 1882, p. LXXXIV), having received a great variety of living oysters from Europe (Report for 1881, p. LVI). This strongly suggests that imported oysters were the means of introduction of *Gonionemus* into Woods Hole, and since the species has not been found elsewhere on the American east coast, except in localities near Woods Hole, it points to Europe as the likely source, though introduction from the Pacific coast cannot entirely be ruled out. Support for the view that the species may be of European, rather than Pacific, origin is perhaps provided by the differences detected between the Pacific *Gonionemus vertens* and the form *G. murbachi* found at Woods Hole (Bigelow, 1909, p. 106; Mayer, 1910, pp. 342–4). The question of racial, subspecific or specific differences in the genus will be discussed later in this paper. Although the medusa was not found at Woods Hole until 1894, the species may have been introduced some years earlier, gradually becoming established in the polyp stage. The sudden appearance of the medusa at Woods Hole, its extremely restricted distribution on the American east coast and its sudden disappearance and probable extinction there in 1931 all strongly suggest that it was not native to the east coast but was accidently introduced into a locality temporarily suited to it.

V. Occurrence and Distribution of *Gonionemus* in Europe

Discussion of the possible significance for the distribution of *Gonionemus* in Europe of the transport and transplantation of oysters requires an outline account of the history of the oyster trade and industry in Europe during the past 120 years or so. In the following summary I have been greatly helped by the comprehensive account of Yonge (1960) and by the excellent library resources at this laboratory.

Because of the greatly increasing demand for oysters in the nineteenth century by the rapidly rising populations of the towns, the extensive natural beds of *Ostrea edulis* that formerly existed in many places along the European coasts became rapidly depleted and in many cases exhausted by overfishing. On the Channel and Atlantic coasts of France, especially in Brittany and southwards along the western shores, there had been many rich beds. Overfishing produced quick and devastating results, most of the beds being destroyed or made economically unworkable by the 1850s. This prompted immediate government action. From the work of de Bon and particularly of P. Coste there developed by about 1870 a very large, highly organized and extremely productive oyster industry, especially in the Morbihan district in southern Brittany, in the Charente area, and at Arcachon. The Arcachon region had been the most severely affected by the depletion of the natural beds, and the revival of the industry there was achieved by the protection of the few remaining natural beds, by the importation of large numbers of young *O. edulis* from England and the north coast of Spain, and by the development of very successful methods for the collection of oyster spat and for the rearing, fattening and conditioning of the oysters. A major change in the French oyster industry was also brought about by the introduction of the Portuguese oyster, *Crassostrea angulata*, from the mouth of the River Tagus. This species was imported from 1867 in relatively small numbers for relaying on the depleted beds at Arcachon. In 1868 a ship laden with them was stormbound and damaged in the mouth of the Gironde and the cargo of oysters, thought to be dead or dying, was dumped on a bank. The surviving oysters flourished and spread remarkably quickly, establishing natural beds for 25 km along the left bank of the Gironde near Le Verdon. The species thence colonized the coast of the Charente Maritime and northwards (the natural beds of *Ostrea edulis* being extinct in this region), the spread being finally stayed south of the Loire by unsuitable temperatures to the north. The Portuguese oyster grew and reproduced well at Arcachon, where it increasingly contributed to the success of the industry, eventually greatly exceeding *O. edulis* in quantities commercially produced. Experiments were also made in the cultivation of the American oyster, *Crassostrea virginica*, at Arcachon: some spat survived the journey from America and grew well, but, because the species was not well thought of, its rearing was abandoned (Bouchon-Brandely, 1877). In Brittany and on the Normandy coast the cultivation of *Crassostrea angulata* has been allowed in certain areas. In the early years of the efforts to rebuild the French oyster industry some American oysters, *C. virginica*, were planted at Hougue de Saint-Vaast in Normandy (de Broca, 1865), but the attempts to cultivate this species here and elsewhere in France were not followed up. (References consulted on the oyster industry in France include: de Bon, 1875; Bouchon-Brandely, 1877, 1878; Calvet, 1910; Coste, 1861; Dean, 1892; Yonge, 1960.)

Gonionemus medusae have been repeatedly found at Callot, near Roscoff, on the north coast of Brittany by Teissier (1930, 1932, 1950, 1965), who collected them, attached to algae in the *Himanthalia* zone, in 1929, 1931, 1945 and 1961–64. *Gonionemus* polyps have been found on stones at Callot and in intertidal pools under stones at Roscoff (Teissier, 1965, p. 30). The species is evidently resident there, but we do not know how long before 1929 it was established there. Teissier (1932, 1950) thought that the species had been introduced into France by medusae carried across the Atlantic from the east coast of North America entangled in seaweed attached to a ship's hull. I consider this most improbable. He was mistaken in thinking that the 1929 record from Callot indicated the first European establishment of the species: it had been found at Cette, Cullercoats, the northern Adriatic, Oslofjord, Port Erin and Gullmarfjord at various dates from 1876 to 1925 (Table II, p. 272). Because the species was noticed at Woods Hole in 1894, there arose the belief that it is native to the North American east coast. I have, however, already given reasons for considering that it is not native to that coast and that it was probably introduced there from Europe late last century.

In the British Isles extensive beds of *Ostrea edulis* formerly existing in many places around the coasts increasingly suffered depletion during the nineteenth century through overfishing. To meet the ever-growing demand, American east-coast oysters, *Crassostrea virginica*, were imported live in large numbers from about 1870 onwards, Liverpool being the chief port of importation (Ingersoll, 1887), this trade having been conducted on a small scale since 1861 (Stevenson, 1899). Surplus oysters, not wanted immediately for market, were laid in the sea in Liverpool Bay until required. Similarly, Irish beds in Counties Louth and Dublin were stocked with American oysters destined chiefly for the Liverpool and Manchester markets (Went, 1962). Young American oysters, too, were laid in large numbers in English and Irish beds. The oyster-growers of south-east England, to replenish their beds, not only went to Ireland, Wales and Scotland for stock from native beds there, but imported large quantities of spat and brood oysters from Holland and France. The oysters from France included not only *Ostrea edulis* but also, and increasingly, the Portuguese oyster *Crassostrea angulata*.

In considering the occurrences of *Gonionemus* in Britain I have suggested the likelihood of its introduction with oysters, and in discussing its occurrence in eastern North America I have concluded that it is highly probable that the species was introduced there on oysters from Europe in the 1880s or early 1890s. If *Gonionemus* was introduced with oysters into these two regions, the circumstances suggest a common source and, moreover, a common event. Now *Ostrea edulis*, the native oyster of western Europe, had long been transported by British and continental growers and merchants. *Gonionemus*, however, was not found in Europe until late last century and in the British Isles not until early this century, though I have already suggested that the introduction into Great Britain may well have been some years earlier: at Port Erin, for instance, the introduction may be attributable to importation of oysters in 1896–99. Having carefully weighed all the circumstances, I consider that it is very probable that the original source of introduction of *Gonionemus* into western Europe, the British Isles and eastern North America was *Crassostrea angulata* and the important event was the importation of *C. angulata* into France from Portugal from 1867 onwards. That is not to say that *Gonionemus* has not since been carried on *Ostrea edulis*: once *Gonionemus* was established in French waters it could probably be transported on any species of oyster. *Crassostrea angulata* not only became very thoroughly established in western France from Arcachon to the Loire but, as already mentioned, it has been cultivated in Brittany. Furthermore, there has been continued importation of *C. angulata* from Portugal to France and, moreover, some importation from Portugal direct to other countries, including the British Isles, so that, if *Gonionemus* has been introduced from Portugal with *C. angulata*, it may have been introduced many times since 1867 and not solely to France. Since, however, France became a major exporter of oysters to other countries, it is likely to have been the chief source of spread of *Gonionemus*. (The subject of the origin of the population of *Gonionemus* in Portugal will be discussed later.)

In the Netherlands the natural beds of *Ostrea edulis* had, as elsewhere in Europe, been largely exhausted by the second half of the nineteenth century. From surviving stocks in the Oosterschelde (a former arm of the delta of the Schelde severed by a viaduct embankment), by closure of free fishing and state supervision from 1870, and by introduction of quantities of *O. edulis* from the natural beds in the Firth of Forth, the oyster industry was re-established in the Oosterschelde, where it became a highly organized and thriving one (Havinga, 1932). In 1920–21 a widespread epidemic oyster-disease in Europe, particularly in France and England, caused enormous mortality on natural and cultivated beds of *O. edulis*. In the years after 1921 Dutch oysters were exported in large numbers to various European countries, especially Brittany and England, to re-establish the beds there (Korringa, 1947). The Dutch beds were themselves endangered in 1930 and the following years by " Dutch shell disease ", a fungal disease traced to vast quantities of old cockle shells that had been laid on the beds over a long period as cultch for spat settlement (Korringa, 1952). During 1936–49, after the beds had been cleaned, large quantities of French oysters were laid in Dutch waters to re-establish the beds (Korringa, 1951). In Germany there were formerly extensive and famous natural beds of *O. edulis* in the Wattenmeer off Schleswig-Holstein (Möbius, 1877), which, by careful control of the fishing, long escaped destruction by overfishing. Eventually these were exhausted by the 1930s, and attempts were made to revive them by introduction of Dutch seed oysters and, to a smaller extent, of French oysters. Dutch oysters were planted on the Ellenbogen bank off northern Sylt, first in 1914 and then in 1925–39, on the Hörnum bank off southern Sylt in 1925–35, near Oland in 1938–39, and off the south-east coast of Föhr in 1944 (Hagmeier, 1941; Werner, 1948). French oysters were laid on the Ellenbogen bank in 1937. The bed off Hörnum was destroyed by soil-dredging in 1935–36 during the construction of an embankment for sealing off Rantumbucht, Sylt, to form the Rantumbecken, a seaplane base (see below).

Gonionemus has, as already stated, been resident near Roscoff, Brittany from 1929, and it had probably been there for many years before. It is likely that the species has been established in other French localities. Oysters from Brittany and Arcachon have been extensively exported to Great Britain, Holland, Belgium, Germany and other European countries for relaying, and, as explained above, Dutch oysters, from beds re-established with French oysters, have been used for relaying on German and other beds. If *Gonionemus* has been transported with oysters, this trade in exporting oysters, of direct or indirect French origin, for relaying can explain the occurrence and distribution of *Gonionemus* in Europe. In Germany the species was found in 1947–49 in the Rantumbecken on Sylt (Werner, 1950a). Werner suggested that *Gonionemus* may have been introduced in the polyp stage by shipping or seaplanes during the 1939–45 war from some other locality where it was already established. I consider that, because of the shallow inshore weedy habitat favoured by the species, it is unlikely to settle on ships' bottoms. I think also that the polyp or the medusa is unlikely to survive aerial transport on the hull of a seaplane. Because of the relaying of foreign oysters on the German beds (particularly on the Hörnum bed, which was destroyed during the construction of the Rantumbecken), the introduction of *Gonionemus* into German waters on oysters is strongly indicated. In the Netherlands *Gonionemus* was found in 1960 at Ritthem, Walcheren (Kramp, 1961, p. 445). *Gonionemus* could well have been introduced into Dutch waters on oysters from France. It is perhaps significant that in the mid-1950s oysters were transplanted from the overcrowded beds in the Oosterschelde to other areas in Zealand waters, including the Zandkreek between North and South Beveland (Korringa, 1956, 1957, 1958). *Gonionemus* may have been spread with the relaid oysters, and the Ritthem record may be accounted for in this way.

In Belgium the Bassin de Chasse at Ostend has been used from 1938 for the fattening of oysters, at first *Ostrea edulis*. Leloup (1948) states that some American oysters, *Crassostrea virginica*, were laid there in 1939–40. Since 1947 the basin has been used for relaying imported *O. edulis* and *Crassostrea angulata* (Halewyck and Leloup, 1951). *Gonionemus* was found in the Bassin de Chasse in 1946 and 1947, and Leloup (1948) considered that it had been introduced on the American oysters laid there in 1939–40. Presumably he was led to this conclusion by the belief that *Gonionemus* is an eastern North American species. I have given my reasons for thinking that *Gonionemus* is not native to the North American east coast, where it was probably introduced from Europe. It is much more probable that *Gonionemus* was introduced into the Bassin de Chasse on oysters from other European beds. Incidentally, *Gonionemus* was not found in a survey of the fauna of the Bassin de Chasse made in 1937–38 (Leloup and Miller, 1940).

In Denmark extensive natural beds of *Ostrea edulis* formerly existed in the Limfjord in northern Jutland. The oyster stocks became very low during 1925–46 and large numbers of Dutch, French and Norwegian young oysters were put in the Limfjord during the years 1925–51 (Spärck, 1951). Korringa (1969) says that these formerly rich beds are almost barren and that oysters are now almost entirely grown there by relaying imported stock. The occurrence of *Gonionemus* at Frederikshavn in 1960 (Kramp, 1961, p. 445) may be related to this laying of foreign oysters.

Gonionemus is well established in the Gullmarfjord. It was detected in 1923 and 1930 (Lönneberg, 1930, p. 173), and in more recent years it has been found abundantly (Jägersten and Nilsson, 1961, p. 182). There are natural beds of *Ostrea edulis* in the fjord, but laying of imported oysters there was carried out over many years from before 1917 (letter from Dr Armin Lindquist, Institute of Marine Research,

Lysekil). In Norway there are the following records: one medusa at Dröbak, Oslofjord, 1921 (Kramp, 1922); several medusae at Espevaer near the mouth of the Hardangerfjord and one at Sandvik, on Halsnöy, in the outer part of the fjord, 1958 (Tambs-Lyche, 1964); one medusa in the Trondheimfjord, 1969 (Gulliksen, 1971). Oyster culture in Norway is based on native *Ostrea edulis.* However, Mr P. Soleim, Fisheries Museum, Bergen, has found from the records that there was some importation of oysters before the First World War from the Netherlands to an oyster poll on Tysnes island in the outer part of the Hardangerfjord and says that oysters for consumption have sometimes been imported to the Bergen area from the Netherlands (personal communication). Tambs-Lyche (1964) ascribed the presence of *Gonionemus* in western Norway to transport on ships' hulls. Although this is possible, I think, in the light of the above information and of the strong circumstantial evidence for the spread of the species elsewhere in Europe through movement of oysters, that the species has been introduced into Sweden and Norway with oysters.

The occurrence of *Gonionemus* in the Mediterranean Sea will now be considered. On the French coast, following the recommendations of Coste (1861), attempts were made in 1860 to establish beds of *Ostrea edulis* in the Étang de Thau at Cette and in the roadstead of Toulon. 3 000 oysters were collected from natural beds in the Gulf of Lions and 40 000 were brought from England (Bon, 1875; Calvet, 1910). In 1865–66 further oysters were put in the Étang de Thau (Bouchon-Brandely, 1878). Coste attempted similarly to establish beds at Saint Tropez (Bouchon-Brandely, 1878). These attempts failed, but the very successful establishment of the oyster industry in the Bay of Arcachon from 1870 onwards suggested the possibility of raising Arcachon oysters in the Étang de Thau. The first concession was granted in 1875 on the shore of the Eaux Blanches, the shallow eastern part of the Étang adjacent to Cette. During 1880–84 seven further concessions were granted along the canals of Cette. These were used for the rearing and fattening of oysters (both *Ostrea edulis* and *Crassostrea angulata*) from the Bay of Arcachon and for the temporary holding of oysters from various parts of the French Atlantic coast (Calvet, 1910). Similarly, oysters from Arcachon were successfully cultivated in the Golfe de Giens, near Toulon (Bouchon-Brandely, 1878).

In 1876 *Gonionemus* medusae were found in a brackish weedy canal leading from the salt-works at Villeroy, near Cette, to the Étang de Thau. (The medusae were reported by Du Plessis, 1879, as *Cosmetira salinarum*, and were reidentified by Picard (1951) as *Gonionemus vertens* from Du Plessis's description and figure. Although there are certain deficiencies in the description, I agree with Picard's identification.) The discovery of the medusae in such an unusual locality excited interest. Despite long-continued studies of the fauna of the Étang de Thau by scientists of the University of Montpellier, which for many years maintained a biological station at Cette, the medusa has never been reported there again. This isolated occurrence in 1876 in such unusual conditions, so soon after the commencement of importation there of oysters from Arcachon, suggests accidental introduction of the species with oysters from Arcachon.

The only other record of *Gonionemus* from the south of France is by Picard (1951, 1955), who found medusae in 1950 and 1951 at Villefranche-sur-Mer in the beds of the sea-grass *Posidonia*. He considered that their occurrence indicated the permanent presence of polyps, and he supported the hypothesis of Teissier (1950), with which I concur, that the species may persist in a locality by asexual reproduction of the polyps. Because the species has so rarely been found in the Mediterranean, a region so long and so assiduously studied by marine zoologists, I consider that its presence at Villefranche indicates the probable local establishment of an introduced species and suggests its introduction with oysters from western France, which have been laid at various places along the south coast.

Gonionemus has been found in aquaria supplied with material from the north-eastern Adriatic. Schaudinn (1894) discovered the polyp in the Berlin Aquarium, which had been stocked from Rovigno. He described it as a new hydroid, *Haleremita cumulans*, and reported the asexual production of frustules, but he did not see the sexual phase and did not know its relationship to *Gonionemus*. Joseph (1918) found medusae in 1917 in the aquarium of the Zoological Institute of Vienna, which had been stocked from the Gulf of Trieste; and in 1924 he found polyps and worked out the life-history (Joseph, 1924, 1925). In 1923 Cori found medusae in the aquarium of the Zoological Institute of the German University of Prague, this aquarium, too, having been supplied from the Gulf of Trieste (Joseph, 1924). Joseph concluded that *Gonionemus* was native to the Adriatic, but these reports are equally consistent with its possible introduction there last century. The Gulf of Trieste and the Canal of Leme, near Rovigno, have long been important oyster-producing areas. Following the successful re-establishment of the oyster industry in western France in the 1870s efforts were made during the last decades of the nineteenth century at various places in the northern Adriatic, including the Lagoon of Grado, the Bay of Muggia (both are in or near the Gulf of Trieste) and Pola and the Canal of Leme, in Istria, to introduce rational systems of culturing the native oyster, *Ostrea edulis* (Allodi, 1906). I have found no mention of the introduction of oysters from elsewhere and can only point out the coincidence of efforts here in the late nineteenth century to redevelop oyster culture and the discovery of *Gonionemus*.

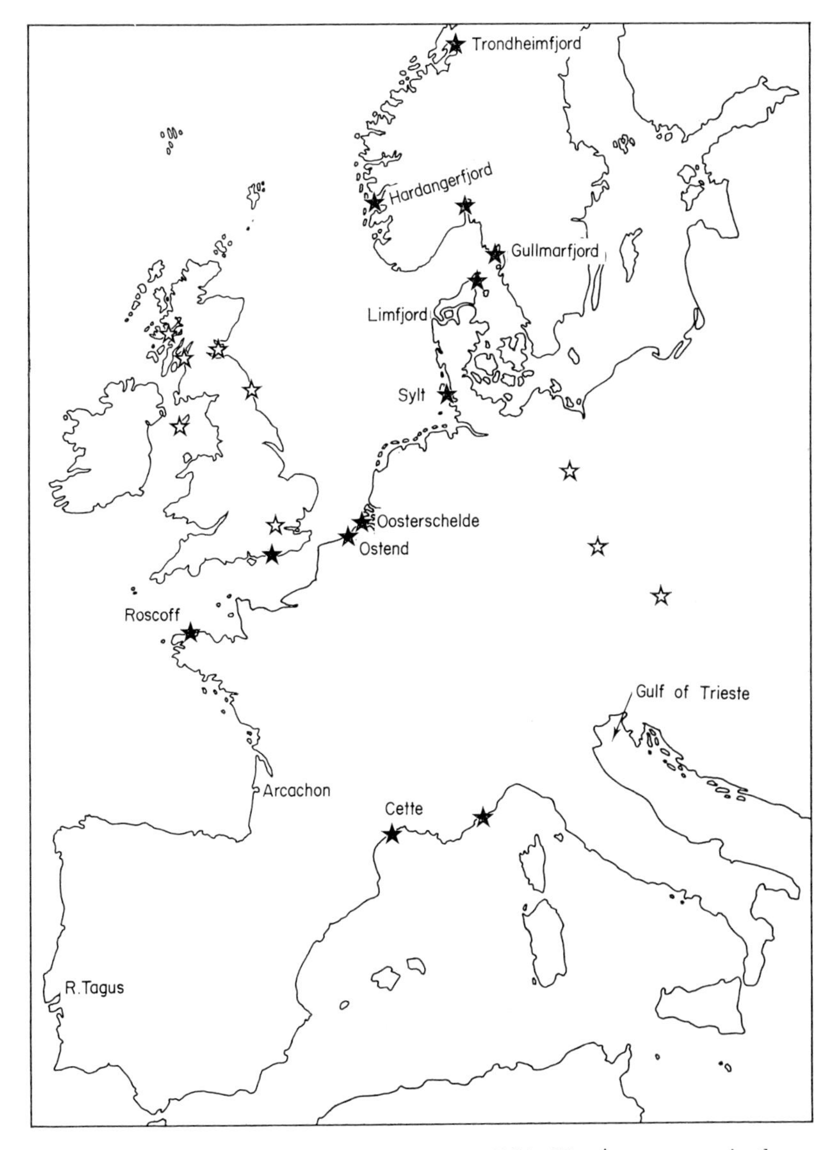

FIG. 2. Occurrence of *Gonionemus* in Europe (see Table II): ★ occurrence in the sea; ☆ occurrence in an aquarium. Some significant oyster localities are named.

Gonionemus can persist in an area in the hydroid stage. If the species is introduced into aquaria, the hydroid can persist for long periods and may produce medusae from time to time under the special conditions of temperature and feeding there. This explains the occurrence of medusae in aquaria at Vienna, Prague, Cullercoats, Port Erin, Aberdour and Dunstaffnage.

In summary, the occurrences, distribution and apparent spread of *Gonionemus* in Europe during the past hundred years are all consistent with the supposition that it has been transported on oysters, especially *Crassostrea angulata*, for which there is much circumstantial evidence (Fig. 2).

VI. On Species and Forms of *Gonionemus*

A number of species and local forms of *Gonionemus* have been described, distinguished by the shape of the bell, the thickness of the jelly, and the numbers of tentacles and statocysts (Fig. 3). *Gonionemus vertens*, from the Pacific coast of North America, has a bell rather greater in height than in diameter, up to 70 tentacles, and about the same number of statocysts, whereas *Gonionemus murbachi*, from Woods Hole, is somewhat shallower than a hemisphere, has 60–80 tentacles, at least half as many statocysts, and a stomach smaller than that of *G. vertens* (Mayer, 1910, pp. 342–4). *Gonionemus agassizi*, from Unalaska, Aleutian Islands, was stated to have more tentacles than *G. vertens* (probably up to 100) and to be shallower than that form (Murbach and Shearer, 1903). *Gonionemus depressum*, from southern Japan, has up to 64 tentacles, about twice as many statocysts, and a bell shallower than a hemisphere (Goto, 1903). *Gonionemus oshoro*, from northern Japan, has up to 80 tentacles and about the same number of statocysts, is very shallow with thin jelly, and has fewer gonadial folds than *G. murbachi* (Uchida, 1925, 1939). *Gonionemus vindobonensis*, found in aquaria at Vienna and Prague, was described solely from young medusae (Joseph, 1918, 1924, 1925).

Mayer (1910, p. 346) says that the statocysts in *G. murbachi* finally become nearly as numerous as the tentacles. Thomas (1921) found that some specimens of *G. murbachi* from Woods Hole had a greater number of statocysts than of tentacles and, from a comparative study of *G. murbachi* and *G. vertens*, concluded that the number and arrangement of statocysts and tentacles in *Gonionemus* are variable. Evidently these two species are not clearly separable on the numbers of tentacles and statocysts. Bigelow (1909) said that the differences between them are slight, consisting of a higher bell in *G. vertens* and stouter tentacles

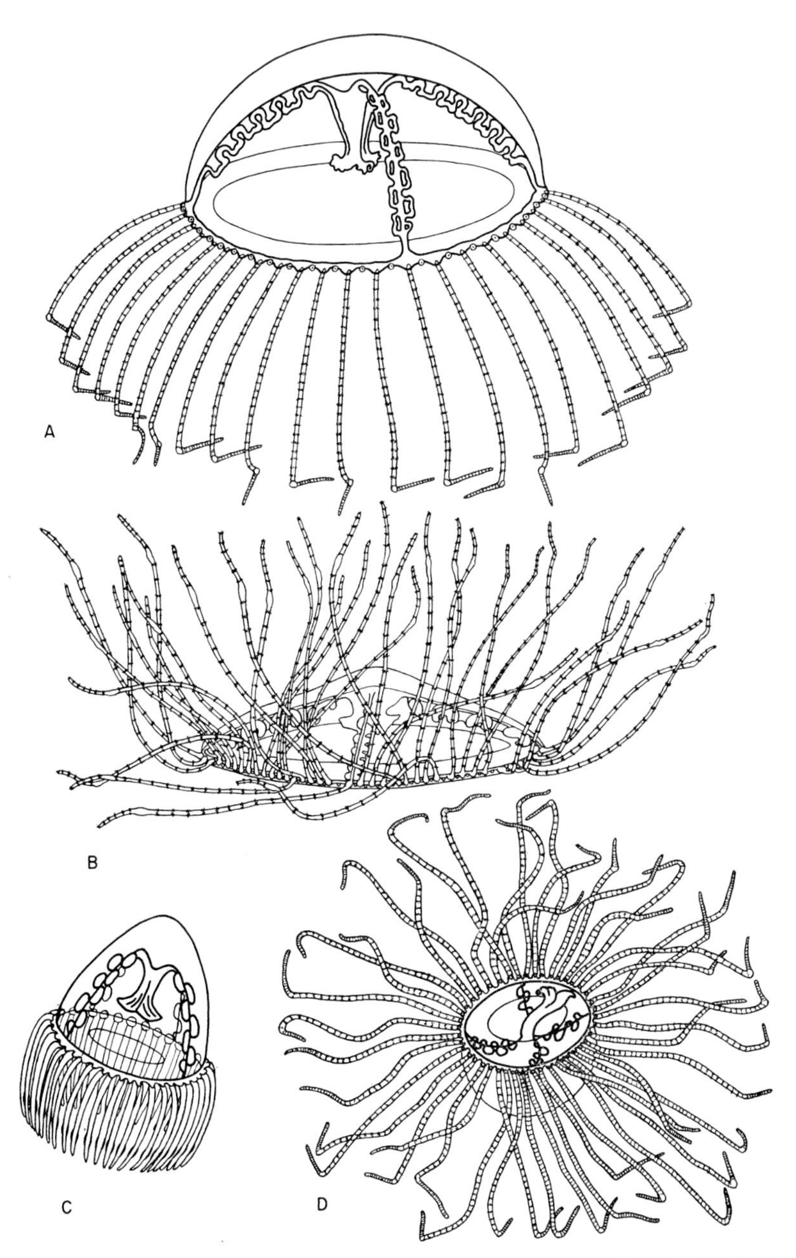

Fig. 3. *Gonionemus* medusae. A, *Gonionemus murbachi*, Woods Hole, Massachusetts (redrawn from Mayer, 1910, Vol. 2, pl. 45); B, *Gonionemus oshoro*, northern Japan (redrawn from Uchida, 1925, Fig. 18); C. *Gonionemus vertens*, Gulf of Georgia, swimming upwards; D. *G. vertens*, Gulf of Georgia, resting attached to seaweed. (C and D redrawn from A. Agassiz, 1865, Figs. 198, 197.)

TABLE II. RECORDS OF THE MEDUSA *Gonionemus* AND ITS HYDROID

Locality	*Date*	*Reference*
Gulf of Georgia, Washington, U.S.A.	1859	A. Agassiz: in L. Agassiz, 1862, p. 350; A. Agassiz, 1865, pp. 128–30 (medusa)
Canal connected with l'Étang de Thau, near Cette, south coast of France	1876	Du Plessis, 1879: as *Cosmetira salinarum* (medusa); reidentified by Picard, 1951, p. 41
Berlin Aquarium (*a*)	—	Schaudinn, 1894: hydroid *Haleremita cumulans*
Woods Hole, Mass , U.S.A.	1894–1930	Murbach, 1895; further references and details in Mayer, 1910, pp. 343–8; Perkins, 1903; Rugh, 1930; Werner, 1950a, p. 498
Noank, Conn. and Muskegat Island, Mass., U.S.A.	—	Murbach: see Mayer, 1910, p. 344 (medusa)
Victoria Harbour, Vancouver Island, Canada	1900	Murbach and Shearer, 1903, p. 184 (medusa)
Puget Sound, Washington, U.S.A.	—	Murbach and Shearer, 1903, p. 184 (medusa)
Unalaska, Aleutian Islands	—	Murbach and Shearer, 1903, p. 186 (medusa)
Many localities throughout Japan	1903 onwards	Goto, 1903; Kirkpatrick, 1903; Uchida, 1925, 1929, 1938, 1940 (medusa)
Dove Marine Laboratory, Cullercoats, north-east England	1913, 1914, 1951	Robson, 1913, p. 27, pl. 2; 1914, p. 90 (as *Cladonema* sp.), identified as *Gonionemus* by Joseph, 1924, p. 129; 1925, p. 377; Russell, 1953, pp. 402–3, pl. 35 (medusa)
Vienna: Aquarium of Zoological Institute (*b*)	1917	Joseph, 1918 (medusa)
Dröbak, Oslofjord, Norway	1921	Kramp, 1922 (medusa)
Strait of Georgia, British Columbia: Ballinac Is., Departure Bay, Mudge Is., Round Is., Cardale Point	—	Foerster, 1923 (medusa)
Prague: aquarium of Zoological Institute (*b*)	1923	Joseph, 1924, p. 129; 1925, p. 377 (medusa)
Marine Biological Station, Port Erin, Isle of Man	1924	Russell, 1953, p. 399 (medusa)
Vienna: Aquarium of Zoological Institute (*b*)	1924	Joseph, 1924, 1925 (hydroid and young medusa)

(*a*) Aquarium stocked from Rovigno, northern Adriatic.
(*b*) Aquarium stocked from Gulf of Trieste, northern Adriatic.

TABLE II.—*contd.*

Locality	*Date*	*Reference*
Gullmarfjord, Sweden	1923, 1930 and later	Lönneberg, 1930, p. 173 (medusa); Jägersten and Nilsson, 1961, p. 182
Roscoff, Brittany, France	1929, 1931, 1945, 1961–64	Teissier, 1930, pp. 185–6; 1932; 1950; 1965, p. 37 (medusa); 1965, p. 30 (hydroid)
Chengshan, China	1936	Ling, 1937, p. 358 (medusa)
Ostend, Belgium	1946, 1947	Leloup, 1948 (medusa)
Sylt, North Friesian Islands, Germany	1947–49	Werner, 1950a (medusa and hydroid)
Villefranche-sur-Mer, south coast of France	1950, 1951	Picard, 1951, 1955 (medusa)
China: several localities on Shantung and south-east coasts	1954–57	Chiu, 1954; Chow and Huang, 1958 (medusa)
Espevaer and Sandvik, Hardangerfjord, Norway	1958	Tambs-Lyche, 1964 (medusa)
Aberdour, Fife, Scotland	1959	Howe, 1959 (medusa)
Sea of Japan, near Vladivostok and Olga Bay; Tatar Strait, near coast of southern Sakhalin; southern Kurile Islands	—	Naumov, 1960 (medusa)
Frederikshavn, Denmark	1960	Kramp, 1961, p. 445 (medusa)
Ritthem, Walcheren, Netherlands	1960	Kramp, 1961, p. 445 (medusa)
Cuckmere Haven, Sussex, England	1963	Russell, 1970, p. 264; Edwards, this paper (medusa)
Santa Barbara, California	1964, 1965	Todd, *et al.*, 1966 (medusa)
Millport, Firth of Clyde, Scotland	1966	Edwards, this paper (medusa)
Vietnam	—	Kramp, 1968, p. 137 (medusa)
Borgenfjord, Trondheimfjord, Norway	1969	Gulliksen, 1971 (medusa)
Dunstaffnage, Argyll, Scotland	1974, 1975	Edwards, this paper (medusa and hydroid)
London: British Museum (Natural History) (*c*)	1975	Edwards, this paper (medusa and hydroid)

(*c*) Aquarium stocked from shell-gravel grounds, off Plymouth, England.

in *G. murbachi*. Picard (1951) considered that the various described species of *Gonionemus* are all referable to one species, by priority called *Gonionemus vertens*. Kramp (1961, 1968) puts all the species, including probably *G. vindobonensis*, under *G. vertens*. Naumov (1960) treats them as forms of *G. vertens*, with two subspecies, *G. vertens vertens* and *G. vertens murbachi*. (Two further species may be mentioned. *Scolionema suvaense*, formerly included in *Gonionemus*, has characters considered worthy of generic distinction (Kramp, 1961, 1968). *Gonionemus hamatus* (Kramp, 1965), known from a single specimen from Australia, differs in several respects from the other species of *Gonionemus*, but has been included in that genus for convenience to avoid undue multiplication of genera. These two species have no bearing on the subject of this paper, and are not considered further.)

Picard (1951), noting that *Gonionemus depressum* is distinguished by its high number of statocysts and its greater size, attributed this to a stage of development attained only in particularly favourable conditions, as found in southern Japan. He considered that the differences reported between *Gonionemus vertens*, *G. murbachi* and *G. oshoro* might have been due to the state of contraction or expansion in life, or the state of maturity, or the state of preservation of fixed specimens. Alternatively, if they are constant differences, he ascribed to them the characters of local races. Concerning *Gonionemus vindobonensis*, described solely from young medusae without gonads, Picard says that among his specimens of *G. vertens* from Villefranche the youngest ones matched Joseph's description, whereas the mature ones with gonads agreed with *G. vertens*. He supported Teissier's (1950) belief that sporadic appearances of *Gonionemus* medusae indicate the permanent presence, in certain localities, of populations of polyps reproducing asexually; and he suggested that small differences reported between medusae from different places are such as may be expected of medusae produced by local clones of polyps. He further emphasized that, where the species is established in a particularly suitable locality, regular production of medusae occurs. I agree with Teissier and Picard that locally persistent populations can be established by asexual reproduction of the polyps, and in favourable places can regularly produce medusae. A clone, however, is a population produced asexually from a single organism. I consider it highly unlikely that a single polyp would be transported to or would survive in an area and there give rise to a population. Accordingly, I think the establishment of clones is an improbable explanation of the variations between populations of medusae. In populations of *Gonionemus* long established in various areas with different environmental conditions it is likely that genetic differences exist. In the Pacific the reported differences between the forms found in various regions are very probably in part of genetic origin, the forms being adapted to local environments and perhaps worthy of subspecific rank. However, medusae are very subject during development to modification by environmental conditions, especially temperature, salinity and food supply. Accordingly, much of the variation found between medusae may be environmentally induced. It is probable that both genetic constitution and environment contribute to the production of local forms, such as the different forms found in northern and southern Japan. Medusae reared in aquaria are particularly liable to variation caused by the abnormal conditions there: those reared by me were produced and kept at low temperature and they attained small size and few vesicles, and those reared at the Plymouth Laboratory by Russell (1953) were fairly small, though mature. Picard (1951) remarked on the much greater size attained by *Gonionemus* medusae at Roscoff than at Villefranche. The medusae in the Gullmarfjord, where *Gonionemus* is well established (Jägersten and Nilsson, 1961), are large. The conditions at Roscoff and in the Gullmarfjord are evidently well suited to the species. If *Gonionemus* has been introduced from the Pacific into the Atlantic, the species would be expected to have varying success in establishing itself and in producing medusae in different places, and variations between medusae from different localities could be attributed to environmental causes. *Gonionemus* has evidently long been endemic in the Pacific and is likely to have undergone some genetic differentiation in various parts of that region. The Atlantic populations are probably derived from a Japanese or other western Pacific form.

VII. General Discussion and Conclusions

The world distribution of *Gonionemus* is peculiar and demands explanation. Picard (1951, 1955) reviewed the distribution, showing that the genus is restricted to the temperate and warm-temperate regions of the northern hemisphere. Since publication of his papers there have been further reports of the species (Table II, p. 272), but the general picture remains little affected. *Gonionemus* is strictly littoral and confined to very shallow water, especially among weeds. It occurs around the North Pacific and adjacent seas, in north-western Europe, in the Mediterranean and the Adriatic, and very locally in eastern North America at Woods Hole and nearby localities (where it may now be extinct). It has not been reported from the Indian Ocean. Picard (1951) considered that the various species reported from different regions are forms of one species, *Gonionemus vertens*, the forms perhaps being " local races ". The outstanding peculiarity of this world distribution is that, whereas the species occurs in both the North Pacific and the North Atlantic, it is absent from the arctic and tropical zones. Such a distribution is very unusual and is not readily explicable on geographical, hydrographic and climatic grounds. Species common to the temperate North Pacific and North Atlantic usually occur also in the subarctic or arctic regions, and have a continuous circum-arctic distribution or may be presumed to have had such a distribution in the fairly recent past. The tropical and arctic zones are barriers to the passage of purely temperate species between the Pacific and the Atlantic (except deep-water species that may find suitable temperatures at depth). That the disjunct distribution of *Gonionemus* is probably man-made is accordingly indicated.

The next point to emphasize is that, whereas *Gonionemus* is abundant and evidently indigenous in many places in the North Pacific, and indeed in Japan is so well known as to have a vernacular name, in the North Atlantic region, where the marine faunas of Europe and eastern North America have long been studied, it is of sporadic and often rare occurrence and was not found before 1876, since which date it has been progressively reported from various European countries. Further, as I have shown, there is strong indication that its distribution in Europe and eastern North America accords with the history of the oyster industry and oyster research and with the transport of oysters. I have suggested that the spread of the species during the past hundred years originated with the importation of *Crassostrea angulata* into France from Portugal from 1867 onwards. The absence of records of *Gonionemus* from Portugal itself I attribute to lack of observers there.

The distribution of *Crassostrea angulata* is also remarkable. Before its artificial spread during the past hundred years or so the species was confined to the coast of Portugal, the south and east coasts of Spain and the Atlantic coast of Morocco. When it was introduced into western France it quickly established itself there, whereas it had not done so before its introduction by man. The failure naturally to extend its range northwards is explicable by its larval ecology and by the current systems: the prevailing surface-currents along the Portuguese and Moroccan coasts are southerly and along the south coast of Spain easterly. The spread of the species farther into the Mediterranean and down the West African coast has probably been prevented by warmer conditions and by salinity: according to Ranson (1948) *C. angulata* requires temperatures of 18–25°C and salinities of 18–23‰ for breeding. Currents, temperatures and salinities probably account for its localization, but they do not explain how the species evolved or arrived in this very restricted area. Such a limited distribution is very unusual and indeed suspect. Many species of *Crassostrea* have been described from various parts of the world. Ranson (1948, 1960) concludes that *C. angulata* is the same species as the Japanese oyster, *C. gigas*, because they have similar temperature and salinity requirements for breeding and because their larval shells are morphologically indistinguishable. He considers two possible explanations of this disjunct distribution, accidental transport on the hulls of ships coming from the Far East and former direct sea connection between the two areas in the Tertiary. He, however, points out that the modern species were not in existence in the Tertiary and that for the second explanation it is necessary to suppose that the common ancestral species gave rise to two almost identical modern species in two widely separated areas. Ranson thinks that the species has been introduced into Portugal accidentally on ships' hulls. He points out that those who have studied the Miocene, Pliocene and Quaternary beds of Portugal have not reported the species, and gives reasons for not accepting a report of the species in Gallo-Roman kitchen refuse at the mouth of the Gironde. Korringa (1952, p. 341) appeared to accept the probability of Ranson's views, but said that proof must await breeding experiments. Imai and Sakai (1961) found that *Crassostrea angulata* crossed well with *C. gigas*, whereas attempts to cross-breed *C. gigas* with *C. virginica*, *C. echinata* and *C. rivularis* failed. In the cross between *C. angulata* and *C. gigas* fertilization took place with a high degree of success, the larvae developed and grew normally, and many spat were obtained and reared to adult size. Menzel (1974) in experiments in 1967–68 had 95% success in reciprocal crosses between *C. angulata* and *C. gigas*, as high a percentage as is achieved when each species is selfed, and he reared hybrid spat to over a year old, some becoming sexually mature and spawning. He found normal meiosis and mitosis in the hybrids, and concluded that the ease of hybridization and the normal chromosomal behaviour in the hybrids support Ranson's view that the two oysters are the same species. I cannot, however, accept Menzel's theory that the Portuguese introduced the oyster into Japan. Having quoted Cahn's (1950) account of the legends of the start of Japanese oyster-culture in the seventeenth century, Menzel says that no evidence has been found in the literature that *C. gigas* occurred in Japan before this. In fact, in Cahn's accounts of the legends and of the discovery of *C. gigas* shells in prehistoric kitchen-middens in Japan there is direct evidence that *C. gigas* is native to Japan and that Japanese oyster-culture is indigenous there.

Stenzel (1971, p. N1082) says that *Crassostrea angulata* and *C. gigas*, as well as the Indian species *C. cattuckensis*, are all derived from the

Miocene species *C. gryphoides*, which occurred over an extensive area from Portugal to Japan at a time when a continuous warm Tethys Sea existed. As Menzel (1974) says, Stenzel's hypothesis requires that *C. angulata* and *C. gigas* have remained morphologically identical and genetically compatible during many millions of years of isolation. I think this is untenable: the time scale, the geographical, climatic and hydrographic changes and the genetic improbability are too great. Even in the comparatively short period of glacial, interglacial and post-glacial times the hydrographic and climatic changes in the Mediterranean and its Atlantic approaches and the accompanying faunal changes (Mars, 1963) have been so great that *C. angulata* could not have been living in Portugal and Spain very long before the present.

Transport of *Gonionemus* polyps from the Pacific to Europe on ships' bottoms is possible: cases of the carriage of temperate species through tropical seas in this way are authenticated. It is, moreover, possible that the polyps were carried on oysters that had settled on the hulls. (Ranson thought that the Pacific oyster had been introduced accidentally on ships' hulls.) However, *Gonionemus* lives in very shallow weedy waters unsuited to ships of sizeable draught, and it is, therefore, less likely to settle on hulls. There is an alternative explanation. The siting of the beds of *Crassostrea angulata* in Portugal, particularly in the estuary of the Tagus, and the evidence for its probable identity with *Crassostrea gigas* suggest the intentional introduction of the Pacific oyster by the Portuguese in their great seafaring days. The coincidence of two North Pacific animals (*Crassostrea gigas* and *Gonionemus*), both abundant in Japan, having apparently been introduced into Europe, and the strong circumstantial evidence given in this paper for the spread of *Gonionemus* in Europe during the past hundred years being due to movement of oysters, particularly *Crassostrea angulata*, seem to point to their simultaneous introduction into Europe. I am inclined to the view that the Portuguese, some centuries ago, found an established oyster-culture in the Pacific, or, at least, natural beds used by the native people, and that they brought back oysters either as meat for the voyage (a well-known practice in sailing days) or with the intention of attempting to start oyster cultivation in Portugal, or both. The transport of live oysters involving journeys of many weeks or even months has been a well-proved practice during the past hundred years. Hydroid polyps are hardy and *Gonionemus* polyps would survive the journey on oysters. The reasons for believing that *Crassostrea gigas* has been introduced into Portugal from the Pacific are independent of those advanced for suggesting that *Gonionemus* has been accidentally introduced into Europe from the Pacific:

this is not a circular argument. *Crassostrea gigas* would, however, provide a very suitable vehicle for transport of *Gonionemus*, and it is most probable that *Gonionemus* was brought to Europe in this way. It is possible that *Crassostrea gigas*, together with *Gonionemus*, was brought from China or Korea rather than Japan: both species occur there (Imai and Sakai, 1961; Stenzel, 1971; Table II, p. 273), and oyster cultivation was practised in China earlier than in Japan.

VIII. Acknowledgements

I wish to express my gratitude to the following, to whom I am much indebted for information: Miss P. L. Cook, Dr P. F. S. Cornelius, Professor P. Korringa, Dr A. Lindquist, Dr R. H. Millar, Dr S. Munch-Petersen, Sir Frederick Russell, Mr P. Soleim, Mr D. T. Streeter, Dr H. Tambs-Lyche, Mr R. C. Vernon, Dr P. R. Walne, Dr B. Werner, Dr K. F. Wiborg and Sir Maurice Yonge.

IX. References

Agassiz, A. (1865). North American Acalephae. *Illustrated Catalogue of the Museum of Comparative Zoology at Harvard College*, **2**, 234 pp.

Agassiz, L. (1862). " Contributions to the Natural History of the United States of America ", Vol. **4**, 380 pp. Little, Brown and Company, Boston.

Allodi, R. (1906). Über die Austernzucht an der nordöstlichen Küste des Adriatischen Meeres. *Stenographisches Protokoll über die Verhandlungen des Internationalen Fischerei-Kongresses, Wien* 1905, 282–96.

Barrett, E. M. (1963). The California oyster industry. *Fish Bulletin, California Fish and Game Commission*, **123**, 103 pp.

Bigelow, H. B. (1909). Reports on the scientific results of the expedition to the eastern tropical Pacific, in charge of Alexander Agassiz, by the U.S. Fish Commission steamer "Albatross" from October, 1904, to March, 1905, Lieut. Commander L. M. Garrett commanding. XVI. The Medusae. *Memoirs of the Museum of Comparative Zoology at Harvard College*, **37**, 243 pp., 48 pls.

Bon, De (1875). *Notice sur la situation de l'ostréiculture en 1875, precedée d'un rapport addressé au Ministre de la Marine et des Colonies.* 27 pp. Paris: Berger-Levrault. (Consulted in English translation: *Report of the Commissioner, United States Commission of Fish and Fisheries*, **1880**, 1883, 885–906.)

Bouchon-Brandely, G. (1877). *Rapport au Ministre de la Marine relatif à l'ostréiculture sur le littoral de la Manche et de l'Océan.* Paris: A. Wittersheim et Cie. (Consulted in English translation: *Report of the Commissioner, United States Commission of Fish and Fisheries*, **1882**, 1884, 673–724.)

Bouchon-Brandely, G. (1878). *Rapport au Ministre de l'Instruction publique sur la pisciculture en France et de l'ostréiculture dans la Mediterranée.* 103 pp. Paris: A. Wittersheim et Cie. (Consulted in English translation: A report on oyster culture in the Mediterranean. *Report of the Commissioner, United States Commission of Fish and Fisheries*, **1880**, 1883, 907–30.)

Broca, P. De (1865). *Études sur l'industrie huitrière des États-Unis.* 266 pp. Paris : Challaune. (Consulted in English translation : On the oyster industries of the United States. *Report of the Commissioner, United States Commission of Fish and Fisheries*, **1873–75,** 1876, 271–319.)

Broch, H. (1924). Trachylina. *In* " Handbuch der Zoologie ", (W. Kükenthal and T. Krumbach, eds.), Vol. 1, pp. 459–84. Walter de Gruyter, Berlin and Leipzig.

Broch, H. (1929). Craspedote Medusen. Teil II : Trachylinen (Trachymedusen und Narcomedusen). *Nordisches Plankton*, **12,** Teil 2, Lief. 21, 481–540.

Bulstrode, T. H. (1896). " Report on an enquiry into the conditions under which oysters, and certain other edible molluscs, are cultivated and stored along the coast of England and Wales". 108 pp. Local Government Board, Public Health Report, London.

Cahn, A. R. (1950). Oyster culture in Japan. *United States Fish and Wildlife Service, Fishery Leaflet* **383,** 80 pp.

Calvet, L. (1910). L'Ostréiculture à Cette et dans la région de l'Étang de Thau. *Travaux de l'Institut de Zoologie de l'Université de Montpellier et de la Station Zoologique de Cette*, Sér. 2, **20,** 104 pp.

Chiu, S. T. (1954). Studies on the medusa fauna of south-eastern China coast, with notes on their geographical distribution. (In Chinese.) *Acta Zoologica Sinica*, **6,** 49–57. (Quoted from Kramp, 1961.)

Chow, T. and Huang, M. (1958). A study on Hydromedusae of Chefoo. *Acta Zoologica Sinica*, **10,** 173–97, pls. 1–5.

Cole, H. A. (1942). The American Whelk Tingle, *Urosalpinx cinerea* (Say), on British oyster beds. *Journal of the Marine Biological Association of the United Kingdom*, **25,** 477–508.

Cole, H. A. (1956) " Oyster cultivation in Britain ", 43 pp. H.M. Stationery Office, London.

Coste, P. (1861). Voyage d'Exploration sur le littoral de la France et de l'Italie. Second edition. 297 pp. Paris. (Consulted in English translation, in part : Report on the oyster and mussel industries of France and Italy. *Report of the Commissioner, United States Commission of Fish and Fisheries*, **1880,** 1883, 825–83, pls. 1–13.)

Dean, B. (1892). The present methods of oyster-culture in France. *Bulletin of the United States Fish Commission*, **10,** 363–88.

Du Plessis, G. (1879). Étude sur la *Cosmetira salinarum* nouvelle méduse paludicole des environs de Cette. *Bulletin de la Société Vaudoise des Sciences Naturelles*, Sér. 2, **16,** 39–45, pl. 2.

Foerster, R. E. (1923). The hydromedusae of the west coast of North America, with special reference to those of the Vancouver Island region. *Contributions to Canadian Biology*, New Series, **1,** 219–77, pls. 1–5.

Goto, S. (1903). The craspedote medusa *Olindias* and some of its natural allies. *In* " Mark Anniversary Volume " (G. H. Parker, ed.), pp. 1–22, pls. 1–3. Henry Holt, New York.

Gulliksen, B. (1971). A new record of *Gonionemus vertens* Agassiz (Limnomedusae) in Norway. *Det Kongelige Norske Videnskabers Selskab Skrifter*, **12,** 1–4.

Haas, F. (1938). Bivalvia. *Bronns Klassen und Ordnungen des Tierreichs*, **3,** Abt. 3, Teil 2, Lief. 2, 209–466.

Hagmeier, A. (1941). Die intensive Nutzung des nordfriesischen Wattenmeeres durch Austern- und Muschelkultur, mit Bericht über die 1932 bis 1940 ausgeführten Untersuchungen der fiskalischen Austernbänke. *Zeitschrift für Fischerei und deren Hilfwissenschaften*, **39,** 105–65.

Halewyck, R. and Leloup, E. (1951). La situation de l'ostréiculture dans le Bassin de Chasse d'Ostende de 1939 à 1948. *Rapports et Procès-Verbaux des Réunions, Conseil Permanent International pour l'Exploration de la Mer*, **128,** (2), 19.

Hargitt, C. W. (1905). The medusae of the Woods Hole region. *Bulletin of the Bureau of Fisheries, Washington*, **24,** 21–79, pls. 1–7.

Havinga, B. (1932). Austern- und Muschelkulter. *Handbuch der Seefischerei Nordeuropas*, **7,** Heft 5, 64 pp.

Herdman, W. A. and Boyce, R. (1899). Oysters and disease. *Lancashire Sea-Fisheries Memoir*, **1,** 60 pp., 8 pls.

Howe, D. C. (1959). A rare hydromedusa. *Nature, London*, **184,** 1963.

Imai, T. and Sakai, S. (1961). Study of breeding of Japanese oyster, *Crassostrea gigas*. *Tohoku Journal of Agricultural Research*, **12,** 125–71.

Ingersoll, E. (1887). The oyster, scallop, clam, mussel and abalone industries. *In* " The Fisheries and Fishery Industries of the United States." (G. B. Goode, ed). Section V. History and Methods of the Fisheries, Vol. **2,** Part 20, pp. 505–626.

Jägersten, G. and Nilsson, L., 1961. " Life in the Sea ", 184 pp. G. T. Foulis, London.

Joseph, H. (1918). Ein *Gonionemus* aus der Adria. *Sitzungsberichte der Kaiserlichen Akademie der Wissenschaften in Wien*, Mathem.-naturwiss. Klasse, Abt. 1, **127,** 95–158, 1 pl.

Joseph, H. (1924). Über *Haleremita*, *Gonionemus* und den Begriff der Trachomedusen. *Verhandlungen der Deutschen Zoologischen Gesellschaft*, **29,** 129–33.

Joseph, H. (1925). Zur Morphologie und Entwicklungsgeschichte von *Haleremita* und *Gonionemus*. Ein Beitrag zur systematischen Beurteilung der Trachymedusen. *Zeitschrift für wissenschaftliche Zoologie*, **125,** 374–434, pl. 8.

Kirkpatrick, R. (1903). Notes on some medusae from Japan. *Annals and Magazine of Natural History*, Ser. 7, **12,** 615–21, pl. 23.

Korringa, P. (1947). Les vicissitudes de l'ostréiculture Hollandaise élucidées par la science ostréicole moderne. *Ostréiculture-Cultures Marines, Bulletin du Comité Interprofessionel de la Conchyliculture et du Syndicat National de l'Ostréiculture et des Cultures Marines, Paris*, **16,** (3), 3–9.

Korringa, P. (1951). The shell of *Ostrea edulis* as a habitat. *Archives Néerlandaises de Zoologie*, **10,** 32–152.

Korringa, P. (1952). Recent advances in oyster biology. *Quarterly Review of Biology*, **27,** 266–308, 339–65.

Korringa, P. (1956). The quality of marketable oysters from the Zealand waters in 1954. *Annales Biologiques, Copenhague*, **11,** 180–2.

Korringa, P. (1957). The quality of marketable oysters from the Zealand waters in 1955. *Annales Biologiques, Copenhague*, **12,** 225–8.

Korringa, P. (1958). The quality of marketable oysters from the Zealand waters in 1956. *Annales Biologiques, Copenhague*, **13,** 238–41.

Korringa, P. (1969). Shellfish of the North Sea. *Serial Atlas of the Marine Environment*, American Geographical Society, **17,** 6 pp., 9 pls.

Kramp, P. L. (1922). Demonstration of *Gonionemus murbachii* from Kristiania-fjord, Norway. *Videnskabelige Meddelelser fra Dansk Naturhistorisk Forening i Kjøbenhavn*, **74,** xi.

Kramp, P. L. (1961). Synopsis of the medusae of the world. *Journal of the Marine Biological Association of the United Kingdom*, **40,** 5–469.

Kramp, P. L. (1965). Some medusae (mainly Scyphomedusae) from Australian coastal waters. *Transactions of the Royal Society of South Australia*, **89,** 257–78, pls. 1–3.

Kramp, P. L. (1968). The Hydromedusae of the Pacific and Indian Oceans, Sections II and III. *Dana-Report*, **72,** 200 pp.

Leloup, E. (1948). Présence de la trachyméduse *Gonionemus murbachi* Mayer, 1901, à la côte belge. *Bulletin du Musée Royal d'Histoire Naturelle de Belgique*, **24,** (27), 4 pp.

Leloup, E. and Miller, O (1940). La flore et la faune du Bassin de Chasse d'Ostende (1937–1938). *Mémoires du Musée Royal d'Histoire Naturelle de Belgique*, **94,** 123 pp., 3 pls.

Ling, S. W. (1937). Studies on Chinese Hydrozoa. I. On some Hydromedusae from the Chekiang coast. *Peking Natural History Bulletin*, **11,** 351–65.

Lönneberg, E. (1930). Några notiser från Kristinebergs Zoologiska station sommaren 1930. *Fauna och Flora, Uppsala*, **4,** 165–74.

Mars, P. (1963). Les faunes et la stratigraphie du Quaternaire méditerranéen. *Recueil des Travaux de la Station Marine d'Endoume*, **28,** 61–97.

Mayer, A. G. (1910). "Medusae of the World". Vol. 1, pp. 1–230; Vol. 2, pp. 231–498; Vol. 3, pp. 499–728. Carnegie Institution, Washington, D.C.

Menzel, R. W. (1974). Portuguese and Japanese oysters are the same species. *Journal of the Fisheries Research Board of Canada*, **31,** 453–6.

Millar, R. H. (1961). Scottish oyster investigations. *Marine Research*, 1961, (3), 76 pp.

Möbius, K. (1877). "Die Auster und die Austernwirthschaft", 126 pp. Wiegandt, Hempel und Parey, Berlin. (Also consulted in English translation: *Report of the Commissioner, United States Commission of Fish and Fisheries*, **1880**, 1883, 683–751.)

Murbach, L. (1895). Preliminary note on the life-history of *Gonionemus*. *Journal of Morphology*, **11,** 493–6.

Murbach, L. and Shearer, C. (1903). On medusae from the coast of British Columbia and Alaska. *Proceedings of the Zoological Society of London*, **1903**, **2,** 164–92, pls 17–22.

Naumov, D. V. (1960). Gidroidy i gidromeduzy morskikh, solonovatovodnykh i presnovodnykh basseinov SSSR. *Opredeliteli po Faune SSSR*, **70,** 585 pp., 30 pls. (Also consulted in English translation: *Hydroids and Hydromedusae of the USSR*, 1969, Jerusalem.)

Northumberland Sea Fisheries Committee (1899). *Report, Northumberland Sea Fisheries Committee*, 1899, 68 pp.

Northumberland Sea Fisheries Committee (1906). *Report on the Scientific Investigations, Northumberland Sea Fisheries Committee*, 1906, 47 pp.

Orton, J. H. (1912). An account of the natural history of the slipper-limpet (*Crepidula fornicata*) with some remarks on its occurrence on the oyster grounds on the Essex coast. *Journal of the Marine Biological Association of the United Kingdom*, **9,** 437–43.

Orton, J. H. and Winckworth, R. (1928). The occurrence of the American oyster pest *Urosalpinx cinerea* (Say) on English oyster beds. *Nature, London*, **122,** 241.

Perkins, H. F. (1903). The development of *Gonionema murbachii. Proceedings of the Academy of Natural Sciences of Philadelphia*, **54,** 750–90, pls. 31–34.

Picard, J. (1951). Notes sur les hydroméduses méditerranéennes de la famille des Olindiadidae. *Archives de Zoologie Expérimentale et Générale, Paris*, **88,** Notes et Revue, 1, 39–48.

Picard, J. (1955). Nouvelles recherches sur les hydroméduses des herbiers méditerranéens de Posidonies. *Recueil des Travaux de la Station Marine d'Endoume*, **15,** 59–71.

Purchon, R. D. (1968). " The Biology of the Mollusca ", 560 pp. Pergamon Press, Oxford.

Ranson, G. (1948). Écologie et répartition géographique des Ostréidés vivants. *Revue Scientifique, Paris*, **86,** 469–73.

Ranson, G. (1960). Les prodissoconques (coquilles larvaires) des Ostréidés vivants. *Bulletin de l'Institut Océanographique, Monaco*, **1183,** 41 pp.

Ricketts, E. F. and Calvin, J. (1952). " Between Pacific Tides ", (3rd Edn.), 502 pp. Stanford University Press, Stanford, California.

Robson, J. H. (1913). Hydroida not previously recorded for the district. *Report, Dove Marine Laboratory*, N.S., **2,** 25–33, pls. 1–4.

Robson, J. H. (1914). Catalogue of the Hydrozoa of the northeast coast (Northumberland and Durham). *Report, Dove Marine Laboratory*, N.S., **3,** 87–103, pl. 3.

Rugh, R. (1930). Variations in *Gonionemus murbachii. American Naturalist*, **64,** 93–95.

Russell, F. S. (1953). *The Medusae of the British Isles: I. Anthomedusae, Leptomedusae, Limnomedusae, Trachymedusae and Narcomedusae.* 530 pp. Cambridge University Press.

Russell, F. S. (1970). *The Medusae of the British Isles. II. Pelagic Scyphozoa, with a Supplement to the First Volume on Hydromedusae.* 284 pp. Cambridge University Press.

Schaudinn, F. (1894). *Haleremita cumulans* n.g. n.sp., einen neuen marinen Hydroidpolypen. *Sitzungsberichte der Gesellschaft Naturforschender Freunde zu Berlin*, 1894, 226–34.

Spärck, R. (1951). Fluctuations in the stock of oyster (*Ostrea edulis*) in the Limfjord in recent time. *Rapports et Procès-Verbaux des Réunions, Conseil Permanent International pour l'Exploration de la Mer*, **128,** (2), 27–29.

Stearns, R. E. C. (1883). The edible clams of the Pacific coast and a proposed method of transplanting them to the Atlantic coast. *Bulletin of the United States Fish Commission*, **3,** 353–62.

Stenzel, H. B. (1971). Oysters. *Treatise on Invertebrate Paleontology*, Part N, Vol. 3 (of three), Mollusca 6, Bivalvia, pp. N953–N1224. Geological Society of America.

Stevenson, C. H. (1899). The preservation of fishery products for food. *Bulletin of the United States Fish Commission*, **18,** 335–576.

Tambs-Lyche, H. (1964). *Gonionemus vertens* L. Agassiz (Limnomedusae), a zoogeographical puzzle. *Sarsia*, **15,** 1–8.

Teissier, G. (1930). Notes sur la faune marine de la région de Roscoff. I. Hydraires, Trachyméduses, Cirripèdes. *Travaux de la Station Biologique de Roscoff*, **8**, 183–6.

Teissier, G. (1932). Existence de *Gonionemus murbachi* sur les côtes de Bretagne. *Travaux de la Station Biologique de Roscoff*, **10**, 115–6.

Teissier, G. (1950). Notes sur quelques Hydrozoaires de Roscoff. *Archives de Zoologie Expérimentale et Générale, Paris*, **87**, Notes et Revue, 1, 1–10.

Teissier, G. (1965). Inventaire de la faune marine de Roscoff. Cnidaires et Cténaires. *Travaux de la Station Biologique de Roscoff*, N.S., **16**, 64 pp.

Thomas, L. J. (1921). Morphology and orientation of the otocysts of *Gonionemus*. *Biological Bulletin*, **40**, 287–98, pl. 1.

Todd, E. S., Kier, A. and Ebeling, A. W. (1966). *Gonionemus vertens* L. Agassiz (Hydrozoa: Limnomedusae) in southern California. *Bulletin of the Southern California Academy of Sciences*, **65**, 205–10.

Uchida, T. (1925). Some Hydromedusae from northern Japan. *Japanese Journal of Zoology*, **1**, 77–100.

Uchida, T. (1929). Studies on Japanese Hydromedusae. 3. Olindiadae. *Annotationes Zoologicae Japonenses*, **12**, 351–71, pl. 1.

Uchida, T. (1938). Report of the biological survey of Mutsu Bay. 32. Medusae from Mutsu Bay (revised report). *Scientific Reports of the Tôhoku Imperial University*, Ser. 4, Biology, **13**, 37–46.

Uchida, T. (1940). The fauna of Akkeshi Bay. XI. Medusae. *Journal of the Faculty of Science, Hokkaido Imperial University*, Ser. 6, Zoology, **7**, 277–97.

United States Commission of Fish and Fisheries (1884a). *Report of the Commissioner, United States Commission of Fish and Fisheries*, 1881.

United States Commission of Fish and Fisheries (1884b). *Report of the Commissioner, United States Commission of Fish and Fisheries*, 1882.

United States Commission of Fish and Fisheries (1886). *Report of the Commissioner, United States Commission of Fish and Fisheries*, 1884.

United States Commission of Fish and Fisheries (1887). *Report of the Commissioner, United States Commission of Fish and Fisheries*, 1885.

Went, A. E. J. (1962). Historical notes on the oyster fisheries of Ireland. *Proceedings of the Royal Irish Academy*, **62**, c, 195–223, pl. 46.

Werner, B. (1948). Die amerikanische Pantoffelschnecke *Crepidula fornicata* L. im Nordfriesischen Wattenmeer. *Zoologische Jahrbücher (Systematik)*, **77**, 449–88.

Werner, B., (1950a). Die Meduse *Gonionemus murbachi* Mayer im Sylter Wattenmeer. *Zoologische Jahrbücher (Systematik)*, **78**, 471–505.

Werner, B. (1950b). Weitere Beobachtungen über das Auftreten der Meduse *Gonionemus murbachi* Mayer im Sylter Wattenmeer und ihre Entwicklungsgeschichte. *Verhandlungen der Deutschen Zoologen in Mainz*, 1949. 138–51.

Yonge, C. M. (1960). " Oysters ". 209 pp. Collins, London.

Adv. mar. Biol., Vol. 14, 1976, pp. 285–443

PHYSIOLOGY AND ECOLOGY OF MARINE BRYOZOANS

J. S. RYLAND

Department of Zoology, University College of Swansea, Wales

I. Introduction

At one time the Bryozoa was regarded as a " minor phylum " which could safely be ignored in all but the most esoteric studies. Slowly, however, this view has changed and bryozoans are beginning to receive attention more in proportion to their abundance in both the world's oceans and its fossil-bearing strata. During the mid-nineteenth century some ten or a dozen papers annually were published on bryozoans, this number rising quite suddenly to forty or fifty between 1870 and 1880. This level was held, with minor fluctuations, until about 1950. The fifties saw an upswing in the amount of research conducted, with the annual number of publications doubling during the decade (Schopf, 1967) and doubling again during the sixties. The two standard textbooks (Hyman, 1959; Brien, 1960) were written largely before the upsurge of interest but contain between them (for neither is complete on its own) the solid foundation of our knowledge of the structure, physiology, ecology and systematics of bryozoans. The annual output of papers wholly or largely devoted to the biology of bryozoans is now so substantial, and major advances made so rapidly, that reviews of progress have become a necessity (Ryland, 1967a).

The bryozoans have had a long geological history, but until recently the studies of living species and of fossils have been essentially separate disciplines. Fortunately this is no longer so, as three successful international symposia testify. These conference volumes (Annoscia, 1968; Larwood, 1973; Pouyet 1975, 1976) span the whole spectrum of bryozoological research. It remains true, however, that bryozoans, which tend to be small, complex, numerous and diverse, are poorly understood by many even though their colonies may readily be recognized as such. My book " Bryozoans " (Ryland, 1970) was an attempt to bring together our knowledge of living and fossil bryozoans, and to explain some basic aspects of their biology to a wider public: its Chapters 1 and 2 can provide any introduction that may be needed for the appreciation of this article.

Following the development and refinement first of transmission and then of scanning electron microscopy, great progress has been made in our understanding of bryozoan zooidal and colonial skeletal morphology. This research is of unqualified importance to students of the phylum. However, partly because its interest is primarily to bryozoologists and partly because there are those far better qualified to review this field than I, subjects such as calcification and wall structure, including ascus formation and interzooidal communications, are not considered in this paper. These topics, together with others such as polymorphism, population genetics, culture methods and questions of phylogeny, will receive detailed treatment from specialists in a forthcoming multi-author volume, the " Biology of Bryozoans " (eds. R. F. Woollacott and R. L. Zimmer), which is currently in the press.

In choosing to review physiology and ecology I am both amplifying and updating the corresponding sections of my book (Ryland, 1970, Ch. 3 and 4), whilst presenting the latest research in these fields to marine biologists generally rather than to bryozoologists particularly. Furthermore, some of the topics covered here are not dealt with at all in the " Biology of Bryozoans ".

The classification followed in this paper is the one first proposed by Borg (1926) and now—but only in recent years—widely accepted. Three classes are recognized: Phylactolaemata, Stenolaemata and Gymnolaemata. The first of these is confined to fresh waters; the second comprises those bryozoans with slender, tubular, calcified walls; the third is the dominant and most diverse marine class. There is only one surviving stenolaemate order, Cyclostomata, and two gymnolaemate orders, Ctenostomata and Cheilostomata. The Ctenostomata are non-calcareous with cylindrical or squat zooids; the Cheilostomata have walls that are more or less strongly calcified, and variably shaped, but usually squat, zooids. It is least confusing, and perhaps wisest in a period of active discovery and taxonomic uncertainty, to classify cheilostomes into two suborders only: the Anasca in which the frontal surface is membranous and directly involved with the eversion of the lophophore; and the Ascophora in which the frontal surface is rigid, either as a covering over a still existing membrane or as a wall overlying a distensible ascus (compensation sac).

II. Growth

A. *The ancestrula and early astogeny*

A new bryozoan colony normally arises from a sexually produced larva. In stenolaemates the larva metamorphoses into the pro-ancestrula, a hemispherical body which later elongates to form a tubular extension (which might perhaps be termed the " opisthancestrula ") housing the polypide. In gymnolaemates the entire ancestrula, which is occasionally a double or triple zooid (see pp. 362, 377), is the product of metamorphosis. Occasionally, as in some stoloniferans and in one species of *Hippothoa*, it is a kenozooid (i.e., it lacks a polypide); in anascans it may resemble the later zooids in appearance, while in ascophorans it is frequently a " tata " (i.e., *Callopora*-like, Fig. 1), though in others it is not very different from the later zooids. The

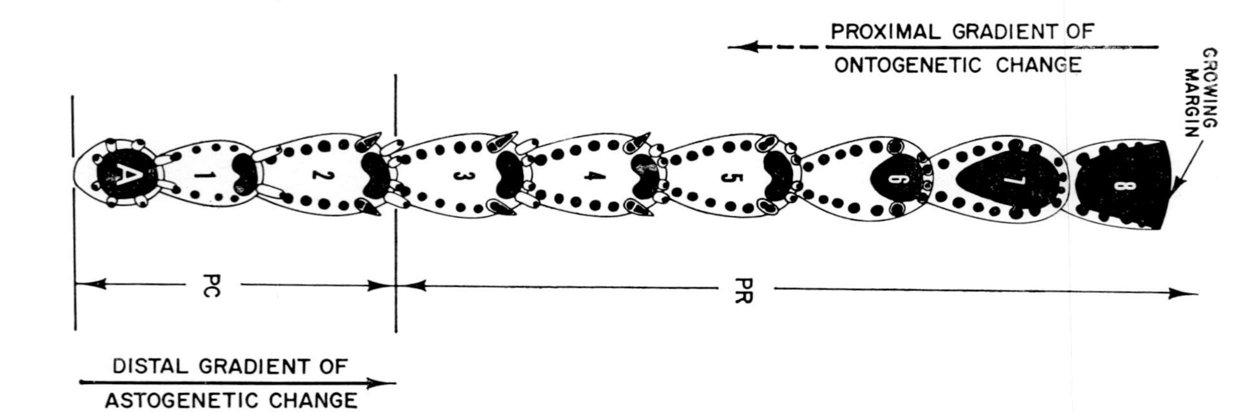

Fig. 1. Diagrammatic representation of astogenetic and ontogenetic change in the colony of a cheilostome bryozoan. The ascending sequence depicts astogenetic change from the ancestrula (A), through eight zooid generations, to the colony margin. During the phase of change (PC) the zooids reach ephebic size and complexity; during the phase of repetition (PR) their appearance is stable. The descending sequence depicts ontogenetic change, or the development and changing structure of a zooid with age. (From Boardman, Cheetham and Cook, 1969.)

phylogenetic significance of ancestrular form is difficult to assess, particularly when—as in *Hippothoa*—kenozooidal, tatiform and adult-type ancestrulae are to be found in the same genus (Ryland and Gordon, 1976); but several authors in the past have opined that the tata represents an ancestral form from which many ascophorans may have been derived.

The ancestrula is usually smaller than the later zooids, as well as frequently differing in morphology. The first phase of astogeny (colony development, from *asty*, town) is, therefore, characterized by a succession of zooids which change in size and shape (Fig. 1). Boardman and Cheetham (1969 and—in depth—1973) have discussed the trends and principles underlying the growth and differentiation of a colony. Following Boardman (1968), they describe the initial period of colony formation and zooid elaboration as the *phase of astogenetic change.* This is succeeded by a (generally much longer) period during which identical zooids, groups of zooids or recurrent zones of zooids are produced by standardized budding procedures (Boardman *et al.*, 1970). This constitutes the *phase of astogenetic repetition.* Cumings (1904) introduced a series of elaborate terms to describe the successive astogenetic phases characterizing a bryozoan colony, but it seems necessary to recognize only the two defined by Boardman. There is advantage, however, in restoring to use two of Cumings' words, viz. "neanastic" and "ephebastic", preferably simplified to neanic and ephebic (Ryland, 1970). The adjectives young and old cannot be used to distinguish the first-formed and later zooids of a colony, since they properly refer to the actual age of the zooids (and the youngest zooids are the latest formed, as shown diagrammatically in Fig. 1). Neanic (*neanikos*, youthful) and ephebic (*ephebos*, adult), however, can be applied unambiguously to zooids laid down during the phases of astogenetic change and repetition respectively. Most of the studies of stenolaemates quoted by Boardman and Cheetham (1969, 1973) refer to Palaeozoic fossils although, belatedly, several studies on cyclostome astogeny have recently been initiated; our present discussions, therefore, concern mainly gymnolaemates and particularly cheilostomes.

The form of the colony may or may not be reflected in the shape of the ancestrula. Thus when the colony is to be erect and tufted the ancestrula itself may be tall and upright, as in *Bugula* and the cyclostome *Crisia*; but other bushy forms, such as *Flustra*, are initially crustose and only later grow free from the substratum. Usually the ancestrula buds one or more daughter zooids to initiate astogeny, but in *Metrarabdotos unguiculatum** the first three buds (Cook, 1973), and in

* Authorities for bryozoan species names are given in the Appendix (p. 425).

Smittina papulifera the first five (Stach, 1938), are formed simultaneously with the ancestrula. There are many distinctive patterns of early astogeny: thus, open colony forms (repent or erect) may commence as a single distal bud from the ancestrula (*Pyripora*, *Bicellariella*); flabellate colonies commonly arise either from paired distolateral buds (*Hippothoa* spp.) or from a triad of buds, one distomedial and two distolateral, e.g., *Fenestrulina malusii* var. *thyreophora* in which the median bud forms first (Gordon, 1971a) and *Electra pilosa* in which the lateral buds appear first (Barrois, 1877). A single distal bud in *Conopeum* is followed by a single proximal bud, so that the young colony is biflabellate (Cook and Hayward, 1966; Dudley, 1973). In *Hippothoa hyalina* there is a single unilateral bud and subsequent budding proceeds in a spiral around the ancestrula (Barrois, 1877), while in *Setosellina goesi* bilateral buds give rise to a double spiral (Boardman and Cheetham, 1969). (It is interesting to note that the

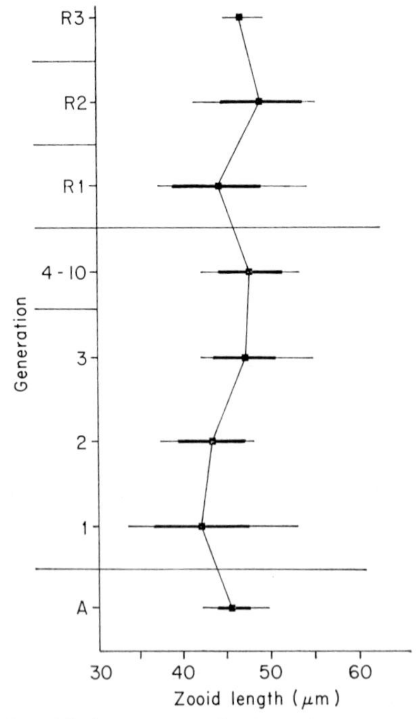

FIG. 2. Dependence of zooid size on growth phase in *Hippoporina porosa*. Ordinate shows growth phase: A, ancestrula; 1–10, initial zooid generations one to ten; R1–R3, regenerative generations one to three. Diagram shows mean (centre point), standard deviation (heavy bar) and range (thin line). Generations 1, 2 and R1 are neanic, 3+ and R2+ ephebic (R3 based on very small sample). (After Abbott, *in* "Animal Colonies" (R. S. Boardman *et al.*, eds.) © Dowden, Hutchinson and Rass, 1973.)

apparently spiral astogeny of *Cupuladria* is an illusion [Lagaaij, 1963]: budding commences with a distal triad.) Five initial buds are usual in smittinids (Soule and Soule, 1972). In *Membranipora* the first daughter zooid arises medially, between the twin ancestrulae (see p. 362), the second and third are lateral to the first, and others form rapidly in a periancestrular crescent (O'Donoghue, 1926; Atkins, 1955a). The triple ancestrula of *Stylopoma duboisii* produces a complete ring of zooid buds (Cook, 1973).

Neanic and ephebic zooids may differ in size, shape, structural complexity, polymorphism and budding pattern. These differences are ordinarily expressed as gradients extending from the ancestrula through the first four to eight generations of zooids. The most consistently evident change is in size. Boardman and Cheetham (1969, Fig. 6) found that the gradient of change may be linear, declining or stepped until the zone of astogenetic repetition is reached. A thorough analysis of morphological change during the astogeny of *Hippoporina* spp. has been made by Abbott (1973) (see Fig. 2). The polypide increases in size with the zooid, as reflected by a rise in the number of tentacles comprising the lophophore, e.g. 8 → 12 in *Conopeum tenuissimum* (Dudley, 1973), 20 → 24–26 in *Smittipora levinseni*, 16–17 → 18 in *Metrarabdotos unguiculatum* and 17 → 18–20 in *Stylopoma duboisii* (Cook, 1973). When colonies regenerate, after damage or seasonal regression, the new zooids frequently revert to neanic form, as indicated in Fig. 2 (Boardman *et al.*, 1970; Abbott, 1973).

B. *The growing edge and budding*

Basic studies on the growing edge are those of Schneider (1957, 1959, 1963) on *Bugula* and of Lutaud (1961) on *Membranipora*. More recently these have been extended to other species: e.g., by Banta (1968, 1969) and Tavener-Smith and Williams (1972). Growth in *Bugula* takes place at the ramus apices and astogeny involves repeated branching; in *Membranipora* growth proceeds on a broad front and lateral budding is suppressed. An important discovery was that the columnar epithelium of the apex (Fig. 3), which in *Bugula* is replaced by a more or less isolated group of cells, extends the cuticle by intussusception—that is by the interpolation of new material among the cuticular elements already present. The apical mass of secretory cells displays regular side to side oscillations, which appear to be correlated with this mode of cuticle formation. The cells do not divide, but are pushed forward by a growth zone behind them. New secretory cells may, however, be recruited from the epithelium in the growth zone.

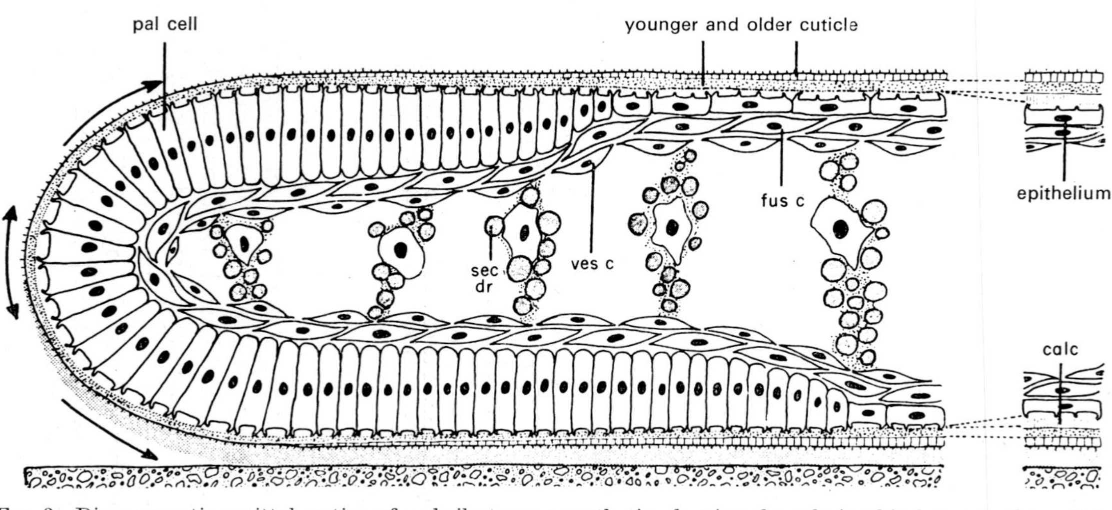

FIG. 3. Diagrammatic sagittal section of a cheilostome growth tip showing the relationship between the cuticle and columnar epithelium of the growing edge, and the skeletal cover and squamous epithelium of the older zooid. The fine stippling between the basal cuticle and the substratum represents mucin. *calc*, calcification; *fus c*, fusiform cell; *pal cell*, pallisade cell; *sec dr*, secretory droplet; *ves c*, vesicular cell. (From Tavener-Smith and Williams, Philosophical Transactions of the Royal Society, B **264**, 1972.)

In effect the cuticle is " stretched " to allow the colony it contains to spread. Mitotic counts and the use of marks of vital stain have shown that in *Membranipora* cell division and growth of the colony occur together in the abapical part of the marginal zone, behind the line of cuticular expansion.

The *Membranipora* colony does not grow regularly all around its periphery. On the line of an emergent axis, the growing edge becomes demarcated as a distinct whitish fringe, functionally a series of giant buds (Fig. 4A). The more rapid the growth, the wider the fringe becomes. Existing longitudinal walls are continued throughout the marginal zone, but a new wall, at the point of division of a zooid row, is initiated as a cuticular ingrowth among the apical columnar cells (Fig. 4B). These cells are differentiated into two distinct zones: the

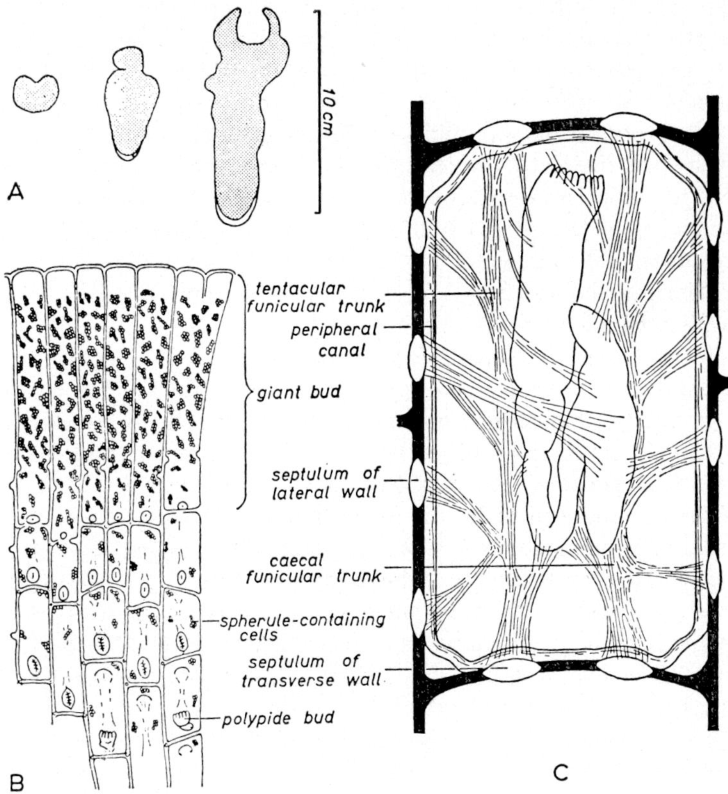

FIG. 4. Growth in *Membranipora membranacea*. A, colony growth, showing emergence of polarity and formation of a fringe (giant buds); B, giant buds at the growing edge, and zooid differentiation; C, septula (pore-plates) and the funicular system. (From Ryland, Oceanography and Marine Biology, an Annual Review, **5,** 1967, © George Allen and Unwin Ltd, after Lutaud.)

inner four-fifths is composed of a dense basophil cytoplasm rich in ribonucleins, and is abruptly demarcated from the outer zone which contains acid mucopolysaccharides and is associated with secretion of the cuticle. At the apex of the giant bud the peritoneum comprises a network of undifferentiated cells. This primordial peritoneum gives rise to two lamellae: an outer layer of fusiform cells and an inner layer of cells distended with lipid globules and inclusions of polysaccharide.

As morphogenesis proceeds, some cells of the inner basal peritoneal layer are utilized in the formation of the two principal funicular cords (Fig. 4C), whilst the corresponding cells from the frontal peritoneum contribute to the developing tentacle sheath. Zooid development continues with the formation of the transverse septum behind the giant bud. The septum is initiated at the base of each lateral wall and infolds as an annular invagination of the epithelium, which subsequently secretes a cuticle. Peritoneal cells migrating inwards from the basal wall contribute to an aggregation which is associated with morphogenesis of the pore-plate. Simultaneously to, and independently of, this process, the polypide rudiment is developing in a median position just distal to the invagination commencing from the frontal wall (Fig. 4B). Development of the polypide at this time is a normal consequence of transverse wall differentiation. The pore-plates of the longitudinal walls have meanwhile developed from unilateral lenticular formations in the distal half of each new zooid, and an existing wall is accordingly broken down at these points.

The giant bud of *Membranipora* is an unusual feature of the growing edge associated with particularly rapid colony growth, but the histological and cytological components of the apex seem to be essentially similar in a number of other species (Tavener-Smith and Williams, 1972) and so perhaps throughout the phylum. Abapically the epithelial cells become reduced in height (eventually squamous) and of changed internal organization. Where the growing zone is hypertrophied, as in *Membranipora*, or at least well developed, as in many British smittinids and schizoporellids, the transverse septa develop as ingrowths either from the four longitudinal walls or from the two lateral walls alone, as Banta (1969) has described for *Watersipora arcuata*. As each septum closes, the zooid proximal to it becomes an entity.

In *Metrarabdotos* the marginal growth zone never exceeds the length of one zooid, and the production of transverse septa apparently proceeds a little differently (Cheetham, 1968). The bud is wedge-shaped, and behind its advancing tip the frontal, basal and lateral walls expand and steadily complete their development. When a bud has reached the appropriate length a transverse septum is produced, initially by upgrowth of the basal wall, while a new bud is pushed forwards (text-fig. 4 in Boardman and Cheetham, 1969). Pore-plates again form where the funicular strands transect the developing septum.

The method of budding in *Electra* and *Callopora* (Silén, 1944b) and *Fenestrulina* and *Micropora* (Gordon, 1971b) is rather distinct from the serial production of zooids by means of transverse septa. Each peripheral zooid terminates in a distal pore-chamber, from which a membranous " window " balloons outwards as the incipient daughter zooid (*op. cit.*, also Fig. 12 in Ryland, 1970).

Lateral buds develop from pore-chambers (used here in the sense of Gordon and Hastings [in prep.] to include all dietellae and septular chambers) in the side walls by the " ballooning " method in all species (Banta, 1969).

The nodular or multilaminar colonies typical of the genera *Celleporaria* and *Turbicellepora*, for example, and found in certain species of *Hippothoa* and *Schizoporella*, are formed when an additional layer or indefinite mass of zooids is added by frontal budding (Pouyet, 1971; Banta, 1972).

C. *The cuticle and colony adhesion*

The structure of the bryozoan cuticle has been investigated by several authors, notably Schneider (1957), Banta (1968), Bobin and Prenant (1968) and Tavener-Smith and Williams (1972). A major obstacle in comparing the accounts is the different descriptive terminology employed by these authors, and I propose here to retain the usage of my book (Ryland, 1970). The whole covering layer secreted by the epidermis is thus the " cuticle " (" ectocyst " of Banta, 1968, etc., and of Bobin and Prenant, 1968). The outermost, organic layer of the cuticle I, like Bobin and Prenant, term the " pellicle ". This is the " epitheca " of Banta (1970) and " periostracum " of Tavener-Smith and Williams (1972). While sympathizing with the latter over their difficulties in applying terms (cuticle and ectocyst) so inconsistently used by previous writers, I doubt whether the problem can be resolved by the introduction of a nomenclature alien to most bryozoan workers.

The pellicle is secreted by the microvillous plasmalemma of the palisade cells and has an initial thickness of 1·5–2·5 μm in *Electra* and *Membranipora* but only of about 0·7 μm in *Bowerbankia*. There is some structural variation, for in the cheilostomes *Electra*, *Membranipora*, *Schizoporella*, *Umbonula*, *Celleporella* and *Cupuladria* and in the cyclostomes *Berenicea* and *Lichenopora* the pellicle is bounded externally by a trilamellar coat some 13–14 nm in thickness. The surface bears a brush of filaments in which, on the basal surface, is entangled a film of mucin. In the ctenostome *Bowerbankia* and in the cyclostome *Crisidia* the trilamellar coat is replaced by a homogeneous electron-dense layer of granular texture, and filaments are rare (Tavener-Smith and Williams, 1972). These are both noncrustose bryozoans.

The cuticle is composed predominantly of protein but also contains mucopolysaccharide. In a frontal membrane or superficial layer of a calcified wall the cuticle remains as a hyaline pellicle; deeper in the wall the cuticle is fibrillar and provides the matrix on which the calcium carbonate crystals are deposited; finally, innermost and adjoining the epidermis, is another thin homogeneous layer (Schneider, 1957, Banta, 1968; Bobin and Prenant, 1968). In *Electra pilosa* Bobin and Prenant were able to demonstrate a build-up of calcium salts in the matrix before any crystallization could be detected. The calcium carbonate is deposited in one or other of the crystal forms calcite or aragonite (rarely both at once): the circumstances deciding which are not understood, but it is not impossible that the organic nature (i.e., amino-acid content) of the cuticle, or dissolved organic materials in the reaction space, influence crystal form.

Primary calcification in cheilostomes usually takes the form of calcite. The manner of deposition may differ between the ancestrula and the subsequent zooids. In the former it must be added to an otherwise fully formed wall and has the nature of a thin layer of crystal spherites secreted below the pellicle (Lutaud, 1953). In budded zooids the calcite can be incorporated within the wall as it develops behind the growing edge (Schneider, 1957; Banta, 1968; Bobin and Prenant, 1968). The complex subject of frontal wall structure will not be considered further in this review, and the reader is referred to the major publication by Tavener-Smith and Williams (1972) and to the chapter by Sandberg in the "Biology of Bryozoans".

It is apparent from inspection that, during the life of the colony, encrusting bryozoans use two different mechanisms for adhesion to the substratum. The ancestrula is cemented down by a secretion from the larval pyriform gland (see p. 369). In phytophilous species the algal substratum may be discoloured (as by *Hippothoa* sp. on *Sargassum sinclarii* Hooker et Harvey) or affected in such a way that the centre of the colony later dissociates from the alga and drops away (*Membranipora membranacea* on *Laminaria* spp.). The rest of the colony remains attached through the adhesive properties of a mucopolysaccharide film associated with the basal surface of the cuticle (Tavener-Smith and Williams, 1972; J. D. Soule, 1973). As it forms at the growing edge the cuticle is extremely ductile and virtually flows over the substratum, perfectly following the microscopic detail of its contours (Soule and

Soule, 1974). As it hardens, the cuticle with its film anchors the colony to the substratum, be it hard like coral or soft like an algal frond. It appears that the adhesive is not a cured cement, since these generally depend upon a protein-polyphenol oxidase system of bonding (as, probably, in the byssus of *Mytilus*), but a Stefan type adhesive, the viscosity of which holds the two surfaces in apposition (as in barnacles; Crisp, 1973). As the differentiation of zooids proceeds, the basal cuticle may become calcified (though often with a central membranous "window" remaining).

D. *The pattern of colony growth*

It seems clear, even from the limited research so far undertaken, that the number of zooids present in a growing bryozoan colony not subjected to environmental constraints increases logarithmically during its growing season. (Such constraints might include lack of space [Hayward, 1973] and seasonal factors such as falling temperature and shortage of food. There are few data for tropical species, and it is not known to what extent colonies suffer from ageing.) The relationship between log zooid number and time is therefore typically linear. This was most successfully established by Bushnell (1966) in the freshwater *Plumatella repens*. In the same paper he stated that *Bugula turrita*, studied at Woods Hole, displayed geometric growth as convincing as that of *Plumatella repens*. The encrusting *Cryptosula pallasiana*, however, showed "a slower and less perfect geometric growth". Growth in *Alcyonidium hirsutum* (Hayward, 1973; Hayward and Ryland, 1975), discussed below, was geometric only for a limited period of time (Fig. 5); and this was also true for *Conopeum tenuissimum* in Chesapeake Bay, where growth rate declines "as limiting factors such as competition for space, predation, etc., begin to take effect" (Dudley, 1973). Menon (1972a), working on three North Sea species of encrusting habit (*Electra pilosa*, *Membranipora membranacea* and *Conopeum reticulum*) likewise found that colony area (proportional to number of zooids) plotted as $\log_{10}$ against time reduced to a straight line. It should also be noted that a recent study on the growth of compound ascidians (Yamaguchi, 1975) has revealed the same relationship. We may therefore (following Bushnell, 1966) summarize that, under unrestricted conditions of growth, the number of zooids N present in a colony at time t is given by

$$N_t = N_o \exp(rt),$$

where N_o is the initial number of zooids and r is the instantaneous rate of growth. In four tropical species (temperatures 27·1 — 31·6°C) observed to an age of 45 days on settlement panels at Cochin, south-west India, Menon (1972b) found that a linear relationship was obtained by plotting log colony area against log time (rather than against time as in the earlier examples discussed).

The growth of *Alcyonidium hirsutum* colonies established on *Fucus serratus* L. on a shore in South Wales has been studied over a three-year period by Hayward and Harvey (1974b) and Hayward and Ryland (1975). The latter (actually the earlier paper) recorded settlement in February–March followed by a pattern of growth that was broadly similar in all three years. As shown in a plot of log colony area against time (Fig. 5), growth is slow in March–April but it accelerates in late spring and then steadies during the summer. This steady rate, which was similar in all three years, does not continue however, for, as the

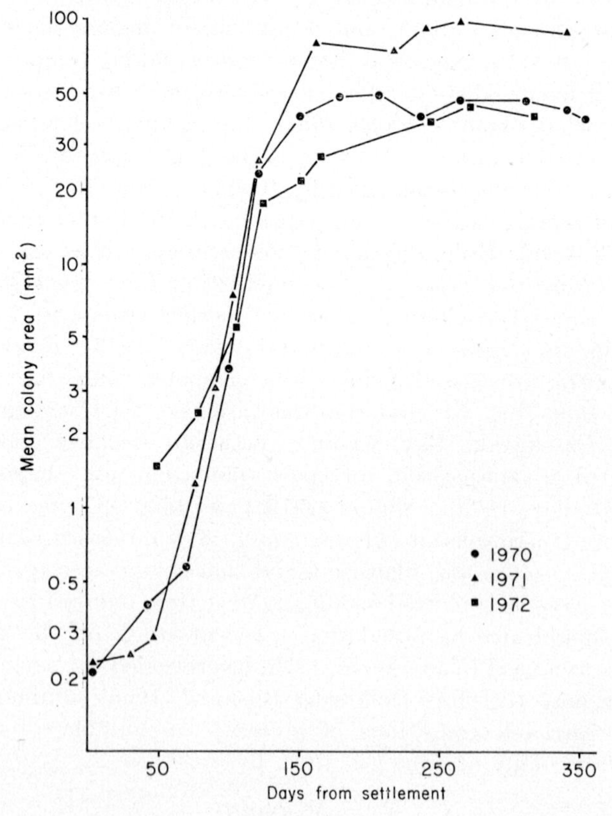

FIG. 5. Growth of *Alcyonidium hirsutum* on a shore in south-west Wales over three seasons. (From Hayward and Ryland, 1975.)

available space on the *Fucus* fronds diminishes, growth slows. It was found that the timing of both larval release and the commencement of accelerating growth varied from year to year, presumably being regulated by environmental factors such as sea temperature. The point from which growth began to slow and the ultimate size of the colonies, on the other hand, seem more likely to depend upon the density of settlement rather than on extraneous environmental factors.

These aspects, and also mortality, were considered by Hayward and Harvey (1974b) in a comparison of growth rates in what they had previously established (1974a) were high- and low-density regions of a *Fucus serratus* frond. There was no evidence of density-dependent mortality—indeed overall mortality rates were very low. By comparing colony size in the two regions month by month, it was found that the colonies did not grow more quickly or reach a larger size in the low-density regions. Since Hayward and Ryland had found that larger colonies resulted from a low-density settlement than from a higher one, larger colonies might also have been expected on the low-density areas of frond. Hayward and Ryland were comparing different years, which introduces an additional variable, but the most probable explanation

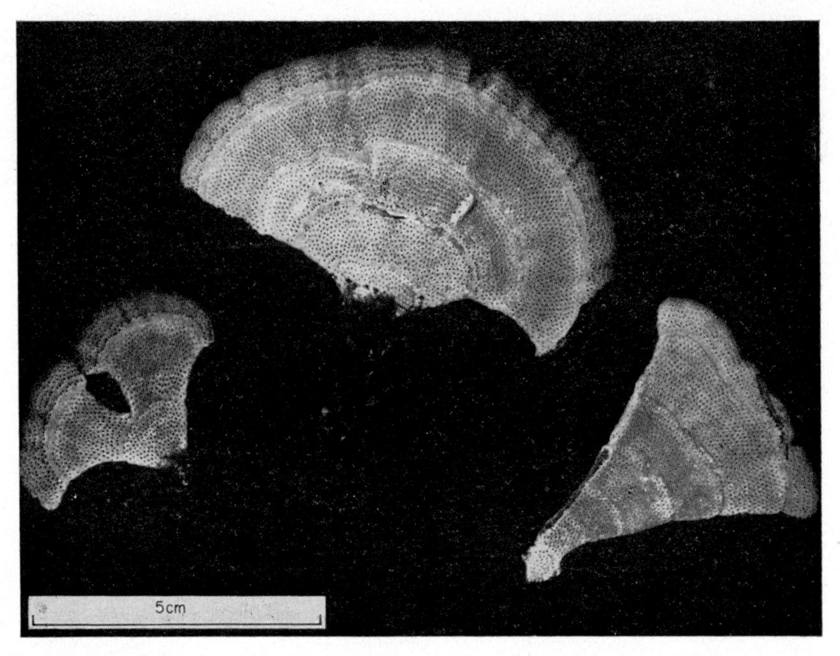

Fig. 6. Frondose antarctic bryozoan *Cellarinella roydsi* showing the conspicuous (and presumably annual) growth check-lines.

for this not occurring is that the low-density areas on a *Fucus* frond represent a sub-optimal habitat, so that colonies do not grow as freely as in the high-density, optimal areas (Hayward and Harvey, 1974b; see also p. 399).

A bryozoon not subjected to spatial constraints on growth is *Flustra foliacea*, a foliose perennial studied by Stebbing (1971a). The larvae settle during February–March and a crustose colony base is established. A bilaminar frond arises from this early in the second year (Eggleston, 1972a).

Seasonal growth off the South Wales coast starts in March, accelerates through spring and ceases, quite abruptly, in November. When the following year's new growth commences, a winter check-line is seen to have been formed across the frond (see Fig. 41, p. 410). (Even more conspicuous check-lines are visible in the fronds of the antarctic species *Cellarinella roydsi* [Fig. 6]. It seems highly probable that they have formed during winter.) Using the winter rings Stebbing was able to show that *F. foliacea* fronds commonly survive and grow for a period of six years, and occasional colonies were 12 years old. Each frond carries a record of its growth through the previous years and, by back measurements, Stebbing found that mean growth rates are exponential over the life of the frond (Fig. 7). The increments were determined in terms of weight of frond formed during each year of life. For the first year this certainly introduces a source of error, for the zooids bud frontally (the phenomenon mentioned on p. 295), so that

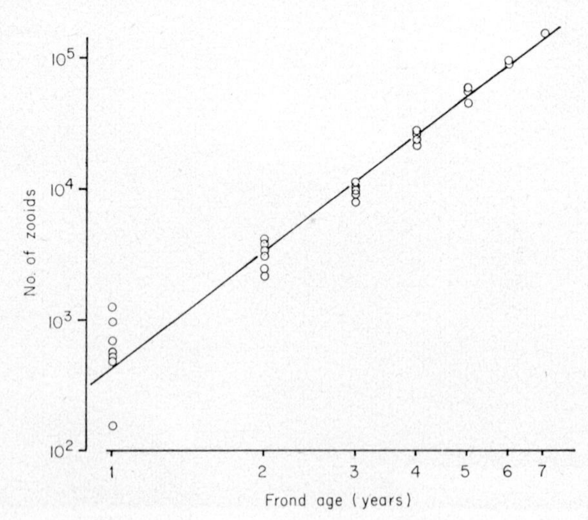

FIG. 7. Growth of *Flustra foliacea*. Ordinate shows estimated number of zooids determined by back measurements. (From Stebbing, *Marine Biology*, **9**, 1971.)

superposed layers of zooids are added around the proximal part of each frond, producing, in effect, a strengthened stem. (Fig. 7 in Stebbing, 1971a, shows that the relation of the number of zooid layers y to frond height x cm is given by $y = 1{\cdot}4 + 1{\cdot}02\,x$). The oldest fronds therefore tend to have the highest number of zooids when backplotted to year 1 (Fig. 7), and the calculated regression line will reflect this. In fact, the lowest point in Fig. 7 actually derives from 4-year-old fronds.

Growth in temperate latitudes is essentially a summer phenomenon, but there are few clear experimental data showing the effect of temperature on growth. Menon (1972a) studied *Membranipora membranacea*, *Electra pilosa* and *Conopeum reticulum* under laboratory conditions.

TABLE I. AFFECT OF TEMPERATURE ON ZOOIDAL DIMENSIONS (MEANS AND STANDARD DEVIATIONS OF 50-ZOOID SAMPLES) (Menon, 1972a)

Temperature (°C)	6°	12°	18°	22°
Electra pilosa:				
Length (μm)	686 (± 175)	596 (± 42)	586 (± 30)	577 (± 37)
Width (μm)	304 (± 36)	312 (± 25)	314 (± 33)	314 (± 51)
Conopeum reticulum:				
Length (μm)	—	558 (± 51)	519 (± 44)	500 (± 49)
Width (μm)	—	315 (± 27)	285 (± 47)	314 (± 35)

Despite evidence of sub-optimal conditions, because *E. pilosa* and *C. reticulum* colonies grew slower than in the sea, the results are useful comparatively. Maintaining *M. membranacea* at constant temperatures of 6°, 12° and 18°C, *E. pilosa* at 6°, 12°, 18° and 22°C, and *C. reticulum* at 12°, 18° and 22°C, all clearly showed increase in growth rate with temperature. However, in culture, growth rate (at all temperatures) fell from an initial high rate, which was maintained only for 2–3 weeks, to a much lower rate after about four weeks.

Menon also investigated lethal temperatures and the effect of temperature on zooid size. His data (Table I) confirm, as observations on wide ranging species suggest, that zooidal dimensions are inversely correlated with temperature.

E. *Orientated colony growth*

1. *Rheotropisms*

Lutaud (1961) was the first to record the proximally orientated growth of *Membranipora membranacea* on *Laminaria* fronds. Ryland (1967a) pointed out the ecological advantage of such growth which

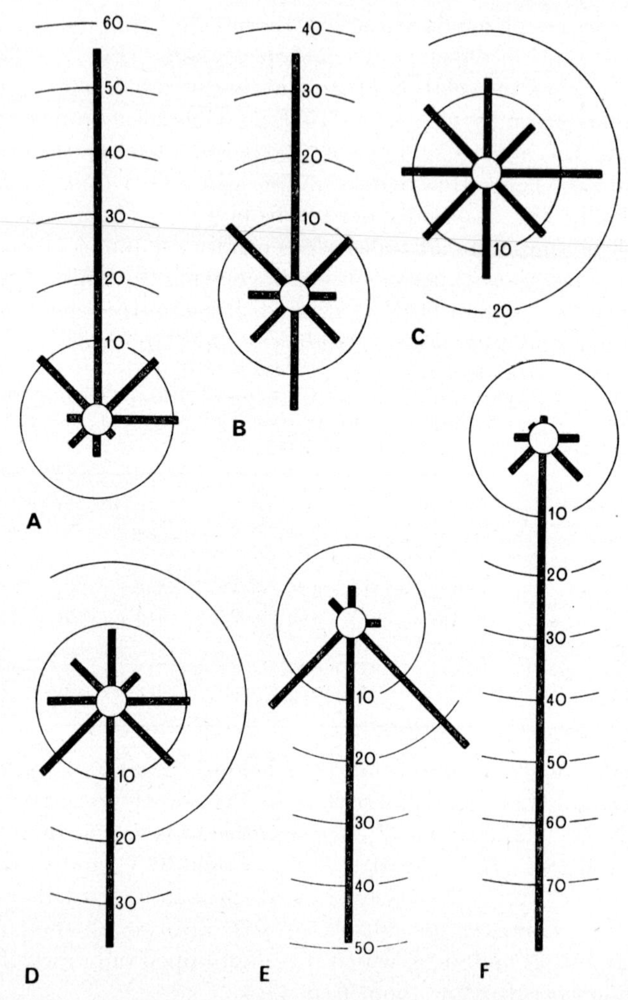

FIG. 8. Orientated growth in bryozoans, showing percentage growth direction by octants. A, *Scrupocellaria scruposa* on *Flustra foliacea*; B, *Electra pilosa* on *Fucus serratus*; C–F, *Membranipora membranacea* on *Laminaria hyperborea*: C, random orientation of ancestrulae; D, emerging growth axis in colonies < 35 mm diameter; E, orientation of growth in colonies 15–35 mm diameter; F, orientation of growth using all colonies. (From Ryland, 1976b, based on Ryland and Stebbing, 1971.)

ensures that the colony spreads continuously towards the youngest part of the frond (see p. 409), and a full investigation was made by Ryland and Stebbing (1971). They first quantitatively confirmed Lutaud's observations. Dominant growth axes were determined and recorded in terms of the octants of a circle, and displayed as eight bars on a circular histogram (Fig. 8). It was found that orientated growth on *Laminaria hyperborea* (Gunn.) Fosl. fronds evolved from randomly orientated settlement (Fig. 8C). A preferentially orientated growth axis emerged early in the astogeny of the colony, often while it consisted of only six to 15 zooids. Indeed, even the first or second rows of daughter zooids may have their long axes skewed towards the algal stipe (Fig. 8D) and by the time the colonies have reached a diameter of 15–35 mm most have acquired a roughly proximal growth orientation (Fig. 8E). When *Membranipora* colonies of all sizes were considered about 80% exhibited growth towards the stipe (Fig. 8F). *M. membranacea* is not particularly common on *Fucus serratus* but a predominantly proximal orientation of growth was noted on this alga also.

Ryland and Stebbing considered as possible causative stimuli light, gravity, water current and polarity of the surface, quickly eliminating the first two despite their known involvement in orientated growth in other bryozoans in different circumstances. It was assumed that algae living in the turbulent waters of the intertidal zone and just below would be in constant motion. The fronds would be swept to and fro by the waves, always streaming in the direction of water flow, which is thus from the base towards the extremities of the fronds. Water passing over a *Laminaria* frond follows the physiological age gradient; but water moving over a *Fucus* thallus is flowing contrary to the age gradient. It thus seemed that the only factor in common was water movement, and it was concluded that the observed orientated growth in *M. membranacea* was indicative of positive rheotropism. This has since neatly been confirmed by Norton (1973). *Saccorhiza polyschides* (Lightf.) Batt. is a kelp in which the stipe is broad and flat. As in *Laminaria* the plant grows from an intercalary meristem where the stipe merges into the frond. If orientated growth were due to rheotropism, *Membranipora* colonies on both frond and stipe would grow towards the holdfast; if, on the other hand, the response were dependent upon the recognition of a physiological age gradient in the alga, colonies on both frond and stipe would grow towards the meristem. Norton showed quite clearly that growth on the stipe was orientated downwards, towards the holdfast, not upwards towards the meristem.

Positive rheotropism had been earlier demonstrated in an unrelated species, *Farrella repens*, by Marcus (1926). He observed an upstream growth of stolons after exposure to a water current of about 1 cm s^{-1}. Curiously, however, Jebram (1970) has obtained rather different results. Using colonies established on glass slides, no orientated stolonal growth was observed at current speeds of either 3 or 15 cm s^{-1}. There seems no obvious explanation for this contrary result.

Jebram also sought rheotropic responses in other species. None was found in *Conopeum reticulum* in a water current of 3 cm s^{-1}. In *Electra monostachys* no response was detected at a velocity of 15 cm s^{-1}, but at 3 cm s^{-1} the compact colony started to produce monoserial zooid rows predominantly in an upstream direction.

Electra pilosa was the subject of further experiments. At current velocities of 3 and 5 cm s^{-1}, Jebram found that: " The main parts of these colonies began to grow against the current direction " and concluded: " Both of my experiments show a clearly positive rheotropic growth reaction under the influence of slow water current speeds." The present writer, however, is unable to see this in the evidence provided (Jebram, 1970, Fig. 11 A–D). At 15 cm s^{-1} Jebram's colonies " showed no special reaction ". Ryland and Stebbing (1971) found a preponderance of apically directed growth of *E. pilosa* on *Fucus serratus* fronds (Fig. 8B). In view of the characteristic early astogeny of *E. pilosa*, in which straight lines of zooids extend the ancestrular axis (see Barrois, 1877, Pl. XV, Figs. 5–9 and Marcus, 1926), it was considered that this stemmed directly from the orientation of the ancestrula. There was certainly no evidence for the emergence of a proximally directed growth axis as happens in *Membranipora membranacea* on the same alga. Additionally, when well-grown colonies were found on the swivelling panel used for the study of ancestrular orientation (see p. 362; Ryland, 1976a), they had the regular stellate form illustrated by Marcus (1926, Fig. 12). It is concluded that, for the present, there is no satisfactory evidence for rheotropic growth in *Electra pilosa*.

2. *Phototropisms*

The occurrence of phototropism in *Bugula* was briefly first reported by Schneider (1955) and Aymes (1956) independently. Aymes studied *B. neritina*; Schneider named his species *B. avicularia*, but Ryland (1967a) thought that it was probably *B. stolonifera*. Schneider established that branches of autozooids were positively phototropic and that the attaching rhizoids were negatively phototropic. There was no response to gravity for, when Schneider (1959) kept his *Bugula* in total darkness, upwards and downwards were no longer distinguished and the branches and rhizoids grew into a disorientated tangle. He also found that closely related species may vary in their responses. Thus light is necessary to *B.* [?] *stolonifera* for typical growth, but *B. neritina* will grow normally in the dark.

The cellular mechanism producing the tropism was also described (Schneider, 1960). As mentioned earlier (p. 291), the branches (and rhizoids) grow from apical buds, in the tip of which lies a plate of dense, spherical cells. The cells display co-ordinated activity, such that the plate rhythmically pulses and makes other movements. If the branch tip is lit from one side, the apical plate shifts towards the source of illumination, commencing its movement between 5 and 30 min after the onset of illumination. With the apical plate of cells relocated, the bud naturally continues its growth in the new direction. At the growing edge of *Membranipora membranacea* the comparable cell group displays regular side to side oscillations (Lutaud, 1961). Presumably a change in growth direction is similarly effected through a lateral shift of these cells.

Scrupocellaria reptans is commonly found growing epizoically on the fronds of *Flustra foliacea* (Fig. 41 and p. 409). Its colonies also display orientated growth (Fig. 8A), predominantly towards the periphery of the *Flustra* fronds (Ryland and Stebbing, 1971). Schneider (1959) noted that the ancestrular bud of *S. reptans* grew towards the light, thereby determining the initial growth axis of the colony. Unfortunately, this was not further clarified, leaving it uncertain how the bud came to face the light in the first instance. It may be that there is an orientating response by the settling larvae, as in *Bugula neritina* (see p. 378), with or without phototropic colony growth as in *B.* [*?*] *stolonifera*.

Bryozoan tropisms have been discussed in greater detail in another article (Ryland, 1976b).

F. *Effect of environment on colony form*

Colony form in erect, branching species is known to be labile. Harmelin (1973c) has studied the relationships between the branching pattern of the cyclostome "*Idmonea*" *atlantica* [= *Idmidronea*] and environmental factors. In the neighbourhood of Marseille, *I. atlantica* is a common species in depths below 5 m (the circa-littoral zone of Pérès, 1967), extending into the "semi-obscure" part of caves, and reaching its maximum abundance in the coralligenous biocoenosis (Pérès, 1967). Harmelin's analysis was based on nine quantifiable variables. The most typical form, exemplified by colonies from the near-optimal conditions of the coralligenous biocoenosis, is a sturdy colony, the branches of which bifurcate after an average of 6·56 series of zooids at an angle of 51° (Table II). They are disposed like spokes from the centre of the colony, which acquires near radial symmetry, as in Fig. 9D. In the caves, under weak illumination and slight water movement, the colonies are slender, attenuate and weakly branched (Fig. 9A–C); in the precoralligenous biocoenosis, where light

TABLE II. CORRELATION OF COLONY VARIABLES WITH HABITAT IN *Idmidronea atlantica* (Harmelin, 1973c). Data are means and standard deviations

Variable	*Caves*		*Coralligenous biocoenosis*		*Precoralligenous biocoenosis*	
No. of zooids per series	3·17	(± 0·69)	3·92	(± 0·62)	4·33	(± 0·81)
Zooid width (μm)	96	(± 10)	102	(± 6)	104	(± 6)
No. of series of zooids between bifurcations	9.61	(±6·97)	6·56	(± 3·29)	6·60	(±2·55)
Branch width (μm)	349	(± 66)	466	(± 57)	481	(± 132)
Bifurcation angle (degrees)	47	(± 14·5)	51	(± 30·0)	58	(± 17·3)

is strong and water movement vigorous, the colonies are stout and well branched (Fig. 9E, F). The implications from the taxonomic viewpoint were also discussed. Some further examples have been considered in a second paper (Harmelin, 1976).

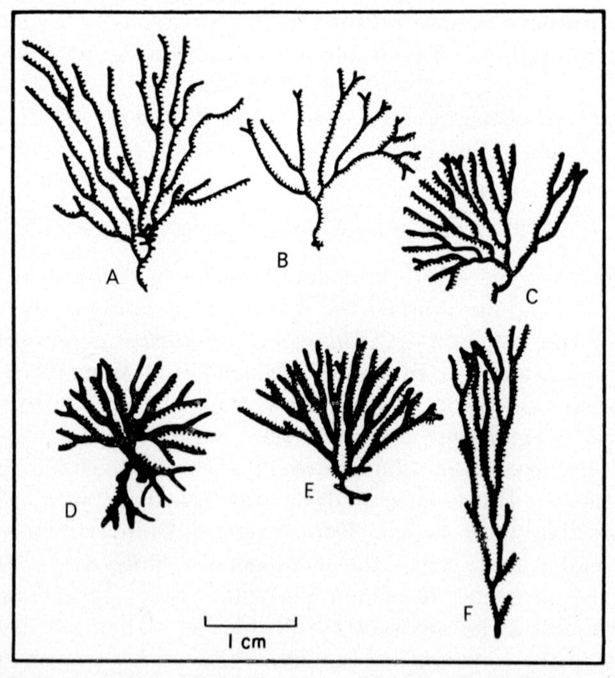

FIG. 9. *Idmidronea atlantica*: variation in colony form according to habitat. A, from the obscure zone of a cave; B, C, from the semi-obscure zone of another cave; D, from deep coastal detritic sand (70 m); E, F, from precoralligenous rocks. (From Harmelin, 1973c.)

G. *Polypide regression and brown body formation*

Colony longevity in bryozoans varies greatly from species to species (Eggleston, 1972a). Among those with short life spans are *Celleporella* (= *Hippothoa*) *hyalina* and *Callopora lineata*, epibionts of algal fronds which themselves disintegrate after a few months. Many colonies are annual or biennial, and a few others live for several years. Examples of such perennial species in European waters are *Pentapora foliacea* and *Flustra foliacea*, the latter occasionally surviving to the age of twelve (Stebbing, 1971a). Individual zooids probably remain alive as long as the colony, or a surviving part of a colony, does. A polypide, on the other hand, has a life of a few weeks only (Table III) after which time it regresses into an ovoid residual mass known as a brown body.

TABLE III. POLYPIDE LONGEVITY AND DURATION OF REGRESSION (DAYS) IN SOME BRYOZOANS (mainly from Gordon, 1976)

Species	*Polypide longevity*	*Duration of regression*	*Authority*
Flustrellidra hispida	21–28[1]	6–7[2]	1 Rey, 1927 2 Joliet, 1877
Cryptosula pallasiana	15–72	6–17	Gordon, 1976
Electra pilosa	6–33	3–8	Gordon, 1976
Eurystomella foraminigera	20–60	5–15	Gordon, 1976
Fenestrulina malusii var. *thyreophora*	35–42	8–10	Gordon, 1976
Bugula flabellata	–	7–9	Joliet, 1877
Bugula neritina	–	2	Harmer, 1891
Carbasea papyrea	–	12	Harmer, 1891
Conopeum tenuissimum	10–14	–	Dudley, 1973
Plumatella casmiana	2–31	2–4	Wood, 1973

The regression phenomenon has been most closely studied in the fouling bryozoon *Cryptosula pallasiana*, using transmission electron microscopy (Gordon, 1973; 1976). In *Cryptosula* and, it appears, in most other bryozoans, regression begins first in the lophophore, starting distally. Ciliated epidermal cells slough off from the axial basement membrane, but individual cells do not appear to regress at the same rate. Cytoplasmic changes typically involve swelling of endoplasmic cisternae and mitochondria, and autophagic ingestion of cell components. Some cells lose their cilia by abscission, whereas others resorb their cilia such that axonemes are encountered within the cells. Nuclei become pyknotic as chromatin condenses, and may also fragment into clusters of smaller " nuclei ", a process called karyorrhexis. There is some ingestion of cells by neighbours, while many others rupture, liberating their contents.

The basement membrane collapses, some of it being incorporated by heterophagy into epidermal cells. A similar fate befalls the tentacle musculature as filaments cease to lie in orderly arrays.

These changes take place over a period of days during which the lophophore condenses by autophagy, heterophagy and concomitant cell movement.

In the gut, different parts regress at different rates. The pharynx and rectum regress concurrently with the lophophore while the stomach maintains its integrity slightly longer. Cytoplasmic changes in the pharynx are much as in the lophophore and the fluid of the large vacuoles increases in density owing to the appearance of small granules and membranous structures. In regressing rectal cells products of absorption that have accumulated during the life of the polypide are a prominent feature. These comprise vacuoles containing homogeneous contents which may be polysaccharide. Other vacuoles (secondary lysosomes) contain membranous lamellae, dense bodies and granular material.

The entire stomach, between cardiac and pyloric sphincters, regresses as a unit. A prominent and characteristic feature of stomach cells is the population of orange-brown inclusions, which are secondary lysosomes and residual bodies, that respond to certain stains and ultraviolet light in the manner of lipofuscin (so-called ageing pigment). These inclusions accumulate during the life of the polypide, being derived from ingested food and possibly also from autophagy of cell organelles (see p. 337). They come to constitute a major part of the brown body proper. Through condensation and digestion these give the dark-brown coloration to the brown body.

Much of the remains of the lophophore, pharynx and rectum becomes distributed throughout the body cavity, evidently among or within mesenchyme cells, or remains as a loose " cap " of necrotic debris at the distal end of the brown body proper which is derived exclusively from the stomach.

The brown body of bryozoans thus represents a huge residual body, comprising necrotic stomach cells, enlarged through heterophagic activity, each containing a central mass of residue surrounded by a narrow zone of cytoplasm which shortly disappears. During regression, rosettes of alpha glycogen accumulate in cells of the mesenchyme and under the floor of the compensation sac.

In *Cryptosula* the whole process of regression and brown body

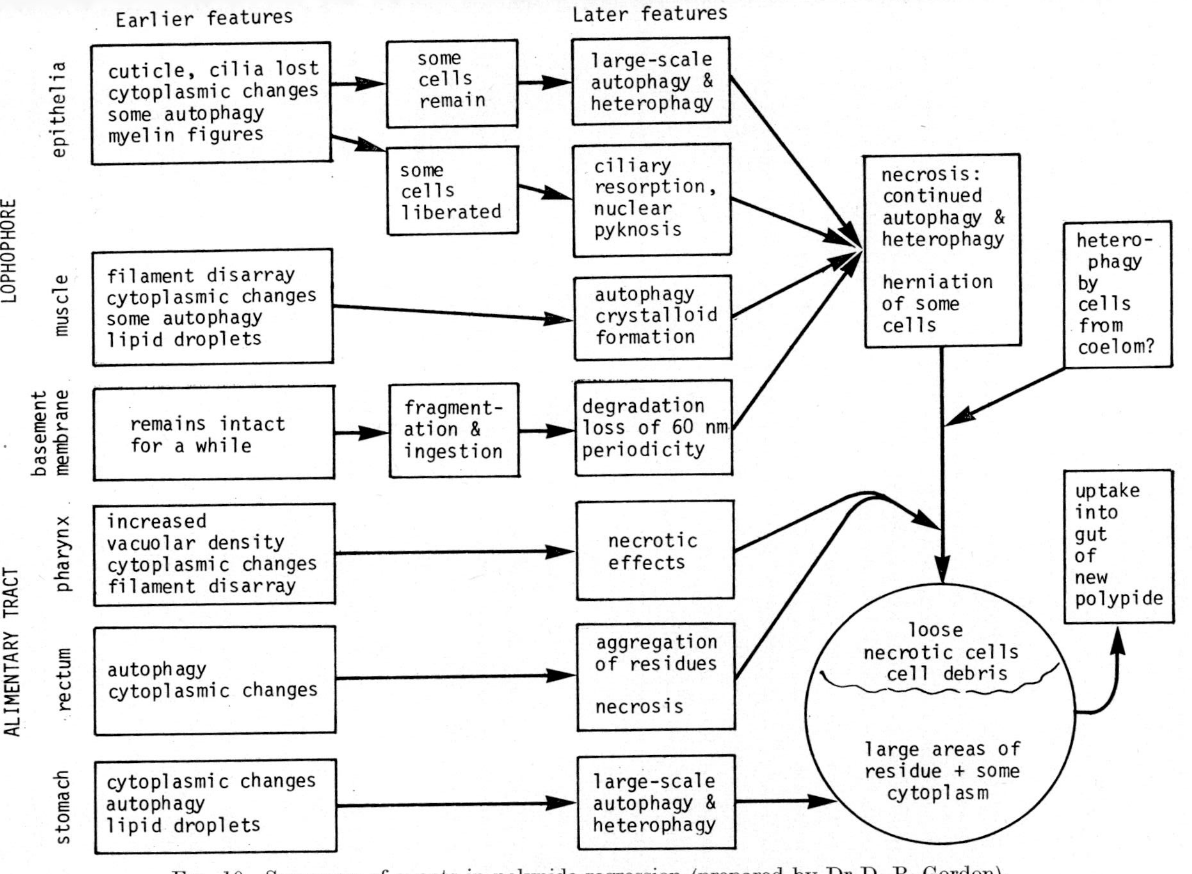

FIG. 10. Summary of events in polypide regression (prepared by Dr D. P. Gordon).

formation (summarized in Fig. 10) takes 6–17 days, but much less (2 days) in *Bugula neritina* according to Harmer (1891).

In many ctenostomes and cellularines (Bugulidae, Scrupocellariidae, etc.) the brown body remains thereafter in the body cavity, but in *Cryptosula* and many other cheilostomes it becomes incorporated into the developing stomach of a new polypide that arises from the epithelia of the distal body wall. When the new polypide defaecates for the first time the brown body is ejected.

While polypide regression is known to be induced by adverse environmental conditions, it does, nonetheless, seem to be the inevitable consequence of the accumulation of inert residues in the stomach cells during its lifetime. Since these cells are not replaced, the life of the polypide must be related to the lives of the stomach cells and hence the availability of food. One would expect that bryozoans living in conditions of a plenitude of food would undergo regression and regeneration with greater frequency than those living under conditions of reduced food supply, but this has yet to be demonstrated experimentally. The topic of polypide regression is discussed in greater detail by Gordon (1976), whose notes provided the basis of the preceding account.

III. The Lophophore and Feeding

The lophophore was defined by Hyman (1959, p. 229) as "a tentaculated extension of the mesosome that embraces the mouth but not the anus and has a coelomic lumen". She elaborated: "By some the term lophophore is limited to the basal ridge that bears the tentacles, but this appears an artificial distinction and the whole structure will be called lophophore or tentacular crown". Hyman's definition is accepted here.

The general form of the lophophore in marine bryozoa is well known and described in the textbooks (e.g. Hyman, 1959; Ryland, 1970). Briefly, the expanded crown has the form of an almost radially symmetrical funnel of slender ciliated tentacles, with the mouth located at its vertex. The mouth and lophophore are situated at the free end of an introvert, the tentacle sheath. The zooid in Cyclostomata is basically cylindrical, protected by its tubular calcified wall. Eversion of the tentacular funnel is by a displacement of fluid from the distal end to the proximal end of the coelom (Borg, 1926). The introvert is unable to evaginate beyond the orifice of the skeletal tube; indeed, while feeding, the mouth seems generally to lie within the orifice (Fig. 11A). In Gymnolaemata protrusion of the tentacles is by deformation of the zooid walls (or part of them or their derivative) and the introvert

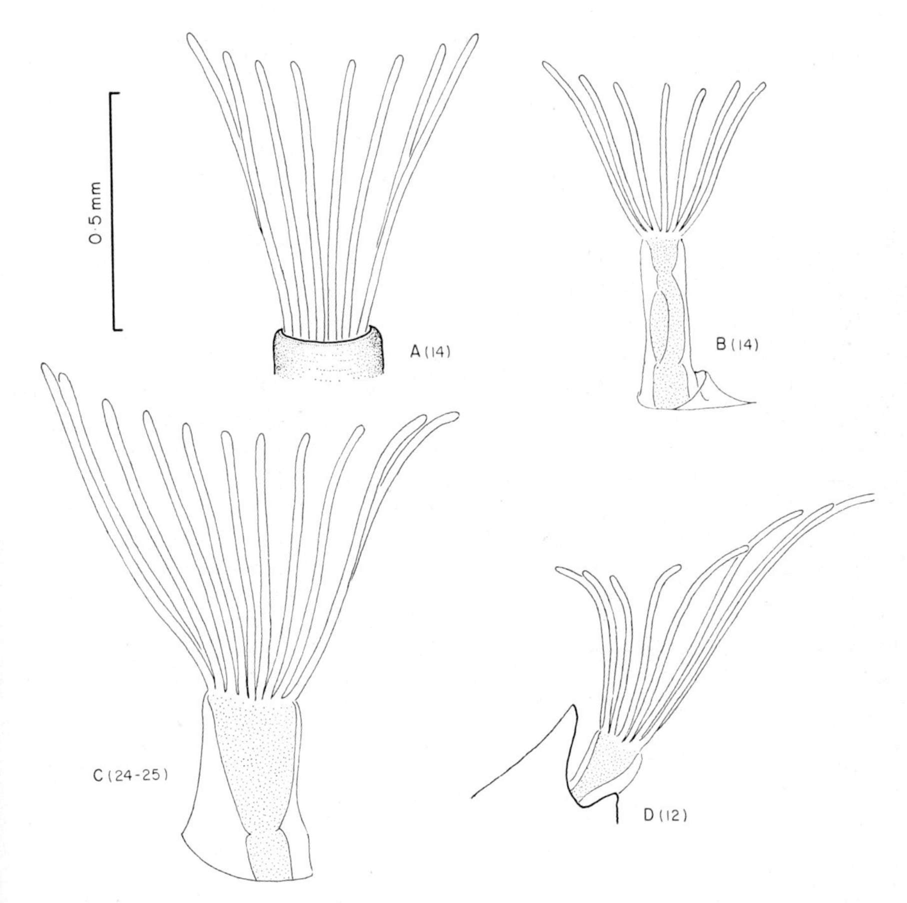

FIG. 11. Representations of the bryozoan tentacular funnel to show variations in size and symmetry. Half the complement of tentacles have been drawn, the modal number being indicated parenthetically. A, *Diaperoecia* sp; B, *Retevirgula acuta*; C, *Steginoporalla neozelanica*; D, *Rhynchozoon rostratum*. (Redrawn from Ryland, 1975.)

evaginates until its entire length may lie outside the orifice (Fig. 11B). The tentacular funnel is thus raised well above the frontal surface of the zooid.

The first satisfactory account of the structure and function of the lophophore in Cyclostomata was given by Borg (1926), while Atkins (1932) followed with a description of feeding in *Flustrellidra hispida* (Ctenostomata). Water, driven by the tentacular lateral cilia, enters the funnel from above and passes downwards and outwards between the tentacles (Fig. 17A, p. 326). The manner in which particles in suspension are captured has proved difficult to determine and two theories are discussed later (p. 322 *et seq.*).

A. *Structure and movements of the lophophore*

1. *Morphometry of the tentacular funnel*

Many authors have counted the number of tentacles in the lophophores of various species (see, e.g., descriptions in Hincks, 1880, and tabulated data for 44 Mediterranean species on p. 138 of Calvet, 1900). The minimum number is eight, in both Stenolaemata and Gymnolaemata, and a low number appears to be primitive. A more or less constant number (cf. Dudley, 1973, and Jebram, 1973, however, who report circumstances under which variation may occur) in the range 12–17 seems most usual in Gymnolaemata with the highest number—usually 28–29 but up to 40 according to Brien (1960)—found in *Flus-*

TABLE IV. MORPHOMETRY OF THE BRYOZOAN LOPHOPHORE
Data (mean, ± standard deviation) for some New Zealand species from Ryland (1975) and for two British species (asterisked) supplied by Mr P. Dyrynda

Species	*Tentacle number,* N	*Tentacle length,* L (μm)	*Funnel top diameter* D (μm)	*Intertentacular tip distance* (μm)
Cheilostomata (Anasca)				
Scruparia ambigua	10·00 ± 0·39	210 ± 21	283 ± 30	87
Crassimarginatella papulifera	14·58 ± 0·84	422 ± 52	437 ± 59	93
*Electra pilosa**	13·18 ± 0·56	418 ± 39	496 ± 51	118
Steginoporella neozelanica	24·47 ± 1·36	862 ± 104	842 ± 87	108
Micropora mortenseni	13·11 ± 0·74	399 ± 44	443 ± 49	105
Cheilostomata (Ascophora)				
Exochella tricuspis	11·69 ± 0·48	333 ± 51	350 ± 54	93
Escharoides angela	16·22 ± 0·55	528 ± 52	488 ± 66	94
Schizomavella immersa	13·70 ± 0·91	408 ± 30	421 ± 105	96
Fenestrulina thyreophora	12·10 ± 0·77	441 ± 56	428 ± 74	110
Watersipora arcuata	19·50 ± 0·53	—	750 ± 56	120
Ctenostomata				
Bowerbankia gracilis	8·00 ± 0·00	274 ± 41	298 ± 48	114
B. imbricata	10·00 ± 0·00	383 ± 70	460 ± 77	142
Elzerina binderi	23·10 ± 0·99	641 ± 41	691 ± 47	94
*Flustrellidra hispida**	27·48 ± 0·73	865 ± 61	1 228 ± 85	140

trellidra hispida. Only recently, however, has there been any comprehensive study of lophophoral morphometry with an attempt to assess the interrelationships between funnel size, tentacle number and other variables.

The smallest funnels generally have fewest tentacles and the largest funnels have most. Thus, in a crown of eight the tentacle length is about 200 μm, for 15 it is about 500 μm and for 25 about 900 μm (Ryland, 1975). Some specific examples are given in Table IV. That the variables of tentacle length, funnel diameter and tentacle number should be positively correlated (Dudley, 1970) is hardly surprising but the very high degree of correlation possibly is: it certainly calls for examination.

In a study of some New Zealand species, Ryland (1975) regarded the funnel as closely approximating to an inverted apically truncated cone. The truncated cone and its constituent tentacles can be fully defined in terms of the top and bottom diameters, and the length of the

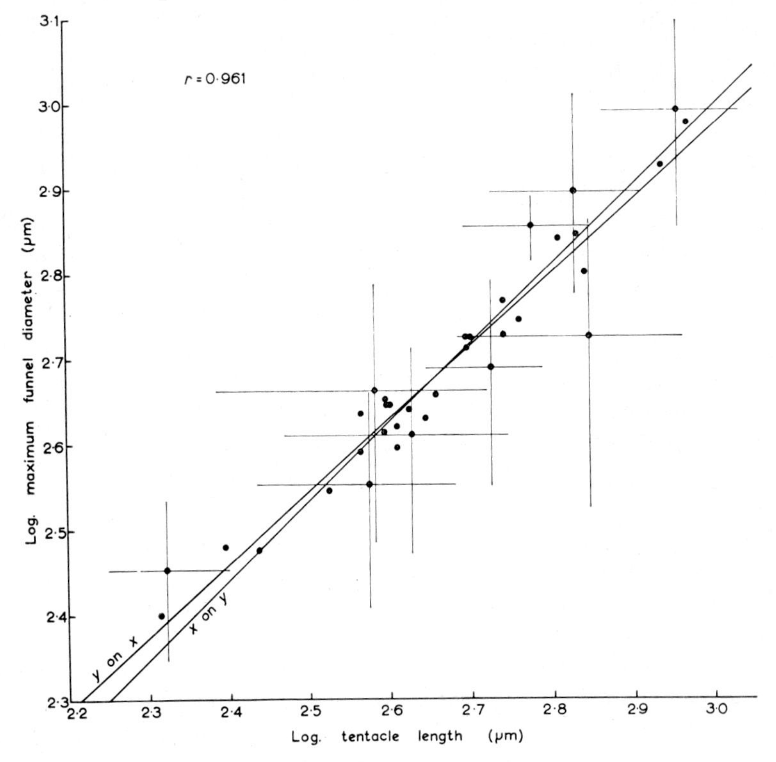

Fig. 12. The relationship between maximum funnel diameter and tentacle length: crossed lines indicate $\pm$ 2 standard deviations. (From Ryland, 1975.)

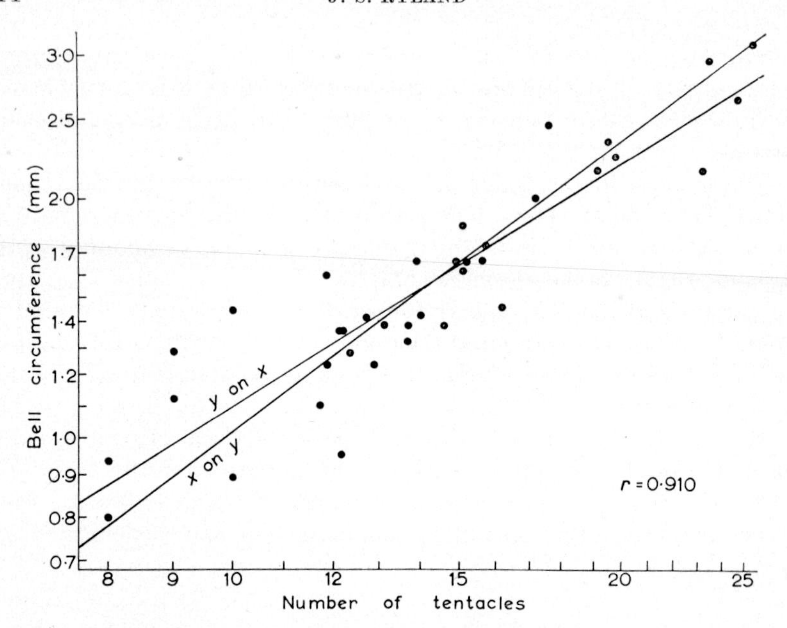

FIG. 13. Relationship between funnel circumference and number of tentacles. (From Ryland, 1975.)

tentacles and their number. Other variables, such as the angle of divergence of the tentacles from the vertical can be calculated. Using statistical methods (partial correlation techniques and principal components analysis), Ryland drew from his data certain conclusions about the basic form of the lophophore in marine bryozoa and the extent of interspecific variability. The top diameter of the funnel is highly correlated with tentacle length ($r = 0{\cdot}96$; Fig. 12) and tentacle number in turn is highly correlated with top diameter ($r = 0{\cdot}91$; Fig. 13), but there is no correlation between tentacle length and number if the effect of diameter is excluded ($r = 0{\cdot}01$).

Principal components analysis showed that almost all the observed variation between the lophophores of species in the sample related to size (93% of the variance was attributable to the first principal axis, which was very highly correlated with tentacle length). The shape of the cone was much less variable (only 5% of the variance was attributable to the second principal axis, considered to be a function of shape and well correlated with the angle of tentacle spread). An important conclusion therefore emerged that, as the size of the funnel changes, so tentacle length, bell diameter and tentacle number maintain stable interrelationships (as in Figs 12 and 13). This is emphasized by the fact that the distance between the tips of adjacent tentacles (given in

Table IV) does not vary in any consistent manner with tentacle number or funnel size but remains roughly constant at 110 μm (s.d. $\pm$ 15 μm). It may be added that even the tentacle bells of the European *Flustrellidra hispida*, substantially larger than any in Ryland's sample, conform quite closely to this rule (a funnel of 1·2 mm diameter with 29 tentacles has a tip distance of 130 μm). It follows from these relationships that, since the more numerous the tentacles the longer they are, the intertentacular angle narrows as the tentacle number increases. In compensation as the funnel size increases, a change in shape from conical to basally convex helps to widen the intertentacular axils, thereby promoting efficient flow. Funnel shape, though fairly stable, in fact changes slightly with size, the cone becoming somewhat more acute with increasing tentacle length. A possible explanation for these stable interrelationships is considered later (p. 327).

The matter of lophophoral symmetry may also be considered further. Perfect radial symmetry was rare in the species examined by Ryland, for the ventral (abanal) tentacles tended to be both longer and less outcurved than the dorsal (anal) tentacles. This discrepancy was most marked in *Rhynchozoon rostratum*, in which the dorsomedial tentacles were only 55% of the length of the ventromedial pair (360 and 660 μm respectively). Asymmetry of this nature was remarked upon long ago by Hincks (1880, p. 550, text-fig. 33), who wrote that the dorsomedial pair of tentacles out of the eight present in *Walkeria* and related genera " are universally bent outwards towards the side, so that the tentacular wreath does not form a perfect circle; the remaining six stand erect in the usual way ". The breaking of radial symmetry may reflect the specialized role of the dorsomedial tentacles in the release of spermatozoa (see p. 349) or the presence, as in *Flustrellidra hispida*, of a ventromedial rejection tract (Atkins, 1932; see p. 329): it is not known, however, how widespread either feature is. Asymmetry has also been reported in association with exhalent " chimneys " (see p. 322). None of these circumstances seems really to explain either the slight departure from radial symmetry seen in so many gymnolaemate lophophores (Fig. 11*b*, *c*) or the more pronounced bilateral symmetry of *Rhynchozoon rostratum* (Fig. 11*d*).

2. *Musculature of the lophophore base*

Turning from gross to fine morphology, recent papers have again contributed greatly to our knowledge, particularly of tentacle ultrastructure. Gordon (1974), however, has studied the microarchitecture of the entire lophophore, discovering a number of previously undescribed features. He worked with the cosmopolitan species *Cryptosula pallasiana*, which has a lophophore of 17 tentacles, length 650–1 200 μm, displaying the weak bilateral symmetry already remarked (Gordon, 1974, Fig. 1, **1**). The lophophore base (Fig. 14), he points out, is structurally the most complex part of the polypide. Here the tentacles unite and their lumina become confluent with the ring-coelom (together constituting the mesocoel). The basement membrane of the epidermal cells is a particularly conspicuous and important element.

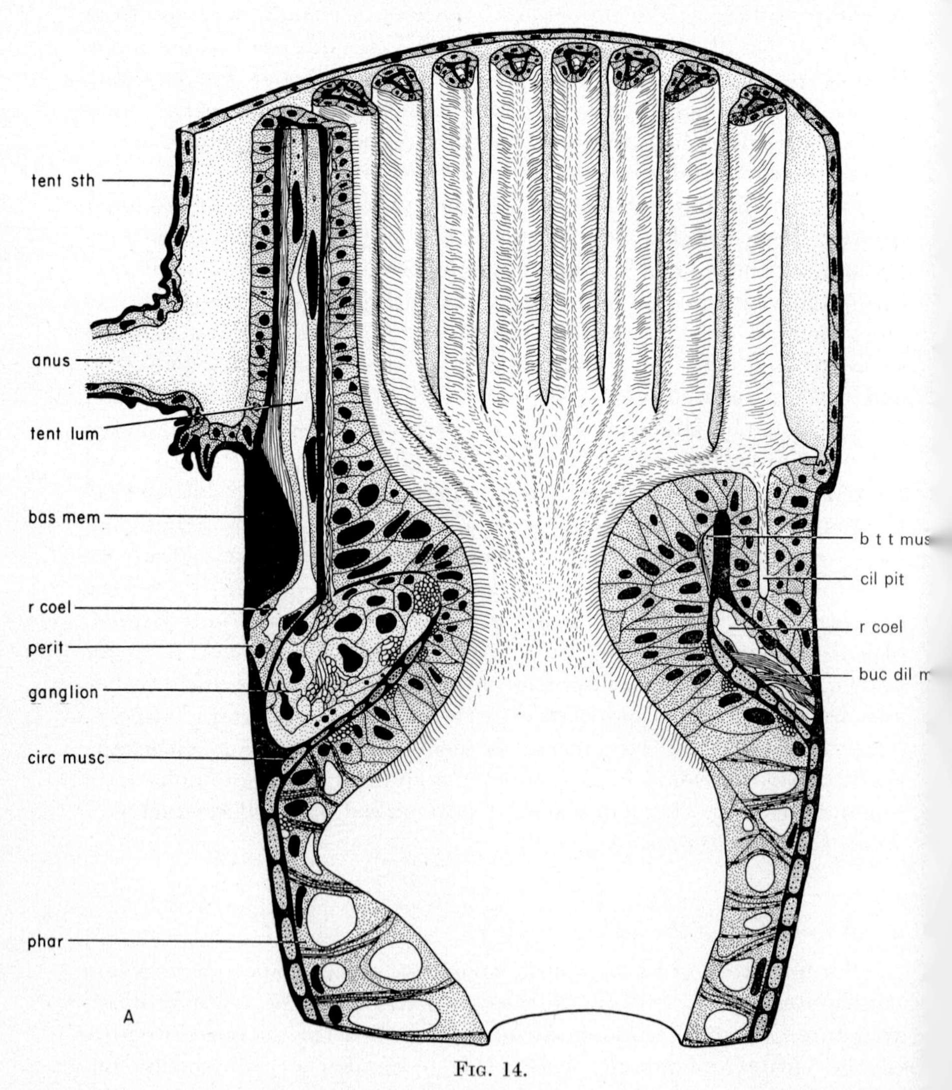

FIG. 14.

It becomes thickened for the insertion of the lophophore retractor muscles and for the outer and inner attachments of the buccal dilator muscles (see below), while its flexile strength supports bending movements of the entire tentacle crown: " It is striking to watch intertidal bryozoans at ebb tide. Contrary to expectations, lophophores are not retracted as small waves wash over the colonies, but with every passing wave lophophores are flung back and forth until finally, with the passage of no more water over the colony surface, they are withdrawn" (Gordon, 1974, p. 156). The epidermis, unilaminar throughout the rest of the polypide, is up to three cells thick in the mouth region and two cells thick along the attachment of the tentacle sheath.

The paired lophophore retractor muscles are inserted laterally on the basement membrane at the level of the ring-coelom and comprise, in *C. pallasiana*, 18–29 cells on each side. There has long been controversy as to whether or not the retractor muscles are striated. Gordon has confirmed the interpretation of Marcus (1939) that the muscle is smooth and that a rippling of the sarcolemma, visible under phase contrast, results in a pseudo-striated appearance. Physiologically this muscle is quite remarkable for, in *Membranipora membranacea*, Thorpe *et al.* (1975) record a peak contraction rate in excess of twenty times its

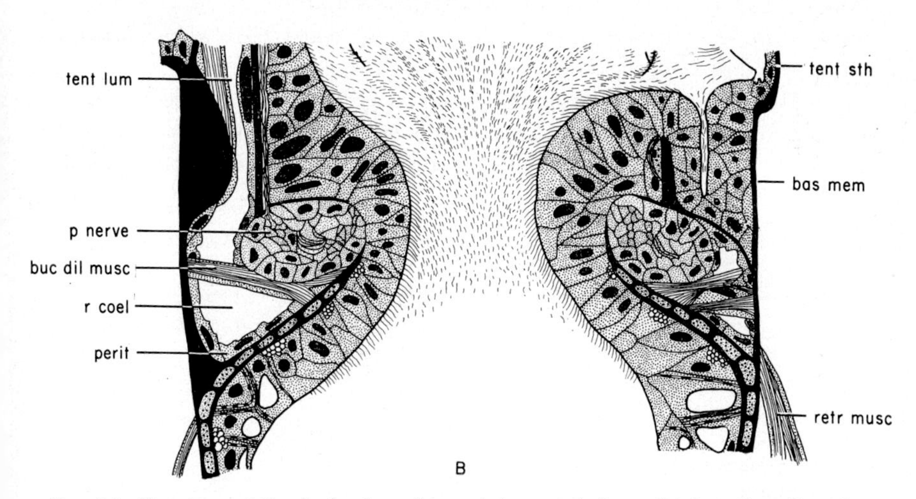

FIG. 14. Structure of the lophophore base. A (*opposite*), in sagittal section (dorsal to the left); B in longitudinal section orthogonal to A. *b t t musc*, basal transverse muscle of the tentacles; *bas mem*, basement membrane; *buc dil musc*, buccal dilator muscle; *cil pit*, ciliated pit; *circ musc*, circular (constrictor) muscles of the mouth and pharynx; *p nerve*, peripharyngeal nerve plexus; *perit*, peritoneal cell; *phar*, pharyngeal cell; *r coel*, ring coelom; *retr musc*, retractor muscles; *tent lum*, coelomic lumen of tentacle; *tent sth*, tentacle sheath. (A, redrawn from Gordon, *Marine Biology*, **27**, 1974; B, from Gordon, unpublished.)

own length (up to 910 μm in *Cryptosula*) per second. This has been claimed as the fastest contracting muscle in the animal kingdom, a fact which—coupled with the well-developed innervation (see p. 338 *et seq.*)—accounts for the high-speed "escape" withdrawal (60–80 ms in *Membranipora*) characteristic of the expanded lophophore.

The buccal dilators are slender, radially disposed, smooth muscles spanning the ring-coelom at intervals. The buccal constrictor comprises a series of striated circular muscles subtending the ring-coelom and serially continuous with the circular muscles of the alimentary canal (Fig. 14). The mouth is not normally closed when relaxed but remains open at a diameter of 25–30 μm; it is fully opened to about 36 μm diameter by contraction of the buccal dilator muscles. Complete closure, achieved by maximum contraction of the buccal constrictor muscles, is infrequent.

Nielsen (1970, p. 231) has amplified and corrected Borg's (1926) description of the lophophore base in the cyclostomatous species *Crisia eburnea*, which is now seen to be similar to that of cheilostomes. The basement membrane again forms a thickened skeletal ring around the mouth, and the retractor muscles are inserted on its lower side. Buccal dilators (the "musculi constrictores canalis annularis" of Borg) originate at the base of each tentacle and are inserted onto the epithelium around the mouth. The buccal constrictor muscles were again noted as being striated.

In the lophophore base of *Cryptosula*, Gordon discovered a fourth set of muscles, which he named the basal transverse muscles of the tentacles. Each consists of a single cell, up to 10 μm in length, stretch-

TABLE V. CHARACTERISTICS OF SOME MUSCLES OF *Cryptosula pallasiana* (Gordon, 1974)

Muscle	*Type*	*Thick filament diameter (nm)*	*Thin filament diameter (nm)*
Retractors of the lophophore	Smooth	13–31	3–7
Transverse parietals	Paramyosin-like	30–72	(?)
Buccal dilators	Smooth	20–24	approx. 3
Buccal constrictors	Striated	20–24	approx. 3
Basal transverse muscle of tentacles	Paramyosin-like	13–75	approx. 3
Tentacular (intrinsic longitudinal) muscle	Smooth [striated in *Flustrellidra*]	22–42	4·5–7·5

ing from tentacle base to tentacle base below each intertentacular axil. The possible function of these muscles will be considered (p. 322) after the ultrastructure of the tentacles has been described.

The characteristics of some bryozoan muscles are summarized in Table V.

B. *Structure and movements of the tentacles*

1. *Ultrastucture of the tentacles*

Lutaud (1955) made a valuable light microscope study on the tentacles of *Pentapora* (as *Hippodiplosia*) *foliacea*, but still more revealing recent studies have been based on transmission electron microscopy (Smith, 1973, on *Flustrellidra hispida*; Lutaud, 1973a, on *Electra pilosa*; Gordon, 1974, on *Cryptosula pallasiana*). As a result, the fine anatomy of the tentacles is now relatively well established.

In transverse section the tentacles are wedge or bluntly T-shaped (Figs. 14, 15), the narrower end being frontal (facing into the funnel). A conspicuous feature of the section is the collagenous tube, strictly the thickened basement membrane of the epidermal cells. This collagen layer, together with the tentacular extension of the mesocoel which it surrounds, constitutes the axial skeleton. Smith (1973) first established the collagenous nature of this tube: the characteristic cross-striations of the fibrils have a periodicity of 55–64 nm in *Cryptosula* and 66 nm in *Flustrellidra*. Two columns of longitudinally overlapping myoepithelial cells, one frontal and one abfrontal, extend through the length of the lumen. The frontal muscle is the larger, comprising (in *Flustrellidra*) about ten small cells when seen in transverse section; the abfrontal block comprises about five cells. The muscle is smooth in *Cryptosula*, striated in *Flustrellidra*, and the filaments within the cells insert in dense bodies on the collagen layer.

The epidermis consists of about nine series of cells: two abfrontal, two pairs lateral and three frontal cells. There is also a discontinuous tenth series medio-abfrontally. The number of cell rows is reduced as the free end of the tentacle is approached, until the summit itself comprises just a quartet of epithelial cells. The shared angles of these four cells slightly gape, leaving a minute opening through which the coelomic lumen and the exterior are in communication (Calvet, 1900, p. 34, Pl. I, fig. 6). The existence of this pore is not a recent discovery, but its vital role in the life of bryozoans is (pp. 345 and 349).

The lateral cilia arise from the two series of cells on each side of the tentacle (Fig. 15). In *Cryptosula* these cilia are about 25 μm long, a value which conforms well with the lengths of 20–25 μm given for

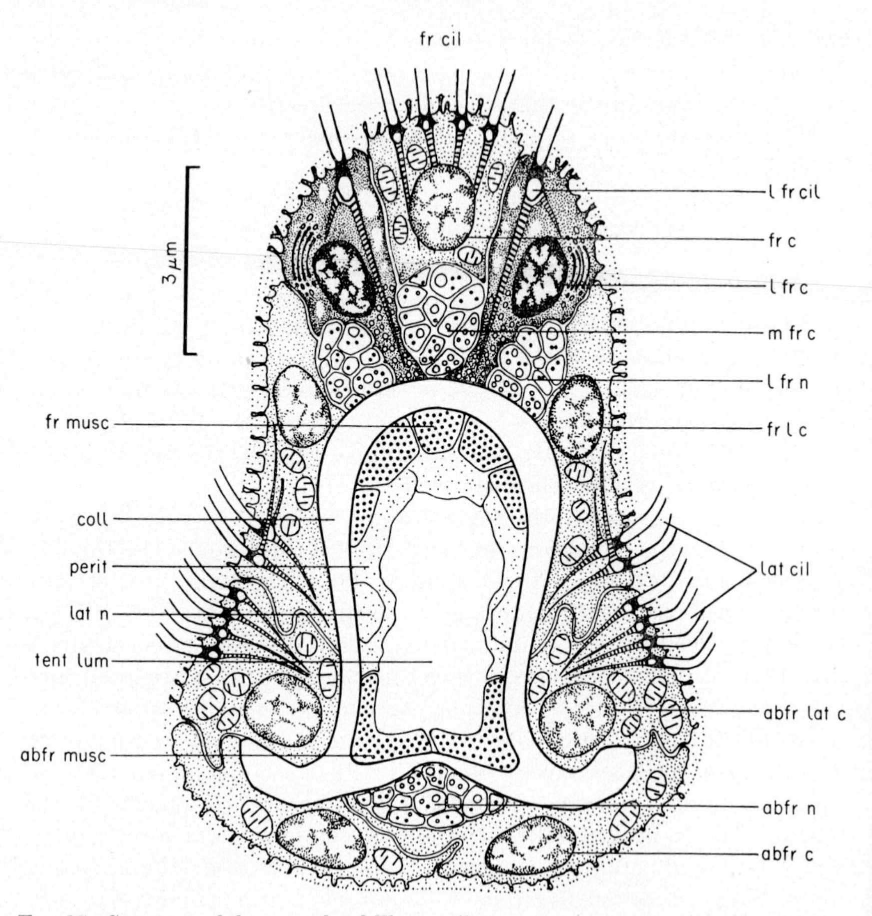

FIG. 15. Structure of the tentacle of *Electra pilosa* as seen in transverse section; *abfr c*, abfrontal cell; *abfr lat c*, abfrontal lateral cell; *abfr musc*, abfrontal muscle fibres; *abfr n*, abfrontal nerve; *coll*, collagen tube; *fr c*, frontal cell; *fr cil*, frontal cilia; *fr l c*, frontal lateral cell; *fr musc*, frontal muscle fibres; *l fr c*, laterofrontal cell; *l fr cil*, laterofrontal cilium; *l fr n*, laterofrontal nerve; *lat cil*, lateral cilia; *lat n*, lateral nerve; *m fr n*, median frontal nerve; *perit*, peritoneal cells; *tent lum*, mesocoelic lumen. (From Lutaud, *Zeitschrift für Zeuforschung*, **140,** 1973.)

Membranipora villosa and *Schizoporella* "*unicornis*" by Strathmann (1973). The frontolateral cell is larger than the abfrontolateral cell but bears fewer cilia (averages of 48 and 63 in *Cryptosula* are given by Gordon, 1974). The cilia are restricted to the abfrontal end of the frontolateral cells, so that the cilia of the two cells combine to form a continuous tract divided only by the cell membrane. The cilia of the two cells are distinguished internally by their rootlets. The cilia of the abfrontal cell have single rootlets which pass through the cell, converging as they approach the basement membrane; those of the frontolateral cell have two rootlets, one of which passes to the basement membrane as in the adjacent cell, while the other proceeds frontally just under, and parallel to, the cell surface for the full length of the cell. These horizontal rootlets are very long, up to 13 μm in *Cryptosula*. As the rootlets are not orthogonal to the longitudinal axis of the tentacle (Smith, 1973), their length is somewhat greater than the transverse length of the cell. The latter is about 9·5 μm in *Electra pilosa* (Lutaud, 1973, Fig. 1, noting that the scale bar should read 3 μm *not* 30 μm) and about 10·5 μm in *Flustrellidra hispida* (Smith, 1973, Fig. 1, although the long axis of this drawing apparently amounts to 23 μm against a figure of 35 μm stated in the text). Smith believed that the lateral rootlets act as anchors resisting shearing stress during the retraction of the lophophore—a process of almost violent rapidity as already noted.

The frontal cilia in *Cryptosula* are of two kinds. Those forming the median tract are about 15 μm in length and possess a single rootlet which passes straight through the cell to the basement membrane. The laterofrontal cells bear single cilia about 20 μm long at intervals of approximately 5 μm. These cilia, which have been presumed tactile and reported from species in all three orders (Lutaud, 1955; Bullivant, 1968b; Nielsen, 1970; Strathmann, 1973) appear rigid but are capable of flicking movements. In addition to the axial rootlet, they have a short lateral rootlet (Lutaud, 1973, Fig. 1; Gordon, 1974). Toward the base of the tentacle the laterofrontal cells bear additional cilia resembling those of the frontal tract. A few cilia may also be present on the abfrontal cells near the base of the tentacle (Brien, 1960).

The disposition of nerves in the tentacle will be described later (p. 338).

2. *The mesocoel and the movement of fluid*

The role of metacoelic fluid, as a hydrostatic skeleton activated by the parietal musculature, in eversion of the lophophore is well known (e.g., Ryland, 1970, pp. 19–22). Gordon (1974) has now emphasized the importance of mesocoelic fluid in the movements both of the lophophore as a whole and of the tentacles individually. At the tentacle bases the axial lumina unite with the ring-coelom. The latter in turn communicates dorsally with the metacoel, via a pore which is incompletely obstructed by the nerve ganglion (Fig. 14).

In sections of withdrawn polypides, the tentacular lumina are occluded by the contracted intrinsic muscles, and the tentacles must be shortened; upon eversion of the lophophore stretching of the tentacles can be observed. This must be achieved by the upward displacement of coelomic fluid, presumably coming from the metacoel, accompanied by the relaxation of the intrinsic longitudinal muscles.

During feeding the buccal dilator muscles are partially contracted. In addition to holding the mouth open, they will keep the floor of the ring-coelom raised, thereby plugging the opening of each tentacular lumen with peritoneal cells. If the mouth is then suddenly closed by the buccal constrictors, simultaneous contraction of the basal transverse muscles perhaps maintains effective closure of the lumina. Tentacles are also capable of individual movements, such as the adoption of an incurved posture or a flicking of the tip (see p. 328). Such movements involve the frontal and abfrontal intrinsic muscles, together with a constriction of the opening of the tentacular lumen as just explained.

C. *Food capture and feeding*

1. *The capture of food particles*

As established by Borg (1926) the lateral cilia generate the water current. They beat somewhat obliquely to the long axis of the tentacle and display laeoplectic metachronism (Knight–Jones, 1954)—the waves pass up the left side of the tentacles (viewed from inside the funnel) and down the right side, so that the effective beat is to the left of the direction of travel of the waves. Water enters the funnel from the top and is propelled outwards and downwards between the tentacles (Fig. 17A). Observation shows that the tentacles of expanded lophophores interdigitate and form a filtration network which may, at times, cover substantial areas of the colony. Below the network of funnels is a space some 300–500 μm in height, depending on the degree of eversion of the tentacle sheath (Fig. 11B, C). As water is driven through the funnels it must occupy this space, from which there must be a corresponding exhalent flow. Where does the water go?

Banta *et al.* (1974) have examined feeding colonies of *Membranipora membranacea* and observe: "When active, most of the colony is covered by extruded polypides, but numerous blank spaces, called chimneys here, are formed because lophophores of zooids located beneath the chimneys lean away from the area. Chimneys are consistent in location; if the colony is disturbed, lophophores re-expand in the same pattern. Lophophores bordering the chimneys are like other lophophores, except that they have longer necks [everted tentacle sheaths] and two or three tentacles bordering the chimneys are longer than other tentacles. It is virtually impossible to detect the location of chimneys when the lophophores are retracted. Incurrent water passes between tentacles into the space between lophophores and

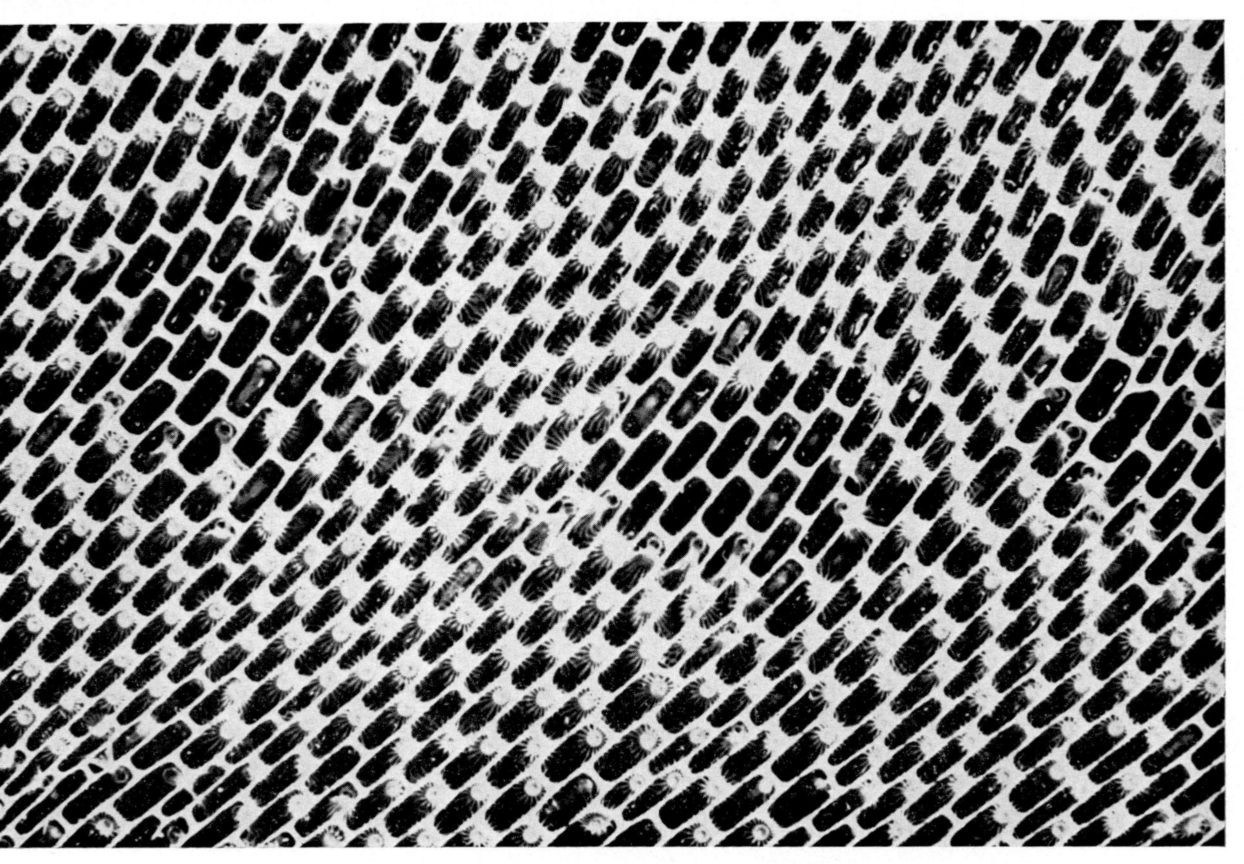

Fig. 16. Part of *Membranipora membranacea* colony showing "chimneys" (see text).

zooecia [the fronts of the zooids], passes peripherally, and is exhaled forcefully at the edge of the colony if an edge is nearby, or through chimneys; small colonies lack chimneys. Chimneys are about 0·8 mm in diameter and are distributed in a rhombic pattern about 1·9 mm from nearest neighbours. Small colonies of a phylactolaemate bryozoan have been found to be more efficient than large colonies in that they have higher clearance rates of food cells per lophophore (Bishop and Bahr, 1973). These observations suggest that lophophores in large, robust colonies interfere with each other unless exhalant channels such as chimneys are present." In *Membranipora membranacea* the present writer's photograph (Fig. 16) shows that there may be a polypide-less zooid near the centre of a chimney, but Nilsson and Jägersten's superb picture (1961, p. 148) clearly shows one where all the central participating zooids have everted funnels strongly inclined away from the chimney.

These perceptive observations by Banta *et al.* have been followed up by Cook (1976). In unilamellar, erect colonies, both flexible (*Bugula*) and rigid (retepore), the water currents pass between the branches from the frontal to the basal side with no restriction. In encrusting colonies three methods of colonial water clearance have been observed or inferred.

The first occurs in small, discoid colonies of cyclostomes (e. g. *Lichenopora*, Fig. 38*a*, p. 402), where the central (neanic) zooids have already undergone ontogenetic elongation of living chambers and lophophores. Their tentacles are thus already longer than those of the recently formed, peripheral zooids. The resultant currents flow centripetally between zooid rows, below the level of the orifices, combining at the centre to form a common exhalent outlet.

The second type is inferred for more massive, nodular, colonies of cyclostomes, cheilostomes and ctenostomes, where raised groups of non-feeding zooids alternate with relatively depressed groups of feeding zooids. The former may consist of " closed ", modified male, or brooding zooids. They constitute areas devoid of inhalent feeding currents which can act as foci for " passive " exhalent currents and occur in a regular pattern throughout the colony.

The third type occurs in gymnolaemate colonies in which one zooid does not feed but is encircled by zooids with modified tentacle crowns. In the cheilostome species observed (*Hippoporina porosa*, *Cleidochasma contractum* and *Parasmittina* sp.) the tentacles of those zooids nearest the non-feeding zooid are greatly elongated, and the funnels often directed away from the central zooid. These circlets form " active " exhalent outlets similar to the " chimneys " described by

Banta *et al.* (1974). Such chimneys are also regularly spaced throughout the colony and are presumed to be maintained throughout cyclic degeneration and regeneration of polypides.

Since frontal cilia are very variably developed on the tentacles of marine bryozoans, being conspicuous in many gymnolaemates but sparse in stenolaemates, the mechanism of food selection is probably largely independent of them. Thus Borg (1926) considered that they played a subordinate role in feeding, and in *Flustrellidra hispida*, in which they are long, Atkins (1932) could see little movement of particles over them; she concluded that the chief function of the frontal cilia was to help produce and direct the main water current toward the mouth (Fig. 17A). Until recently the matter rested there; then Bullivant (1968b) published his theory of impingement feeding.

Impingement separation is a well-known industrial process for the collection of particles from a transporting fluid. The principle is that when a suspension is directed against a baffle or baffles, which cause sharp deflexion of the liquid, the particles are thrown against the baffles by their own momentum. Bullivant suggested that the sharp deflexion of the current as it approaches the mouth causes particles to be thrown towards the mouth or into an eddy of comparatively still water above it. The effectiveness of the process will depend upon the velocity attained by the particles, their specific gravity relative to sea water, their size, the radius of curvature of deflexion and the viscosity of sea water. It is undoubtedly a criticism of Bullivant's proposal that factors such as the velocity attained by, and the specific gravity of, particles were insufficiently considered. In particular, protistans are scarcely denser than sea water and would have insufficient momentum to travel across the flow lines.

Thus, for echinoderm larvae, in which the current paths " invite comparison with the current paths figured by Bullivant (1968b) for a bryozoan ", Strathmann (1971) rejected impingement as the method of collection. He quoted the stopping distance L of a particle as:

$$L = \frac{v_1 \rho D^2}{18\eta},$$

where η is the viscosity of sea water, v_1 is the initial velocity ($\approx 0{\cdot}2$ cm s^{-1}), D is the diameter of the particles and ρ their density ($\leqslant 1{\cdot}2$ g cm^3). The value of L was then 10^{-2} μm. Unless bryozoans can generate a flow several orders of magnitude faster than these larvae, they also would be unable to concentrate particles to any significant degree by impingement.

After several other studies on the manner in which marine organisms

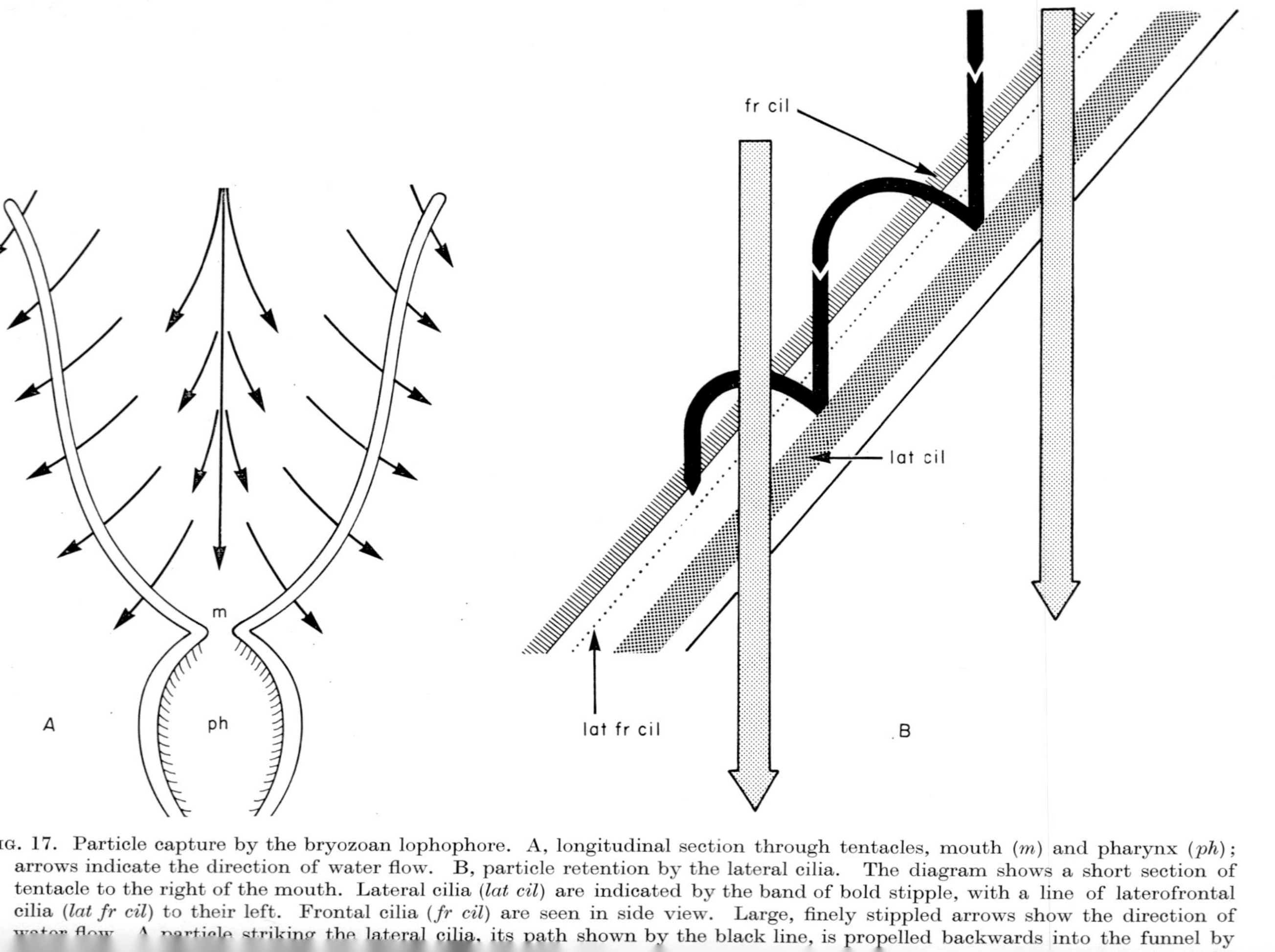

Fig. 17. Particle capture by the bryozoan lophophore. A, longitudinal section through tentacles, mouth (*m*) and pharynx (*ph*); arrows indicate the direction of water flow. B, particle retention by the lateral cilia. The diagram shows a short section of tentacle to the right of the mouth. Lateral cilia (*lat cil*) are indicated by the band of bold stipple, with a line of laterofrontal cilia (*lat fr cil*) to their left. Frontal cilia (*fr cil*) are seen in side view. Large, finely stippled arrows show the direction of water flow. A particle striking the lateral cilia, its path shown by the black line, is propelled backwards into the funnel by

remove particles from suspension, Strathmann (1973) has suggested that bryozoans employ a method entirely different from that proposed by Bullivant. Strathmann has assumed that particles will move with the through-flowing water. Consequently, unless trapped, they will pass between the tentacles and out of the funnel. Many particles—sometimes nearly 100%—are however, retained inside the funnel. Many of the particles were first carried between the tentacles and then moved back towards the frontal surface, for which the only adequate explanation can be that they are driven back by reversal of beat in the lateral cilia (Fig. 17B).

He noted that particles being transported along a tentacle frequently get carried to the side but are then quickly moved back towards the frontal midline; or that they may even oscillate from side to side. The particles often moved in jumps of about 100 μm, a much greater distance than they could be directly pushed by cilia of about 25 μm in length. He suggests that ciliary reversal must be sufficiently strong to produce a local reversal of current: the resultant flow direction would be proximal, towards the vertex of the funnel.

Bullivant (1968b) and Strathmann (1973) are agreed that mucus plays no major role in feeding: i.e., there is no trap or transporting film of mucus. Heavier particles may, as Bullivant suggested, be projected towards the mouth; but mostly it seems that particles are retained within the funnel by local reversal of the lateral cilia and are moved towards the mouth. The reversal must be triggered by particles contacting either the lateral cilia themselves or the laterofrontal cilia, but in either case the reversal mechanism can be turned-off and the particles allowed to continue in the stream of out-flowing water. Since there are so many cilia on each of the lateral cells (in contrast to many other suspension feeders in which there is only one), the co-ordination of local reversal must be fairly complex.

The structure of the laterofrontal cilia ("sensory bristles") has already been described (p. 321). During feeding they are situated upstream from the lateral cilia themselves. Strathmann argues that with upstream sensors a particle could be retained from a faster flowing current than would otherwise be possible and clearance rates could be higher than if the lateral cilia had to function as both sensors and effectors.

Ryland (1975) discussed the significance of the approximately constant intertentacular tip distance in relation to Strathmann's (1973) views on particle capture. If particles are to be retained upstream of the lateral cilia, it follows that the cilia of the opposing lateral tracts should close the intertentacular gap, at least for a substantial part of the funnel height. If ciliary length remains in the order of 25 μm regardless of funnel dimensions, and the tip distance is 110 μm, filtration should be fully effective in approximately the lower half of the funnel (by height) whatever the tentacle length. Variations in tip distance alter this relationship quite considerably (Fig. 18). Filtration will obviously be less effective in the distal part of the funnel, but a supplementary collecting mechanism here comes into use. If a particle contacts the distal third of a tentacle, the latter may flick inwards as described on p. 322 (Borg, 1926; Bullivant, 1968b; Strathmann, 1973). By this means the particle is transferred to the central current and will be carried proximally towards the mouth.

Some relevant observations have been made on *Electra pilosa* (relaxed mouth diameter 27 μm) and *Flustrellidra hispida* (44 μm) by P. E. J. Dyrynda (unpublished). He has noticed, as I inferred, that as particles enter the funnel they may indeed be carried away unimpeded through the upper half of the intertentacular spaces. Once captured, however, they accumulate above the mouth: if small (16 μm in *F. hispida*) they reach the gut by ciliary action aided by occas-

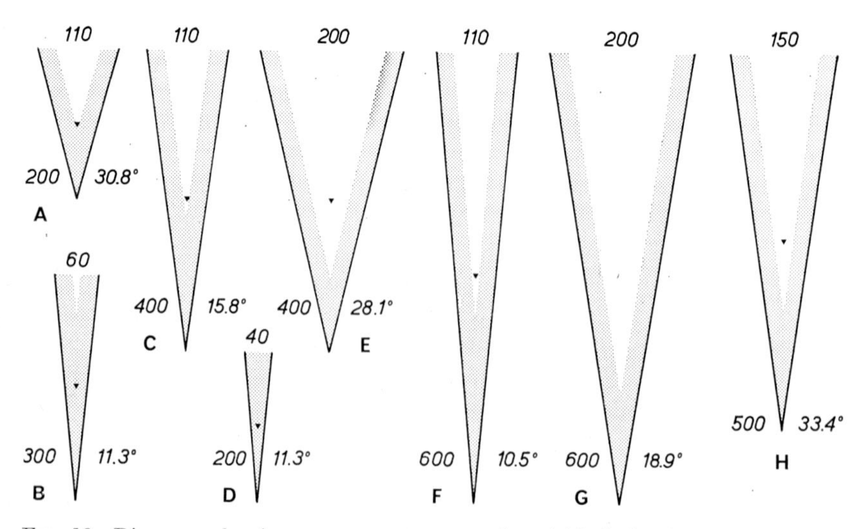

FIG. 18. Diagrams showing some consequences of variable lophophore morphometry on the space between adjacent tentacles. The black lines defining each V represent the opposing sides of the two tentacles. Numerals above each V show the distance apart of the tentacle tips (μm); those to the left show the tentacle length (μm) and these to the right the axillary angle (decimal degrees). Stippled band indicates the spread of the lateral cilia at commencement of the power stroke, assuming a length of 25 μm. The triangular spot marks the vertical midpoint of the intertentacular space. Irrespective of tentacle length and intertentacular angle, the lateral cilia meet near the vertical midpoint only when the tip distance approximates to 110 μm.

ional pharyngeal dilation, but larger particles (26 μm in *F. hispida*) can be engulfed only by the latter method. For the ingestion of still larger particles, of diameter approximating to that of the mouth, buccal dilation must accompany that of the pharynx. Large particles entering the tentacular funnel centrally may be greatly accelerated as they approach the mouth, being then propelled without check directly into the pharynx or even down to the stomach.

The methods for preventing the intake of distasteful particles were enumerated by Borg (1926) and Atkins (1932), viz. (1) complete or partial retraction of the lophophore; (2) approximation of the tentacle tips, effectively closing the mouth of the funnel; (3) flicking of an individual tentacle to remove an adhering particle; (4) closure of the mouth; (5) use of a rejection tract; (6) ejection from the pharynx (see later); and (7) clearing the funnel by first widening it and then contracting it with great rapidity. To these must be added (8) the overriding of the reversal of beat response in the lateral cilia, allowing the particles to be swept from the funnel.

The rejection tract is either of rather unusual occurrence in marine bryozoans or has been widely overlooked. Atkins' (1932) description was based on *Flustrellidra hispida* (not *Farrella* as stated by Bullivant, 1968b, p. 140) and applies equally to the similar *Pherusella tubulosa*, but a comparable tract has been reported in *Sundanella sibogae*, *Watersipora* "*cucullata*" (Marcus, 1941) and *Bugula neritina* (Bullivant, 1968b). These species are not all closely related, but have in common rather large tentacular funnels (*F. hispida*, 28–29 tentacles; *P. tubulosa*, 25–30; *S. sibogae*, 31–34; *Watersipora* (*arcuata*), 20–21; *B. neritina*, 23–24). In *Zoobotryon verticillatum*, which has only eight tentacles, Bullivant is definite that there is no tract, so its presence or absence is perhaps associated with funnel size. The rejection tract is a ciliated groove leading centrifugally between the ventromedial tentacles, but it is best considered after an account has been given of the structure and function of the pharynx (p. 332). Even in the absence of a tract, however, unwanted particles are usually expelled between the adneural tentacles by ciliary action (Bullivant, 1968b for *Zoobotryon verticillatum*; P. E. J. Dyrynda, unpublished, for *Electra pilosa*).

2. *The food of marine bryozoans*

The qualitative nature of the food consumed by bryozoans is still largely unknown, but it is generally assumed to consist mainly of bacteria and phytoplankton such as the flagellates, small diatoms, silicoflagellates, small peridinians and coccolithophores recorded by Hunt (1925) in the guts of *Pentapora* (as *Lepralia*) *foliacea*. No doubt the analysis of faecal pellets, as Wood (1974) has done for freshwater bryozoans, would provide much more information. Most of our knowledge is based upon the foods accepted by bryozoans in the laboratory. Thus Jebram (1968), following Grell (unpublished) and Schneider (1959, 1963) has successfully cultured many species on the heterotrophic dinoflagellate *Oxyrrhis marina* Dujard. (length 22–23 μm) reared on the chlorophycean *Dunaliella tertiolecta* Butcher (8 μm). The latter alone was unsuitable for *Bugula neritina* and Bullivant (1968a) later found this to be true also in *Zoobotryon*; but *Dunaliella* has nevertheless proved useful in experimental work. The chrysophycean flagellate *Monochrysis lutheri* Droop (= *Pavlova lutheri* (Droop) Green) (5·4 μm) and the diatom *Phaeodactylum tricornutum* Bohlin (28·5 μm) were good foods for *Zoobotryon* (which has a gizzard, see p. 334). The coccolithophore *Cricosphaera carterae* (Braarud and Fagerl.) Braarud (11·5 μm) was also consumed. Pollen grains, in the range 14–40 μm diameter, are readily accepted by *Electra pilosa* and *Flustrellidra hispida*, and are being used in work currently in progress at Swansea. Pollen may occasionally be abundant in the intertidal habitat and has been observed naturally in the guts of *Cryptosula pallasiana* (D. P. Gordon, personal communication).

The nature and abundance of the food consumed by bryozoans under laboratory conditions affects the form of both zooids and colonies (Jebram, 1973). In an extensive series of experiments, Jebram (1975) investigated the effect of single species and mixed algal cultures on the growth of *Conopeum seurati*. Growth was found to be negligible when the diatoms *Phaeodactylum tricornutum* or *Cyclotella nana* Houstedt were used alone, but was good with a *Cyclotella* + *Dunaliella* mixture. Early growth was fastest when the *Conopeum* was fed with *Oxyrrhis marina*, but over a longer period (25 days) mixed *Oxyrrhis* + *Dunaliella* produced the best growth rate. Sexual maturity (presence of male and female gametes) was achieved on diets of *Cryptomonas* sp., *Monochrysis lutheri* and *Oxyrrhis* fed singly, and on *Cryptomonas* + *Dunaliella* and *Oxyrrhis* + *Dunaliella*. *Bowerbankia gracilis*, on the other hand, matured only when fed with *Oxyrrhis*. Interestingly, however, when the *Oxyrrhis* had been reared on *Phaeodactylum* the ova were of the normal pink colour, but when the dinoflagellate had been raised on *Dunaliella* the embryos remained whitish.

When mixed cultures were used as food, Jebram found that the bryozoans preferentially selected the larger cells. Thus, in a roughly 100 : 1 *Dunaliella* + *Oxyrrhis* mixture, *Conopeum seurati* consumed more *Oxyrrhis* (12–28 × 15–41 μm) than *Dunaliella* (5–8 × 7–13 μm); and from a roughly 1 : 1 *Cryptomonas* + *Dunaliella* mixture, 90% of the cells selected were *Cryptomonas* (8–15 × 10–15 μm). The relaxed mouth gape (see p. 318) in *C. seurati* is 20–28 μm.

3. *Physiology of feeding in* Zoobotryon

Bullivant (1968a) has carried out the only study so far on the rate at which a bryozoon can clear a phytoplankton suspension. *Zoobotryon verticillatum*, at 23–25°C, was fed with suspensions of various algae while the change in concentration was determined using a Coulter Counter. Satiation concentrations (the lowest concentration of particles at which the animal achieves a maximum ingestion rate) and maximum ingestion rate for four algae are given in Table VI. It has to be appreciated that satiation concentrations of suitable plankton will rarely, if ever, occur in the sea, except perhaps at the times of algal blooms. More usual spring–summer ranges (in cells ml^{-1}), quoted by Bullivant, are: flagellates < 100– > 5 000, diatoms 50–500, dinoflagellates 20–50 and coccolithophores 20–100.

TABLE VI. SATIATION CONCENTRATION AND MAXIMUM INGESTION RATES FOR *Zoobotryon verticillatum* FED ON FOUR SPECIES OF ALGA (Bullivant, 1968a)

Alga	*Cell volume* (μm^3)	*Satiation concentration cells* ml^{-1}	*Max. ingestion rate cells* $zooid^{-1}$ h^{-1}
Monochrysis lutheri	81	7 900	2 872
Phaeodactylum tricornutum	103	6 000	3 486
Dunaliella tertiolecta	257	5 100	1 270
Cricosphaera carterae	796	1 700	1 505

Clearance rates were calculated (necessarily) for experiments in which the concentration of algae was below the satiation level. Quite consistent values were obtained, the means of which are given in Table VII. *Monochrysis* was the best food, and a mean clearance rate of 0·368 ml $zooid^{-1}$ h^{-1} was obtained. Since the dry weight of a *Zoobotryon* zooid was calculated as $10{\cdot}91 \times 10^{-3}$ mg, the mean clearance rate can be expressed as 33·7 ml (mg dry weight)$^{-1}$ h^{-1}. This is a comparatively high rate, roughly comparable to that of oyster veligers and small crustaceans. As a high rate is characteristic of small animals, which have a comparatively high respiratory rate, it appears that metabolically the *Zoobotryon* individual is the zooid (0·5–1·0 × 0·25 mm) rather than the colony (a festoon 0·5 m or more in length). Bullivant suggested that if bryozoans are considered to consume

TABLE VII. CLEARANCE RATE OF *Zoobotryon verticillatum* FED ON FOUR SPECIES OF ALGA (AS IN TABLE VI). Rates are means of three experiments. (Bullivant, 1968a)

Alga	*Initial concentration cells ml^{-1}*	*Clearance rate ml zooid^{-1} h^{-1}*
M. lutheri	4 504	0·352
	6 768	0·384
C. carterae	969	0·794
P. tricornutum	5 024	0·562
D. tertiolecta	4 386	0·224
	2 086	0·327

about $2{\cdot}7 \times 10^{-3}$ ml (mg dry weight)$^{-1}$ h^{-1}, then *Zoobotryon* feeding on *Monochrysis* clears about 11·6 l. water for each ml 0_2 consumed: this corresponds well to the performance of sponges, ascidians, cladocerans and bivalves.

Menon (1974) found that when *Electra pilosa* and *Conopeum reticulum* were acclimated to four different constant temperatures (6°, 12°, 18°, 22°C) and fed with *Cryptomonas* sp., the filtration rates varied significantly as a function of temperature. Colonies acclimated to the higher temperature levels showed the highest rates of filtration.

D. *The alimentary canal*

Bryozoans, like many other sedentary suspension feeders, have a U-formed alimentary canal. Silén's (1944a) revision of the terminology of its component parts is followed here (Fig. 19, p. 336).

1. *Structure and functions of the pharynx*

The pharynx is a short, thick-walled region of the gut lying immediately internal to the mouth. It is an important organ of unusual structure, the detail of which has only recently been established (Bullivant and Bils, 1968; Matricon, 1973; Gordon, 1974, 1975a; Smith, unpublished). The external outline in transverse section is circular, but the lumen is triangular (*Bowerbankia*, *Zoobotryon*) or stellate (*Alcyonidium*, *Flustrellidra*). The epithelial walls between the sulci bulge inwards and are made up from distinctive, tall vacuolated cells apparently with cross-striated walls. The nature of these striations has long been the subject of controversy (Bullivant and Bils, 1968). In *Cryptosula* the epithelial basement membrane incorporates collagen fibrils aligned orthogonally in two layers (Gordon, 1975a). Peripheral to the basement membrane is circular muscle, serially continuous with that of the mouth (Fig. 14), which in turn is covered by peritoneal cells.

The height of the epithelial cells varies, in *Alcyonidium*, from 8–25 μm (Matricon, 1973). Their inner boundary is a microvillous plasmalemma, and all have a large central vacuole comprising some three-quarters of the cell's length and, at maximum, nine-tenths of its width (see Fig. 14). The lateral membranes are subtended by a sleeve of muscle which imperfectly encapsulates the vacuole and is attached at each end to the plasmalemma. Centrifugally the sheath divides into separate myofibrils which originate from the plasmalemma, each anchored by a hemidesmosome. Centripetally small fascicles of myofibrils are inserted by means of a filament reaching the plasmalemma between the microvilli. The striations have now positively been identified as the Z-bands of the myofibrils.

The ciliation from the lophophore base extends through the mouth into the pharynx; in some bryozoans (*Cryptosula*, *Zoobotryon*) it covers only the buccal region, but in *Flustrellidra* and in stenolaemates the epithelium in the entire distal half of the pharynx is ciliated. Since the mouth is maintained partially open during feeding (p. 318), many particles enter directly; those remaining outside the mouth are periodically engulfed by the sudden dilatation of the pharynx, which abruptly assumes the form of a hollow globe and then reverts to its resting shape. The dilatation is clearly achieved by contraction of the epithelial myofibrils, but depends equally on the plasticity of the vacuole and the incompressibility of the vacuolar fluid. The height of the intersulcal cells diminishes and the overall diameter of the pharynx is increased. Relaxation of the myofibrils with contraction of the circular muscle restores the pharynx to its resting condition.

Atkins (1932) established in *Flustrellidra* that unwanted particles could be expelled from the pharynx. Of the longitudinal sulci, the midventral is the deepest. Cilia in this groove, unlike those elsewhere in the pharynx, beat upwards, and the groove is continued through the mouth and between the ventromedial tentacles as a ciliated rejection tract. Long cilia beside the tract beat transversely, towards it, at irregular intervals. In *Bowerbankia* the midventral groove is ciliated (Brien, 1960) even though the remainder of the pharynx is not, and no rejection tract has been reported.

A combination of ciliary and muscular action brings food particles to the lower end of the pharynx, from whence they are conveyed to the stomach by peristalsis.

2. *The gizzard*

A sphincter and valve separate the pharynx from the cardia or descending arm of the stomach. In the Vesicularina, a division of the Ctenostomata Stolonifera, exemplified by *Bowerbankia* and *Zoobotryon*, the distalmost portion of the cardia is differentiated as a gizzard, the structure of which was described at the level of optical microscopy by Bobin and Prenant (1952) and Brien (1960). In the gizzard of *Bowerbankia imbricata* and *B. pustulosa* each modified cell of the gut epithelium bears a single tooth. As described by Bobin and Prenant the teeth are of two kinds, and the larger, harder teeth (up to 40 μm in length and 18 μm in base diameter) are restricted to two patches, one on the dorsal surface and one on the ventral. Gordon (1975b) has examined the structure of the larger teeth by electron microscopy. Each tooth-cell has the shape of a narrow-based pyramid, rhomboidal in cross-section. The luminal surface of the cell is at first studded with long (90–135 nm) microvilli. During tooth development a matrix is secreted which hardens around each microvillus. The microvilli are later withdrawn, leaving the matrix perforated by a series of tubes 18–25 nm in diameter.

Immediately surrounding the epithelium is a basement membrane 65–100 nm in thickness, outside which is a well-developed layer of circular muscle, comprising filaments of two sizes (3–7 and 18–35 nm diameter). This is bounded by another basement membrane, external to which are some widely separated strands of longitudinal muscle. Gordon commented upon the close ultrastructural similarity of gizzard

TABLE VIII. THE OCCURRENCE OF A GIZZARD IN BRYOZOANS

Genus	*Nature of gizzard surface*
Aeverrillia	4 shields
Amathia	many teeth
Avenella	many teeth
Bowerbankia	> 40 teeth
Buskia	
Cryptopolyzoon	2 shields
Hippothoa[1]	> 4 blunt teeth
Hislopia	smooth lining, cell apices keratinized
Penetrantia[2]	smooth lining, no keratinization
Spathipora[2]	many teeth
Terebripora[2]	many teeth
Vesicularia	many teeth
Zoobotryon	30–50 teeth

[1]Gordon, 1975c; [2]Soule and Soule, 1975; the remainder not from recent literature, see Gordon, 1975c.

teeth to the setae and homologous epidermal derivatives of polychaetes, brachiopods and cephalopods (the bryozoan gut is believed to be entirely ectodermal in origin: Woollacott and Zimmer, 1971).

A gizzard has recently been discovered in a cheilostome, a species of *Hippothoa* (Gordon, 1975c). While a description of this *Hippothoa* was in preparation, a gizzard was found by Dr Gordon in another member of the genus. Both species are being described as new (Ryland and Gordon, 1976). In some bryozoan gizzards, including those of *Hippothoa*, the hard-parts are plates or shields of multicellular origin (Table VIII). The " gizzard " of the shell-burrowing genus (*Penetrantia*, considered by Soule and Soule (1969, 1975) as probably belonging to the Cheilostomata (instead of to the Ctenostomata, in which all shell-boring genera have heretofore been placed), is muscular but lacks horny teeth.

3. *The stomach*

In bryozoans lacking a gizzard, the stomach consists of three distinct parts (Fig. 19): the descending arm or cardia, the central stomach which extends into a proximal dilatation (the caecum), and the pylorus. Several uncertainties concerning the ultrastructure of the stomach and the physiology of digestion have been clarified by Gordon (1975a) in a recent study on *Cryptosula*.

The stomach epithelium is of unicellular thickness throughout. In the cardia it comprises columnar cells 12–15 μm tall, with an internal brush border of microvilli about 3 μm long. Cilia are infrequent. The cells contain a substantial amount of rough endoplasmic reticulum and well developed Golgi bodies: secretion is evidently a function of importance. The stomach floor in the boundary region between the cardia and the caecum is strongly ciliated to assist the passage of particulate matter to the latter. The caecum is the main locus for intracellular digestion. Cells are columnar, taller than in the cardia. The apical half of each is concerned with secretion and absorption, and contains Golgi bodies and endocytotic vacuoles. The cell apex is complexly canalized and vesiculated, the vesicles becoming larger by fusion deeper in the cell. Digestion is thus extra-and intra-cellular. In the central stomach outside the caecum, the cell apices are somewhat less microvillous, and cilia are of more frequent occurrence. Food particles leaving the caecum are here aggregated into a food cord—the ergatula of Morton (1960)—which projects from the pylorus.

The epithelial basement membrane is overlain by weak strands of striated circular muscle, from which thin, longitudinally orientated fibrils branch. Cardia, central stomach and caecum undergo

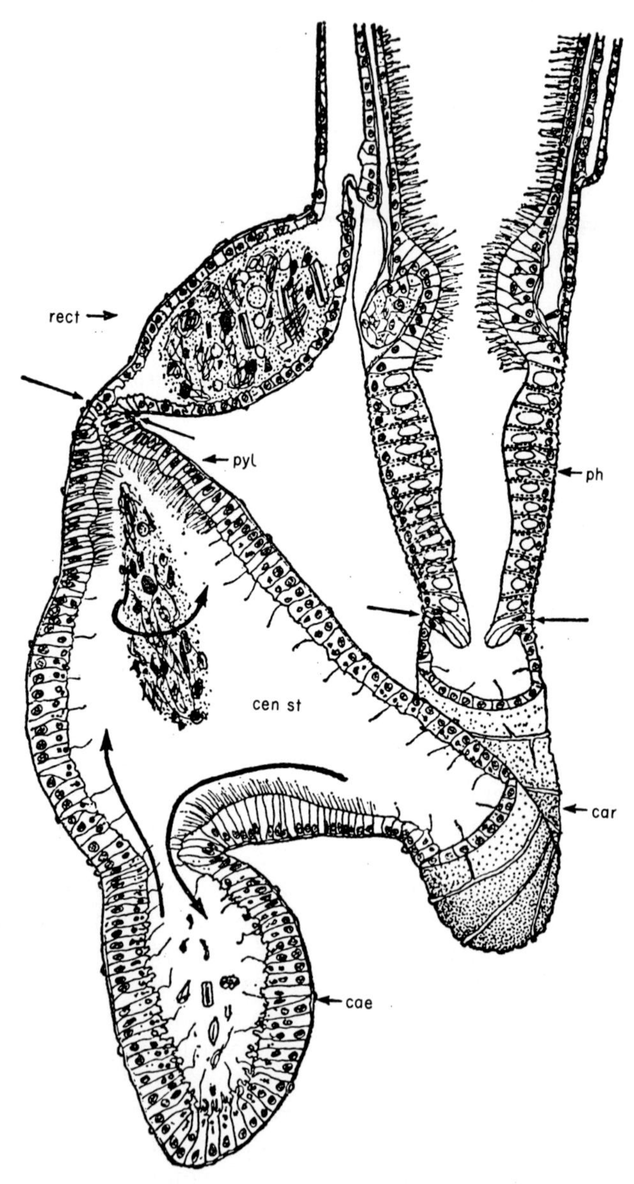

FIG. 19. Diagrammatic representation of the alimentary canal of *Cryptosula pallasiana*. Pairs of thin arrows indicate sphincters. Arrows in the stomach indicate passage of material in and out of the caecum and the rotation of the ergatula in the central stomach; *cae*, caecum; *car*, cardia; *cen st*, central stomach; *ph*, pharynx; *pyl*, pylorus; *rect*, rectum. (From Gordon, 1975a.)

contraction and dilation, evidently more to promote mixing than the transport of particles.

Conspicuous in the stomach cells are the relatively large inclusions (secondary lysosomes and residual bodies) which show as orange-brown granules under the light microscope and which give the stomach its brownish colour. They are absent in newly formed polypides; they accumulate, and enlarge by the addition of smaller vesicles, during the life of the polypide. Incomplete elimination of undigested material from the cells in which intracellular digestion takes place leads to the accumulation of this material.

Gordon's electron-microscope study thus showed that, contrary to the statements in the earlier literature, distinctive cell types (supposedly recognizable by their straining characteristics) do not occur in the stomach. Cells which, from their staining characteristics, have previously been regarded as acidophilic or basophilic, almost certainly differ only as a result of differential fixation. There are obvious differences in the types of cell apex found in the caecum, the central stomach and the ciliated tracts, but there is a gradation of cells where these areas merge.

The pyloric cells differ from the other stomach cells mainly in being densely ciliated. The pylorus is funnel-shaped and its major function seems to be to condense food material and skeletal fragments expelled from the caecum into a compact mass. The beating cilia cause the resulting cord of bacteria and fragments to revolve, its more diffuse end projecting into the central stomach. There is no mucus binding this cord. Of the functions of an ergatula listed by Morton (1960), compaction of particles is probably most important in bryozoans. The compacted material is stored in the rectum. The rectal cells have a well-developed brush border apical to an extensive system of endocytotic channels. These channels fuse into inclusion bodies resembling digestive vacuoles in the caecum. The rectum, then, seems to be essentially absorptive in function. The faecal pellet is moved through and expelled from the rectum by peristalsis.

4. *Food passage rate*

The following average times (in minutes, standard deviations in parentheses) between ingestion and defaecation in colonies fed with *Cryptomonas* sp. were found by Menon (1974).

	6°C	12°C	18°C	22°C
Electra pilosa	85 (±44)	68 (±34)	37 (±18)	29 (±14)
Conopeum reticulum	70 (±39)	62 (±38)	35 (±19)	34 (±22)

IV. Nerves and Co-ordination

A. *The nervous system*

Some of the most interesting recent research on bryozoans concerns the nervous system; indeed, our knowledge of both the disposition of the nerves and the mode of operation of the system depends largely on papers published since 1968. Older work, which frequently yielded dubious or ambiguous results, can also now be better evaluated. Some of the anatomical detail has been revealed by electron microscopical studies of the lophophore (including tentacles), but mostly it stems from the perfection by Dr Geneviève Lutaud of well-established silver and methylene blue staining techniques and their critical application to minute bryozoan zooids. Knowledge from earlier sources has been summarized by Hyman (1959) and by Bullock and Horridge (1965).

The nerve ganglion is dorsally situated in the ring-coelom of the lophophore, adjacent to the pore through which the mesocoel and metacoel communicate (Fig. 14). It constitutes the most developed part of a peripharyngeal nervous ring, or deep crescent (Gordon, 1974). From the ganglion and ring arise nerves which may be considered conveniently as lophophoral, polypidial (supplying the polypide exclusive of the lophophore) and cystidial (supplying the wall structures of the zooid).

1. *Lophophoral nerves*

Graupner (1930) reported that, in *Flustrellidra hispida*, four nerves from the peripharyngeal ring ascended each of the tentacles, two inside the axial tube (claimed as motor from their proximity to the muscles) and two (presumed sensory) outside it. Studies by electron microscopy (Lutaud, 1973a, on *Electra pilosa*; Smith, 1973, on *Flustrellidra hispida*; Gordon, 1974, on *Cryptosula pallasiana*) have shown that the full complement of unsheathed tentacular nerves probably is six (Fig. 15). While Lutaud (1973a) was not convinced that the two intra-axial " filaments " were nerves, Gordon (1974) felt confident that the structures he observed were single axons of variable diameter: their function is unclear, but they are not the motor nerves. Of the remaining extra-axial nerves, one is abfrontal and median (when observed in section) while the others constitute a symmetrical frontal triad, all four being subepidermal and more or less adnate to the axial tube (Lutaud, 1973a; Gordon, 1974). The two median nerves are probably motor, for nerve fibrils were observed between the median frontal nerve and the muscle by Smith (1973, Pl. 1, fig. 4) and between the abfrontal nerve and the muscle by Lutaud (1973a, fig. 2b). The two flanking frontal nerves are assumed to be sensory (Gordon, 1974). The abfrontal nerves link with the lower (proximal) fascicle of the peripharyngeal ring and the frontal nerves join the upper (distal) fascicle. The flanking frontal nerves from adjacent tentacles become approximated in the tentacular axil prior to union with the peripharyngeal ring.

2. *Polypidial nerves*

In *Electra pilosa*, Lutaud (1969, 1973b) has shown that the polypidial (non-lophophoral) nerves have three components (Fig. 20), the first two of which comprise paired nerves arising from the posterior part of the ganglion. The three components are: (1) the paired trifid nerves, with branches to the tentacle sheath, the retractor muscles and the suprapharyngeal ganglion (see below); (2) the paired tentacle sheath nerves which merge with the tentacle sheath branches of the trifid nerves to form the compound tentacle sheath nerves; and (3) the small accessory (suprapharyngeal) ganglion with three visceral nerves to the alimentary canal arising from it. The compound tentacle sheath nerve divides considerably and innervates the diaphragm, the operculum and the body wall distally to the orifice. Then there are two recurrent branches, the first of which reaches the frontal membrane and passes to the insertion of each parietal muscle, and the second, presumably sensory, which ramifies to the mural spines around the frontal membrane. No nerves, apparently, directly underlie the frontal membrane.

3. *Cystidial and " colonial " nerves*

The paired cystidial nerves originate from the anterior part of the ganglion (Fig. 20) and then lie in the tentacle sheath parallel with the paired compound tentacle sheath nerves already mentioned (Lutaud, 1969). Lateral to the orifice/operculum, each of these nerves joins a peripheral ring which encircles the entire cystid (zooid wall), entering in turn each of the interzooidal pore (communication) chambers. This modern study thus fully corroborates the briefly reported findings of Hiller (1939), who illustrated an annular nerve joining the pore chambers. Since the corresponding nerve in all the contiguous zooids also ramifies into the pore chambers, it is by this means that nervous co-ordination between zooids, and thus within the colony as an entity, could be achieved in *Electra pilosa*. Although the course of nerves has not been traced through the lateral pore-plates (multiporous septula) or synapses observed, the physiological evidence (see below) strongly supports the view that the cystidial systems are functionally linked and constitute a " colonial nervous system ".

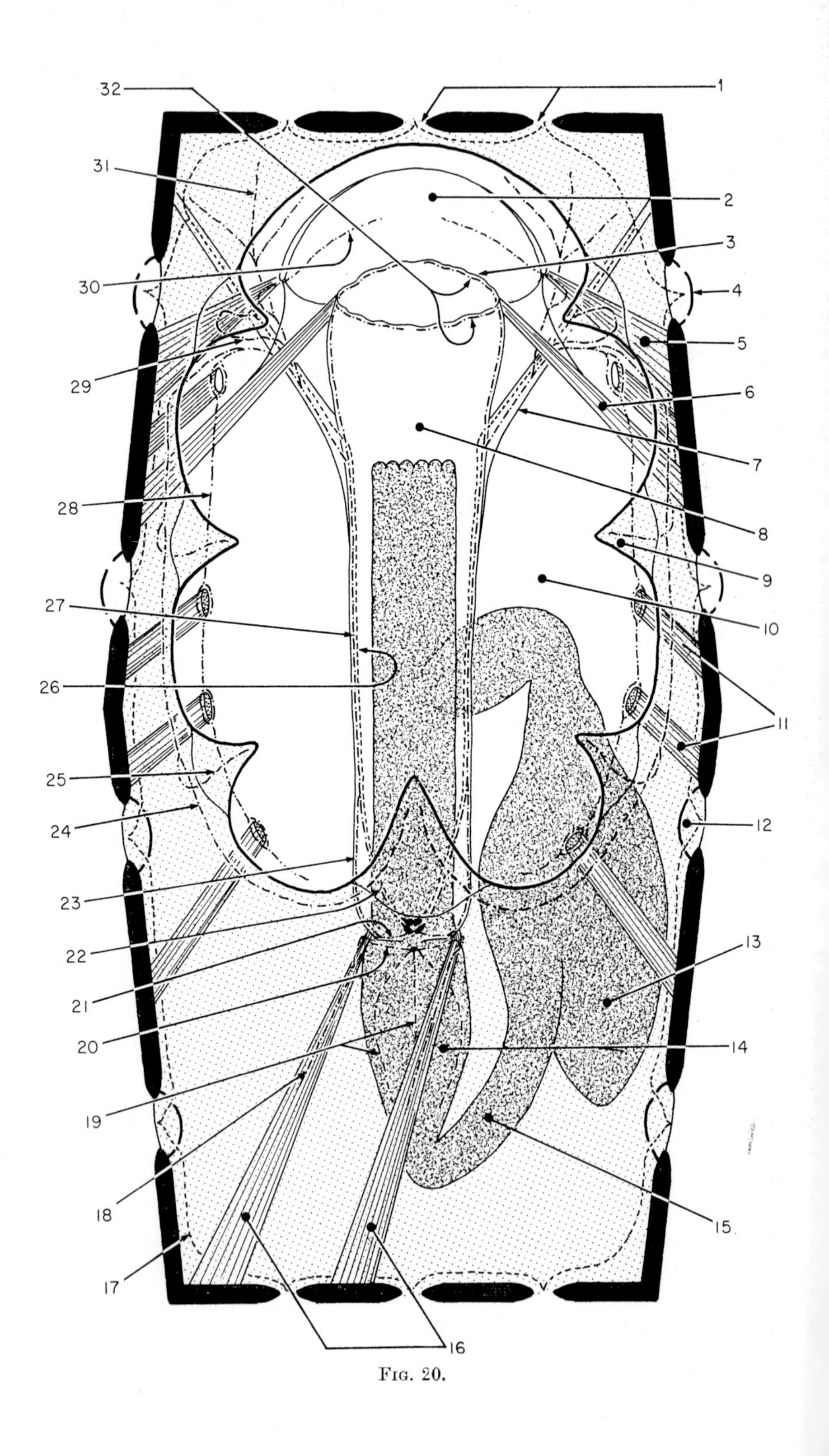
32
1
31
2
3
30
4
5
29
6
7
8
28
9
10
27
26
11
25
12
24
23
22
21
13
20
14
19
18
15
17
16

Fig. 20.

A system of somewhat different form is the network described in the stoloniferans *Farrella repens* (Marcus, 1926) and *Bowerbankia gracilis* (Lutaud, 1974). This comprises a superficial plexus of elongate bipolar cells orientated parallel with the longitudinal axis of both autozooids and stolons. There is no recent confirmation of Bronstein's (1937) claim that, in *Bowerbankia imbricata*, the network is continuous through the autozooidal-stolonal septa (Gordon, 1975d), and it is probable that he was misled by staining artifacts (Lutaud, quoted by Thorpe, 1975). Thus, at the moment, there is no evidence for a colonial nervous system in the Stolonifera.

In *Electra pilosa*, however, it is clear that, in addition to the well-developed nervous system to parts of the polypide and parietal musculature, nerves ramify into the interzooidal pore-chambers. Such an arrangement suggests the possibility of nervous co-ordination between zooids as Hiller (1939) was aware: ". . . polypides frequently react in groups to the stimulation of the frontal cystid wall, especially in the region of the operculum." Marcus (1926, p. 54), on the other hand, found in this and other species that mechanical stimulation of an expanded tentacle crown caused withdrawal of that crown but not of those nearby. Unfortunately it is Marcus' generalization, that each zooid's behaviour is uninfluenced by that of its neighbours, that has been the more widely disseminated and has established an erroneous impression of the level of behavioural co-ordination within a bryozoan colony.

B. *Nervous co-ordination*

Elegant corroboration and elaboration of Hiller's observations has now been provided by Thorpe (1975) and Thorpe *et al.* (1975). For their experiments *Membranipora membranacea* was the species principally employed, although the main results were repeatable with *Electra*

Fig. 20. Diagrammatic representation of the zooidal nervous system in *Electra pilosa*. Nerves are labelled to the left of the diagram, non-nervous structures to the right. The " cystidial " nerve is shown as a pecked line, the " polypidial " nerves as alternating dots and dashes. 1. distal pores, 2. operculum, 3. diaphragm, 4. septulum, 5. occlusor muscle, 6. parieto-diaphragmatic muscle, 7. parieto-vaginal muscle, 8. tentacle sheath, 9. lateral spine, 10. frontal membrane, 11. parietal muscles, 12. pore-chamber, 13. caecum, 14. pharynx, 15. cardia, 16. retractor muscles, 17. pericystidial nerve, 18. branch to retractor muscle, 19. nerves to alimentary canal, 20. ganglionic branch of trifid nerve, 21. trifid nerve, 22. tentacle sheath nerve, 23. tentacle sheath branch of trifid nerve, 24. recurrent marginal branch, 25. branch to spine, 26. cystidial nerve, 27. compound tentacle-sheath nerve, 28. parietal muscle recurrent branch, 29. branch to distal spine, 30. opercular branch, 31. distal superficial branch, 32. diaphragmatic branch. (Compiled from figures in Lutaud, 1969 and 1973b.)

pilosa—and Lutaud's unpublished research (quoted by Thorpe, 1975) has confirmed that *M. membranacea* has cystidial nerves comparable to those of *E. pilosa*. Electrical stimuli were given and recordings made using suction electrodes of 100 μm tip diameter applied to the frontal membrane. It was confirmed that mechanical or electrical stimulation of an expanded lophophore resulted solely in retraction of that lophophore, while similar stimulation of the frontal membrane led to the immediate withdrawal of all the expanded lophophores near to the point of stimulation. The area affected was usually roughly diamond shaped, elongated in correspondence with the longitudinal axis of the zooids, to a maximum extent of about 10×5 cm. Hiller's statement that the region in the vicinity of the orifice was most sensitive was also confirmed.

Using electrical stimulation, near maximal numbers of zooids responded to a stimulus well above an established threshold (1·5 V for 10 ms) or to two or more successive stimuli just above threshold. With mechanical stimulation a force of about 10 mg had to be applied. The latency between stimulus and response was 20–30 ms cm^{-1} with increasing distance of the responding zooid from the point of stimulation. A cut in the colony inhibited propagation of the response, although there was limited spread around the end of the cut. Electrical recording revealed two types of pulse: " type 1 " (T1) and " type 2 " (T2). T1 pulses were diphasic, with short duration ($\sim$ 3 ms) and small amplitude ($\sim$ 10 μV), and were usually produced in bursts with a peak frequency sometimes in excess of 200 Hz. The notable regularity of the T1 pulses suggests the presence of pacemaker cells in the ganglion. The conduction velocity was about 100 cm s^{-1} in directions parallel with the longitudinal axis of the zooid and about half this is at right angles to it. T2 pulses had a longer duration and a high amplitude and were recorded whenever a lophophore was withdrawn or further retracted. They are, in fact, in two parts: the first part is diphasic and briefly displays high amplitude (about 200 μV, 5–10 ms), while the second is long and of low amplitude (10–15 μV, about 100 ms). Thorpe (1975) has suggested that the high amplitude of the first part indicates propagation by a giant axon in the retractor muscle branch of the trifid nerve, while the second part is caused by contraction of the retractor muscle.

Thorpe *et al.* argue that the properties of the T1 system are consistent with those that would be predicted from linkage of the cystidial nerve plexuses demonstrated histologically by Hiller (1939) and Lutaud (1969), although a simple nerve net or neuroid conduction system cannot be absolutely ruled out on physiological grounds alone. T1 pulses evidently travel through a colonial nervous system, which is assumed to show bidirectional polarization, and are conducted across the colony at the same rate as the spread of lophophore retractions. It seems reasonably clear that the cystidial nerve plexuses correspond with the electrophysiologically demonstrated colonial nervous system, although the kind of nerve junction present in the pore-chambers remains to be discovered. The most important sensory nerves are evidently those branches of the compound tentacle sheath nerve to the diaphragm, operculum and distal frontal membrane (Fig. 20). Thorpe (1975) reported that little or no response was obtained by stimulating the spines of *Membranipora membranacea*. (It should be noted that the spines of *Membranipora* (Jebram, 1968) are not comparable to those of *Electra* (Bobin, 1968) which Lutaud (1973b) has shown to be innervated by a branch of the compound tentacle sheath nerve.) It is quite obvious that there is no through conduction from the sensory nerves in the tentacles to the cystidial/colonial system.

Activity in *Membranipora* and many other marine bryozoans consists mainly of movements of the polypide in relation to protrusion and retraction of the lophophore. The co-ordination of such movements is now seen to involve two distinct nervous systems. Sensory nerves within the zooid lead to cells in the posterior of the ganglion. Neighbouring zooids influence the ganglion (and *vice versa*) through the cystidial/colonial system, the component nerves of which join cells in the anterior of each ganglion. The position of the lophophore thus depends upon integration within the ganglion of input from its sensory nerves and from the cystidial/colonial system (T1 pulses). The remarkably rapid (see p. 317) withdrawal of the lophophore must be controlled by the ganglion *via* the trifid nerve, which issues from cells in the posterior of the ganglion and probably incorporates a giant axon. The much slower process of eversion presumably involves the recurrent branch of each compound tentacle sheath nerve: these innervate the insertions of the parietal muscles.

V. Reproduction

A. *The Stenolaemata*

1. *Spermiogenesis and the release of spermatozoa*

The cyclostome colony grows from its periphery or, occasionally, all over its surface, and the germinal cells arise in the growth zone. Spermiogenesis in *Tubulipora liliacea* was described by Franzén (1956) and differs in some ways from that of gymnolaemates. The testis, which is embedded within a cellular membrane, forms at the end of the funiculus in the autozooids. At the division of both primary and secondary spermatocytes the daughter cells remain connected so that the spermatids develop in syncytial tetrads, with their nuclei embedded in a spherical cytophore (Fig. 21, (1) *a–d*). As the spermatid nuclei elongate the syncytium becomes polarized, with the four flagella projecting from the pole opposite to the nuclei (Fig. 21, (1) *e–g*). Four

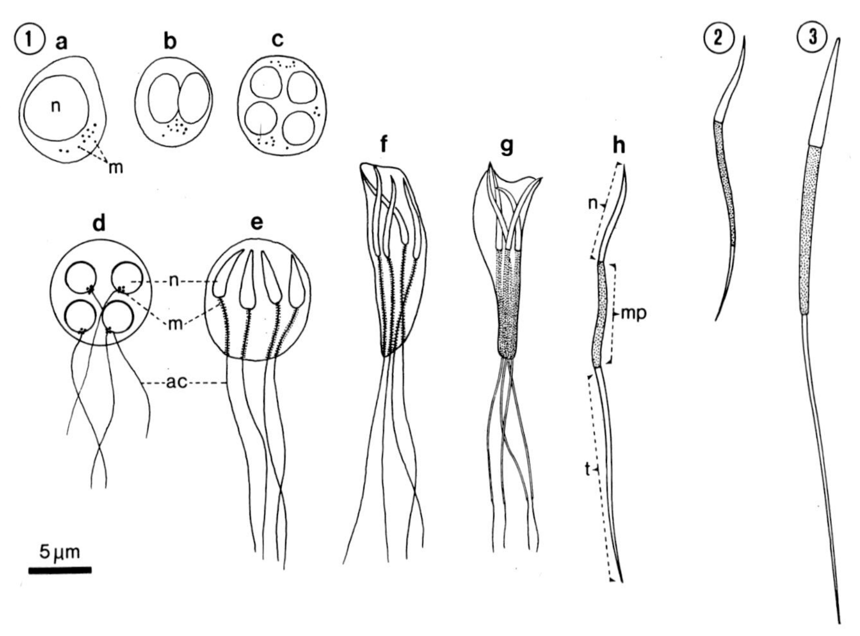

Fig. 21. Spermatogenesis and spermatozoa in stenolaematous Bryozoa. (1) Diagrammatic representation of successive stages of spermatogenesis in *Tubulipora liliacea*: *a*, primary spermatocyte; *b*, secondary spermatocyte; *c*, four spermatid nuclei in a syncytium; *d*, spermatogenesis: chromatin condensation occurring at the anterior poles of the nuclei, with mitochondria grouped in a ring around the base of each flagellum or axial filament complex; *e*, continuing nuclear condensation and elongation with the commencement of caudal displacement of mitochondria; *f*, elongation of whole spermatid syncytium; *g*, nuclei and midpieces in their final shape, the tails not yet morphologically complete: the syncytium persists; *h*, spermatozoon. (2) Spermatozoon of *Berenicea patina*. (3) Spermatozoon of *Diplosolen obelia*. *ac*, axial filament complex; *m*, mitochondria; *mp*, midpiece with mitochondria; *n*, nucleus; *t*, tail. (From Franzén, 1976.)

mitochondria can be distinguished around the proximal part of each flagellum during early maturation (Fig. 21, (1) *d*). These disappear later, being replaced by a thin layer of mitochondrial material extending around the flagellum from the base of the nucleus to the syncytial membrane (the future midpiece of the spermatozoon) (Fig. 21, (1) *e*, *f*). As the cytophore lengthens in the long axis of the spermatids, so the cytoplasm with mitochondria slides backwards along the flagella. The spermatids then separate from each other. The head of the mature spermatozoon (Fig. 21, (1) *h*) is 8 μm long, slender and pointed, and lacks an acrosome; the midpiece is also about 8 μm in length. These proportions vary slightly in other cyclostomes (Fig. 21, (2)–(3) (Franzén, 1956, 1976).

Release of spermatozoa through the tentacles has been observed by Silén (1972) in *Crisia ramosa* and *Lichenopora radiata*, just as in gymnolaemates (p. 349).

2. *Oogenesis and gonozooids*

Oogenesis, like spermiogenesis, takes place at the growing edge of the colony, but most of the resulting ova degenerate. A few only are enveloped by the mesoderm of developing polypides: this mesoderm constitutes a follicle but no proper ovary is formed (Borg, 1926). It appears that most of the ova that have become thus associated with polypides abort, for there are generally very few reproductive female zooids in a cyclostome colony. In general, throughout the order, such a zooid becomes enlarged and constitutes a gonozooid or zooidal brood chamber in which the embryos develop (though it is a colonial cavity in *Lichenopora*). In *Crisia* the gonozooid is pear-shaped and it tends to have this simple form in *Stomatopora*; in *Tubulipora* it becomes irregular as it expands around and between neighbouring autozooids. In almost all instances the brood chambers have highly punctate walls, probably to satisfy the respiratory requirements of the developing embryos (Ryland, 1967b).

Borg (1926) established that the gonozooid polypide developed, but degenerated once the zygote had commenced to cleave. Silén (1972) was able to observe the tentacles of the female polypide expanded through the orifice of the gonozooid in *Crisia ramosa* and *Lichenopora radiata*, so it seems likely that the presence of a protrusible lophophore is necessary to permit the entry of spermatozoa through the supraneural pore, as in gymnolaemates (p. 351).

3. *Embryology*

After fertilization the zygote cleaves to produce an indistinctly hollow ball of blastomeres in which cell layers cannot be distinguished. From this primary embryo, as Harmer (1893) first described, a number of protrusions develop whose tips separate to give rise to secondary embryos. These may in turn constrict off tertiary embryos in the same way until, in some instances, a hundred or more have been produced. At this stage the chamber contains many solid balls of blastomeres, but the cells in each embryo become rearranged in two layers around a central cavity. Our knowledge of the later stages of embryology have been amplified by Nielsen's (1970) recent study. The attachment organ (adhesive sac) develops as an abapical invagination and is bordered internally by high cells with large nuclei containing conspicuous nucleoli (Fig. 22). The apical organ appears first as an unciliated epithelium above an equatorial ring fold which constricts towards the upper pole of the embryo: it never projects in the manner of the

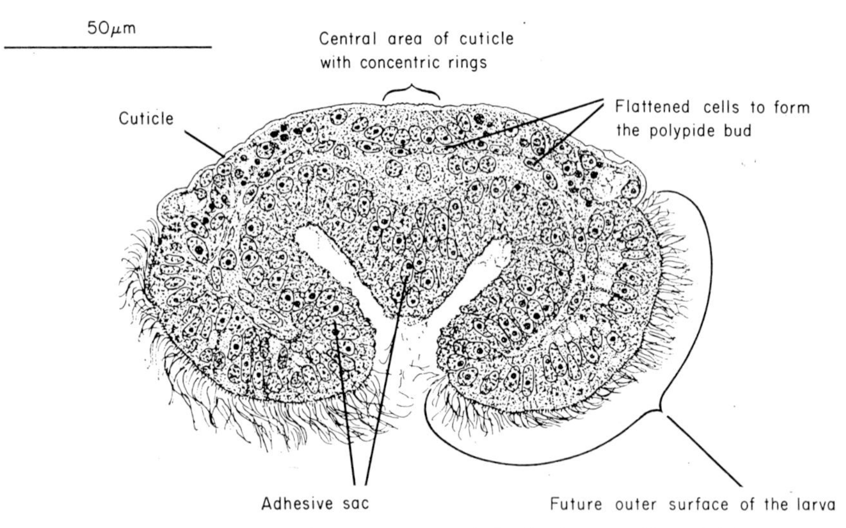

FIG. 22. Median section through a late embryo of *Crisia eburnea*. (From Nielsen, 1971.)

gymnolaemate apical organ. The epithelium between the pore of the adhesive sac and the annulus delimiting the forming apical organ is heavily ciliated. There is no ciliated groove (cf. p. 369).

B. *The Gymnolaemata: gametogenesis*

1. *Sexual polymorphism*

Gymnolaemate zooids are commonly hermaphrodite, though sometimes with slight protandry (Silén, 1966). Certain species, however, are probably monoecious, with the zooids morphologically alike though sexually either male or female. *Hippopodinella adpressa* has distinctive male polypides bearing a reduced number of tentacles (six instead of 15–16) inside zooids which are morphologically indistinguishable from autozooids (Gordon, 1968). *Hippoporidra senegambiensis* has two externally distinguishable zooid morphs: the autozooid has 10–12 tentacles and the modified zooids (which *may* be male, although this was not confirmed) have polypides with six unciliated tentacles (Cook, 1968b).

Structural dimorphism involving female zooids is apparently of wider occurrence, with the orifice being larger in the female zooids of *Bifaxaria* and *Calyptotheca*, for example. Dimorphism has recently been reported in a species of *Haplopoma* in which the female zooids measure about two-thirds the length of the feeding zooids (Silén and Harmelin, 1976). *Cribrilina annulata* produces dwarf female zooids by frontal budding, although zooids of ordinary size in the initial zooid layer may also bear ova and ovicells (Powell, 1967). Colonies of *Hippothoa*, irrespective of form, generally include three operculate zooid morphs (several examples discussed by Ryland and Gordon, 1976). In *H. hyalina* from Brazil, Marcus (1938) established that the polymorphs comprised neuter (feeding), male and female zooids. The feeding zooids were the largest and their polypides bore lophophores of 10 tentacles 200 μm in length; male zooids were smaller, their polypides lacked a functional gut and the lophophore consisted of about six tentacles 18–46 μm in length; female zooids were also small but of different shape and their polypides were still more rudimentary, the lophophore being reduced to 2–3 tentacles 7 μm in length. Recent work, discussed below, has established why it should be necessary for the polypide of sexual zooids to retain some sort of lophophore.

2. *Spermiogenesis*

The testis differentiates from the peritoneum of the developing zooid, being usually situated proximally on the body wall or funiculus. Spermatogonia become free from the testis and give rise in the coelom to spermatocytes clustered in cytophores. In *Membranipora* and *Electra* the mature spermatozoa are aggregated in tightly adherent groups of 32 or 64, orientated with the terminal points of their nuclei in the centre and the midpieces and tails radially projecting (Franzén, 1956; Silén, 1966).

Mitochondria can be seen beside the nucleus from an early stage (Fig. 23, 4*a*), but their number decreases until four are placed symmetrically around the point of origin of the growing flagellum (Fig. 23, 4*b*). According to Reger (1971) mature *Bugula* spermatozoa remain in essentially this condition, except that the head is elongated and the cytoplasm reduced. In most other species (e.g., *Membranipora membranacea*, *Flustra foliacea* and *Triticella korenii*), as described by Franzén (1956, 1976) the mitochondrial material becomes concentrated into two masses which later elongate and become diffuse, spreading

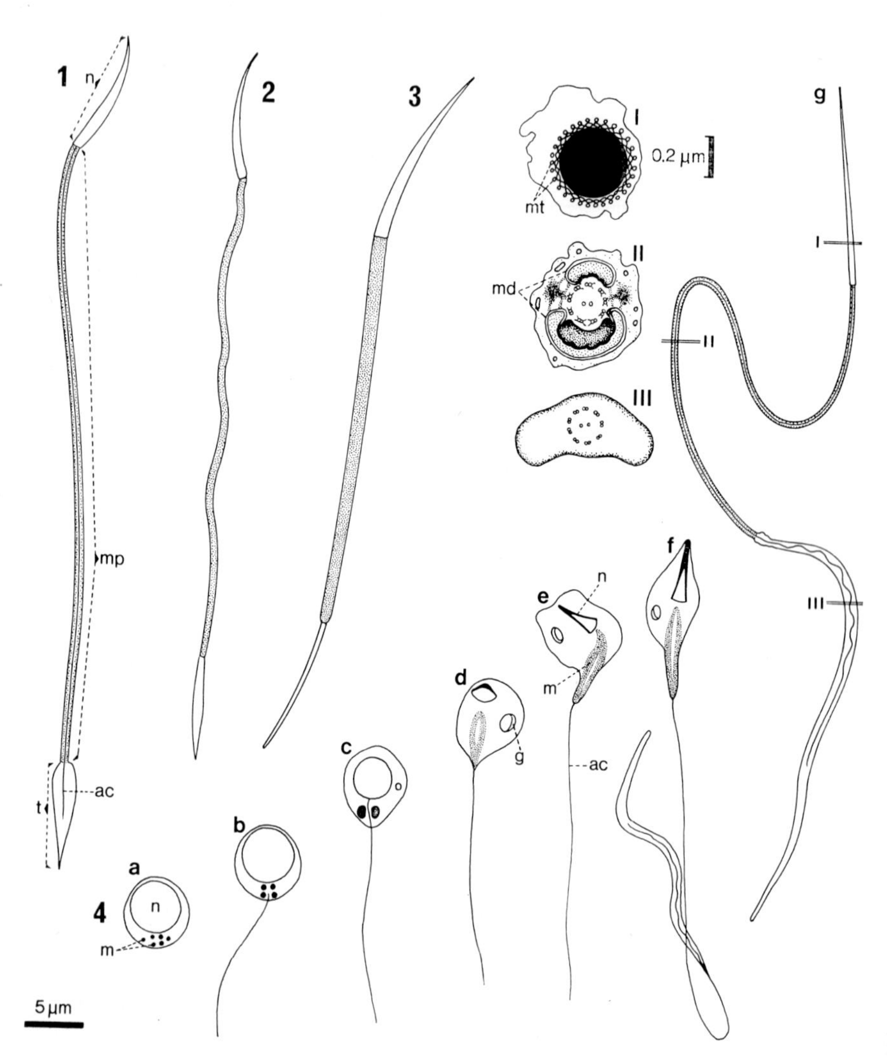

FIG. 23. Spermiogenesis and spermatozoa in gymnolaematous Bryozoa. (1) mature spermatozoon of *Flustra foliacea*; (2) mature spermatazoon of *Scrupocellaria scruposa*; (3) mature spermatazoon of *Bicellariella ciliata*; (4) spermiogenesis and the mature spermatozoon of *Triticella korenii*: *a*, young spermatid, *b*, orientation of four mitochondria around the growing flagellum; *c*, two large mitochondria at the base of the flagellum; *d–f*, successive stages of nuclear condensation and spermatid elongation, with mitochondria forming a pair of dark rods flanking the growing axoneme; *g*, spermatozoon: the sectional planes I, II and III correspond to the drawings illustrating the ultrastructure at different levels: I, nuclear region (head); II, the midpiece; III, the tail, with the axoneme enveloped by a broadly tube-shaped cell membrane. *ac*, axial filament complex; *g*, Golgi body; *m*, mitochondria; *md* mitochondrial derivative; *mp*, midpiece with mitochondria, *mt*, microtubules; *n*, nucleus; *t*, tail. (From Franzén, 1976.)

posteriorly along the developing axoneme of the future midpiece (Fig. 23, 4*c–f*). The nucleus lengthens until it is finely conical. In *Triticella* the head length is 16 μm, the midpiece about 50 μm and the lanceolate tail about 33 μm (Fig. 23, 4*g*). The proportions vary somewhat from species to species. The typical gymnolaemate spermatozoon is thus characterized by the absence of an acrosome, the presence of two mitochondrial rods in the midpiece, and the tubular form of the membrane around the axoneme in the tail (Fig. 23, 1–3). It differs from the stenolaemate spermatozoon in the greater development of the midpiece and corresponding shortness of the tail (in *Triticella* the tail is exceptionally long). Why the *Bugula* spermatozoon should retain a more primitive structure is at present unclear, but it is becoming apparent that *Bugula* is *sui generis* in several aspects of its reproductive biology (see below).

3. *Release of spermatozoa*

Mature spermatozoa break away from the cytophore and accumulate in the metacoel, but their route of exit from the body for a long time was a puzzle. Indeed, the authors of most textbooks (Cori, 1941; Hyman, 1959; Brien, 1960) accepted that self fertilization must be the rule. The problem was finally solved by Silén (1966), studying particularly *Electra posidoniae*. Silén noted that during spermatogenesis the lophophores were rarely protruded, but once the spermatozoa were mature and free, the lophophores were alternately everted and withdrawn for periods of 3–4 h without a break. During the periods of eversion the tentacles did not " exhibit the incessant movements performed by tentacles of the sterile [feeding] zooid ", but remained rather stiff, only occasionally incurving their tips. With the tentacles thus spread, spermatozoa accumulated in the distal part of the metacoel and were seen by Silén wriggling in single file up the lumina of the two dorsomedial tentacles—the pair to which the supraneural coelomopore (or its modification, the intertentacular organ) is axillary. In *E. posidoniae* the rate of evacuation varied considerably, from as low as eight to as high as 50 per min. The liberation of spermatozoa takes place simultaneously not only from neighbouring zooids but from nearby colonies.

The presence in each tentacle of a terminal pore has already been noted (p. 319) and through this the spermatozoa escape to float slowly away from the tentacular funnel. Drifting spermatozoa which pass close to another expanded lophophore are caught by the ingoing current and drawn into the funnel. They are not, however, driven to the mouth as food particles but, in *E. posidoniae*, by their own movements achieve contact with the unciliated abfrontal surface of the tentacles where they remain until ova are discharged. In *Electra crustulenta* the flask-shaped intertentacular organ acts as a receptaculum seminis. Release of spermatozoa in the manner described has now also been established for *Membranipora* as well as *Electra*, and similarly, though through all the tentacles, in *Bugula*, *Cellaria*, *Chorizopora*, *Schizoporella*, *Celleporina*, *Turbicellepora*, *Myriapora* and *Zoobotryon*; and comparable release may be assumed in *Farrella* and *Triticella* (Marcus, 1926; Silén, 1966, 1972; Bullivant, 1967; Ström, 1969). These genera represent widely distant families of Anasca, Ascophora and Ctenostomata, and embrace both brooding and non-brooding embryonic types. The role of the tentacles in the release of spermatozoa makes it clear why some at least are always present on the otherwise reduced polypides of special male zooids.

4. *Oogenesis*

Eggs develop within the ovary, which is simply a cluster of oocytes contained by a peritoneum. It is usually stated that there are two kinds of oogenetic and embryonic development in gymnolaemates but, as Woollacott and Zimmer (1972b, 1975) have pointed out, there are really three, all of which have been described in the earlier literature.

In non-brooding anascans, such as species of *Electra* and *Membranipora*, some 8–20 ovoid eggs are discharged into the coelom, where they rapidly enlarge to a mature size of 60–70 μm. Such ova are greyish, with no coloured yolk. *Triticella korenii* produces a greater number of eggs, up to 60 per zooid being observed by Ström (1969). In this species the oocytes reach a maximum size of 65 μm in the ovary; meiosis commences as the oocytes loosen from the ovary, but its progress is apparently blocked at metaphase of the first division. Maturation occurs within an hour of discharge of the egg through the coelomopore, the ovum being the same size as the fully formed oocyte. By squeezing a zooid gently, Ström was able to release some eggs. A few minutes later neighbouring zooids shed their eggs. It appears that the expelled eggs, or accompanying body fluids, have some kind of chemical effect on other zooids. The eggs do not float away, however, but adhere to the outside of the zooid by their sticky membrane.

Brooding occurs in the genus *Bugula*, and descriptions have been given by Calvet (1900 for *B. simplex*, as *B. sabatieri*), Corrêa (1948, for *B. flabellata*), and Woollacott and Zimmer (1972b and 1975, for *B. neritina*). Oocytes enlarge successively in the ovary and only one at a time reaches full size of about 50 μm diameter, suspended in a pedunculate follicle. About three ova may mature sequentially in the same zooid. Maturation proceeds in the coelom, with the polar bodies appearing by the time the ovum reaches the brood chamber. Calvet stated that the egg was alecithal, but Corrêa seems more accurate in describing it as oligolecithal: ova in the coelom are obviously yellow (in most species, but not in *B. neritina*).

The third pattern of oogenesis is seen in *Callopora dumerilii*, and is considered to occur also in *Escharella immersa* and *Fenestrulina malusii* (Silén, 1945). One oocyte enlarges in the ovary until it reaches a length of about 120 μm. It then bursts into the coelom and there continues to expand until its length measures about 200 μm. A sequence of three to four ova is produced within the zooid, the second oocyte developing in the ovary whilst the first ovum matures in the coelom, and so on. The eggs are telolecithal: the yolk in brooding gymnolaemates is frequently coloured yellow, orange or red with carotenoids, or is occasionally pure white (Marcus, 1938; Silén, 1943, 1945, 1951; Ryland, 1958, 1963b; Eggleston, 1970; Gordon, 1970).

C. *The Gymnolaemata: fertilization*

As established by Silén (1966), fertilization in non-brooding species occurs either in the intertentacular organ as the ova are discharged (*Electra crustulenta*) or in the sea immediately outside the intertentacular organ (*E. posidoniae*). The latter is presumably the case in species which are provided only with a coelomopore, such as *Farrella repens* and *Triticella korenii* (Marcus, 1926; Silén, 1966; Ström, 1969). Marcus (1938, pp. 86–87) tabulated the known occurrences of the intertentacular organ: it is present only in *Conopeum*, *Electra*, *Membranipora* and *Tendra* (Cheilostomata, Electridae and Membraniporidae) and in some species of *Alcyonidium* (Ctenostomata).

In no brooding species has fertilization yet been witnessed, but it presumably occurs either just before or just after transfer of the ovum to the brood chamber. The greatest doubt seems to surround the genus *Bugula*, already noted as having more primitive spermatozoa and smaller eggs than other brooding bryozoans. Calvet (1900) surmised, from the presence of the ovum and of mature sperm in the coelom together, that self fertilization occurred. Corrêa (1948), too, reached that conclusion. However, Silén (1972) has seen spermatozoa of *B. flabellata* " liberated through all tentacles ", so that cross fertilization presumably does occur. Corrêa, implying that his conclusion was based on the examination of microscopical sections, found " the ovocytes of *B. flabellata* precociously inseminated while they are still in the ovary surrounded by the follicle " and stated that fertilization was monospermic. It transpires from Corrêa's account that during early oogenesis the polypide remains functional; by the time the ovum is mature in the coelom the polypide has degenerated and there is no lophophore. *B. flabellata* displays the usual slight protandry so, to conform with his account, release of spermatozoa and their entry through the coelomopore both have to take place before the egg is discharged into the coelom. Possibly Corrêa was mistaken for, by the time the ovum is ready for transfer to the brood chamber, a new polypide has differentiated and entry of spermatozoa is again possible.

It is thus fairly clear now that, throughout the marine Bryozoa, interzooidal fertilization is certainly possible and perhaps usual; but for this to be genetic cross fertilization the zooids concerned must belong to different colonies. Sterility mechanisms to prevent self fertilization may be present but, if so, they have yet to be demonstrated. Cross fertilization will be favoured by the epidemic discharge of spermatozoa reported in *Electra posidoniae* (Silén, 1966) and also by the aggregated settlement frequently found in bryozoans (even in those with planktotrophic larvae) which often results in mosaics of small colonies (Ryland, 1959b, 1972; Hayward, 1973; Hayward and Ryland, 1975; Ryland and Gordon, 1976). However, as Silén recognized, outbreeding does not require that inter-colony fertilization should be the rule, merely that it should happen regularly. Marcus (1926) reported that *Electra pilosa* spermatozoa survived in sea water for only 10 min, but Silén (1966) found in *E. posidoniae* that they could remain motile for much longer. Transference between colonies is therefore by no means unlikely. Finally, Schopf (1973) has demonstrated that gene frequencies in *Bugula stolonifera* and *Schizoporella errata* conform to the Hardy-Weinberg equilibrium, which is evidence that both species are, in fact, routinely outbreeding.

D. *The Gymnolaemata: brooding and embryogenesis*

1. *Structure and formation of ovicells*

Embryonic brood chambers of several kinds exist in the Gymnolaemata, but the most characteristic is the ovicell or ooecium, a more or less globular chamber placed at or near the distal transverse wall of an egg-bearing (maternal) zooid. The terms ovicell and ooecium generally have been regarded as interchangeable, but recent work on *Bugula* (Woollacott and Zimmer, 1972a) suggests that, in *Bugula* at least, a distinction might advantageously be made. The term ooecium, as used here, will simply be the double-walled chamber which encloses the developing embryo: this is in strict accordance with the established practice of naming the inner and outer walls entooecium and ectooecium respectively. (Woollacott and Zimmer call this structure the oecial fold, which is entirely satisfactory within the context of their paper, but unsuitable to use in systematic accounts.) The ooecium is closed by a membranous, or sometimes chitinized, structure generally termed the inner vesicle (ooecial vesicle by Woollacott and Zimmer). The entire structure, ooecium with inner vesicle, will here be termed the ovicell. Silén's (1945) proposal to call the ooecium, as here defined, the

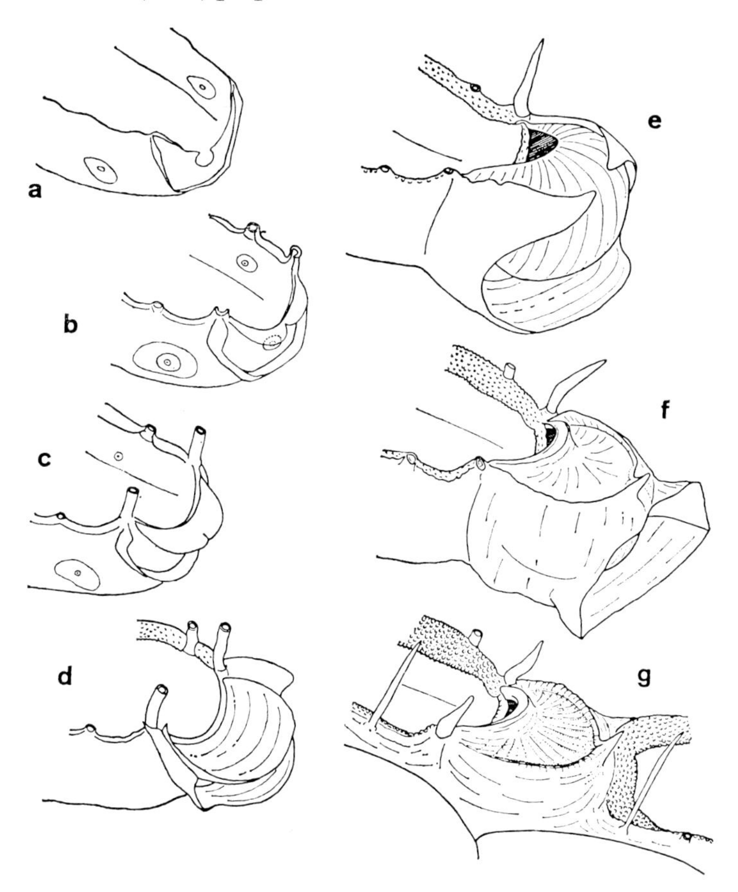

Fig. 24. Formation of ooecia in *Crassimarginatella maderensis*. *a*, the end-wall of the maternal zooid is nearing completion; *b*, appearance of twin ooecial rudiments (and of distal spines); *c*, fusion and enlargement of the rudiments; *d*, saucer-like basal wall of the ectooecium has formed from the rudiments; *e*, *f*, that part of the ooecium derived from rudiments is complete; " wings " from the walls of the distal zooid are overarching the ectooecium; *g*, the complete ooecium, with its " cover " from the distal zooid. (From Harmelin, 1973a; the drawings were based on hypochlorite-cleaned specimens, so that the membranous entooecium was destroyed.)

ectooecium, and the inner vesicle the entooecium is too much at variance with the rest of the literature to be acceptable.

Silén's (1945) account of the development of the ovicell in *Callopora dumerilii* may briefly be summarized. The ooecium starts to form only when a recognizable ovum is enlarging in the ovary. It commences as two hollow knobs above the transverse wall of the maternal zooid; the knobs fuse, and spread as a flat prominence which becomes bowl-shaped on the proximal end of the distal zooid and then overarches to form a hood open toward the maternal zooid. The ooecium, being hollow from the start, is thus bilaminar, with the entooecium and ectooecium separated by a flat lumen. This lumen communicates with the maternal zooid through a pore. Harmelin (1973a) has described essentially similar development in *Crassimarginatella*. The ectooecium again starts as twin rudiments arising from the distal wall (Fig. 24 *a–c*); they meet and overarch as in *Callopora* (Fig. 24 *e–g*). An additional complexity in *Crassimarginatella*, however, is that much of the ecto-oecium becomes overgrown with a calcified layer derived from the distal zooid.

Silén criticized—unjustly it now transpires—Calvet's (1900) account of the structure of the ooecium in *Bugula simplex* (as *B. sabatieri*). Calvet's text-fig. 10 shows the communication pore *beyond* the transverse wall, and thus opening into the distal, rather than the maternal, zooid. (It also has to be noted that Levinsen, 1909, shows a *Callopora* ooecium opening into the distal zooid. It is, however, difficult to reconcile a pore located as shown in his Pl. 24, Fig. 16, with the origin of the rudiments shown in Pl. 9, Fig. 3*a*, and it seems justifiable to regard the former figure as inaccurate.) There is no controversy surrounding the inner vesicle, which evaginates from the distal end of the maternal zooid to close the ooecium. Calvet's figure actually shows the structural relationships of the entire *Bugula* ovicell correctly and clearly.

The vindication of Calvet's interpretation is a new study by Woollacott and Zimmer (1972a) on the origin of the ovicell in *Bugula neritina*. The ooecium arises as a single (not paired) rudiment from the distal zooid (Fig. 25) while the inner vesicle simultaneously evaginates from the maternal zooid. The junction between the ooecium and the distal zooid is a perforated septum plugged with cells: this suggests that the ooecium should be regarded as a highly specialized kenozooid, linked to the distal zooid by a simple communication pore.

At the present time it thus appears that anascan ooecia develop in two quite different ways: (1) from twin rudiments at the distal end of the maternal zooid, and (2) from a single rudiment at the proximal end

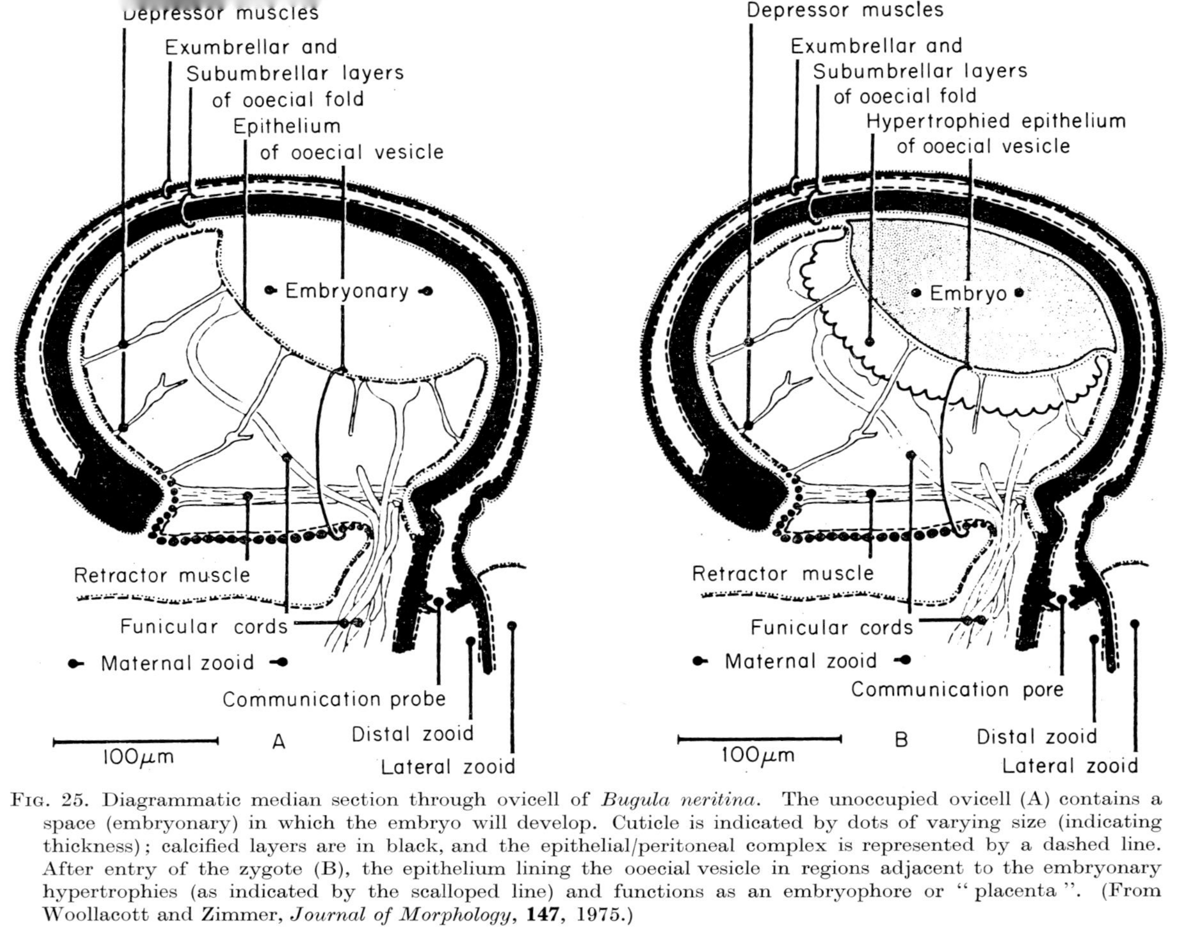

FIG. 25. Diagrammatic median section through ovicell of *Bugula neritina*. The unoccupied ovicell (A) contains a space (embryonary) in which the embryo will develop. Cuticle is indicated by dots of varying size (indicating thickness); calcified layers are in black, and the epithelial/peritoneal complex is represented by a dashed line. After entry of the zygote (B), the epithelium lining the ooecial vesicle in regions adjacent to the embryonary hypertrophies (as indicated by the scalloped line) and functions as an embryophore or "placenta". (From Woollacott and Zimmer, *Journal of Morphology*, **147**, 1975.)

of the distal zooid. The formation of ascophoran ooecia has been less thoroughly investigated in recent years but is certainly variable (D. F. Soule, 1973) and too much generalization on the basis of Levinsen's 1909) work seems unwise.

2. *Transfer of the ovum to the ovicell*

This remarkable process has been witnessed twice only, in *Bugula*, which has a smallish egg, by Gerwerzhagen (1913) and in *Callopora*, which has a large egg, by Silén (1945). The latter's account has, with justification, been much quoted. The polypide emerges and the ovum is forced into the " neck " region just below the lophophore; the polypide is then partly withdrawn, so that the supraneural coelomopore is aligned with the opening of the ovicell. The inner vesicle is lowered by its intrinsic musculature (Fig. 25) and the ovum flows, thread-like, through the coelomopore into the embryonary where it re-forms as a spheroid.

Subsequent development of the embryo is well known, at least for oligolecithal types, and need not be repeated. It appears, however, that there may be important physiological differences between the way in which embryos of oligolecithal genera such as *Bugula* and telolecithal genera such as *Callopora* are nourished (Woollacott and Zimmer, 1972b, 1975).

3. *Embryonic nutrition*

The telolecithal ovum of *Callopora* apparently is supplied with sufficient yolk to meet all requirements of the developing embryos. The colouration of the embryo fades during development and there is no increase in its size, both consistent with the views that yolk is being used up and that there can be little external supply of nutrients during embryogenesis. Silén (1945) believed that the inner vesicle acted as a cushion only, having a mechanical and not a nutritional role. This is echoed by Corrêa (1948) in the statement that " there are many species without nourishing in the ovicell and without placenta ". Nevertheless, Silén found that embryos stop developing if removed from the ovicells. This kind of development, in which the embryo and larva " feed exclusively on the content of yolk within the egg cell from which they originate " is lecithotrophic (Thorson, 1946) in the strict sense.

In *Bugula* the situation is quite different. First, the embryo increases greatly in size, by a factor of about 50 in *B. flabellata* but by as much as 500 times in *B. neritina*, in which the zygote is smaller and the larva larger. Second (see Fig. 25B) that part of the inner vesicle towards the embryo has a hypertrophied epithelium of columnar cells (Calvet, 1900; Marcus, 1938; Corrêa, 1948; Woollacott and Zimmer, 1972a). Marcus and Corrêa termed this a placenta, and it is certainly obvious from the growth of the embryo that an external source of nourishment in involved.

Woollacott and Zimmer (1972a, 1975) based their study on *Bugula neritina*. Prior to transfer of the zygote, the embryonary (the cavity between the entooecium and the inner vesicle) is spacious. The epithelium of the inner vesicle is, at this time, squamous throughout. Outside the ovicell its cuticle is about 2·2 μm in thickness, stratified and externally convoluted; inside, where adjacent to the embryonary, in the region about to differentiate into the placenta or (as Woollacott and Zimmer prefer) embryophore, the cuticle is only 0·2 μm in thickness and externally plane. The cell apices lack microvilli and lie flat against the cuticle. The most conspicuous feature of these cells is the abundance of ribosomes. The bases of many of the cells are sheathed by the expanded distal ends of funicular cells—part of a tissue which essentially links the gut (especially the caecum) to the interzooidal pore-chambers (see Fig. 4C on p. 293), and is believed to serve for the transport of metabolites (Bobin, 1971, 1976). The bases of other cells in the epithelium of the embryophore, however, remain in direct contact with metacoelic fluid.

During early embryogenesis the inner vesicle becomes closely applied to the entooecium, except where it abuts against the embryo, compressing it to a lens-shaped body (Fig. 25B). The epithelial cells of the future embryophore now hypertrophy, increasing in height five-fold. It becomes possible to distinguish in these cells three zones, according to the distribution of organelles: (1) a basal region containing the nucleus, mitochondria and rough endoplasmic reticulum; (2) a centre region with Golgi bodies and numerous (presumably secretory) vesicles; and (3) an apical region largely devoid of conspicuous organelles. The once smooth plasmalemma is now folded into numerous microvilli which do not, however, penetrate or deform the thin cuticle. There are morphological indications of the secretion of vesicles by exocytosis into the extracellular space between the microvilli. The basal plasmalemma of the epithelial cells does not become microvillous but remains smooth in profile where approximated to the covering funicular cells.

Part of the embryonic ectoderm (presumptive internal sac) lies against the embryophore, but the embryonic and maternal tissues remain separated by the cuticle of the embryophore. The cell apices possess numerous infoldings from which pinocytotic channels lead and apparently cut off vesicles into the cytoplasm. These channels and vesicles contain material of similar electron density to that occurring in the extracellular spaces apical to the epithelial cells of the embryophore. Regions of embryonic epithelia remote from the embryophore lack the pinocytotic channels just described. At the conclusion of the active phase of extraembryonic nutrition the uptake epithelium invaginates into the interior of the embryo (to form the internal sac), breaking contact with the embryophore.

The evidence from the growth of the embryo during its development in the embryonary, taken in conjunction with the ultrastructure of the cells in the two approximated epithelia (as established by Woollacott and Zimmer, 1975), demonstrates unequivocally that metabolites must pass from the maternal zooid to the embryo. The involvement of the funicular system seems equally certain, for not only do strands lead to the embryophore but the funicular plexus there enlarges during the period of embryophoral activity. Woollacott and Zimmer suggest that

TABLE IX. CHANGES IN DIMENSIONS OF EMBRYOS DEVELOPING IN OVICELLS

Species	*Diameter of major axis* (μm) *Ovum or zygote*	*Larva*
Callopora dumerilii[4]	200	no increase
Bicellariella ciliata[1]	62	172
Bugula avicularia[3]	50	200
B. flabellata[5]	50–80	180
B. neritina[6]	36	300–400
B. simplex (= *sabatieri*)[2]	about 65	about 200
Kinetoskias smitti[3]	—	no increase
Hippothoa hyalina[3]	40–47	108–140
Hippopodina feegeensis[3]	237	316
Vittaticella (= *Catenicella*) *elegans*[3]	90 [16-cell]	144
V. contei[3]	100	100
Schizoporella carvalhoi[3]	155	150
Celleporina (= *Siniopelta*) *costazii*[3]	155	150
Hippoporella gorgonensis[3]	—	no increase
Hippodiplosia americana[3]	—	no increase
Microporella ciliata[3]	—	no increase
Rhynchozoon phrynoglossum[3, 5]	200	no increase
Celleporaria (= *Holoporella*) *mordax*[3, 5]	220	no increase
Escharella immersa[4]	—	no increase
Fenestrulina malusii[4]	—	no increase

Sources: [1]Nitsche, 1869; [2]Calvet, 1900; [3]Marcus, 1938; [4]Silén, 1945; [5]Corrêa, 1948; [6]Woollacott and Zimmer, 1975.

the funicular tissue provides a pathway for nutrient flow from actively feeding individuals to the functional embryophores (as well as to branch apices, zooids with regenerating polypides, etc.). Since, at least in *B. flabellata* (Corrêa, 1948), activity of the placenta appears more or less to coincide with regression of the maternal polypide, the metabolites passing to the embryo must be either transported from elsewhere in the colony or be degradation byproducts.

The mode of embryonic nutrition has so far been established for very few cheilostomes (Table IX), but both methods evidently occur among related groups (e.g. Cellularina: *Bugula*, *Kinetoskias*) or even within a genus (*Vittaticella*). No obvious systematic or ecological pattern is evident in the method found in the species tabulated (admittedly a small sample). The rediscovery by Woollacott and Zimmer of "placental" nutrition in ooeciferous cheilostomes has thus led not only to a valuable ultrastructural study, but to the posing of fundamental questions about colonial metabolism in bryozoans and about the biological significance of the way in which brooded embryos receive nourishment.

E. *Reproductive ecology*

This topic was considered quite comprehensively in my earlier review and book (Ryland, 1967a, 1970). With regard to reproductive season and annual cycles, data which had then been made available to me in advance have now been published in full (Gordon, 1970; Eggleston 1972a). Médioni (1972) has summarized breeding period and other ecological data for the region of Banyuls-sur-Mer (Mediterranean), and Abbott (1975) has investigated some effects of temperature on the growth pattern and attainment of sexual maturity. Fouling continues to attract research and some recent papers containing useful information on bryozoan settlement are by Meadows (1969, noting that Pl. I D illustrates *Microporella ciliata* not *Cryptosula pallasiana*), Geraci and Relini (1970), Haderlie (1971), Menon and Nair (1971), Humphries (1973), Menon (1973) and Geraci (1976).

VI. Larvae and Metamorphosis

A. *Planktotrophic larvae*

1. *Cyphonautes*

The general form and structure of the large *Membranipora* cyphonautes are quite familiar (see Atkins, 1955a; Ryland, 1964; Mawatari and Itô, 1972) but the smaller larvae of this type are less well known. The latter occur in the membraniporine genera *Conopeum* and *Electra* (descriptions given by Cook, 1960, 1962 and Ryland, 1965) and in the ctenostomatous species *Alcyonidium albidum*, *Farrella repens*, *Hypophorella expansa* and *Triticella korenii*. The larvae of *F. repens* were reared by Marcus (1926) and the same has now been done for *T. korenii* by Ström (1969).

Cyphonautes are triangular, cupuliform or oval in side view, compressed between a bivalved shell in membraniporines, shelled or shell-less and " hat-like " in ctenostomes. The valves, if present, gape at their tip and along the base. There is no shell in *Triticella*. The apex or aboral pole is surmounted by a sensory organ. The base is fusiform or round, formed by the ciliated mantle edge or corona; within the corona lies the mantle cavity. The pyriform gland is anteriorly placed inside the mantle cavity, and associated with it is a tuft of plume cilia. A neuromuscular tract links the apical organ and the plume cilia. The adhesive organ lies posteriorly inside the mantle cavity, just behind the adductor in shelled forms. There is no functional alimentary canal in *Triticella*, but it is present and tripartite in the larger cyphonautes: the midgut forms from entoderm and merges with the stomodaeal and proctodaeal invaginations.

Mawatari and Mawatari (1975) have established in *Membranipora serrilamella* that the shell is derived from the fertilization membrane of the zygote. By about two days after discharge of the egg (at 15°C), the embryo has trapesoid shape and is surrounded by an extensive perivitelline space. Cilia on the developing corona drive the embryo into contact with the oral surface of the membrane, which perforates to allow the cilia to project. The developing larva is not released from the membrane. The embryo elongates as the pyriform gland and adhesive sac develop, until the apical organ makes contact with the aboral surface of the membrane, where the appearance of a pore permits protrusion. While organogenesis proceeds to completion, the larva becomes increasingly compressed until, about three days after discharge, the now flattened egg membrane separates into the paired shell valves. The larva now has the typical triangular profile of a membraniporine cyphonautes and measures about 220 μm across the base, a dimension that reaches about 600 μm with full growth.

The ciliary feeding mechanism of the *Membranipora membranacea* larva was described in some detail by Atkins (1955b) and has recently been re-examined by Strathmann (1973). Food is obtained from a water current circulated through the mantle cavity, which is divided into inhalent and exhalent chambers by paired ciliated ridges. The current is generated by lateral cilia on the ridges, which also bear frontal cilia and a row of laterofrontal cilia. Both authors found that the laterofrontal cilia were stationary for much of the time. Strathmann noted that particles were retained just upstream from the ciliated ridges and moved along the ridges towards the mouth in a series of jumps. Since there was nothing to suggest that the laterofrontal cilia could be responsible for this motion, he believed that retention and transport of particles may often result from a local reversal of the lateral cilia (cf. the role of lateral cilia on the tentacles of the adult lophophore, p. 327). Atkins established that the frontal cilia, which are active only in the presence of food organisms, and the cilia of the funnel-like upper portion of the inhalent chamber transport particles to the mouth. Rejection of unwanted particles is effected by the mouth remaining closed, the material then being carried away into the exhalent chamber. The gut is internally ciliated, and Atkins believed that the cilia would beat with a spiral or rotary movement.

Hyman (1959) regarded the cyphonautes as a modified trochophore, but Jägersten (1972) was cautious about such an interpretation and Zimmer (1973) rejected it.

2. *Behaviour*

Little is known about the free life of cyphonautes, except that they swim in a spiral path with the apical organ in advance (O'Donoghue, 1926; Atkins, 1955b; Cook, 1960; Ström, 1969) or occasionally in a vertical position. The direction of the coronal ciliary beat is from the apex downwards, the laeoplectic metachronal waves passing clockwise when viewed from above (Knight–Jones, 1954).

Cyphonautes appear generally indifferent to light, showing no indication of phototactic or photokinetic behaviour (Silén and Jansson, 1972). Using a dark tube suspended horizontally in the sea, these authors found that the interior surface of the tube was rapidly covered by evenly distributed colonies; the whole exterior surface was also covered in the absence of algal growth, but only the lower face received settlement when the upper side was covered by a dense rug of algae. The larvae appeared to be geonegative, the action of the coronal cilia causing them to rise in the water.

3. *Settlement*

Approaching settlement, the *Electra pilosa* larva glides in an erect position over the substratum, the pyriform gland foremost and the plume cilia beating on, and presumably testing, the surface (Atkins, 1955a). Jägersten (1972) has observed in an unidentified cyphonautes that the pyriform organ can be extended like a tongue and is probed in different directions while the larva crawls about: this confirms an earlier statement by Kupelwieser (1906). Jägersten further established that this organ was acting as a foot and that the cilia clothing the " sole " of the foot propel the creeping larva and, for gliding, seem more important than the cilia of the corona. The larva adheres to the substratum with a secretion from glandular cells of the pyriform organ, and the cilia of the corona gradually become quiescent. Immediately before settlement the cyphonautes move around in circles until, suddenly, as the adhesive sac everts, the valves separate and are pulled down onto the substratum. During this flattening the apical organ comes to lie just behind the pyriform gland. The larva is now strongly attached; the shell valves fall away some time later.

The larval organization completely breaks down during metamorphosis (review by Nielsen, 1971) and, in *Electra*, a new structure, the ancestrula or primary zooid of the colony, arises in its place. *Membranipora* differs in that metamorphosis gives rise to twin primary zooids (O'Donoghue, 1926; Atkins, 1955a). The two zooids are joined proximally, but their distal or blastogenic ends diverge. During metamorphosis, which has been described by Atkins (1955a), there is a reversal of internal polarity: the distal end of the polypide and the blastogenic face of the ancestrula correspond to the posterior of the larva. This correlation is important, for it reveals how the initial direction of colony formation depends on the orientation of the larva at the time of settlement.

The cyphonautes of *Triticella korenii* metamorphoses into a very small, polypideless ancestrula. This gives rise to the stalk of the first, small autozooid from one end and to a stolon segment from the other. The second autozooid is budded from this, as is the next stolon segment; and so on (Ström, 1969).

4. *Orientation at settlement*

Several sets of observations have shown that cyphonautes usually do not settle in a random orientation but respond to water movements or other directional stimuli. As I have reviewed this subject in detail elsewhere (Ryland, 1976b), only a summary of conclusions will be presented here. Ryland and Stebbing (1971) found that *Electra pilosa* larvae on *Fucus serratus* settled with twice as many facing the plant base as faced the frond apices. They assumed that, since the algal fronds would be swept to and fro by the waves, always streaming in the direction of water flow, this was probably a rheotropic response. Since the cyphonautes is rather sail-like, when gliding on its pyriform

" foot " (as described by Jägersten, 1972), its only stable orientation in a water current would be facing into it, pivoting rudder-like about the foot. This interpretation seemed to be confirmed by the preferentially upstream settlement of *E. pilosa* recorded on panels held in a tidal current (Ryland, 1976a). Unfortunately, however, *Membranipora membranacea* larvae settling on *Fucus serratus* tend to be distal-facing (at least on the unusually sheltered and wave free shore studied by Ryland and Stebbing (1971)); whilst orientated settlement of *M. tuberculata* was unexpectedly recorded on the leaflets of gulf-weed *Sargassum fluitans* Börgesen and *S. natans* (L.) Meyen (Ryland, 1974a). Since, in the pleustal habitat, the *Sargassum* clumps should move with the water mass, it is difficult to visualize any consistent pattern of water flow over the leaflets. Generalization thus seems impossible at the present time, and experiments with living larvae are clearly needed to help interpret the observational data.

B. *Stenolaemate larvae*

1. *The larva and settlement*

The first description of cyclostome larvae and their metamorphosis was given by Barrois (1877). Unfortunately, as Nielsen has pointed out, these illustrations (Barrois, 1877, Pl. 3, Figs. 20, 21) have been reproduced upside down and wrongly labelled in the textbooks of Cori (1941, Fig. 520a) and Hyman (1959, Fig. 132H). Moreover, the specimen depicted by Barrois seems to be abnormally elongated. The larvae, which may be liberated naturally or obtained by breaking open the brood chambers, have the simplest larval structure known in marine bryozoans (Nielsen, 1970). The shape is ovoid, the longest axis measuring 100–150 μm according to species, and the whole surface—apart from minute apical and abapical areas—is covered with cilia (Fig. 26A).

Larvae are released during daylight hours, but not necessarily as a response to illumination; in *Tubulipora pulchra* at least they show no clear phototaxis (Mawatari, 1947). The larvae may bear pigmented spots, purplish brown in *T. misakiensis* (Mawatari, 1948), white in *T. plumosa* and *T. phalangea*, but there are no structures corresponding to the so-called eye-spots present in many cheilostome larvae (p. 372). There has so far been no experimental work on the free-swimming or settlement behaviour of the larvae of any cyclostomatous species. They hold, perhaps, greatest potential in genetical studies, since all the larvae derived from a single gonozooid are presumed to be genetically alike.

The larvae swim about rapidly with a spiralling motion, rotating clockwise about the long axis, the apical pole leading; but they soon settle—usually within 15 min in *Crisia eburnea* (Nielsen, 1970) though after 3–5 h in *Tubulipora misakiensis* (Mawatari, 1948). The larvae then start to explore the available substratum, creeping in

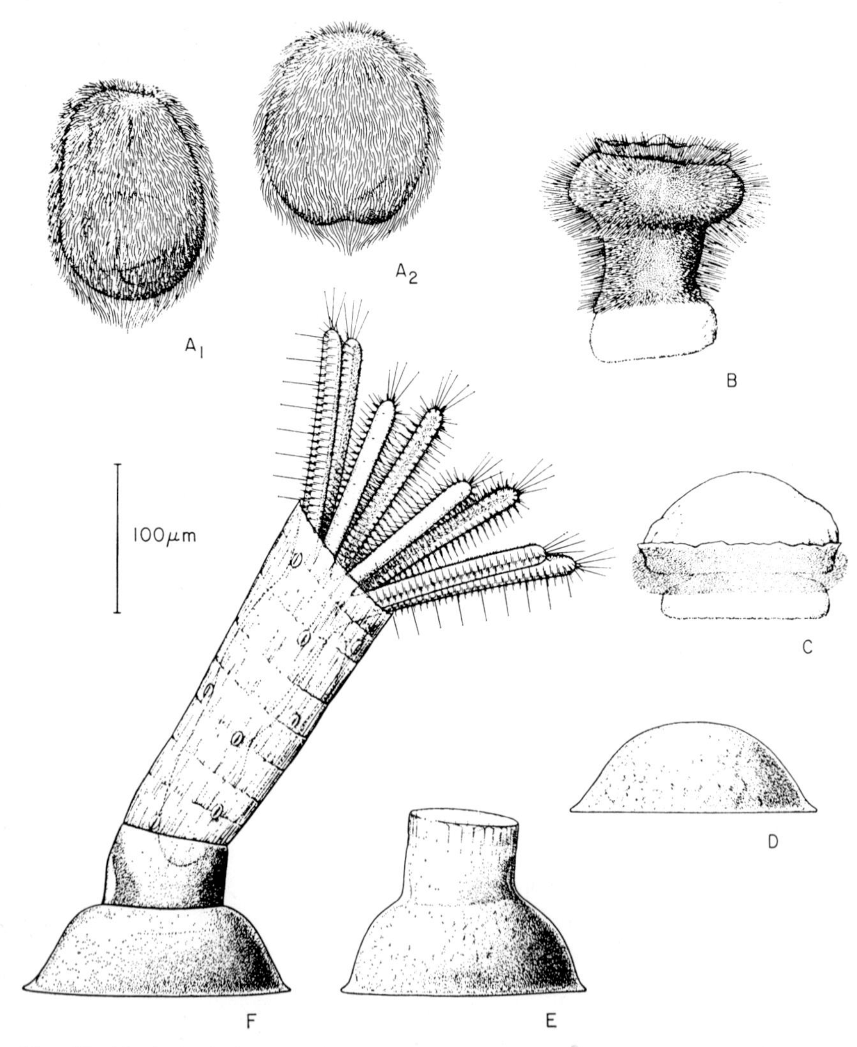

FIG. 26. The larva, metamorphosis and ancestrula of *Crisia eburnea*. A, free-swimming larva seen in profile and from the flattened side; B, early metamorphosis, eversion of the upper and lower invaginations having just started; C, late metamorphosis; D, early primary disc (proancestrula); E, young ancestrula, about two days after the onset of metamorphosis; F, ancestrula with the lophophore protruded, about seven days after settlement. (From Nielsen, 1970.)

different directions, rotating or resting, until the cilia slow down and the larva begins spasmodic pulsations.

As described by Nielsen for *Crisia eburnea* the settling larva turns so that the pore of the adhesive organ comes into contact with the substratum. " Immediately afterwards the adhesive sac everts; the conical central portion comes out first and flattens, and soon afterwards the whole invagination spreads out as a large circular plate which sticks firmly to the substratum " (Fig. 26B, C). The apical invagination also everts and the outer layers of the larva move downwards towards the flattened attachment organ, enclosing an annular cavity. The product is the hemispherical proancestrula or primary disc (Fig. 26D), whose upper surface immediately commences calcification, except for a small circular or subcircular area, centrally located in *Crisia*, eccentric in *Tubulipora*. Within the proancestrula complete reorganization of the tissues takes place. The epithelial cells, originally on the outer surface of the larva but now inside the annular cavity, round off and enclose the cilia in large vacuoles. The cells of the adhesive sac degenerate. The polypide rudiment starts to form, while the cylindrical portion of the ancestrula slowly elongates from the uncalcified area of the primary disc (Fig. 26E). In *Crisia* and related genera a chitinous joint separates the primary disc from the ancestrular cylinder (Fig. 26F). It should be noted that formation of the ancestrula in two stages is a stenolaemate characteristic.

2. *Polypide formation*

The development of the polypide in cyclostomes had not been well described prior to Nielsen's (1970) paper, so this process is now related in some detail.

The cells in the centre of the annulus marking the base of the ancestrular cylinder re-form in two layers. The deep layer sinks towards the middle of the proancestrula, whilst the superficial layer invaginates away from the covering cuticle. The deep layer—the polypide rudiment —becomes cup-shaped, open-ended below the invaginating superficial layer, and bilaminar, probably by transverse division of the single layer of cells. Before the end of the second day the polypide rudiment has elongated and its lower end has turned towards one side of the primary disc. Its outer layer gradually becomes diffuse and the cells flatten. Degenerating ciliated cells are distributed around the polypide rudiment. The distal invagination—the future atrium—is now long, slender and drop-shaped, with the cuticle above it still intact.

The ancestrular cylinder elongates slowly, and the incipient atrium likewise. The distal part of the sacciform polypide rudiment evaginates and forms a thick, bilaminar ring (the lophophore base) around the future mouth. At the lower side of the lophophore base the ganglion forms as a small invagination. Other cells around and in front of the lophophore base give rise to the tentacle sheath. The tentacles develop later, as do the two retractor muscles: each passes from the upper wall of the primary disc to the lophophore base. Meanwhile, the proximal part of the rudiment has elongated and become reflected beneath the lophophore base before differentiating into the various parts of the gut. The outer cell layer of the rudiment is still rather diffuse, but here and there two laminae can be distinguished: the inner is destined to be the gut peritoneum and the outer the membranous sac (an epithelial layer on the inside of a basement membrane, which, in cyclostomes, encloses the polypide).

About six days after settling the ancestrula is almost fully formed and the polypide extends well up the ancestrular cylinder. The tentacles are now long and slender, and both lateral and frontal cilia are appearing. Around the tentacular lumen two cell layers can be distinguished, with a thick membrane between them. These tubular membranes are open proximally where their lateral-abfrontal parts fuse to form a somewhat thickened ring level with the mouth (cf. p. 318). Each of the retractor muscles is inserted on the basal side of this ring, while its origin has gradually shifted upwards into the base of the cylinder (above the joint in *Crisia*). The polypide is now complete.

Nielsen's account was based principally on *Crisia eburnea* but applied equally to *Crisiella producta*. In *Berenicea patina*, *Tubulipora* spp. and *Disporella hispida*, which are encrusting, the atrial invagination initiates in front of the polypide rudiment, rather than above it; and, of course, there is no chitinous ring between the primary disc and the ancestrular cylinder. Otherwise they are very similar. He concluded that polypide development within the Cyclostomata follows a very consistent pattern.

C. *Short-lived gymnolaemate larvae*

1. *The larvae of* Alcyonidium polyoum *and* Bugula neritina

The gross structure of fully developed, brooded larvae is well established, especially by Barrois (1877) and Calvet (1900), and is given in the textbooks (e.g. Hyman, 1959). Nielsen (1971) has tabulated with full documentation those species for which the larvae have been adequately described. However, new studies based on transmission electron microscopy, have been published on the larvae of *Alcyonidium polyoum* (d'Hondt, 1973) and *Bugula neritina* (Woollacott and

Zimmer, 1971). It was noted on p. 360 that the larva of *Alcyonidium albidum* is a cyphonautes; although that of *A. polyoum* is quite different in shape, being disciform and oro-aborally flattened, it nevertheless displays a number of similarities with cyphonautes. Thus, by comparison, the apex is laterally expanded to form a broad, circular apical organ (calotte) fringed with cilia; shell valves are absent and the corona is equatorial (Fig. 27A). The pyriform organ complex remains at the anterior of the oral hemisphere, and the internal sac is posterior—though it is much larger than in a cyphonautes. A rudimentary alimentary canal remains, with a minute oral opening situated between the pyriform organ and the internal sac, and with a " blind " mid-gut above the latter.

The 48 coronal cells are not vertically elongated as in *Bugula* (see below) but are disposed as a narrow band about 30 μm in height. Cilia are restricted to the central one-third of the cell apex and their roots continue into the cell for about two-thirds of its length. Above the corona, and separated from it by a ring of supracoronal cells, a deep annular pallial groove, lined with palisade epithelial cells, delimits the apical organ. The apical organ is composed of biciliated epithelial cells. The sensory cap constitutes only part of the apical organ and is placed eccentrically toward the larval anterior. The pyriform organ is complex: at its centre is a protruding cluster of cells about 20 μm high bearing the plume cilia; around these and extending inwards are grouped the secretory cells; centrifugally again, in face view, and also continuing inwards around the glandular cells, are large (about 50 × 15 μm) cubical, vacuolated cells. Immediately aboral to the vacuolated cells is an infracoronal annulus of ciliated cells. The remaining larval epithelium consists of cells about 6·5 μm in height, apically microvillous but non-ciliated.

The internal sac is substantial, occupying about one-third of the larval volume, conical, with its summit directed towards the sensory cap. Its walls are thick (up to 35 μm) but composed of a single layer of palisade cells; the lumen is flattened anteroposteriorly but crescentric as seen in sagittal section. The posterior extension of the lumen is bifid. Although d'Hondt refers to the " floor " and " roof " of the internal sac, there is no specially differentiated roof of the kind described in *Bugula* (see below), presumably because in *Alcyonidium polyoum* there is no neuromuscular tract connecting the internal sac with the sensory cap. A distinctive group of cells at the opening of the internal sac (" cellules périphérique au débauché du sac interne ") may correspond with the " neck " cells in the *Bugula* larva.

The main nerve trunk is enlarged apically where it underlies the

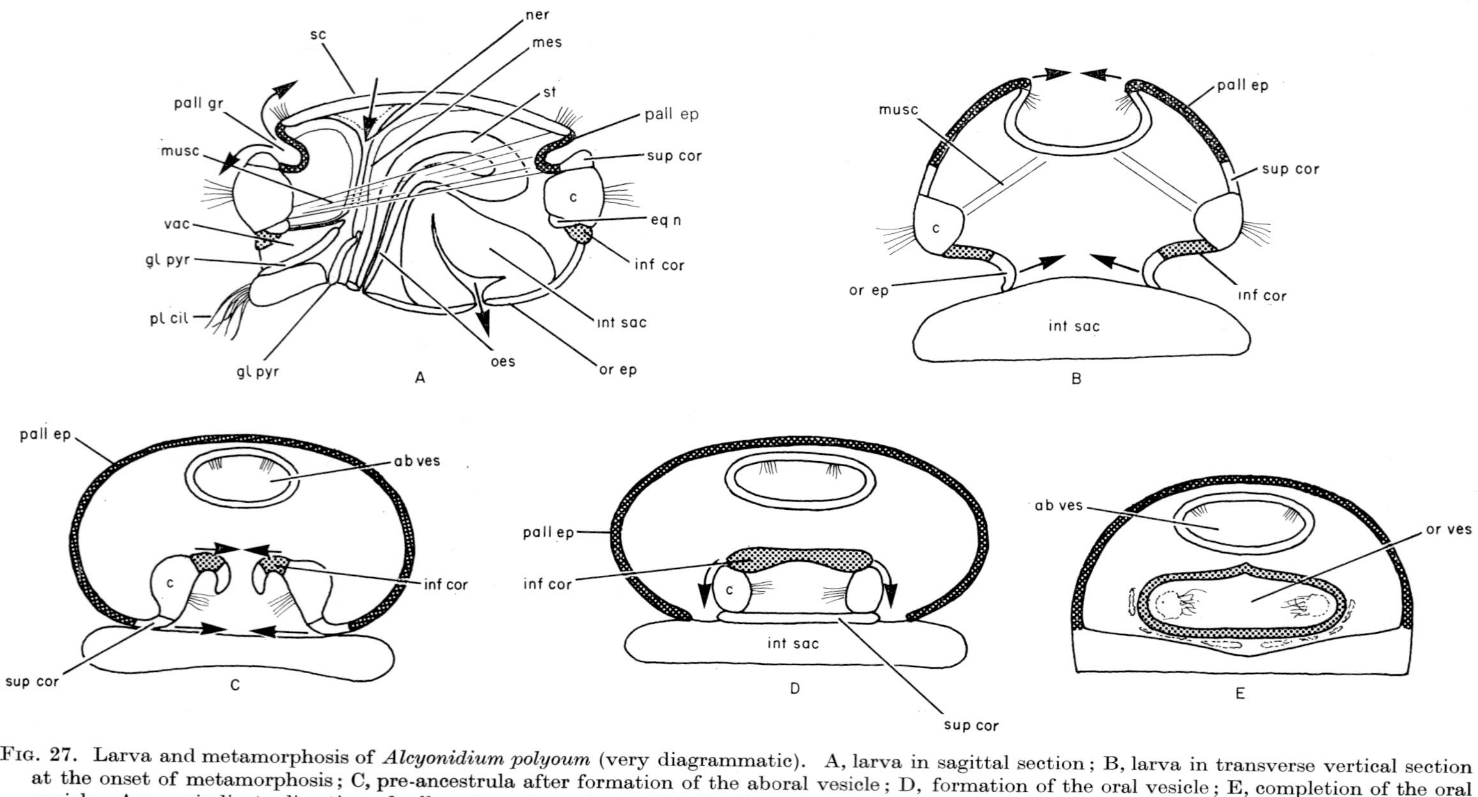

FIG. 27. Larva and metamorphosis of *Alcyonidium polyoum* (very diagrammatic). A, larva in sagittal section; B, larva in transverse vertical section at the onset of metamorphosis; C, pre-ancestrula after formation of the aboral vesicle; D, formation of the oral vesicle; E, completion of the oral vesicle. Arrows indicate direction of cell movements. *ab ves*, aboral vesicle; *c*, coronal cell; *eq n*, equatorial nerve ring; *gl pyr*, glandular cells of pyriform organ; *inf cor*, infracoronal cell; *int sac*, internal sac; *mes*, mesodermal covering; *musc*, muscle; *n*, nerve trunk; *oes*, oesophagus; *or ep*, oral epithelium; *or ves*, oral vesicle; *pall ep*, pallial epithelium; *pall gr*, pallial groove; *pl cil*, plume cilia; *s c*, sensory cap; *st*, stomach; *sup cor*, supracoronal cell; *vac*, vacuolated cells. (Redrawn from d'Hondt, 1972, 1974.)

sensory cap. The " ganglion " comprises a peripheral network of entangled cellular prolongations within which is a mass of cell bodies. The ciliated epithelial cells of the sensory cap are tall (25 μm) and tapered basally: they have synaptic junctions with the nerve cells in the centre of the ganglion. The nerve descends from this dorsal ganglion and trifurcates as it approaches the pyriform organ. The median ramus passes to the ciliated cells, where there is a plexus but no cell bodies; the lateral rami are deflected to form the equatorial nerve ring which lies against the coronal cells, bounded orally by the infracoronal ciliated cells. There are no " eye-spots ".

The musculature comprises: (1) two large fascicles of non-striated fibres that cross the larva from the pyriform organ. Postero-laterally the fibres are attached to desmosomes on the aboral epithelial cells of the pallial groove, while anteriorly they are inserted on desmosomes either on the infracoronal cells or on the ciliated cells of the pyriform organ. These muscles seem to correspond with those in a cyphonautes which extend from the gut to the pyriform organ (Kupelwieser, 1906; Atkins, 1955a, Fig. 5*a*). (2) Axial striated muscle, which in *Alcyonidium* is restricted to the region of the pyriform organ, originating from the infracoronal cells and being inserted on the pallial epithelium. This muscle appears to be important in compressing and distending the pyriform organ at the onset of metamorphosis. The myofibrils are not, in *A. polyoum*, associated with the nerve trunk. (3) A small, but very interesting muscle, which is positioned transversely across the apical organ. This is regarded, quite justifiably, by d'Hondt as the homologue of the adductor in shelled cyphonautes.

The larva of *Bugula neritina* (Fig. 28A) is ovoid, about 350 μm high and 250 μm in diameter (rather larger than most other *Bugula* larvae). The coronal cells, numbering about 230, cover almost the whole surface. They are long and slender, vertically aligned, externally ciliated and covered with microvilli. The nucleus of each cell is situated at the equator. Above the coronal cells and comprising the aboral pole is the disc-shaped apical organ delimited, as in *Alcyonidium*, by the pallial groove. In the centre of the apical organ is the sensory cap, which extends inwards as a neuromuscular cord linking it to the plume cilia of the pyriform organ, the " eye-spots " and the summit of the internal sac.

The pyriform organ is antero-equatorial, rather than antero-oral as in *Alcyonidium*. In descending order it consists of: a superior group of elongated glandular cells, the cells bearing the plume cilia, and the ciliated groove (an orally extending trough lying between two ciliated ridges, absent in *Alcyonidium*). Mawatari (1951), also describing the

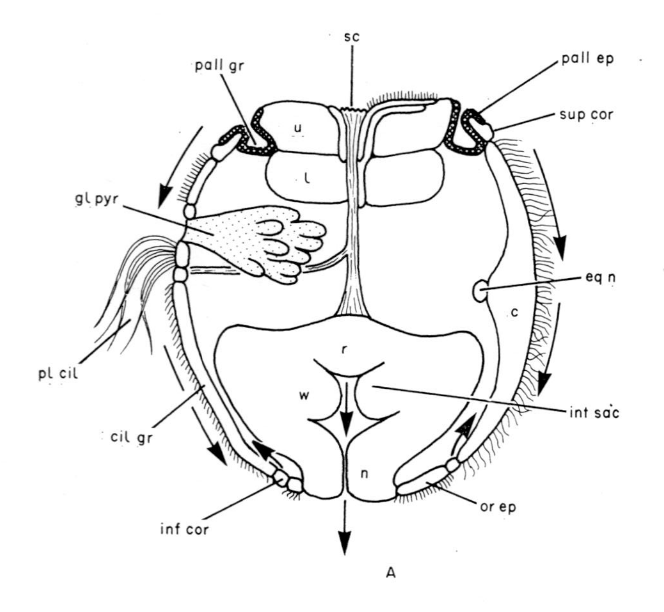
sc
pall gr
pall ep
sup cor
u
l
gl pyr
eq n
c
pl cil
r
w
int sac
cil gr
n
or ep
inf cor
A

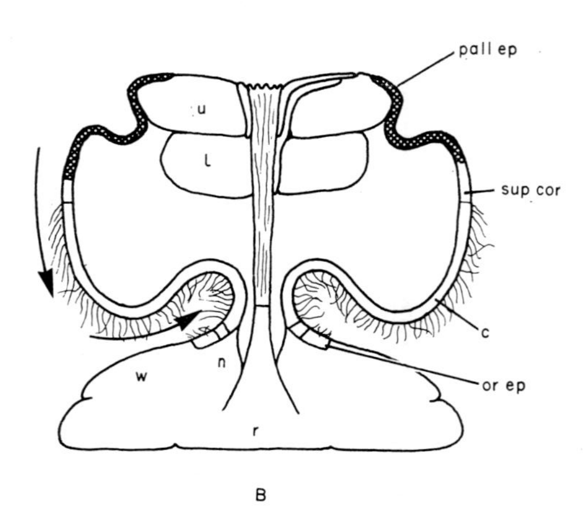
pall ep
u
l
sup cor
c
or ep
w
n
r
B

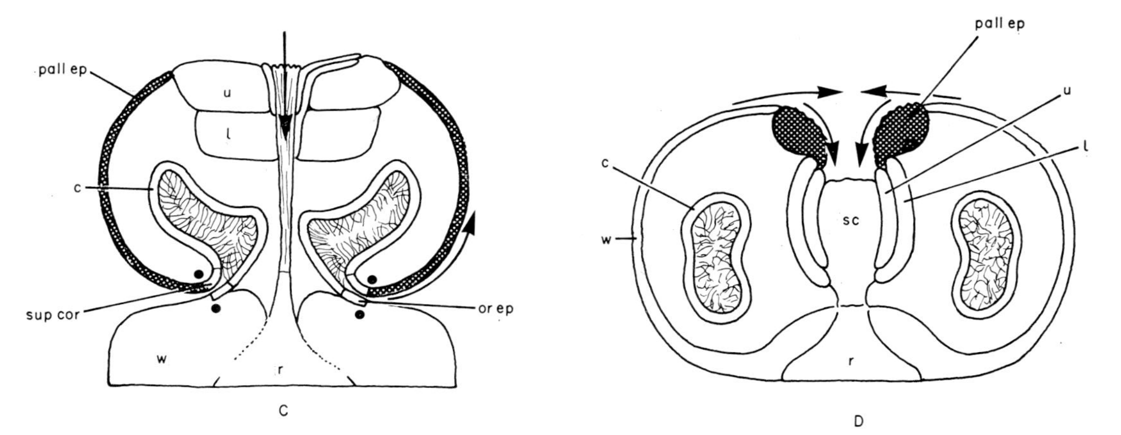

Fig. 28. Larva and metamorphosis of *Bugula neritina* (very diagrammatic). A, larva in sagittal section; B, larva in transverse vertical section at the onset of metamorphosis; C, larva in later metamorphosis before involution of the apical organ; D, pre-ancestrula. Arrows indicate direction of cell movements; paired dots in C mark position of cell layer partition and union of supra- and infra-coronal cells and of pallial epithelium and wall cells. *c*, coronal cell; *cil gr*, ciliated groove; *eq n*, equatorial nerve ring; *gl pyr*, glandular cells of pyriform organ; *inf cor*, infracoronal cell; *int sac*, internal sac; *l*, lower blastema cell; *n*, neck cells of internal sac; *or ep*, (ciliated) oral epithelium; *pall ep*, pallial epithelium; *pall gr*, pallial groove; *pl cil*, plume cilia; *r*, roof of internal sac; *s c*, sensory cap; *sup cor*, supracoronal cell; *u*, upper blastema cell; *w*, wall of internal sac. (A redrawn from Woollacott and Zimmer, 1971, B-D constructed from their descriptions and photographs.)

larva of *Bugula neritina*, showed only an inferior group of glandular cells and most bryozoan larvae have been described as having both superior and inferior groups of cells (Calvet, 1900). The internal sac is an extensive invagination from the oral surface. In sections the sac is seen to be differentiated into three regions: epithelial cells of the neck, filled with large inclusions; columnar epithelium of the wall or centre; and the roof, linked by the neuromuscular strand with the sensory cap.

Nerve fibres also form an equatorial ring, which lies adjacent to the nuclear region of the coronal cells and the base of the plume cilia. The partly ciliated annulus surrounding the sensory cap overlays what Woollacott and Zimmer (1972a) have named the upper and lower blastema cells. Situated between coronal cells, but not actually reaching the surface, are groups of pigment cells. Each of the two larger " eye spots " present in this larva has a small pit in its centre, from which a tuft of cilia projects. At its base this pit comes into close association with the equatorial nerve ring. The organization of these two " eye-spot " complexes and their approximation with the underlying nerve tract suggest a sensory role, which could be photoreception.

2. *Fine structure of the larval " eye-spots "*

Many brooded bryozoan larvae are provided with orange-red pigment spots: for example, *Bugula plumosa* has four, *B. simplex* eight and *B. flabellata* ten. Calvet (1900) described these spots in *B. simplex* (as *B. sabatieri*) as oval, red-vermilion, and supporting a cluster of fused cilia. In the *B. neritina* larva, which is rather differently pigmented: " On the side of the larva opposite the median furrow are two, prominent, black, diamond-shaped eye-spots lying almost on the equator and about 90° apart " (Lynch, 1947).

Woollacott and Zimmer (1972c) have examined the ultrastructure of the " eye-spots " of *B. neritina* (Fig. 29). The pigmented field measures 40–80 μm in length and has at its centre a pit 4–6 μm in depth and 1–2 μm in diameter. The walls of the pit are lined by an epidermis of modified coronal cells, from which arise the cilia which project as a tuft. The pit terminates with a single " basal sensory cell " bordered with microvilli and having a laterally displaced soma. The cilia of the refractile granule are derived from this basal sensory cell and are twisted into a tightly packed mass 1·0–1·5 μm in diameter. The arrangement is such that the ciliary shafts lie at right angles to the plane of incident light. These cilia possess the usual 9 + 2 construction of the axoneme. Subepidermal pigmented cells ensheath the pit and give rise to the underlying black rhombic field. The basal sensory cell abuts against the equatorial nerve ring, and Woollacott and Zimmer were

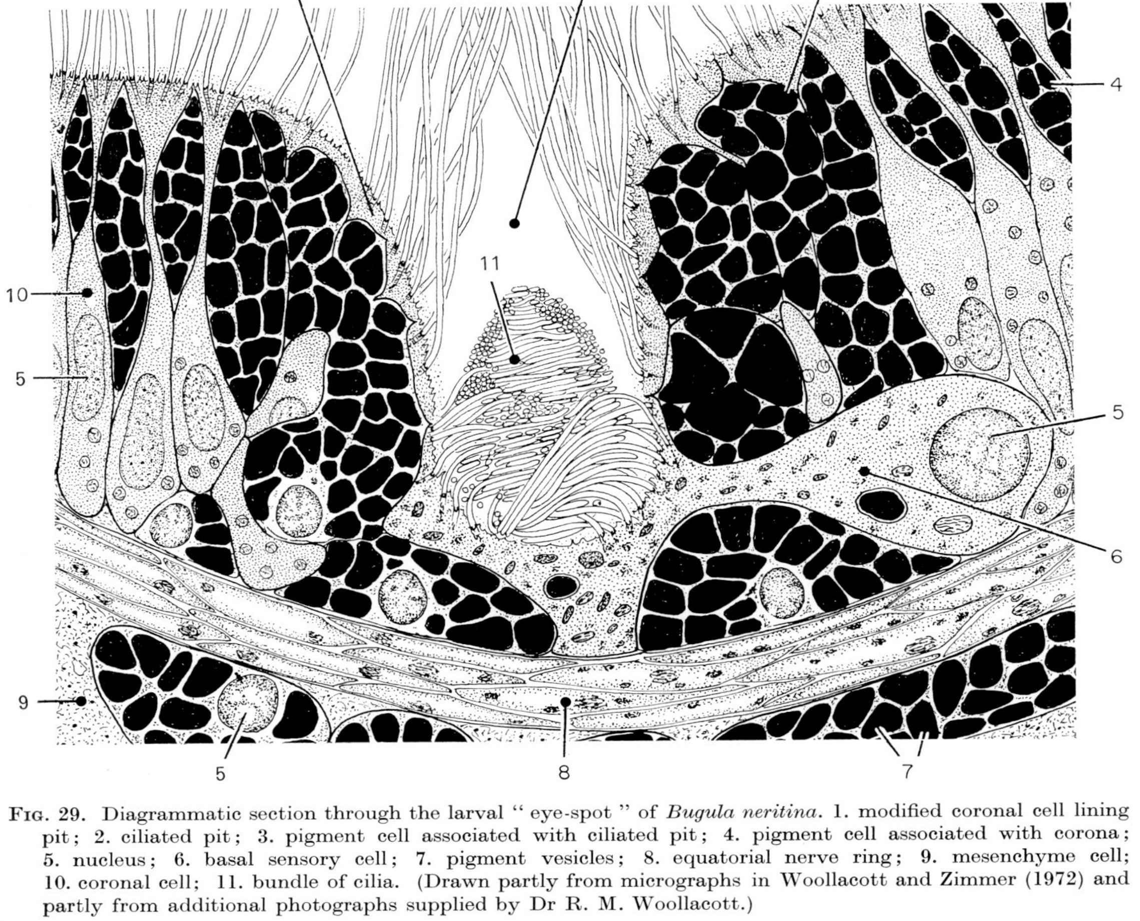

Fig. 29. Diagrammatic section through the larval " eye-spot " of *Bugula neritina*. 1. modified coronal cell lining pit; 2. ciliated pit; 3. pigment cell associated with ciliated pit; 4. pigment cell associated with corona; 5. nucleus; 6. basal sensory cell; 7. pigment vesicles; 8. equatorial nerve ring; 9. mesenchyme cell; 10. coronal cell; 11. bundle of cilia. (Drawn partly from micrographs in Woollacott and Zimmer (1972) and partly from additional photographs supplied by Dr R. M. Woollacott.)

able to find morphological evidence of gap junctions between the sensory cell and neuronal cells suggesting the presence of a pathway for electrotonic conduction. There is no evidence, however, that these cells are linked by chemical synapses.

The presumed photoreceptor is quite remarkable, apparently being of unique construction among invertebrate ocelli, although it does have the basic deuterostome combination of pigmented cells surrounding a cup containing ciliary shafts (Eakin, 1968). Usually, however, such ocelli have a photosensitive surface provided by the modification of membrane ensheathing the axonemes; such proliferations are absent from the pigmented spots of *B. neritina*, and Woollacott and Zimmer suggest that the photosensitive surface is provided by the unique arrangement of tightly packed, transversely orientated ciliary shafts. There is no physiological evidence that the larval " eye-spots " are photoreceptors, though this may be judged likely from what we know of their morphology. However, species such as *Bugula turbinata* and *Alcyonidium hirsutum* show orientation to light yet lack pigmented " eye-spots ". From Calvet's description of the *B. turbinata* larva it may perhaps possess comparable, but unpigmented, structures.

3. *Larval responses to light*

The stimulus triggering the discharge of brooded embryos appears always to be light, at least in the intertidal and shallow water species that have been the subject of investigation. This is true whether or not the larvae subsequently display phototactic responses. Ström (1969), however, believed that there is " probably a chemical factor also influencing hatching of larvae in brood-protecting bryozoans such as *Alcyonidium polyoum*. When one zooid has delivered its larvae, the larvae of others swarm out, too." The normal technique for obtaining larvae is simply to keep the colonies in the dark, transferring them to a lighted place when liberation is required.

Once released most brooded larvae are strongly photopositive but at settlement generally show a preference for surfaces shaded from the light. The experimental work on this subject has been reviewed elsewhere (Ryland, 1976b) and will only be summarized here. Exposed to directional illumination, larvae swim either toward or away from the light source, rotating by ciliary action as they go. Their response to directional illumination thus appears to depend upon movement of the receptors, which can thus make comparison of intensities at successive points in time: it is therefore a photoklinotaxis in the sense of Fraenkel and Gunn (1940). Ryland (1960) obtained four types of response: (a) Initally photopositive, becoming photonegative (the commonest type); (b) Initially photopositive, with a partial change to photonegative (e.g. *Alcyonidium hirsutum*). Perhaps rather than a part only of the population becoming permanently photonegative, the larvae develop a changing response in which positive and negative periods alternate; (c) Photopositive throughout (e.g., *Flustrellidra hispida*); (d) Indifferent to light; i.e., displaying no response (e.g. *Alcyonidium polyoum*). The large number of species displaying type (a) responses or known to be photopositive on release have been listed by Ryland (1976b). The transition from photopositive to photonegative appears to be induced by metabolic progress or change, its onset being influenced by temperature but not by light itself (Ryland, 1960, 1962b, 1976b).

The oft-recorded phenomenon of both brooded and planktotrophic larvae settling preferentially on lower or shaded surfaces (see Ryland, 1976b) has sometimes been attributed to skotopositive behaviour. Thus Grave (1930) stated that *Bugula simplex* (as *B. flabellata*) larvae in their photonegative phase " quickly turn toward a shadow ". Ryland (1960) thought that aggregation in shaded places was best explained in terms of high photokinesis; that is to say, activity is high in the light and low in the dark, so that larvae having arrived in a shaded place tend to stay there. It is also evident that not all of this behaviour can necessarily be explained in terms of responses to light. Ryland (1976b) has summarized the known effects of other stimuli and finally concludes that a re-investigation of the responses of free-swimming bryozoan larvae would be of value.

4. *Metamorphosis*

The attachment process and subsequent metamorphosis of *Bugula* larvae are fairly well known through the studies of Barrois (1877), Calvet (1900), Grave (1930), Lynch (1947) and Mawatari (1951) (with a review by Nielsen, 1971), but several details have recently been clarified by Woollacott and Zimmer (1971). Prospecting *B. neritina* larvae explore the surface with their plume cilia, while rotating counterclockwise about an axis perpendicular to the opening of the pyriform gland: the apical organ therefore faces horizontally. As with cyphonautes, secretion from the pyriform gland appears to hold the larva against the selected surface before it settles. At the moment of fixation the median groove grasps the substratum: then the adhesive sac is suddenly everted, spreading as a disc beneath the larva. The median groove releases its hold, and the apical organ returns to a more or less dorsal position. The larva then rotates about its perpendicular axis until the " eye-spots " are located on the lighted side. This orientation is accomplished after eversion of the internal sac (Lynch, 1947).

When the internal sac evaginates, it is its roof which creates the contact zone between the settling larva and the substratum. The wall region of the sac is not involved in the anchoring process and the cells of the neck region secrete a (presumably protective) coating which envelops the metamorphosing mass. There is concurrently a down-folding and involution of the surface coverings into the interior of the larva—the " umbrella " stage—which also results in the creation of an internal annular cavity (Fig. 28B, p. 370). As this infolding proceeds, the pallial groove epithelium spreads out as a sheet which progressively covers the metamorphosing larva (Fig. 28B, C). The reflected wall portion of the internal sac then unfolds, and is drawn externally towards the apical organ complex (Fig. 28C, D). Next, the apical organ, and the pallial groove epithelium with it, is withdrawn as an invagination until it is enclosed within the walls of the internal sac (Fig. 28D). As the pallial groove epithelium infolds, the walls of the internal sac are carried upwards, close apically, and differentiate into the epidermis of the ancestrular body wall. The digestive tract and lophophore develop from the now invaginated upper blastema; while some at least of the adult peritoneum develops from the lower blastema. The larval tissues are histolysed, and the annular space becomes merged in the developing coelom of the ancestrula (Woollacott and Zimmer, 1971). The polypide is soon fully formed and budding commences.

Metamorphosis in *Alcyonidium polyoum* has been re-investigated recently by d'Hondt (1974) and apparently differs in major respects from that described in other gymnolaemates. Attachment, as usual, is by eversion of the internal sac. Following this, the apical organ invaginates to form an aboral or upper vesicle (Fig. 27B, C, p. 368). This constitutes the presumptive tentacle sheath of the ancestrular polypide. A consequence of this invagination is that the pallial groove epithelium spreads to cover all the upper hemisphere of the pre-ancestrula. Peripheral involution takes place immediately above the extruded internal sac, now spread as a disc (Fig. 27C). The sequence is that the infracoronal cells precede, and the supracoronal cells follow, the corona into the lower hemisphere of the larva. This inwards migration causes the pallial groove epithelium to extend orally until it covers all but the attachment disc of the pre-ancestrula. The pallial groove cells thus form the epithelium of the ancestrula.

As the pre-ancestrula continues to flatten, the peripherally involuting cells close together as a second or oral vesicle. Closure is achieved by a centripetal movement of the infracoronal cells leading the involution and a similar movement of the supracoronal cells concluding it (Fig. 27C, D). The vesicle thus comprises a roof of infracoronal cells, sides of coronal cells (cilia on the luminal face), and a floor of supracoronal cells. However, all but the former infracoronal cells degenerate, while the latter spread, replacing first the sides and then the bottom of the oral vesicle (Fig. 27E): this is the presumptive polypide.

From d'Hondt's (1974) account it is evident that metamorphosis differs not only from that of *Bugula neritina* but from both gymnolaemates generally and stenolaemates (p. 365) in that the polypide forms largely from the orally involuted infracoronal cells instead of from apically invaginating or delaminating cells. There is also the clear difference from *Bugula* that the presumptive ancestrular epidermis is the pallial groove epithelium rather than the wall cells of the internal sac. It is obvious that more studies of this kind, tracing the cell movements of metamorphosis by transmission electron microscopy, are required before a satisfactory synthesis of metamorphic patterns in the various major taxa of Bryozoa will be possible.

In a brief comparison between the larvae of *Alcyonidium polyoum*, *Bowerbankia imbricata* and *Flustrellidra hispida* (the last being bivalved and cyphonautes-like), d'Hondt (1975) notes that the infracoronal cells—important in polypide formation in both *A. polyoum* and *B. imbricata*—partly degenerate during metamorphosis in *F. hispida* and do not contribute to the polypide. In this species it is a band of cells aboral to the pallial groove that gives rise to both the tentacle sheath and the ectodermal moiety of the polypide. In d'Hondt's view, the larval type of *Alcyonidium* is the most similar to the archetype: the larva of *Flustrellidra* can be derived by elongation of the supracoronal epithelia (which bear the shell valves), whereas that of *Bowerbankia* (as that of *Bugula*) displays elongation of the coronal cells. As a conclusion he proposes separation of *Alcyonidium* and allies from *Flustrellidra* and relatives (heretofore regarded as closely related) at the level of super family.

It seems generally to be true that encrusting species with a recumbent ancestrula have more or less dorsoventrally flattened larvae, e.g. *Alcyonidium*, *Cryptosula*, *Hippothoa*; while in those with an erect ancestrula, like *Bugula*, *Scrupocellaria* and *Amathia*, the larva is dorsally elongated. The ancestrula is usually a single zooid (cf. *Membranipora*, p. 362), but triple ancestrulae are formed in *Hippopodina feegeensis* (Eitan, 1972) and *Stylopoma duboisii* (Cook, 1973). The distal (and usually blastogenic) end of a single ancestrula must be that closest to the apical organ-internal sac complex, which means that, as during metamorphosis of cyphonautes, there is a reversal of polarity.

In *Bugula neritina* the settling larva orientates with its " eye-spots " towards the light and the pyriform gland away from it (Lynch, 1947). Since the ancestrula stands upright, its frontal surface should face the same way as the pyriform gland at the fulfilment of metamorphosis. Confirmation that this is so is found in the study by McDougall (1943), that: ". . . the first, and therefore all subsequent zooids, face away from the source of illumination." McDougall found that colonies on vertical surfaces faced downwards, and he was able to induce the reverse orientation in an experimental container lit only from below. There seem to be no observations on other species.

VII. Settlement, Distribution Patterns and Competition

A. *The location and choice of substrata*

1. *Availability of support*

The relationship between benthic animals and the substratum in or on which they live is one of fundamental importance. There is increasing evidence that larval settlement is far from haphazard, and that larvae search for and metamorphose after finding a suitable biotope. This is not to imply that habitat selection is necessarily the overriding factor in explaining the distribution of bryozoans, but rather that " given suitable opportunity (phenological and hydrographical), behavioural selection in conjunction with habitat availability on a scale related to the size of the organism, initially determines distributtion within the arena of the individual's power of effective independent locomotion " (Moore, 1975).

Nearly all bryozoans require a firm support for their settlement and continued existence—exceptions include the sand-faunal species with discoid or conical colonies (reviewed by Ryland, 1967a), the meiofaunal species of *Monobryozoon* (Franzén, 1960; Gray, 1971), and those like *Kinetoskias* spp. and certain deep water ctenostomes (d'Hondt, 1976) which can anchor themselves in fine sediments. Some bryozoans are highly stenotopic. Three examples mentioned by Eggleston (1972b) are *Triticella korenii* which is epizoic on burrowing crustaceans such as *Calocaris macandreae* Bell, *Goneplax rhomboides* (L.) and *Nephrops norvegicus* (L.); *Arachnidium hippothoides*, found only on *Ascidiella* tests; and *Hippoporida edax*, which off the Isle of Man occurs only on shells inhabited by the hermit crab *Pagurus cuanensis* Bell. In *Triticella* synchronization of the breeding season with the moult of the crustacean host is important (Ström, 1969; Eggleston, 1971). Cook (1968) discussed the association of bryozoans and hermit crabs, mainly occupying *Turritella* shells, in the Gulf of Guinea. She categorized the association as ranging from (1) probably fortuitous, to (2) those simply attaching to shells, dead or alive, to (3) commensals of gastropods or pagurids (e.g. *Membranipora commensale*), concluding with (4) the obligate pagurid commensals of the genus *Hippoporidra*.

The importance of " opportunity " is well documented in the studies by Cook (1968) in the Gulf of Guinea, and Eggleston (1972b) off the south of the Isle of Man. In the latter area even the more eurytopic species are dependent on the distribution of stones, shells and hydroids in what, from a bryozoan " viewpoint ", is an uninhabitable sandy waste. As Eggleston succinctly stated: "Abundance of support is the main factor controlling the distribution of ectoprocts in Manx waters ". In the sands of the Gulf of Guinea the available substrata are: sand grains and small foraminiferans; the large colonial foraminiferan *Jullienella foetida* Schlumberger; hydroids, which can apparently colonize the sea bed directly; and gastropod shells, mainly of *Turritella* spp. Since live *Turritella* burrow in the sand, their shells would not be available for colonization prior to the death of the animal; if most dead shells have been occupied by hermit crabs, it is easy to comprehend how bryozoan-pagurid commensalism could have arisen in an area like this by chance. Cook made the same point in regard to *Jullienella*, the arenaceous test of which is so important as a bryozoan support. Some species encrust it exclusively, but *J. foetida* is not necessarily the preferred substratum for such bryozoans, it is simply the only one available.

The abundance and distribution of the main support types off the Isle of Man are related to the pattern and strength of the water movements. As most support is available where the current is strong, the abundance of bryozoans becomes positively correlated with current strength. The more abundant the support, the more abundant the bryozoans (per unit area of suitable support, not merely per unit area of sea bed) in both species and number (Fig. 30). (It would seem not unreasonable to suppose, however, that wastage—as represented by that proportion of the larval output that is carried away to the muddier areas—is considerable.) Table X compares the number of colonies per 100 cm^2 of shell collected from coarse sand and muddy sand respectively.

It is important to remember that most bryozoans have brooded embryos and short-lived larvae. Delayed metamorphosis, of the type familiar in the planktotrophic larvae of benthic animals, can hardly occur. As Eggleston emphasizes, the distances travelled by the larvae

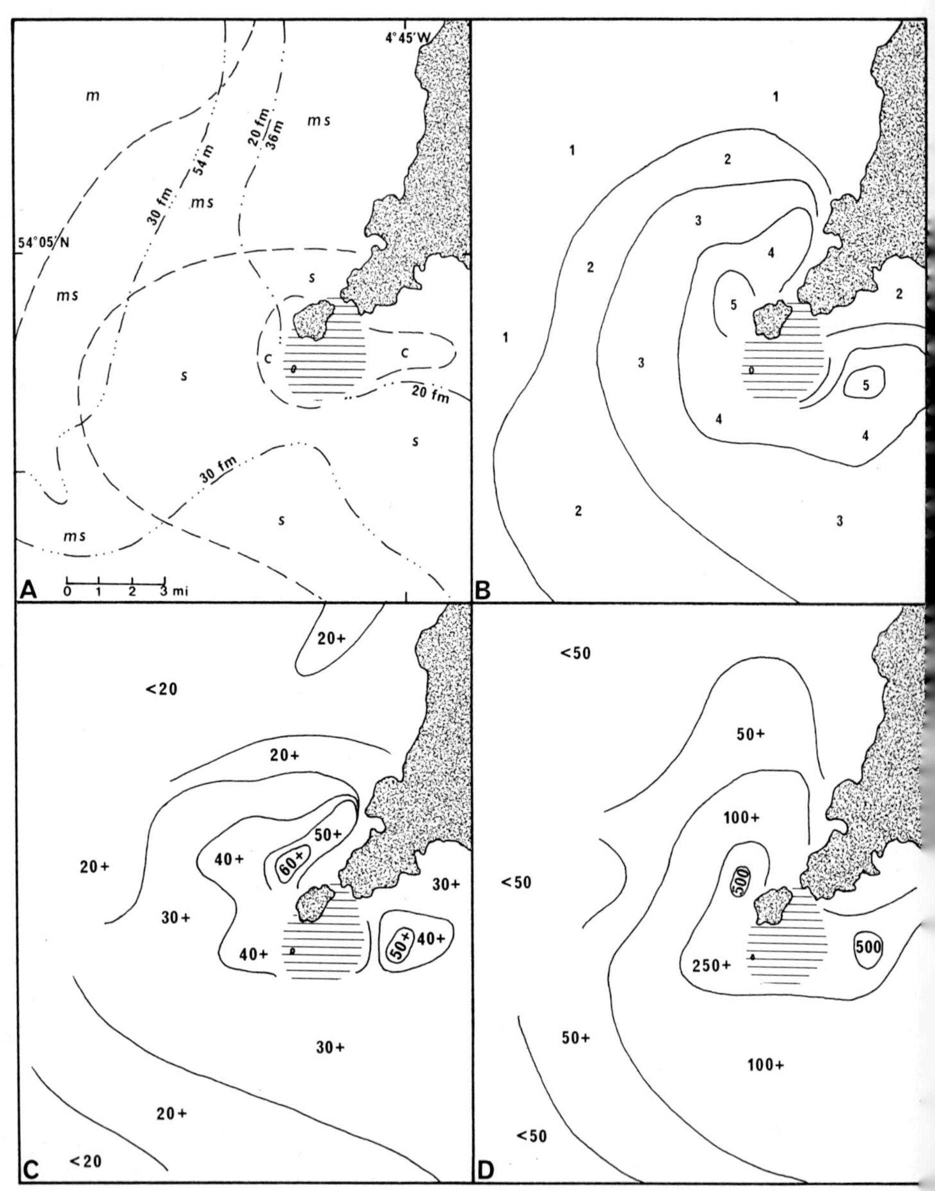

FIG. 30. Distribution and abundance of bryozoans off the south-western tip of the Isle of Man, showing their dependence on the availability of suitable support. A, survey area showing isobaths and type of bottom (*m*, mud; *ms*, muddy sand; *s*, sand; *c*, coarse). The hatched area is too rough to sample by dredging. B, abundance of suitable bryozoan support (arbitrary scale 1→5). C, number of bryozoan species recorded. D, number of colonies per 1 000 cm² of support. (Redrawn after Eggleston, *Journal of Natural History*, **6**, 1972.)

TABLE X. NUMBER OF BRYOZOAN COLONIES PRESENT ON SHELLS FROM MUDDY SAND GROUNDS (WHERE SHELL IS SCARCE) AND COARSE SAND GROUNDS (WHERE SHELL IS ABUNDANT), EXPRESSED AS COLONIES 0·1 m^{-2} (Eggleston, 1972)

Species	*Muddy sand*	*Coarse sand*
Alderina imbellis	2	—
Cellaria sinuosa	1	—
Chorizopora brongniartii	—	13
Diastopora suborbicularis	3	11
Disporella hispida	9	—
Electra pilosa	—	11
Escharella immersa	4	52
E. ventricosa	2	—
Fenestrulina malusii	—	124
Hippoporina pertusa	1	—
Microporella ciliata	—	110
Parasmittina trispinosa	—	14
Porella concinna	4	14
Schizomavella auriculata	10	12
Scrupocellaria scruposa	—	11
Tubulipora spp.	2	—
Others	11	120
Total	49	492

in search of support must be rather small; when suitable substrata become too scarce, larvae will be unable to find them during their free life (see also the discussion on yolk reserves and energetics in Crisp, 1976). I have drawn attention elsewhere (Ryland, 1967a) to the fact that significant settlement is found only near to a reservoir of breeding colonies.

2. *Selective settlement*

Notwithstanding the consideration of availability, it remains true that the larvae of many non-related invertebrate groups are highly selective at settlement, and that their behaviour at this time is remarkably similar (reviews by Meadows and Campbell, 1972 and Crisp, 1974). So far as bryozoans are concerned, most attention has so far been paid to the species dwelling on algae. These range from the eurytopic to the highly specific in their choice of algae (Prenant and Teissier, 1924; Prenant, 1927; Rogick and Croasdale, 1949; Gautier, 1962; Ryland, 1959b, 1962a, 1974a; Pinter, 1969; Woollacott and North, 1971; Moyano and Bustos, 1974; with additional observations scattered through a vast literature). Withers *et al.* (1975) have recently recorded nine bryozoan species on *Sargassum muticum* (Yendo) Fensholt, an immigrant to British waters. Other species are associated with the Mediterranean sea grass *Posidonia oceanica* (L.) Delile (Gautier, 1954, 1962; Kerneis, 1960; Harmelin, 1973b; Hayward, 1976).

If the combination of physical factors and appropriate behaviour brings the free-swimming larvae into (or retains them in) the proximity of their appropriate biotope, the physico-chemical structure of the investigated surfaces, and their established biota, will largely influence subsequent behaviour. This may proceed through the phases of "broad exploration", "close exploration" and detailed "inspection", culminating with fixation and metamorphosis (Crisp, 1974).

The nature of the surface is probably of prime importance, and many bryozoans prefer a smooth or even a glossy finish. Several experiments have been carried out using Tufnol (a paper laminate impregnated with phenol formaldehyde resin), Perspex (acrylic plastic) or glass, comparing the surface as manufactured with that in which the original glaze has been abraded to a matt finish with fine carborundum powder. *Bugula neritina*, *Electra pilosa* and *Hippothoa hyalina* all show statistically significant preferences for the glazed finish (Ryland, 1976a). There was some indication with *B. neritina* that a heavy bacterial film made the matt surface somewhat more attractive. Experiments with *Bugula flabellata* have also demonstrated a clear preference for smooth surfaces (Crisp and Ryland, 1960). Wisely (1962) found that larvae of two species of *Bugula* settled readily even on surfaces coated with polytetrafluorethylene, which has a particularly low coefficient of friction. Only one experiment of those discussed by Ryland showed a contrary result: *Alcyonidium polyoum* settled preferentially on matt Perspex, but the panels had been treated with *Fucus serratus* extract (see below) which had perhaps been better adsorbed onto the slightly roughened surface. Eggleston (1972b) found that a majority of bryozoans are commoner on the inner, smooth surface of *Chlamys opercularis* (L.) and *Pecten maximus* (L.) shells, although species such as *Turbicellepora avicularis* and *Aetea sica* prefer the outer, rough surface.

The moulding or contour of the surface is also important. Many bryozoans settle in depressions on a crinkly algal frond, e.g. *Hippothoa* species on *Laminaria saccharina* (L.) Lamour. and *Macrocystis pyrifera* (L.) C. A. Ag. (Ryland, 1959b; Ryland and Gordon, 1976), or in the grooves beside a midrib, e.g. *Alcyonidium hirsutum* on *Fucus serratus* (Hayward, 1973). Ryland (1959b) showed that the larvae of four bryozoans settled overwhelmingly (227 : 30) in the channelled side of *Pelvetia canaliculata* (L.) Dcne. and Thurs. and *Gigartina stellata* (Stackh.) Batt. fronds. This is not, however, in any way an absolute rule. Ancestrulae of *Membranipora tuberculata* are found as commonly (per unit area) on the bladders of *Sargassum natans* as on the leaflets (Ryland, 1974a), and Morton and Miller (1968), describing the epibionts of *Sargassum sinclairii*, observed that *Lichenopora novaezelandiae* is generally found fitting over the top of a bladder. Such species as *Bicellariella ciliata* and *Turbicellepora avicularis*, which are often found on hydroids, seem to prefer convex surfaces (Eggleston, 1972b).

The association of particular bryozoans with specific algae has already been mentioned. It should be noted, however, that clear-cut preferences in one geographical region may be replaced by others elsewhere. *Cryptosula pallasiana* is abundant on intertidal *Fucus vesiculosus* L. at Espegrend, western Norway (Ryland, 1962b) but it is a rock encrusting species in Wales; *Haplopoma graniferum* inhabits stones and rock surfaces around the British Isles (Ryland, 1963a) but *Posidonia* in the Mediterranean and Aegean (Harmelin, 1973b; Hayward, 1976). *Haplopoma impressum*, found characteristically on red algae in British waters (Ryland, 1963a), occurs on *Posidonia* blades in the Mediterranean and Aegean seas (Gautier, 1962; Harmelin, 1973b; Hayward, 1974). That the mechanism underlying such preferences is active selection was demonstrated by Ryland (1959b), who showed that larvae of four intertidal bryozoans settled preferentially on exactly the algae on which they are most commonly found on the shore. Thus larvae of *Alcyonidium hirsutum*, *A. polyoum* and *Flustrellidra hispida* all settled most heavily on *Fucus serratus*, their normal algal host, when offered a choice of eleven species. A significant development from this work was the demonstration that the inert and unfavourable surface of Tufnol could be made attractive to *A. polyoum* larvae by filming with *Fucus* extract (Crisp and Williams, 1960).

Ryland's (1959b) experiments also established that not all parts of an algal frond are necessarily equally attractive to settling larvae. *Alcyonidium polyoum* larvae preferred the frond tips, and settlement declined with increasing age of the frond; *Flustrellidra hispida* larvae appeared to be less particular, though they avoided the frond bases. This is in accord with a recent study which shows that, on *Fucus serratus* plants, *Flustrellidra* colonies reach their greatest abundance nearer the holdfast than *Alcyonidium* colonies (Boaden *et al.*, 1975). Stebbing (1972) found that the density of settlement of *Scrupocellaria reptans* larvae on discs cut from the lamina of *Laminaria digitata* (Huds.) Lamour. followed the age gradient of the alga, with the youngest part being most favourable. It is also evident from observation that *S. reptans* settles also on the youngest parts of the frondose bryozoon *Flustra foliacea* (Stebbing, 1971b).

Hayward and Harvey (1974a) have analysed the distribution of *Alcyonidium hirsutum* ancestrulae settled on *Fucus serratus* fronds. Although they confirmed Ryland's (1959b) earlier finding, for choice experiments with *A. polyoum*, that not all regions of the frond were equally attractive, their results differed in detail. They considered both the components of the frond (midrib, midrib groove and frond flat) and region in relation to age. Settlement on the rib was low throughout; elsewhere there were marked differences that related to the age of the frond. Density was lowest on the basal regions, increased to a peak in the middle regions and declined again (contrary to the results obtained in Ryland's choice experiments) towards the growing tips. This is in agreement with the study of Boaden *et al.* (1975), which showed that *A. hirsutum* most successfully colonizes the central regions of a *Fucus serratus* plant. Hayward and Harvey remark that the high density area is generally in the middle of the main vegetative axis of the plant—a distribution of notable adaptive significance in that the *Fucus* plants tend to lose foliage by basal erosion together with peripheral denudation of tips that have fruited.

Cyphonautes of *Membranipora tuberculata* settling on *Sargassum natans* evidently also eschew the young thallus (Ryland, 1974a). In *S. muticum*, Withers *et al.* (1975) note that " epibionts were absent not only from the growing tips but also for a considerable length of frond proximal to the tip ". The absence appeared greater than the rapid growth rate of *S. muticum* alone could account for. Selectivity for frond age, therefore, is widespread; and it is not always the youngest growth which is chosen. In an analogous situation, Moore (1973) noted the faunal paucity on the youngest whorls of haptera in *Laminaria hyperborea* holdfasts.

Micro-organisms, and the coating they produce on submerged objects, are so important to settling larvae that the view seems widespread that a bacterial film is indispensible for larval attachment. In bryozoans, at least, this is not true. Scheer (1945) found that bacterial coating favoured the settlement of hydroids but not of bryozoans. Larvae of *Hippothoa hyalina* settle readily on unfilmed surfaces (Ryland, 1959a), as do those of *Bugula neritina* (Miller, 1946), although a 24 h film does appear advantageous in this species (Ryland, 1976a). Particularly striking, however, was the 9 : 1 preference for unfilmed over filmed Tufnol shown by *Bugula flabellata* larvae (Crisp and Ryland, 1960). It may well be, as Stebbing (1972) proposes, that it is the microbial surface flora which determines the zone of favourability within the age gradient of algal fronds. Certainly, the absence of settlement near the thallus tips of both *Sargassum natans* and *Fucus vesiculosus* (not, in the latter species, of bryozoans) apparently is correlated with the presence of an age dependent tannin with antibiotic properties (Conover and Sieburth: several papers quoted by Harvey and Hayward, 1974a, and Ryland, 1974a), and Ghelardi (unpublished; quoted by Moore, 1973) suggested that *Macrocystis pyrifera* might secrete bacterial inhibitors.

In another series of experiments it has been shown that the surface of certain British algae, such as *Chondrus crispus* (L.) Stackh. and *Laminaria digitata*, possesses outstanding antibacterial properties (Hornsey and Hide, 1974). Moreover, maximum activity may be associated with either the youngest tissue, as in *C. crispus*, or the oldest, as in *Laminaria saccharina* (Hornsey and Hide, 1976). In *L. digitata* activity decreased in the lamina just beyond the meristem but then steadily increased towards the free end of the plant. This is possibly significant in relation to Stebbing's (1972) experiments mentioned above. Little antimicrobial activity was reported by Hornsey and Hide (1974) in *Fucus serratus*. This work seems clearly to indicate the potentiality of comparisons between the antibacterial properties of algae with the pattern of their colonization by bryozoan and other epibionts.

B. *Aggregation*

1. *Major habitats*

Bryozoan colonies are commonly found in aggregations. At a macro level this implies high overall abundance and/or diversity in habitats particularly suited to filter feeders; at a micro level it reflects the very precise selection of habitat referred to in the previous section.

Most studies on bryozoan ecology still deal with intertidal habitats (exceptions include the work of Eggleston off the Isle of Man (p. 379), of Stebbing on *Flustra foliacea* communities (pp. 300 and 409) and, particularly, the effective use of SCUBA by Dr J. G. Harmelin in his work on Mediterranean cyclostomes (1974 and p. 305). That particular shores are remarkable for their concentrations of bryozoans is well known, but it is less easy to identify precisely those ecological factors responsible for the favourability. One remarkable shore is the Echinoderm Reef Flat, close to the Leigh Marine Laboratory, Auckland, New Zealand. The reef comprises an almost level, boulder-strewn terrace exposed only at spring tides. The rock is sparsely covered with algae, mainly corallines, and a lithothamnion " paint " spreads down the sides onto the lower surface of the boulders: it is here that the bryozoans abound. Ryland (1975) suggested three factors with which the faunal richness of the shore is probably correlated. (1) The degree of exposure to wave

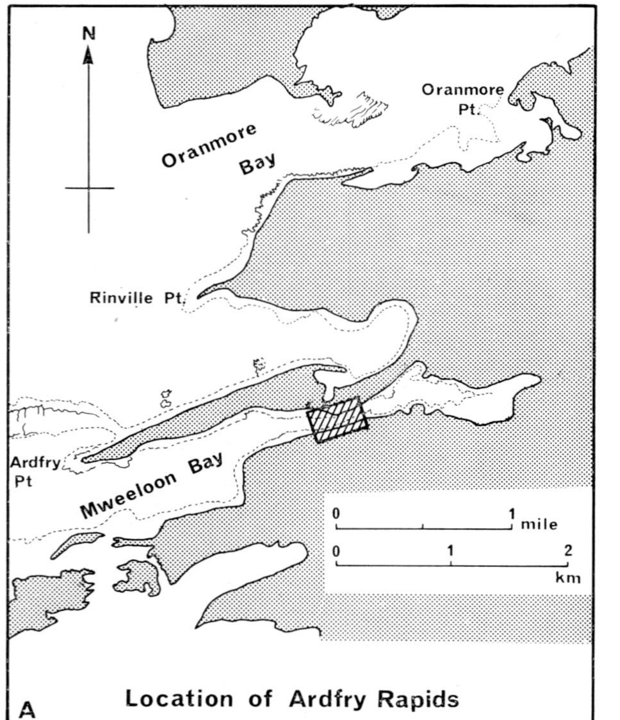

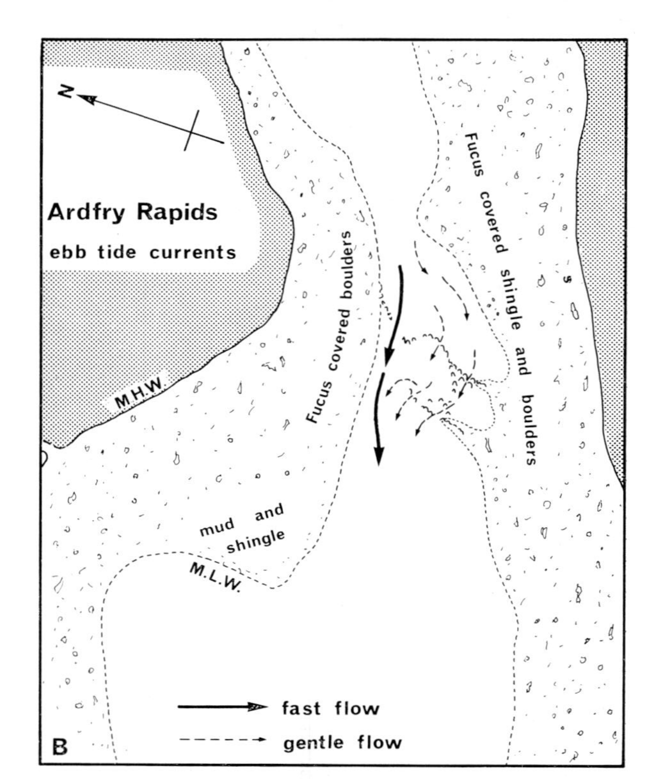

FIG. 31. The Ardfry rapids in Galway Bay: location and ebb tide currents. (From Ryland and Nelson-Smith, *Proceedings of the Royal Irish Academy*, B **75**, 1975.)

action, which is apparently optimal. Destructive waves are rare but water movement is sufficient to keep the shore free of sediment. (2) The structure and configuration of the bedrock, which produces flat boulders of moderate size, stable in a breaking swell. These tend to be trapped in runnels and shallows so that their lower surface does not lie in silt. (3) The clarity of the water, in contrast to the turbid conditions elsewhere in the Auckland area.

The importance of sea-water turbidity as a factor affecting the abundance of bryozoans has been assessed by Moore (1973), as part of a larger study on the fauna of *Laminaria hyperborea* holdfasts on the north-east coast of England. The southern part of this coast (Yorkshire, Durham and south Northumberland) suffers high coastal erosion, with the consequence that the water is turbid; in contrast, the water off the northern part (north Northumberland and south Berwickshire) is clear. The highest bryozoan diversity is associated with the northern region, with species such as *Celleporaria pumicosa*, *Escharina spinifera*, *Escharoides coccinea* and *Membraniporella nitida* being restricted to this part of the coast. Other species, including *Alcyonidium* spp., *Callopora lineata*, *Electra pilosa*, *Hippothoa hyalina* and *Umbonula littoralis*, however, are more widely distributed. *Alcyonidium* "seems to flourish in the turbid water and was the dominant bryozoan on kelp holdfasts in this [southern] area".

In parts of the Menai Strait in North Wales, in contrast to the Reef Flat at Leigh, the shores are covered with fucoids and the under-boulder habitat is rendered uninhabitable by mud. Despite the high turbidity, a relatively small number of species (*Alcyonidium* spp., *Bowerbankia imbricata*, *Flustrellidra hispida* and *Hippothoa hyalina*) reach great abundance, while several others are relatively common. The essential factors here seem to be (1) absence of wave action, and (2) high mass transport of water through the Strait ensuring a rich supply of food. The few species that can tolerate the turbidity obviously flourish.

It is the combination of these same two characteristics that makes "rapids" systems so suitable for bryozoans, as was demonstrated by the group studying Lough Ine, in southern Ireland (see Kitching and Ebling, 1967). More recently two smaller rapids in Ireland have been examined, in a fairly preliminary way, by a Swansea University group (Ryland and Nelson-Smith, 1975). One of these localities, the Ardfry rapids at the head of Galway Bay (Fig. 31) provides an instructive comparison with Lough Ine.

In both localities water funnels through a strait, narrowest at a sill in its middle region. Discharging water flows turbulently over the sill,

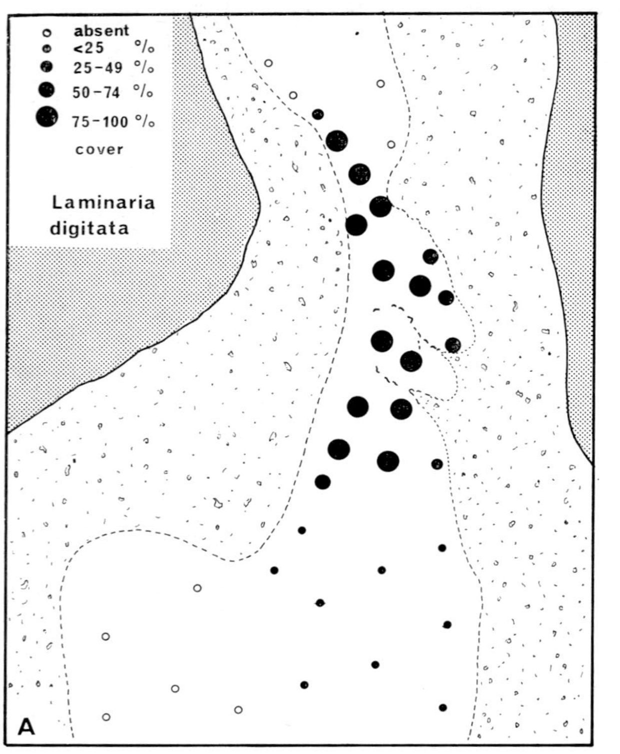

absent
<25 %
25-49 %
50-74 %
75-100 %
cover
Laminaria digitata
A

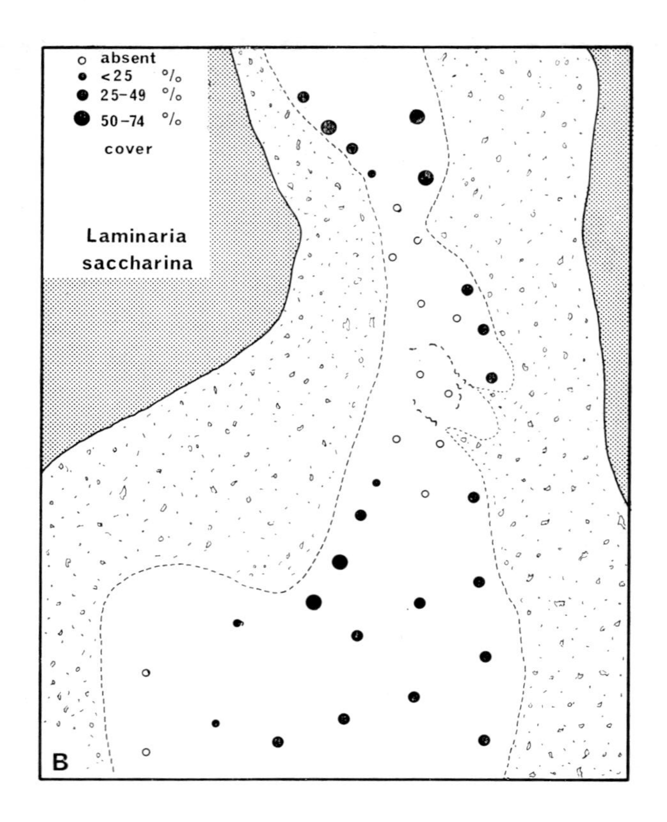

absent
<25 %
25-49 %
50-74 %
cover
Laminaria saccharina
B

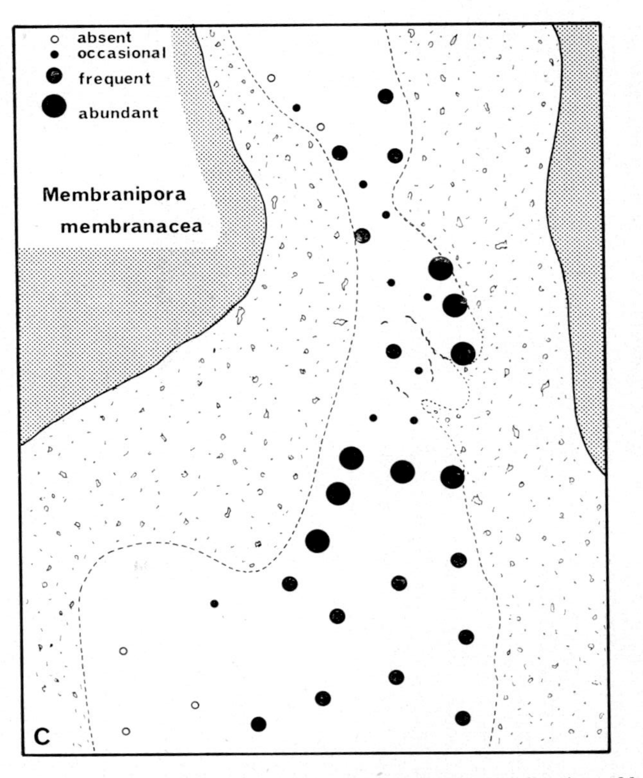

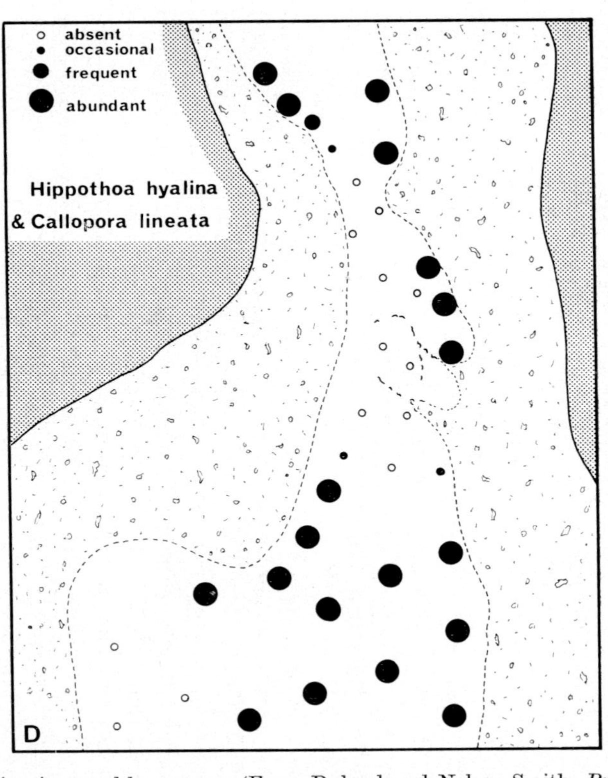

FIG. 32. The Ardfry rapids in Galway Bay: distribution of laminarians and bryozoans. (From Ryland and Nelson-Smith, *Proceedings of the Royal Irish Academy*, B **75**, 1975.)

reaching 2·5 m s^{-1} or more in the L. Ine rapids (Bassindale *et al.*, 1948). The dominant large alga is *Saccorhiza polyschides* in the L. Ine rapids but *Laminaria digitata* at Ardfry. *Membranipora membranacea* colonizes the fronds of both algae. At both localities it is virtually absent from the turbulent central section, but reaches great abundance towards each end of the rapids, where, at L. Ine, the currents are in the order of 1·0–1·5 m s^{-1} (Kitching and Ebling, 1967, Figs. 8 and 19). As current velocity drops still further and, at Ardfry, *Laminaria saccharina* replaces *L. digitata* (Fig. 32A, B), the bryozoan fauna also changes. Sheets of *Membranipora* (Fig. 32C, D) are replaced by the clustered scabby colonies of *Hippothoa hyalina* (and *Callopora lineata* at Ardfry), and the current is " slight " (Ebling *et al.*, 1948; Ryland and Nelson-Smith, 1975). The ways in which the current could affect the biota were discussed by Ebling *et al.* (1948) and Lilly *et al.* (1953). The weak current supplies abundant oxygen and food but permits some sedimentation—apparently enough to eliminate *Membranipora membranacea*. The stronger current may interfere with both settlement and feeding, and the observations suggest that the subspherical larvae of *Hippothoa* and *Callopora* are far less successful in attaching in strongly flowing water than are the cyphonautes of *Membranipora*.

A second rapids system, located at Cashla Bay on the northern side of Galway Bay, was also described by Ryland and Nelson-Smith. Here, where the water is cleaner and more oceanic, the bryozoan fauna is more varied than at Ardfry. Scattered clumps of *Laminaria digitata* and *L. saccharina* grow in the main stream and in the shallows, their fronds well covered with tufts of *Scrupocellaria reptans* together with encrusting species such as *Hippothoa hyalina, Callopora lineata and Microporella ciliata*. *Scrupocellaria reptans*, although bushy in growth habit, is well adapted to survive in flowing water, for its frequently divided branches are anchored and held close to the frond by numerous rhizoids (cf. Figs. 8 and 23 in Kitching and Ebling, 1967). Some studies on this fauna are mentioned later.

2. *Microhabitats*

The importance of support availability for bryozoans on sublittoral sands has already been mentioned. When associated with hard substrata, as on rocky shores and coral reefs, the restricted availability of surfaces suitable for colonization by bryozoans again results in the formation of aggregations. Of great importance is the settlement behaviour of larvae which results in their attaching to downward-facing surfaces (full discussions in Ryland, 1976a, b). Such local habitats, intertidally, are protected from insolation and desiccation, while subtidally they will not accumulate a smothering deposit of sediment or a mat of competing algae. Thus lithophilic bryozoans can be found concentrated on the roofs of caves (Riedl, 1966; Harmelin, 1969a; Norton *et al.*, 1971), below overhangs and on the underside of boulders. The favourability of the last depends on the extent to which water can circulate freely past the surface, so that there is no trap for silt and organic debris (Morton and Chapman, 1968; Ryland, 1975). In the western Mediterranean (Banyuls-sur-Mer), bryozoans are essentially restricted to overhangs in the superficial 5 m, additionally colonize north-facing vertical walls in the range 5–20 m, and are found on surfaces of any orientation at depths of 20–40 m (Médioni, 1972; Fiala-Médioni, 1974).

Intertidal areas of coral reef, as characterize Australia's Great Barrier Reef, offer essentially the same under boulder microhabitats as do rocky sea shores. Ryland (1974b) mentioned the high species diversity on coral boulders at Low Isles and Green Island. As might be expected, boulders on the sediment-rich reef flat support no bryozoans, but they become common towards the edge of the reef at low water. Boulders derived from *Acropora hyacinthus* (Dana), now upturned and with a concave lower face, offer a sheltered silt-free habitat for delicate branching species such as *Reteporella graeffei* as well as more robust clumps of *Margaretta triplex* and numerous encrusting species. Sublittorally, bryozoans on the reef occur in cavernous areas (Ross, 1974) and on the lower surface of plate corals such as *Montipora*. Soule and Soule (1974) report nearly 200 species on corals at Hawaii. In the West Indies the bryozoans colonize dead *Millepora* and *Acropora palmata* (Lamarck) rubble on the outer reef slope and are uncommon in the back-reef zone (Ryland, 1974b; Schopf, 1974). Here it is again the underside of corals, such as *Agaricia*, and *Diploria* in Bermuda (Cuffey, 1973), which attract most bryozoans. Thus on rocky shores and in coral reefs localized areas of hard substratum become densely covered with bryozoan colonies (and other encrusters in greater or less amounts, of course) in what is presumably a very competitive situation (see Jackson and Buss, 1975).

Even on a relatively homogeneous surface, such as a rock face or a kelp frond, there are really many microhabitats. Below a boulder there will be a gradient of light, certainly influencing distribution (Gordon, 1972), while the rock surface itself is irregular, with pits, crevices and hollows affecting the behaviour of settling larvae. Study of microdistribution patterns on boulders presents formidable practical difficulties and algal fronds offer a more satisfactory alternative.

Ryland (1972) collected *Laminaria digitata* fronds from the Cashla

Bay rapids (mentioned above) and analysed the occurrence of species in each cm^2 of 10 cm × 10 cm quadrats laid on the frond. The dispersions on the frond were non-random, although (the well-known weakness of discrete sampling methods) the degree of non-randomness demonstrated varied with quadrat size. The pattern found was dominated by local clumping, with the colonies occupying and exploiting the most favourable parts of the frond (probably in terms of exposure to water flow).

Aggregations, irrespective of scale, involve mainly larvae with brooded development and limited powers of dispersal. The demonstration in the same paper (Ryland, 1972) that colonies of *Electra pilosa*, which has a cyphonautes, were aggregated on *Fucus serratus*, however, cannot in any way reflect restricted dispersal from the breeding stock. In this study the *F. serratus* fronds in a random sample were treated as " quadrats " of variable area. The method, based on computer simulation, permitted comparison of the observed frequency distribution with a Poisson series (Fig. 33), and clearly established that the colonies were clumped on particular fronds. It did not, of course, resolve the mechanism underlying the aggregation.

In acorn barnacles and spirorbid tubeworms " gregariousness ", in which searching larvae are stimulated to settle by contact with adults of the same species, plays a major role in the development of aggregations (see Knight-Jones and Moyse, 1961, and Crisp, 1974, for reviews). Gregariousness, however, has been rarely reported in bryozoans, and for some years *Watersipora arcuata* provided the only definite example (Wisely, 1958, as *W. cucullata*). Ryland (1976a) demonstrated gregariousness in *Bugula neritina* and *Hippothoa hyalina* where the density of individuals in the offered settlement did not exceed $2{\cdot}5\ cm^{-2}$. At higher densities there was no gregariousness in these species, nor was any demonstrable in *Alcyonidium hirsutum* and *Flustrellidra hispida*. When *Alcyonidium polyoum* larvae were offered a plane and uniformly abraided Perspex surface filmed with *Fucus* extract, the resulting settlement was usually aggregated, at least for resultant densities up to about $12\ cm^{-2}$ (Harvey *et al.*, 1976). It is always possible, however, that the formation of clumps reflects heterogeneities in the adsorbed film: it cannot be taken as definite evidence for gregariousness.

Harmelin (1974) discussed aggregation in cyclostomatous bryozoans. For example, large colonies of *Plagioecia patina* on the inner face of *Pinna* shells may be surrounded for a distance of 2–3 cm by numerous small colonies. He believes that such clumping is a manifestation of limited dispersal rather than of gregariousness. This, in turn, implies an

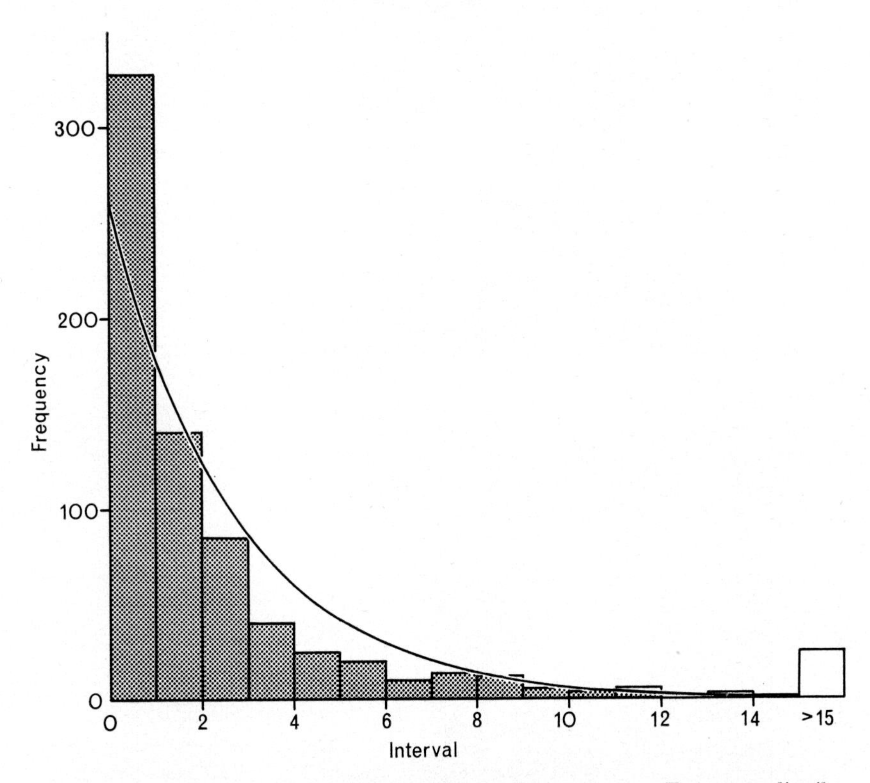

Fig. 33. Aggregation of *Electra pilosa* colonies on *Fucus serratus*. Frequency distribution of intervallic distances in a linear computer simulation of the dispersal of *E. pilosa* on *F. serratus*, and (smooth curve) the random distribution having the same mean. The observed class frequencies differ significantly from expected ($P(\chi^2) < 0{\cdot}001$), and the excess in the lowest and highest (combined) classes clearly indicate aggregation. (From Ryland, *Journal of Experimental Marine Biology and Ecology*, **8**, 1972.)

extremely short larval life (cf. p. 364), at least for larvae not leaving the shelter of the *Pinna* valves.

The consensus from the analytical studies performed so far is that small-scale clumping of bryozoan colonies results from the detection by prospecting larvae of irregular surface contours and/or uneven physico-chemical attractiveness of the surface. Gregariousness, if of consequence at all, plays a minor role in contrast to its undoubted importance to barnacles and spirorbids. The biological advantages of aggregation, however induced, are (1) that the population is concentrated into those regions in which survival, growth and reproduction are likely to be optimal, and (2) that it favours genetic cross-fertilization (p. 351), and establishes populations which are routinely outbreeding (p. 352). Its

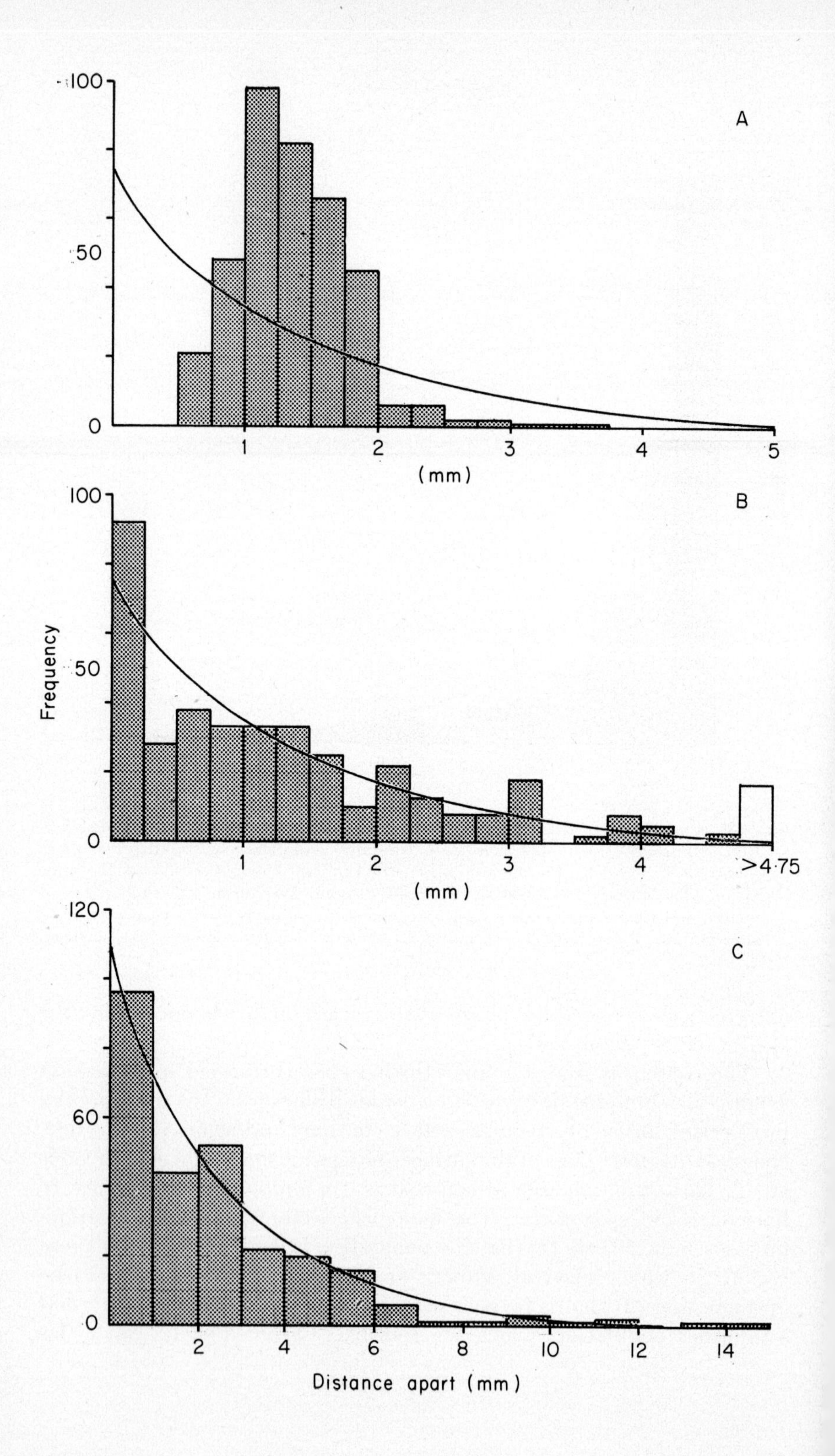

100
50
0
A
1
2
3
4
5
(mm)
100
50
0
B
Frequency
1
2
3
4
>4·75
(mm)
120
60
0
C
2
4
6
8
10
12
14
Distance apart (mm)

major biological disadvantage might seem, *a priori*, to be increased competition within restricted microhabitats.

C. *Spatial competition*

1. *Intraspecific competition: the absence of spacing apart*

Sessile invertebrates which aggregate by gregariousness may alleviate the affects of overcrowding by spacing apart at settlement (Wisely, 1960; Crisp, 1961). In spirorbids the function of spacing apart is perhaps to allow room for the individuals to reach a mature size; but in barnacles, which can produce eggs even at high densities, the main advantage is evidently the acquisition of sufficient space for the young individual to become established and start growing (literature discussed by Hayward and Ryland, 1975). As bryozoan larvae may also settle in clumps, do they also exhibit spacing apart behaviour?

The question was approached by Hayward (1973) and Ryland (1973) by the same method that had enabled Wisely (1960) to discover spacing apart in spirorbids. The rugophilic larvae of *Alcyonidium hirsutum*, like those of *Spirorbis*, settle in the grooves alongside the midrib on *Fucus serratus* fronds (p. 382) and thereby become arranged in single file. This presents a simple situation for analysis (Crisp, 1961). The intervals between successive ancestrulae are measured and the mean interval calculated. The results are divided between about 15 frequency classes of convenient length. Since, in a random settlement, the frequency distribution of intervals will be exponential, the numbers expected in each class can readily be calculated and compared with the observed frequencies by means of χ^2. Examination of Wisely's data in this way clearly confirmed his supposition that the larvae spaced apart, often by about 1 mm (Fig. 34A). The distribution of *Alcyonidium hirsutum* ancestrulae, by contrast, did not depart significantly from random (Fig. 34 B, C).

In another situation, the dispersal of *Crisia eburnea* colonies on selected growth zones of *Flustra foliacea* fronds (pp. 300 and 410) was examined. The analysis performed was of nearest-neighbour distances (Clark and Evans, 1954) whereby, for each *Crisia* colony present in the

FIG. 34. Settlement alongside the midrib of *Fucus serratus*. The frequency histograms show distance apart, z, of newly settled spat of (A) the tubeworm *Spirorbis spirorbis* and (B, C) *Alcyonidium hirsutum* compared with an exponential distribution (smooth curve). In A and B the mean distance apart is the same ($\bar{z} = 1{\cdot}36$ mm); C is a less dense population ($\bar{z} = 2{\cdot}62$ mm). Spacing out is clearly displayed by *Spirorbis* but not by *Alcyonidium*. (A, B from Ryland, 1973, C redrawn from Hayward, 1973.)

sample, the distance was measured to its nearest conspecific neighbour. The frequency distribution of distances in a population randomly dispersed in two dimensions is not exponential but, starting from zero, rises to a peak and declines to an asymptote (Fig. 35). An excess of observed values below the peak indicates aggregation (Fig. 35B), while spacing apart would be marked by a corresponding shortfall in low values. Ryland (1973) found that the densest populations displayed aggregation, while the distribution of less dense populations did not differ significantly from random (Fig. 35A). Neither gave evidence of spacing apart.

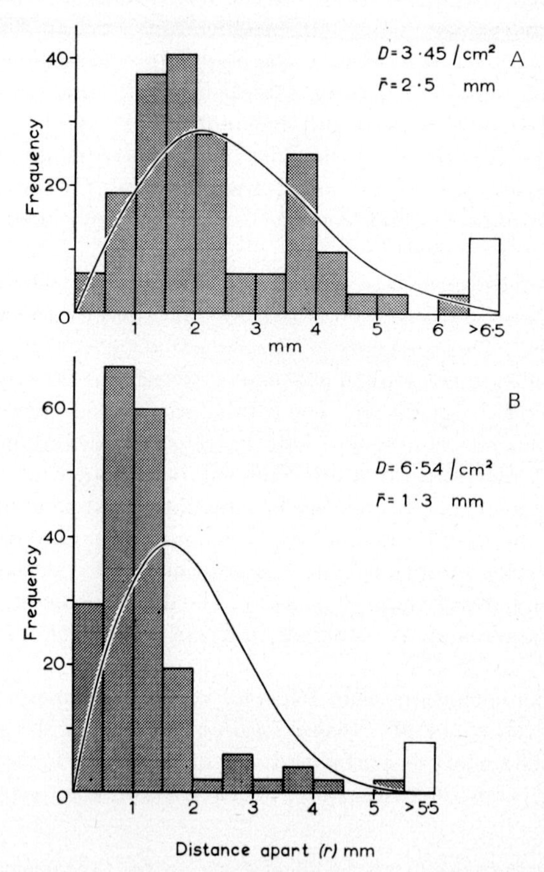

FIG. 35. Frequency histograms showing nearest neighbour distance, r, in two populations of *Crisia eburnea* growing on *Flustra foliacea* compared with random distributions (smooth curves) for the same density. A is the less dense population ($D = 3{\cdot}45$ cm^{-2}; $\bar{r} = 2{\cdot}5$ mm) and the distribution does not depart significantly from random; B is denser ($D = 6{\cdot}54$ cm^{-2}; $\bar{r} = 1{\cdot}3$ mm) and the distribution is contagious (showing aggregation). (From Ryland, 1973.)

While the above examples were based on the direct analysis of nearest-neighbour distances, there are advantages in employing the square of distance as variate (Skellam, 1952), one being that the frequency distribution for a randomly dispersed population is again exponential. The change, however, accentuates another problem. When a population is patchily dispersed, the accumulated nearest-neighbour data comprise many short, within-clump distances and a smaller number of longer distances relating to those individuals which have settled outside the clumps. The length interval selected for the boundaries of the frequency classes must be sufficiently large to include the longer distances within the analysed distribution. In the above-mentioned examples the boundaries were in the range of 0·25–1·0 mm apart. If spacing out were present, but for a distance less than that covered by the first class, it might go undetected. This, of course, is particularly true when distances less than one are squared. Pielou (1962) suggested that if evidence for spacing apart were being sought, the frequency distribution could be truncated, and subsequent analysis restricted to within-clump distances of some predetermined value. She derived distribution functions for this purpose.

To check whether small-scale spacing apart of this kind occurred in *Alcyonidium*, Harvey *et al.* (1976b) obtained settlements of *A. polyoum* on matt Perspex discs filmed with *Fucus* extract at densities ranging from about 1 to > 30 cm^{-2}. Similar settlements of *Spirorbis spirorbis* (L.) were used as a control. Nearest-neighbour distances were measured from every ancestrula (except for those near the edge of the disc) and squared. The data were then expressed as cumulative distributions (Fig. 36) and compared with those expected under each of two null hypotheses.

The first null hypothesis was that a larva encountering an ancestrula would settle immediately adjacent to it (maximum aggregation); the second was that after such an encounter the larva would move away and settle randomly. When the observed and expected distributions are compared, a positive deviation (i.e. above the expected curve, as with the lower line in Fig. 36C) is taken to indicate aggregation, while a negative deviation (as in Fig. 36B) is indicative of spacing apart. When the observed distribution has sigmoid form, showing first a negative deviation and then a positive one, there is both aggregation and spacing apart within the aggregations (Fig. 36C, relative to the upper line). The authors' results demonstrate aggregation over a wide range of densities (at least up to 12 cm^{-2}, which is probably as dense as any natural settlement of *A. polyoum* on *Fucus serratus*.) At higher densities *Spirorbis* was spaced apart under both null hypotheses

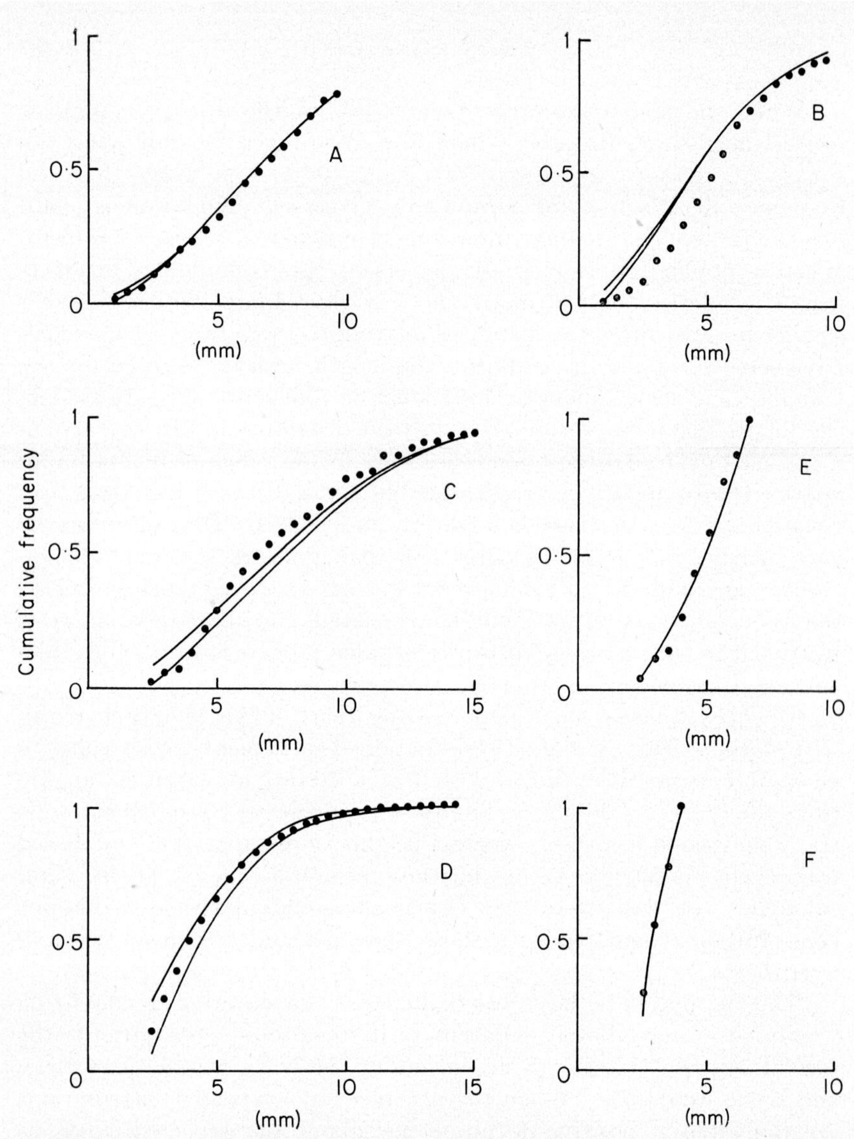

Fig. 36. Cumulative probability curves for various settlements. Results are shown as dots, expected curves as continuous lines: the upper line represents the null hypothesis that a larva encountering a set individual settles immediately adjacent to it, while the lower line represents the null hypothesis that a larva encountering a set individual moves away and settles randomly elsewhere. Abscissae are nearest neighbour measurements in mm from enlarged ($\times$ 6) photographs of the settlement. A, settlement of the tubeworm *Spirorbis spirorbis* (density $D = 18{\cdot}22$ cm^{-2}); B, high density settlement of *S. spirorbis* ($D = 37{\cdot}59$ cm^{-2}) showing spacing out; C, settlement of *Alcyonidium polyoum* ($D = 13{\cdot}18$ cm^{-2}); D, high density settlement of *A. polyoum* ($D = 45{\cdot}61$ cm^{-2}); E, truncated curve corresponding to C; continuous line represents settlement expected from the second null hypothesis; F, truncated curve corresponding to D; continuous line represents settlement expected from the second null hypothesis. E and F show no indication of spacing out. (From Harvey *et al.*, *Journal of Experimental Marine Biology and Ecology*, **21**, 1976).

(Fig. 36B), while the *Alcyonidium* results suggested spacing apart under the first, but aggregation under the second, null hypothesis (Fig. 36C, D). The aggregation implied under the second hypothesis was then investigated by Pielou's (1962) truncation method, with the cut-off at the point of maximum deviation between expected and observed curves. A good fit was found between the observed and expected (truncated) distributions (Fig. 36E, F), giving no evidence of spacing apart under the second null hypothesis.

The experiments showed that both *Alcyonidium* and *Spirorbis* settlements are aggregated at low densities, and that the latter are spaced apart at high, but natural densities (> 18 cm^{-2}) under both null hypotheses (Fig. 36B). *Alcyonidium* larvae produce spatial distributions at high densities which conform with the second null hypothesis—viz., that if a larva encounters an ancestrula it will move away and settle randomly (Fig. 36E, F). In terms of the first null hypothesis, that contact with an ancestrula stimulates a larva to settle immediately adjacent to that ancestrula, *Alcyonidium* larvae can be said to space apart at high densities (> 13 cm^{-2}). It is apparent, therefore, that when a prospecting larva encounters an ancestrula its reaction is to move off, and it subsequently settles randomly and not deliberately spaced away from that ancestrula. Why should bryozoan larvae differ in this respect of their behaviour from barnacle cyprids and spirorbid metatrochs?

Harvey *et al.* (1976b) conclude that spacing apart is a behavioural phenomenon associated with aggregation by gregariousness. There is little or no gregariousness in *Alcyonidium:* settlement densities are not quite as high as in barnacles and spirorbids, and spacing apart is unnecessary. (Experiments with *Bugula neritina* and *Watersipora* spp. might be interesting, since these are gregarious.) Hayward and Harvey (1974b), and Hayward and Ryland (1975) considered respectively the effect of aggregation on the growth rate and fecundity of *Alcyonidium hirsutum*. Comparisons of growth rate were made between populations occupying high density and low density regions of *Fucus serratus* fronds (Hayward and Harvey, 1974a: see p. 384). No consistent differences were found, suggesting that any disadvantages conferred by low settlement density were outweighed by the effects of being situated on a less favourable region of the frond.

Fecundity (number of gonozooids) was found to be linearly proportional to the surface area of the colony (Fig. 37) and small colonies were at no relative disadvantage. Only very small colonies, of area $< 3{\cdot}5$ mm^2, did not produce gonozooids. Hayward and Ryland similarly found for a species of *Hippothoa* from Portobello, New

Zealand, with very aggregated colonies (Ryland and Gordon, 1976), that number of ovicells was proportional to colony area above about 2·7 mm^2.

Obviously a colony must have a nucleus of feeding zooids before it can reproduce (the female zooids in both *Alcyonidium* and *Hippothoa* are non-feeding), but breeding usually commences while the colony is still very small. Nevertheless, the minimum area for maturation is still greater than in spirorbids, the larvae of which space apart. Undoubtedly the significant difference between the growth of the two

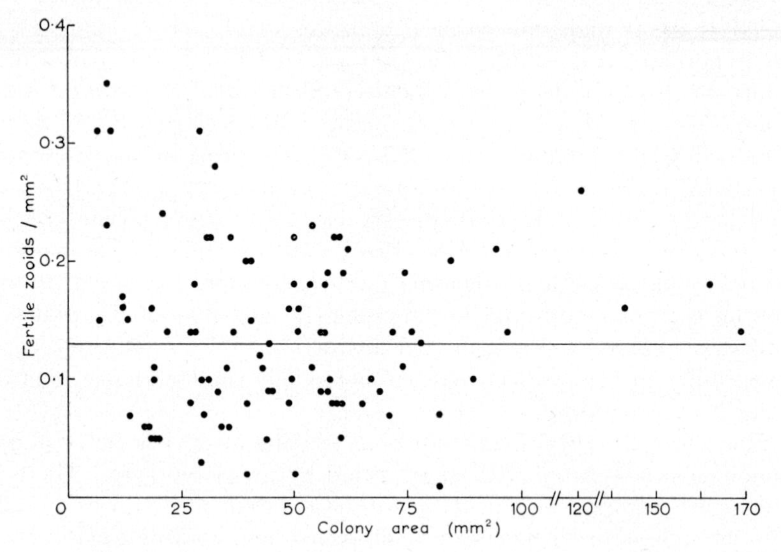

Fig. 37. Relationship between fecundity and colony size in *Alcyonidium hirsutum*. (From Hayward and Ryland, 1975.)

organisms is the lability found in the bryozoan. When a colony is obstructed on one side, it continues to grow just as fast in another direction. Hayward and Ryland (1975) and Harvey *et al.* (1976b) are in full agreement with Stebbing (1973a) that it is this property of bryozoans that makes spacing apart unnecessary. The result, not only for *Alcyonidium* but in *Flustrellidra*, *Hippothoa*, *Membranipora* and probably many other encrusting species, is an interlocking mosaic of variably sized and irregularly shaped colonies making maximal use of the available substratum. Space, then, is frequently a limited resource for which, by definition, there must be considerable competition.

According to O'Connor *et al.* (1975): " intraspecific competition should result in an increase in the range of a resource spectrum used by a species, as at high population levels the advantages to any individual of being at the competition-free optimum of a resource gradient are offset by the intense intraspecific competition found there". This they sought to verify using populations of *Alcyonidium hirsutum* on *Fucus serratus* (the situation for which Hayward and Harvey (1974a) had defined the population optimum). Unfortunately this claim to have demonstrated intraspecific competition in this situation is open to objection (Harvey *et al.*, 1976a). In particular, increase in range with population density is not a test for intraspecific competition. Even in the absence of such competition, if the settling larvae chose the resource optimum with an associated variance about a normal distribution (of, say, zero mean and unit variance), then range increases simply as a function of density. Thus, from Table XX in Fisher and Yates (1963), the expected range for five individuals would be 2·32, for twenty-five, 3·94, and for fifty, 4·50. Moreover, it is not necessary that the population be normally distributed to expect range to increase with population density. Attention should also be drawn (see p. 399) to the conclusion of Hayward and Harvey (1974b) that growth rate was not greater in colonies distant from the resource optimum.

2. *Inter- and intraspecific competition: colony fusion and overgrowth*

Knight-Jones and Moyse (1961) reviewed the subject of intraspecific competition in marine invertebrates. Much of their article was concerned with barnacles and tubeworms, and showed rather clearly how little was then known about competition in colonial organisms. Fortunately, some instructive studies have recently been carried out with bryozoans. Fundamental to the concept of spatial competition are the reactions of any two colonies which achieve close proximity or actually come into contact. Overt interspecific aggressiveness, as occurs in scleractinians (Land, 1973), is absent from bryozoans but a variety of other adaptive responses, inter- and intraspecific, have been reported.

Knight-Jones and Moyse introduced two useful terms: "autosyndrome" which is the fusion between parts of the same colony and "homosyndrome" which is the fusion between separate colonies of the same species. Autosyndrome is of common occurrence among bryozoans (Silén, 1944b, p. 454). It may be seen, for example, in "dendritic" encrusting forms like *Electra monostachys* and certain species of *Hippothoa* (Ryland and Gordon, 1976) in which chains of zooids repeatedly diverge only to meet and join up to other chains. Fusion of buds may even be an essential part of zooid formation, as in *Beania discodermae* (Silén, 1944b). *Alcyonidium hirsutum* and *Flustrellidra hispida*, both species with compact colonies, are commonly found

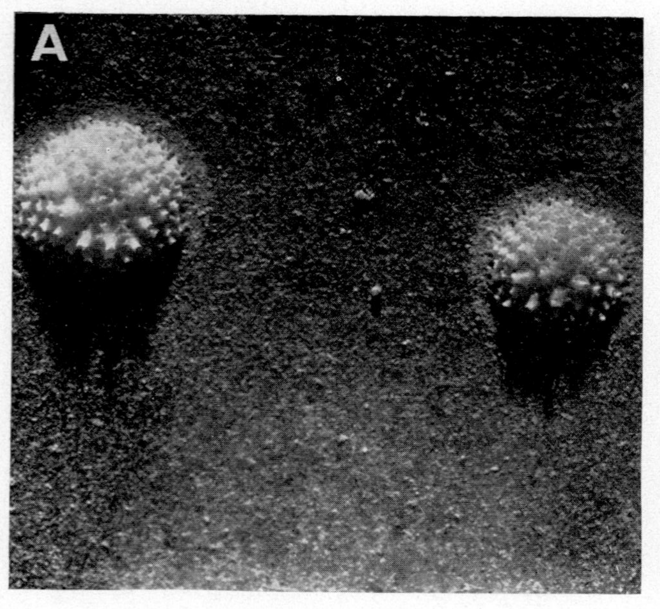

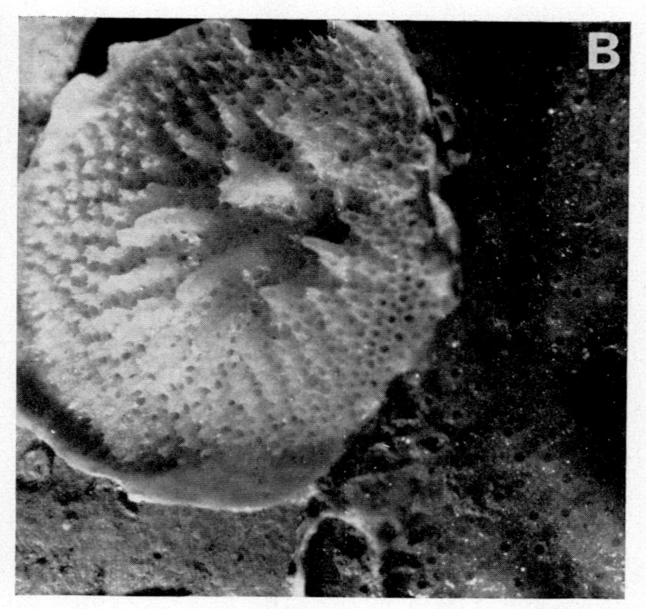

FIG. 38.

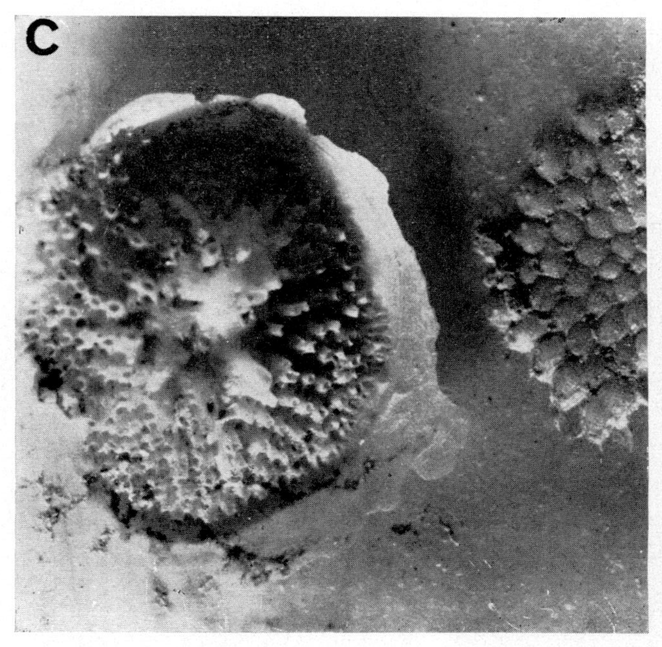

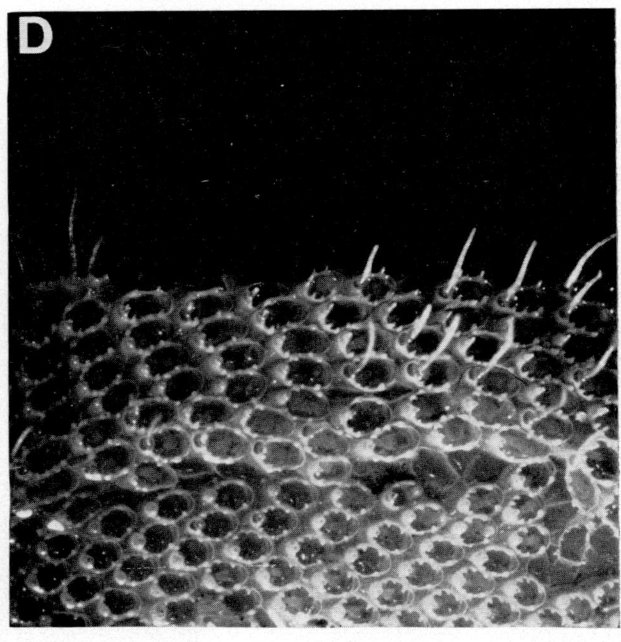

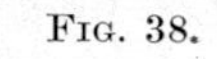

Fig. 38.

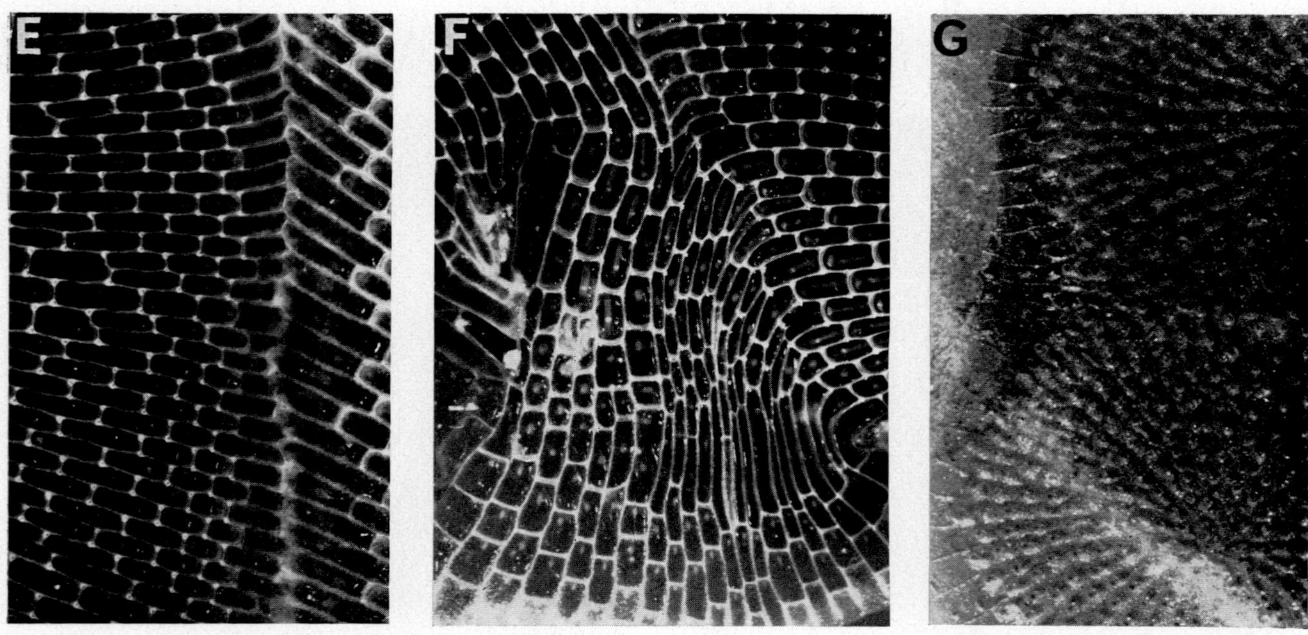

FIG. 38. Some apparently defensive strategies in bryozoan colonies. A, *Disporella hispida*: isolated colonies showing the adherent growing edge (× 8·25); B, C, *Disporella hispida*: colonies showing elevated growing edge when overgrowth is threatened (× 8·25); D, *Electra pilosa*: colony showing some zooids with a proximal denticle, others with a proximal bristle (× 9·0); E, *Membranipora membranacea*: cessation of growth where two colonies have met (× 10·0); F, homosyndrome in *Membranipora membranacea*: two colonies have met at the top of the photograph; as growth has been redirected, so that the zooids of each colony are parallel, fusion has occurred and the two colonies have a common growing edge (× 10·0); G, homosyndrome in *Parasmittina trispinosa*: two colonies have met and produced a common growing edge (× 6·75). (Photographs from Stebbing, 1973b.)

to have surrounded the lower part of *Gigartina stellata* fronds as tubular encrustations. Fusion is not simply a coming together of zooids. The tissues of zooids within a colony are linked through special connecting structures best referred to as pore-chambers (Silén, 1944b; Banta, 1969; Gordon and Hastings, in prep.). Fusion can be said to have taken place only when adjacent zooids are united in this manner.

Homosyndrome is apparently not uncommon in sponges and cnidarians, at least in young colonies, but is very rare in bryozoans. Most usually, when colonies of the same species meet, both stop growing along the line of contact (Fig. 38E) and expand in other directions. This results in the formation of mosaics (Stebbing, 1973a; Hayward, 1973). Stebbing recorded overgrowth in *Electra pilosa*, but few zooids were involved. Homosyndrome was first reported by Moyano (1967) in *Membranipora hyadesi*: unfortunately, however, his paper contains no unequivocal statement that interconnecting pore-chambers had formed. Stebbing (1973b) did demonstrate the presence of pore-chambers between the zooids of two *Membranipora membranacea* colonies that had grown together (Fig. 38F), and suspected the occurrence of homosyndrome in ten colony pairs of *Parasmittina trispinosa* (Fig. 38G). It may be relevant that zooid formation occurs rather similarly in these two species, with cross walls arising serially from a very wide growing edge (see p. 294). In the case of *Parasmittina* on dead shells it is not unlikely that the colony pairs were siblings in every instance; but that this should have been true for *M. membranacea*, with its planktotrophic larvae, is highly improbable. In this species non-fusion is undoubtedly general, and I have seen *Laminaria* fronds covered with a terrazo of rather regular colonies about 5 cm across.

In the compound ascidian *Botryllus primigenus* Oka autosyndrome is normal, and homosyndrome may or may not occur. Oka (1970), summarizing many series of experiments, concluded that the genes controlling fusibility are heterozygous multiple alleles. Colonies sharing at least one common gene are mutually fusible, otherwise different colonies are non-fusible. The same alleles control compatability: fusible colonies are incompatible and *vice versa*. The mechanism underlying fusibility in bryozoans is quite unknown but, if it were similar to that of *B. primigenus*, homosyndrome among sibling colonies might be expected. Homosyndrome has not been recorded in any *Alcyonidium* mosaics but Harmelin (1974) has observed fusions between colonies of roughly the same size in *Diplosolen obelia*. The similarity in size implies a closeness of age, so the colony pairs may well have been derived from the same parent. Both species are cyclostomes, in which larvae develop by polyembryony (p. 345), so that the possibility of the colony pairs actually being identical twins cannot be excluded. There is obviously considerable scope for experimentation in this field using larvae of known maternal parentage. Any genetically controlled individuality mechanism of the type found in *Botryllus* involves only a limited number of genotypes, and it is to be expected that occasional adjacent colonies should, by chance, be capable of fusion.

Gordon (1972) enumerated the types of boundary response found when colonies of different species meet. Mutual cessation of growth is common, though perhaps with the opposed edges upthrust as a crest. Overgrowth also occurs, especially of a thin by a thicker colony. Sometimes, when certain species pairs make contact, there is neither a

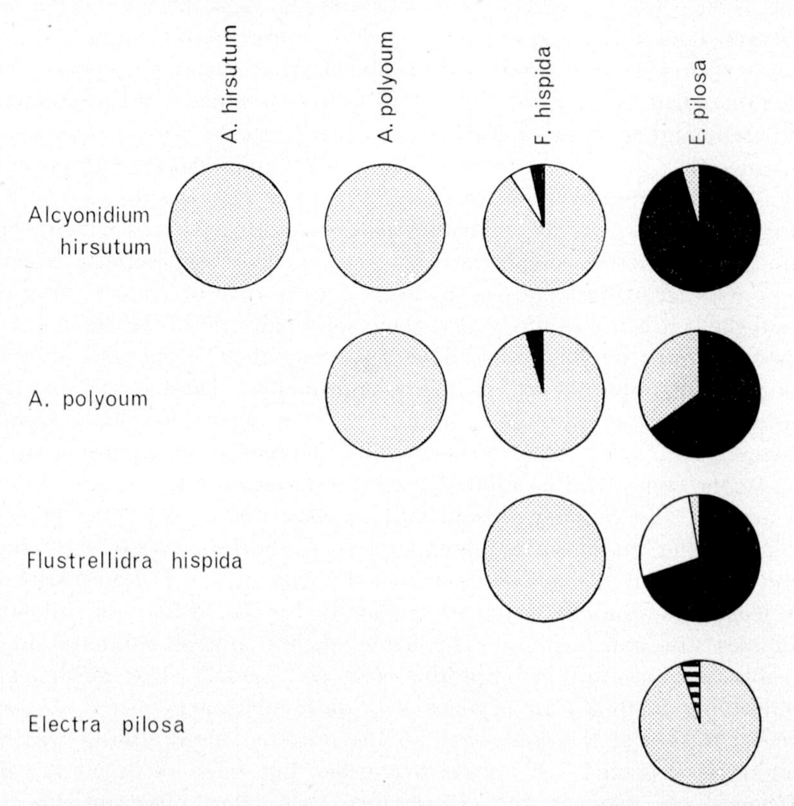

FIG. 39. Intra- and interspecific competition between colonies of four bryozoans living on *Fucus serratus*. The sector diagrams summarize the responses noted when the growing edges of two colonies met: *grey*, mutual cessation of growth; *black*, interspecific overgrowth by the species named at left; *white*, interspecific overgrowth by species named above; *stripes*, intraspecific overgrowth. (Data from Stebbing, 1973a.)

definite cessation of growth nor manifest overgrowth: instead, stolonic outgrowths are formed which spread, as though probing, some way over the surface of the opposing colony.

Stebbing (1973a) also investigated interspecific boundary effects. In the epiphytes of *Fucus serratus*, for example, four species of encrusting bryozoa frequently cohabit the same frond. Between certain pairs—*Flustrellidra hispida: Alcyonidium hirsutum, F. hispida: A polyoum,* and *A. hirsutum: A polyoum*—the almost invariable reaction is a cessation of growth by both; but where any of these meet *Electra pilosa* the commonest result (66–95%) is that the *Electra* gets overgrown (Fig. 39). Zooidal morphology in *E. pilosa* is variable, both in the number of marginal spines (3–12) and in the degree of development of the proximal spine: it may be a denticle like the others or it may be produced as a tall cuticular bristle (Fig. 38D). It has been known since the work of Marcus (1926) that the presence of bristles tends to be associated with growth on a convex substratum, but Stebbing (1973b) has suggested that their presence confers benefit by resisting overgrowth by competitors. Perhaps his most intriguing observation was the apparently " defensive response " of *Electra* to the " threat " of being overgrown. " Often spines occurred sparsely over whole colonies, but where *E. pilosa* and *Alcyonidium polyoum* grew together and touched, spines were often restricted to that part of the colony that was about to be overgrown by *A. polyoum*. Sometimes colonies would be found like that illustrated [Fig. 40]. At one end the new growth of an *Electra pilosa* colony has come into contact with a colony of *Alcyonidium polyoum*. Long-spined zooids are found at the point threatened with overgrowth, but in addition to these there are also long spines at the other end of the colony, where other new zooids have been laid down. Sometimes, star-shaped colonies were found with growth occurring at the tip of each arm. When one arm of such a colony has encountered a colony of *A. polyoum*, the zooids constituting the tips of all the other arms may include some zooids with long spines. It appears that here is an example of interzooidal communication of a ' message ' of functional significance for the colony as a whole ".

Of course, it may simply have been that the bristles started to develop on zooids just at that particular stage in colony formation; but Stebbing evidently felt satisfied that their appearance was a response to the proximity of an *Alcyonidium polyoum* colony, and that a hormonal substance was produced by the *Electra* which affected the differentation process (ontogeny as applied to bryozoans) in all developing zooids. This should be followed up.

Another defensive response recorded by Stebbing (1973b) was

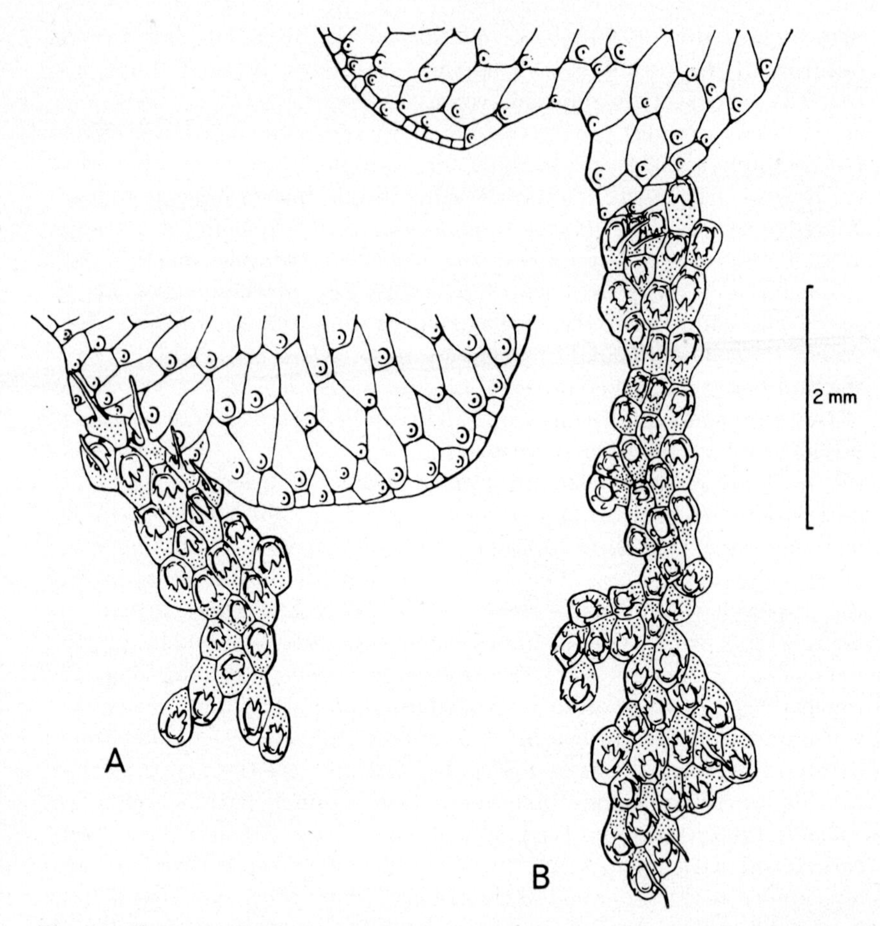

Fig. 40. Encounters between colonies of *Alcyonidium polyoum* and *Electra pilosa*. A shows how long-spined zooids occur typically at the point where overgrowth is occurring and frequently appear to stop it. B shows a similar encounter: long-spined zooids of *E. pilosa* can be seen at the point where overgrowth appears imminent, but they are also present at the other extremity of the colony, where new zooids are being formed. (From Stebbing, 1973b.)

upcurling of the basal lamina in discoid cyclostomes such as *Disporella hispida* (Fig. 38A–C). Harmelin (1974) has recorded the same phenomenon in *Plagioecia patina* and *Diplosolen obelia*. Discoid species with a basal lamina can overgrow almost all other encrusting cyclostomes and many cheilostomes as well; the raised margin may even prevent overgrowth by sponges and ascidians. When bryozoans meet non-bryozoan competitors they frequently fare the less well, however, although on *Fucus serratus* they overgrew the stolons of the hydroid

Dynamena pumila (L.) (Stebbing, 1973a). In the tropics sponges and sometimes compound ascidians often smother the bryozoans. Jackson and Buss (1975) have shown that chemicals produced by certain sponges and compound ascidians are capable of killing zooids of the bryozoan *Stylopoma spongites* and of causing deterioration in others. They suggest that such " allelochemicals " may have considerable competitive importance in cryptic coral reef communities. In New Zealand " Competition between bryozoans and ascidians is very onesided being always in the favour of the latter "; but the ascidians are sometimes short-lived and the bryozoan colonies survive their temporary burial. Sponges may be much more devastating. These and some other biological relationships were discussed by Gordon (1972).

3. *Inter- and intraspecific competition: the partition of resources*

Highly selective settlement behaviour, while sometimes perhaps exacerbating spatial competition (p. 393), in other circumstances diminishes it. Stebbing (1971b) noted how the epibionts *Bugula flabellata* and *Scrupocellaria reptans* settled on the youngest (and cleanest) part of a *Flustra foliacea* frond, which is its periphery. Later (Stebbing, 1972), he showed how *S. reptans* selectively settled on the youngest part of a *Laminaria digitata* frond, this being its base. He pointed out that settlement away from the concentrations of adults was contrary to "gregariousness" and enabled the young colonies to become established on that region of the frond having the least intense competition for space.

Also of apparent adaptive significance in avoiding competition are some of the orientated growth responses described earlier (p. 301). *Flustra foliacea* is a bryozoon with broad flat fronds which grows by peripheral extension over a period of years. As the fronds age they often become densely populated by epizoites, one of which is *Scrupocellaria reptans*. Ryland and Stebbing (1971) found that 78% of the colonies of this species were orientated distally, so that their growth was directed away from the epizoites clustered near the base of the frond towards the young, clean marginal zone.

A second example of orientated growth is that displayed by *Membranipora membranacea* on species of *Laminaria* (p. 301). The lamina of these kelps forms from an intercalary zone near to the junction with the stipe, so that it is youngest and most free from epiphytes proximally. *Membranipora* colonies, virtually without exception, grow towards the base of the fronds (Fig. 8, p. 302). There is, however, an adaptive significance to this beyond that of avoiding competition. In both *L. hyperborea* (the species discussed by Ryland and Stebbing) and *L. digitata* (Pérez, 1969) the growth of the new frond is greatest during the early part of the year, while autumn is characterized by peripheral defoliation. *Membranipora* larvae settle during the summer, and proximally orientated growth by the young colonies will help to ensure their survival through autumn and winter, and enhance their prospects for rapidly colonizing the new lamina in the spring.

Stebbing's (1971b) detailed studies on the epizoic fauna of *Flustra foliacea* revealed some other remarkable situations, including the complementary distributions on the host of two species pairs, *Crisia eburnea* and *C. aculeata*, and *Scrupocellaria reptans* and *S. scruposa* (Figs 41 and 42). In each genus, the first mentioned species was found predominantly on the youngest, peripheral region of the fronds, while the second species was commoner on the older, basal regions. It seems that in both pairs of species, the one found more peripherally is the better adapted to exposed situations and greater water movement (e.g., *S. reptans* produces more numerous and shorter attachment rhizoids than *S. scruposa* and—as its name implies—has the more

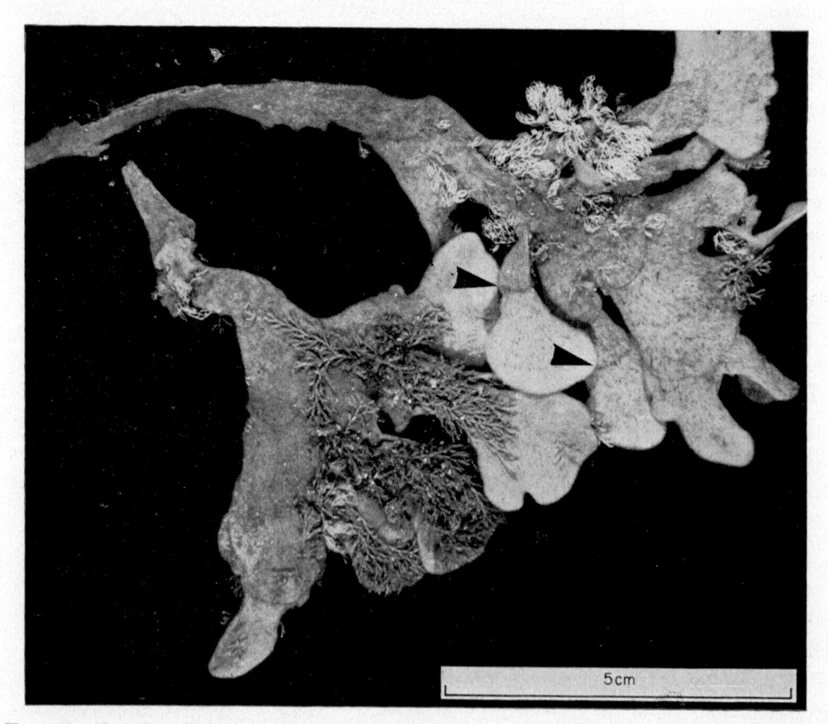

FIG. 41. Fronds of *Flustra foliacea* showing (arrowed) growth check-lines (see p. 300) and bryozoan epizoites *Crisia eburnea* (fine white tufts) and *Scrupocellaria reptans* (coarser, adherent colonies).

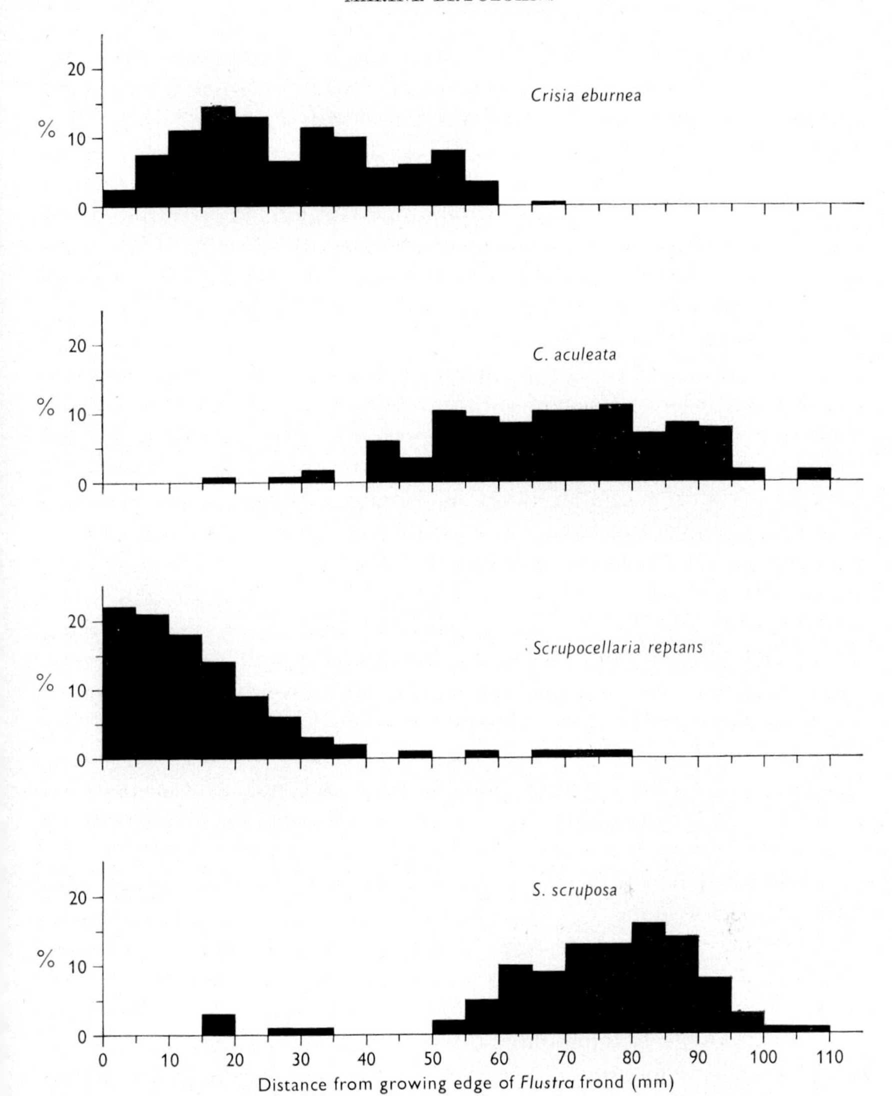

FIG. 42. The distribution of epizoites on *Flustra foliacea* fronds. Two species of *Crisia* and two of *Scrupocellaria* show essentially discrete distributions in relation to the age of the substratum. (From Stebbing, *Journal of the Marine Biological Association of the United Kingdom*, **51**, 1971.)

creeping habit; *S. scruposa* is more bushy). Whatever the physiological mechanisms and direct cause, the result is an interspecific partition of available space apparently in accordance with Gause's hypothesis that two species with identical ecological requirements do not occupy the same niche.

But does Gause's hypothesis really apply to sessile colonial invertebrates? And is there competition for any resource other than space? An interesting situation concerns two species of *Hippothoa*, with similar-sized lophophores, recently found together on red algal fronds in Jersey, Channel Islands. One, presently known in the literature as *H. divaricata* var. *conferta* (see Ryland and Gordon, 1976), differs from the common *H. hyalina* in possessing a gizzard (Gordon, 1975c; see p. 335). Does the presence of a gizzard in one and its absence in the other perhaps indicate that these two species are selectively feeding on different kinds of food particle?

It was mentioned earlier (p. 383) that there was evidence for resource partitioning when *Alcyonidium hirsutum* and *Flustrellidra hispida* occur together on *Fucus serratus*. Moreover, O'Connor *et al.* (1975) have shown that the greater the number of species present on a *F. serratus* plant, the smaller is the niche width (range of the resource spectrum occupied) of each. But, of the species normally common on *F. serratus*, some at least differ greatly in the size of their lophophores, thus: *Electra pilosa*, 11–15 tentacles, funnel diameter 0·5 mm; *Alcyonidium polyoum*, 19–20 tentacles, 0·8 mm; *Flustrellidra hispida*, 28–40 tentacles, 1·2 mm (see also Table IV, p. 312). This suggests that there may be a resource partitioning of both space and food, but it leaves quite unresolved the apparently anomalous situation involving *Alcyonidium hirsutum* and *A. polyoum*. These species may occur together on *Fucus serratus* (though they did not in the material studied by O'Connor *et al.*) and their lophophores are closely similar.

Similarly I had noticed the large size range of zooids in cheilostomatous species found below coral plates in the Caribbean and under boulders on the shore in New Zealand. In both regions the large zooids of *Steginoporella* spp. were conspicuous among the other species. A full survey at Leigh, New Zealand (see p. 385) confirmed that there was a wide range of lophophore sizes present among bryozoans in the under boulder community (see Fig. 11, p. 311). When, however, the relative abundance of all the contributing species was assessed, the situation revealed was somewhat unexpected, for all the common species had lophophores of roughly the same size (Ryland, 1975). Fig. 43A shows the percentage of species having a mean tentacle number approximating to each of the integers 8 to 21, with numbers above 21 grouped. Bryozoans with 11 tentacles are rare, otherwise the distribution approximates to normal with a high variance ($\bar{x} = 13{\cdot}9$; var $= 19{\cdot}95$). When the frequency histogram is compiled from the occurrences of species (Fig. 43B), however, the distribution becomes

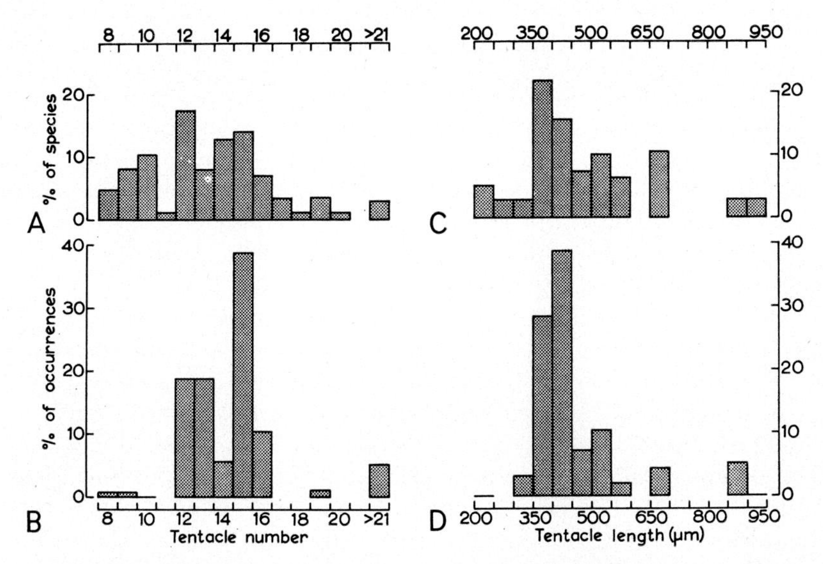

FIG. 43. Frequency histograms showing the abundance of bryozoans on the shore at Leigh, New Zealand, in relation to morphometry of the lophophore: A, species against tentacle number; B, occurrences against tentacle number; C, species against tentacle length; D, occurrences against tentacle length. (From Ryland, 1975.)

more peaked and the variance is greatly reduced ($\bar{x} = 14{\cdot}6$; var $= 7{\cdot}40$). The data for tentacle length show the same trend, with a wide spread in Fig. 43C ($\bar{x} = 475$ μm; var $= 28\,548$). When occurrences are considered (Fig. 43D) there is a tall, narrow peak and relatively little spread ($\bar{x} = 460$ μm; var $= 15\,125$): the variance is again roughly halved.

The four histograms summarize a clear picture. (The large discrepancy between the 14 and 15 tentacle classes in Fig. 43B is because mean values have been scored and the very abundant *Crassimarginatella papulifera* with 14·58 tentacles just falls into class 15). The commonest species are those in the centre of the distribution—those with 12–16 tentacles of length 350–450 μm. Those species in the community which possess a much smaller or a much larger tentacular funnel tend, almost without exception, to be rare (which is not to imply, of course, that all the species in the middle of the size range are common; they are not.) Ryland concluded that all the common species in this community were likely to have been consuming particles in the same size range, but that a different situation might prevail in other habitats or different geographical locations. This work is currently being continued.

VIII. Geographical Distribution

The geographical distribution of bryozoans is a neglected subject, and likely to remain so until the precision of taxonomy is considerably improved. However, there are several substantial areas of ocean from which the bryozoan fauna is reasonably well-known, e.g., west European coasts of the Atlantic, the Mediterranean, the Arctic and, increasingly in recent years, the Antarctic (see Rogick, 1965; Bullivant, 1969; Androsova, 1973 for further references).

One outstanding study which deserves to be better known is that of Maturo (1968) on the distribution of bryozoans to the north and south of Cape Hatteras (North Carolina)—a well-established zoogeographic boundary (Briggs, 1974). The Atlantic seaboard of America between Cape Cod, Massachusetts, and Key West, Florida, has been divided by zoogeographers into three marine provinces: Virginian, Carolinian and Tropical. The Virginian Province extends from Cape Cod to Cape Hatteras and the Carolinian Province embraces the inner shelf from Cape Hatteras to Cape Kennedy, Florida. Cape Hatteras thus marks a boundary between the cooler northern water and warmer southern water. As the Gulf Stream follows the edge of the continental shelf northwards, passing close to Cape Hatteras before swinging away from shelf and coast, the Tropical Province is considered to comprise the outer shelf along the south-eastern coast and to extend somewhat north of Cape Hatteras.

In the work-up of samples collected by dredge and other means, Maturo identified 246 species of Bryozoa (including a few Entoprocta). The distribution patterns were as follows: Ranging extensively north and south of C. Hatteras (cosmopolitan): 12%; Restricted to the Virginian Province: 8%; Ranging from C. Hatteras southwards: 67%; Ranging from Georgia southwards: 7%; Recorded from Florida only: 6%.

The bryozoans provided no evidence in favour of separate Carolinian (inner shelf) and Tropical (outer shelf) provinces; rather their colonies " seem to be widespread over the shelf wherever suitable substrates can be found " (see p. 378 *et seq.*) and also " occur in the Gulf of Mexico, Straits of Florida, and the Caribbean Sea." In other respects the bryozoan distributions (percentages of species assigned to defined areas) tally remarkably closely with those of other invertebrates (Cerame-Vivas and Gray, 1966). Maturo's paper contains a full tabulation of species and their distribution as well as a number of maps.

In another study, Hayward (1974) described a large collection of shallow water bryozoans from the little studied Aegean Sea. In a zoogeographic appreciation he was able to utilize also the recent results of deep-water dredgings from the same region (Harmelin, 1969b). Pérès (1967), in a general review of the Mediterranean benthos, recognized seven faunal elements: endemic, Atlantic temperate-boreal, circumtropical, cosmopolitan, warm temperate (Central Atlantic), Indo-Pacific, and Senegalian. In Table XI the percentage of

TABLE XI. ZOOGEOGRAPHICAL AFFINITIES OF MEDITERRANEAN BRYOZOA from Hayward (1974)

	Western Mediterranean (*Gautier*, 1962)	*Aegean* (*Harmelin*, 1969b; *Hayward*, 1974)
	%	%
Mediterranean endemic	32·2	36·3
Atlantic temperate-boreal	27·0	23·4
Cosmopolitan	19·5	17·7
Circumtropical	12·1	13·7
Central Atlantic warm-temperate	6·9	4·0
Indo-Pacific/Mediterranean/West African	2·3	4·8
	n = 174	n = 124

species in each category is listed and compared with the corresponding figure from the western Mediterranean (from Gautier, 1962). The three authors considering the bryozoans have combined the last two of Pérès' groups.

The extent of agreement is substantial, with endemics comprising the largest group and the Atlantic temperate-boreal second largest. Considering the invertebrate fauna generally Pérès had found the latter group to be the largest, followed by the Mediterranean endemics, but the proportion of endemics increased as the mobility of the animals decreased. For the ascidians, as for the bryozoans here considered, the endemics constituted the largest group. The numbers of species representing the two smallest groupings in the table are perhaps too low for the difference to be regarded as significant. Hayward found no evidence in the Aegean fauna of genuinely Indo-Pacific species, although four (*Hippopodina feegeensis*, *Thalamoporella gothica indica*, *Hippaliosina acutirostris* and *Celleporaria aperta*) are now present on the coast of Israel, having apparently colonized the Mediterranean *via* the Suez Canal (Powell, 1969). The last mentioned is known to have reached Port Said by 1924 (Hastings, 1927).

IX. Predators of Bryozoa

Bryozoans, at first sight, appear an unsatisfactory source of food for other organisms. In cheilostomes, and especially ascophorans, a very high proportion of the body consists of calcium carbonate and the overall nutritional value is low. Yet bryozoans do form a significant proportion of the diet in more animals than is generally supposed, and a number of successful species are specialized bryozoan predators. At one extreme is the highly specific feeding relationship between the nudibranch *Adalaria proxima* (Alder and Hancock) and *Electra pilosa* (Thompson, 1958); at the other is the probably fortuitous occurrence of *Bugula turrita* in the guts of the smooth dogfish (*Mustelus canis* Mitchell) and puffer (*Sphaerodes maculatus* (Bloch and Schneider)) and of assorted bryozoans in the stomachs of king and Pacific eiders (*Somateria spectabilis* (L.) and *S. mollissima* (L.)) (Osburn, 1921).

Many omnivorous species regularly take substantial quantities of bryozoans. Thus Qasim (1957) found that bryozoans occurred in the guts of one-fifth of the 840 shannies (*Blennius pholis* (L.)) he examined and that the proportion rose to as high as one-half in early summer. Many other grazing fishes (e.g. labrids) include bryozoans in their diet, but echinoids probably constitute the largest group of this kind of omnivore, and numerous species are known to consume bryozoans. Included are *Echinus esculentus* L., which has eaten *Cellaria*, *Crisia*, *Membranipora*, *Pentapora* and *Scrupocellaria* (Hunt, 1925; Stebbing, unpublished thesis, 1970), *Psammechinus miliaris* (Gmelin), which has eaten *Electra* and *Flustra*, scraping the former off *Mytilus* shells (Milligan, 1916), *Strongylocentrotus drobachiensis* (O. F. Müller), *Evechinus chloroticus* (Val.) from New Zealand (Gordon, 1972) and the tropical *Diadema setosum* (Leske), *Echinothrix calamaris* (Pallas), *Stomopneustes variolaris* Lamarck and *Tripneustes gratilla* (L.) (Herring, 1972). Some of these (e.g. *Echinus*, *Evechinus* and *Tripneustes*) are essentially grazing species, but Elmhirst (1922) found that young *Echinus esculentus* browsed "on such limy food as *Membranipora*" and the adults feed "on *Laminaria*, particularly if encrusted with *Membranipora*". Similarly, Sevilla (1961) found that *Strongylocentrotus drobachiensis* preferred *Nereocystis luetkeana* (Mert.) P. and R. fronds when covered with *Membranipora membranacea*. If an urchin browses over 60–70 cm^2 each day a great deal of *Membranipora* must be consumed.

Gordon (1972) recorded that two asteroids, *Patirella regularis* Verrill and *Coscinasterias calamaria* Gray, feeding on rock surface biota incidentally consumed bryozoans. If a feeding specimen of either starfish was removed and found to be on a bryozoan colony, a clean white patch of zooid skeletons marked the place.

Certain omnivorous chitons consume encrusting bryozoans and on one New Zealand shore, near Auckland, Gordon (1972) considered that *Cryptoconchus porosus* Burrow and *Notoplax violaceus* (Quoy and Gaimard) were their most important predators. He dissected individuals whose guts contained only bryozoan fragments. An unusual and apparently selective predator on the Pacific coast of America is the 10 mm long polyclad *Thysanozoon californicum* Hyman. It was recorded on the encrusting ascophoran *Celleporaria brunnea* by Haderlie (1970), who noted that the flatworm is dorsally papillated and closely matches the *Celleporaria* in colour and texture. The *Thysanozoon* was invariably found closely appressed to the bryozoon; when loosened and turned over it was often seen to have the pharynx extended and pressed over the zooids. In many cases such zooids were whitened and with the polypide destroyed. Undoubtedly most significant, however, are two groups of specialist bryozoan predators: pycnogonids and nudibranchs.

A. *Pycnogonid predators*

Of marine arthropods, pycnogonids are the most important consumers of bryozoans. It seems true to say that wherever bryozoan colonies are found, numerous pycnogonids will be in attendance. In the Swansea area *Anoplodactylus pygmaeus* (Hodge) and *A. angulatus* (Dohrn) have been found in clumps of *Bowerbankia*, and *Achelia echinata* Hodge in *Bugula* (see below). *A. echinata* and *Anoplodactylus pygmaeus* have been found on *Laminaria digitata* fronds well covered with *Scrupocellaria reptans*, *Hippothoa hyalina* and other bryozoans (Ryland, 1972). Numerous small pycnogonids were present in clumps of *Elzerina binderi* collected at Kaikoura, New Zealand, and in the mat of cryptofaunal bryozoans (*Nellia oculata*, *Poricellaria rationiensis* and *Synnotum aegyptiacum*) on a tropical Australian shore (Ryland, 1974b). Association does not, of course, necessarily imply that the pycnogonids are predating the bryozoans: thus Wyer and King (1973) recognized three possible levels of association, the bryozoan being (1) simply a substratum; (2) a surface from which algal or detritic food may be collected; or (3) the prey. An example of the second relationship is provided by *Achelia longipes* Hodge associated with *Flustrellidra hispida*. Although the colony surface of many bryozoans remains remarkably free from epizoites, this is by no means always the case. On the Pembrokeshire coast *F. hispida* is found abundantly on the rhodophycean *Gigartina stellata*. Red algal sporelings are in turn found on the *F. hispida* and are consumed by the pycnogonid. The latter " move over the bryozoan colony making no effort to feed on the autozooids, even when they emerge in front of them" (Wyer and King, 1973). Similarly, in the epizoic community on *Flustra foliacea* (see p. 409), *Nymphon rubrum* Hodge feeds on the hydroid *Laomedea angulata* (Hincks) while *Pycnogonum littorale* (Ström) consumes trapped detritus. *Achelia echinata*, however, is a true predator of *Flustra* zooids and its feeding behaviour has been carefully recorded (Wyer and King, 1973).

Achelia echinata is abundant on the *Flustra* and a gallon breffit of the latter may yield up to 200 *Achelia* at certain times of the year. The pycnogonid moves slowly over the frond, with its proboscis lowered towards the surface and the palp tips held just ahead of and below the end of the proboscis. The distal segments of the palps bear sensory bristles which are orientated to face the substratum over which the pycnogonid is walking. If there are sudden movements, as when an avicularium adducts, the palps are swung away sideways.

At intervals the pycnogonid encounters the spines on the distal angles of a zooid: " If the spines curve away . . . the pycnogonid . . . may pause momentarily but soon passes on. If the spines curve towards it, which means that it is facing the side at which the operculum opens, it lowers the cephalon so that the proboscis is at an angle of about 60° to the autozooid and the proboscis tip is between the two large spines and thus in direct line with the operculum " (Fig. 44A). This position may be held for up to 30 min; if the operculum begins to open, the pycnogonid lunges forwards so that the tip of the proboscis penetrates the orifice and enters the zooid. Having gained access the pycnogonid moves forwards so that the angle between the proboscis and the zooid increases to the vertical. This ruptures the frontal membrane and the *Achelia* feeds for about 10 min with its proboscis pushed well into the zooid (Fig. 44B).

Achelia echinata has also been found in the hanging tufts of *Bugula turbinata*, but here it feeds on the detritus accumulated in the older part of the colony. The absence of pycnogonids from the more distal part of the colony, where the autozooids are vigorous and the avicularia are constantly in motion, might be attributed to the activities of the latter. Kaufmann (1971) noted that a pycnogonid of 10 mm leg span could be held by a *Bugula* avicularium. However, Wyer and King (1973) found that any parts of the legs, proboscis or palps that were seized by an avicularium were soon released having suffered no visible damage.

Nymphon gracile Leach, which commonly feeds on actinians and

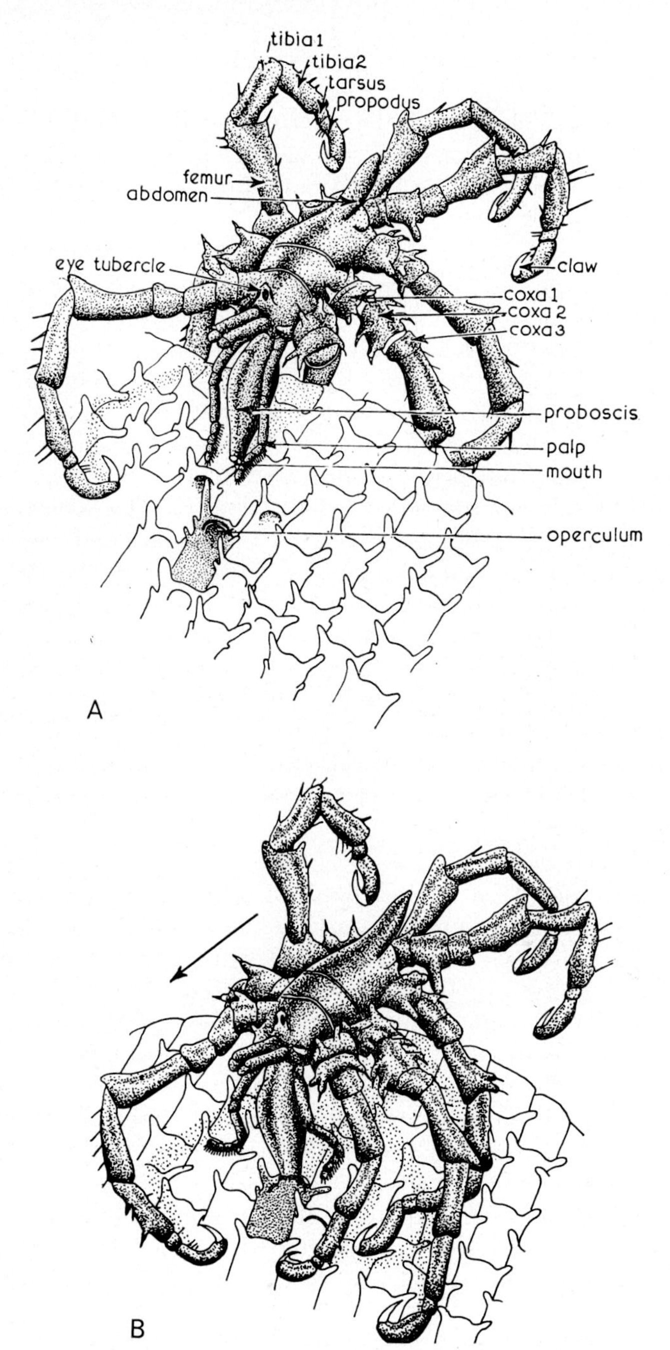

FIG. 44. Pycnogonid *Achelia echinata* preying on *Flustra foliacea*. A, representation of *A. echinata* walking over *Flustra* zooids; note the position of the proboscis and palps in relation to the spines and operculum of the proposed prey zooid (shaded). B, after the *Achelia* has thrust its proboscis through the orifice it moves forward levering the operculum further open. (From Wyer and King, 1973.)

hydroids, has been observed feeding on zooids of *Bowerbankia imbricata*. Individual zooids are grasped by the chelifores and the wall ruptured so that the polypide can be pulled out, shredded by the chelae and ingested (Wyer and King, 1973).

Fry (1965) studied the food preferences of two Antarctic pycnogonids, *Rhynchothorax australis* Hodgson and *Austrodecus glaciale* Hodgson. In his experiments the former rarely associated with the bryozoan colonies offered but *A. glaciale* chose bryozoans on about 50 per cent of occasions. The most favoured species was *Cellarinella roydsi**, a species with rigid flabellate fronds up to about 40 mm across (shown in Fig. 6, p. 299), although the feeding behaviour was unfortunately never witnessed. Fry's detailed consideration of the proboscis and its musculature is outside the scope of this review, but the salient features are important. The proboscis is long (about 2 mm compared with a body length of 3 mm and a leg span of 10 mm), very slender, down-curved and distally flexible; structurally, rings of highly sclerotized wall alternate with relatively soft annuli (W. G. Fry, *in litt.*). It can be moved in a ventral direction by a powerful muscle. *A. glaciale* is apparently highly adapted to feeding on *Cellarinella* species, but how exactly is this accomplished? Fry based his analysis on a model bryozoan zooid which was really rather different from *Cellarinella* and, for this reason, I believe, may not necessarily have reached the correct conclusion.

Although *Cellarinella* has an ascophoran type zooid (see, e.g., Ryland, 1970, fig. 4), it is specialized. The frontal wall is very thick (up to 450 μm) and the pseudopores large (29–79 μm in diameter). The orifice is very complex: the secondary orifice (superficial opening) measures about 290 $\times$ 340 μm and is "guarded" by the "external" avicularium of mandible length about 114 μm. The primary orifice (inner opening) is situated at the bottom of a tunnel-like peristome of length equal to the thickness of the wall; it opens proximally into the zooidal lumen. There is no operculum, but the peristome is partially occluded by an oblique shelf with which is associated an "internal" avicularium of mandible length about 126 μm.

It seems by no means impossible that *A. glaciale* feeds by inserting its long, slender proboscis through the peristome. Its great length, down-curvature and ventrally deflecting muscle would clearly be

* Fry named his species *Cellarinella foveolata*, quoting *C. roydsi* as a possible synonym. The two species are not alike in colony form and it seems improbable that they are synonymous. From the data given by Fry, in his Table IV, I conclude that the species concerned is probably *C. roydsi* Rogick (1956) from whose paper the morphological data here quoted have been taken.

advantageous in probing for and obtaining access through the primary orifice, particularly if the pycnogonid made its attack from the proximal side of the superficial opening. Since there is no operculum, Fry's argument that: " Penetration to the polypide through the operculum is fraught with danger; even if the pycnogonid were able to place its proboscis in the orifice, the closing of the operculum, which is usually very thick, would probably crush it " cannot apply.

Nevertheless, particularly if the softer rings of the proboscis are vulnerable, it remains possible that the two peristomial avicularia successfully protect the primary orifice. If so, Fry may be correct in assuming that the styliform proboscis penetrates the zooid through the pseudopores, though these are probably not the simple openings into the ascus that he assumes. Which of the two methods of attack is correct will be settled only by close observation. It is highly desirable that this should be done in *A. glaciale*, or in one of the other species mentioned by Fry as having a styliform proboscis, for a protective function of such specificity has never been demonstrated for a bryozoan avicularium (though defence, in general terms, has often been suggested).

In his analysis of the functional morphology of the *Bugula* avicularium, Kaufmann (1971) concluded that its best developed faculty was that of grasping objects of diameter about one-half the length of the mandible: i.e., of about 60 μm in *Cellarinella*. For much of its length the proboscis of *A. glaciale* is not much wider than this and there seems little doubt that both the external and internal avicularia should be able to grip it firmly. Whether they would damage the proboscis, or even deter the pycnogonid, is a matter that only future observation can determine. The *Bugula* avicularia observed by Kaufmann were most effective in catching setose arthropods, such as small amphipods.

B. *Nudibranch predators*

The best known, and probably the most destructive, predators of bryozoans are nudibranchs of the families Polyceridae, Goniodoridae and Arminidae (Miller, 1961; Thompson, 1962; with additional unpublished details provided in personal communications by Miss P. L. Cook and Dr M. C. Miller.) As with pycnogonids, however, associated nudibranchs are not necessarily predators. One Ghanaian species found on *Anguinella palmata* was evidently a cleaner, feeding on the bacterial film and other detritus. It was totally specific, reacting violently against the other species in its habitat (*Electra verticillata*, *Membranipora tuberculata*) and refusing to clean large colonies of a

Nolella, which appeared indistinguishable from *Anguinella* with regard to surface flora and fauna (Cook, pers. comm.).

Of European polycerids, the food preferences and feeding techniques of *Polycera quadrilineata* (O. F. Müller) and *Limacia clavigera* (O. F. Müller) are well established. *P. quadrilineata* has been reported on various bryozoans but is particularly associated with *Membranipora membranacea*. It is very destructive, ploughing its way across the colonies and destroying all the zooids in its path. Miller (pers. comm.) recorded 94 mm^2, about 475 zooids, eaten in 15 h at 16°C and, at certain seasons, several specimens may be present on a *Membranipora* encrusted *Laminaria* frond from shallow water. *Polycera dubia* M. Sars is said to feed on ascophorans (Miller, 1961), and Morton and Miller (1968, p. 412) mention a New Zealand species which feeds on *Bugula neritina*. Small specimens attack zooids individually, eating the polypide, but larger slugs (10 mm) destroy the ends of the branches as depicted in Morton and Miller's Fig. 153*c*. *Limacia clavigera* has been recorded on ascophorans but usually feeds on anascans; entire zooids are consumed, leaving only the lateral walls and marginal spines. One 12 mm long specimen observed by Miller ate 165 *Membranipora* zooids in 14 h.

Two well camouflaged polycerids are *Nembrotha morosa* Beigh and *N. kubaryana* Beigh from New Zealand. They are found on the bottle green *Bugula dentata* (Ryland, 1974b), which they match. Both the *Bugula* and the nudibranchs are illustrated in colour by Doak (1971).

The goniodorids are probably the most frequently encountered and best documented nudibranch predators. *Goniodoris nodosa* (Montagu) is a commonly found intertidal species in the British Isles, occurring preferentially on the fleshy ctenostomes *Alcyonidium hirsutum*, *A. polyoum* and *Flustrellidra hispida*. It is a suctorial feeder which appears to aid ingestion by pressing on the frontal surface. *Acanthodoris pilosa* (O. F. Müller) is another suctorial feeder on the *Alcyonidium* species, consuming the polypide without destroying the frontal wall. Miller (pers. comm) studied a 7 mm long specimen which ate 276 *A. polyoum* polypides in three days. The developing embryos are also taken.

A dorid which lives on *Membranipora membranacea*, *Electra pilosa*, other anascans, fleshy ctenostomes and ascophorans such as *Hippothoa hyalina* and *Escharella immersa* is *Onchidoris muricata* (O. F. Müller). It is the commonest nudibranch in kelp holdfasts on the north-east coast of England (Moore, 1973). This again is a suctorial feeder, taking the polypides without destroying the frontal membrane. The slug appears to feel for the orifice with its oral veil and spends much time positioning its mouth. One specimen 7·5 mm long ate eight polypides in five hours. The *Membranipora* apparently habituates to the touch of the slug and the lophophores remain everted when pushed to one side (Miller, pers. comm.). Other suctorial feeders are *Okenia plana* Baba, also on *M. membranacea*, and *O. pellucida* Burn which takes the polypides of *Zoobotryon verticillatum* (Morton and Miller, 1968, p. 412, Fig. 153*a*).

The related dorid *Adalaria proxima* was comprehensively studied by Thompson (1958). It is almost entirely dependent on *Electra pilosa*, and is so adapted to this species that its veligers will metamorphose only after contact with it. Using *Electra* colonies growing on glass, Thompson was able to observe closely how the slug attacked the frontal membrane with its radula, which bears a few rows of large pointed teeth, using short bursts of three to eight strokes at a time. Once the membrane had been punctured, the lips of the dorid were applied to the breach and the polypide sucked out by dilatations of the buccal pump.

On the west coast of North America *Membranipora* is preyed upon by *Corambe pacifica* MacFarland and O'Donoghue. This species is so prevalent that it is rare to find uninjured young colonies of *Membranipora villosa* (O'Donoghue, 1926). When small the slug can be seen inside the zooids from which the polypide has been eaten, but larger specimens attack zooids at the growing edge of the colony where the frontal membrane is particularly thin. *Corambe* is pale grey and closely matches its bryozoan background.

In Ghana a species of *Corambe* is found on *Membranipora tuberculata*, against which it is very well concealed (Cook, pers. comm.). Several others (usually 1–5 mm, with a maximum length of 10 mm), belonging to species still undescribed, have been observed in abundance on Ghanaian bryozoans. All had some form of camouflage, in coloration and/or shape, and were very active. They also were apparently specific in their food preferences; for example, one would consider only *Trematooecia turrita* and *T. magnifica*, and another was confined to an undescribed species of *Alcyonidium*. All were suctorial feeders. The species on *Trematooecia* sucked open the operculum to reach the polypide, but also removed the cuticle, epidermis and tissue in the hypostegal coelom (space above the cryptocystal frontal wall in this ascophoran). In life the hypostegal coelom is turgid and opaque, and presumably nutritious, and is not afforded protection by the frontal wall. Each nudibranch is able to clear 10–15 zooids in 24 h, and about 30 were present on a colony of approximate dimensions 20 × 20 × 10 cm. Avicularia did not appear to be stimulated by contact with the nudibranchs, and the latter were never seen to be deterred by avicularian activity. However, *T. magnifica* buds frontally (see p. 295) at least the equivalent of a new layer of zooids each 60–72 hours (the colony is multilaminar) and this colony had an estimated 160 000 zooids actually at the surface—so while it was actively growing it could easily keep pace with the predation (Cook, pers. comm.).

Dudley (1973) mentions *Doridella obscura* Verrill as a predator of *Conopeum tenuissimum* in Chesapeake Bay.

Miller (1970) has described the feeding of the arminacean *Caldukia rubiginosa* on *Beania magellanica*, the two species occurring together under boulders in the vicinity of Auckland. The zooids—which form a loose network in *B. magellanica*—are attacked singly and torn apart by the large, dentate jaws of the slug. The zooid is usually grasped about the middle, fractured by the large teeth of the jaws closing upon it and drawing it into the buccal cavity. This is repeated many times until the zooid, or part of it, is torn from the colony. The polypide is macerated during this process, while the radula acts as a conveyer belt transporting the zooid fragments to the oesophagus.

Eolid nudibranchs generally feed on coelenterates, so the recent record of one species, *Favorinus ghanensis* Edmunds, apparently subsisting exclusively on *Zoobotryon verticillatum*, is of interest. The eolid is specific in its diet, apparently ignoring two species of *Bugula* which were growing with the *Zoobotryon* (Edmunds, 1975).

From the foregoing account it seems that most of the bryozoans predated by nudibranchs are either anascans or ctenostomes, and that both encrusting and erect colony types are attacked. Ascophorans regularly preyed upon are fewer, although a number, together with their nudibranch predators, were listed by Thompson (1962). In addition to *Polycera dubia*, *Onchidoris muricata* and the un-named predators on *Trematooecia*, all mentioned above as preying on ascophorans, *Onchidoris pusilla* (Alder and Hancock) shows preference for an ascophoran, *Escharella immersa*. Usually the operculum is pushed aside or into the zooid, although it may, on occasion, be torn and swallowed. The balance of observations, therefore, supports the assumption (Ryland, 1970, p. 35) that the unmodified anascan zooid, with its extensive and delicate frontal membrane, is vulnerable to predators. Clearly there have been selective pressures which have favoured the evolution of calcified protective walls, in various guises, which characterize the Cheilostomata, particularly the Ascophora. Nevertheless, the anascans are an abundant and successful group: their defensive strategies seem to depend upon the rapid growth that can accompany light calcification (e.g., in *Membranipora mem-*

branacea the growing edge may advance locally at up to about 1 cm day^{-1} (Lutaud, 1961), and in *Bugula* the bushy tufts, up to 8 cm high, form anew every year). And, as pointed out elsewhere (Ryland, 1970), it is unlikely to be a coincidence that the most elaborate and mobile avicularia have evolved in the Bugulidae and related families which retain an extensive and generally unprotected frontal membrane. The nodding of pedunculate avicularia, through an arc of nearly 180°, greatly increases their chances of contact with a marauding organism (Kaufmann, 1971). Nevertheless, even large avicularia, as occur in *Beania magellanica*, are no defence against many predators; the major role(s) of avicularia remain a subject for speculation.

X. Acknowledgements

My attempt to provide an up-to-date review has been greatly assisted by the kindness of colleagues in supplying manuscripts or notes covering publications in the press or in preparation: particularly, Miss P. L. Cook, Mr. P. E. J. Dyrynda, Dr Å. Franzén, Dr D. P. Gordon, Dr M. C. Miller, Professor Lars Silén, Drs J. D. and D. F. Soule, Dr A. R. D. Stebbing and Dr R. F. Woollacott. The typescript, or parts of it, has been read and commented on by Miss P. L. Cook, Mr W. G. Fry, Dr P. J. Hayward and Mr J. P. Thorpe, although the responsibility for the final content rests solely with me. It gives me great pleasure to thank those mentioned and to express my appreciation to them for their interest and help. I am indebted to Dr C. Nielsen for providing translations of Mawatari's two papers in Japanese (1947, 1948). Finally, I wish to thank the Editors for their great forbearance over several months.

XI. Appendix

List of Species of Bryozoa Mentioned

PHYLACTOLAEMATA

Plumatella casmiana Oka
P. repens (L.)

STENOLAEMATA (Cyclostomata)

Berenicea patina (Lamarck) [syn. *Plagioecia patina*]
Crisia aculeata Hassall
C. eburnea (L.)
C. ramosa Harmer
Crisiella producta (Smitt)
Diastopora suborbicularis Hincks
Diplosolen obelia (Johnston)
Disporella hispida (Fleming)
Idmidronea atlantica (Forbes in Johnston)
Lichenopora novaezelandiae (Busk)
L. radiata (Audouin)
Plagioecia patina (Lamarck) [syn. *Berenicea patina*]
Tubulipora liliacea (Pallas)
T. misakiensis Okada
T. phalangea Couch
T. plumosa Harmer
T. pulchra MacGillivray

GYMNOLAEMATA (Ctenostomata)

Alcyonidium albidum Alder
A. hirsutum (Fleming)
A. polyoum (Hassall)
Anguinella palmata van Beneden
Arachnidium hippothoides Hincks
Bowerbankia gracilis Leidy
B. imbricata (Adams)
B. pustulosa (Ellis and Solander)
Elzerina binderi (Busk)
Farrella repens (Farre)
Flustrellidra hispida (Fabricius)
Hypophorella expansa Ehlers
Pherusella tubulosa (Ellis and Solander)
Sundanella sibogae (Harmer)
Triticella korenii G. O. Sars
Zoobotryon verticillatum (della Chiaje)

Cheilostomata (Anasca)

Aetea sica (Couch)
Alderina imbellis (Hincks)
Beania discodermae (Ortmann)
B. magellanica (Busk)
Bicellariella ciliata (L.)
Bugula avicularia (L.)
B. dentata (Lamouroux)
B. flabellata (Thompson in Gray)
B. neritina (L.)
B. plumosa (Pallas)
B. sabatieri Calvet [= *B. simplex*]
B. simplex Hincks
B. stolonifera Ryland
B. turbinata Alder
B. turrita (Desor)
Callopora dumerilii (Audouin)
C. lineata (L.)
Carbasea papyrea (Pallas)
Cellaria sinuosa (Hassall)
Conopeum reticulum (L.)
C. seurati (Canu)
C. tenuissimum (Canu)
Crassimarginatella maderensis (Waters)
C. papulifera (MacGillivray)
Cribrilina annulata (Fabricius)
Electra crustulenta (Pallas)
E. monostachys (Busk)
E. pilosa (L.)
E. posidoniae Gautier
E. verticillata (Ellis and Solander)
Flustra foliacea (L.)
Kinetoskias smitti (Danielssen)
Membranipora commensale (Kirkpatrick and Metzelaar)
M. hyadesi Jullien
M. membranacea (L.)
M. serrilamella Osburn
M. tuberculata (Bosc)
M. villosa Hincks
Membraniporella nitida (Johnston)
Micropora mortenseni Livingstone
Nellia oculata Busk
Retevirgula acuta (Hincks)
Scruparia ambigua (d'Orbigny)
Scrupocellaria reptans (L.)
S. scruposa (L.)
Setosellina goesi (Silén)
Smittipora levinseni (Canu and Bassler)
Steginoporella neozelanica (Busk)
Synnotum aegyptiacum (Audouin)
Thalamoporella gothica indica (Hincks)

Cheilostomata (Ascophora)

Cellarinella roydsi Rogick
C. foveolata Waters
Celleporaria aperta (Hincks)
C. brunnea (Hincks)
C. mordax (Marcus)
C. pumicosa (Pallas)
Celleporella hyalina (L.) [syn. *Hippothoa hyalina* (L.)]
Celleporina costazii (Audouin)
Chorizopora brongniartii (Audouin)
Cleidochasma contractum (Waters)
Cryptosula pallasiana (Moll)
Escharella immersa (Fleming)
E. ventricosa (Hassall)
Escharina spinifera (Johnston)
Escharoides angela (Hutton)
E. coccinea (Abildgaard)
Eurystomella foraminigera (Hincks)
Exochella tricuspis (Hincks)
Fenestrulina malusii (Audouin)
F. malusii var. *thyreophora* Busk
E. thyreophora (Busk) [syn. *F. malusii* var. *thyreophora*]
Haplopoma graniferum (Johnston)
H. impressum (Audouin)

Hippaliosina acutirostris Canu and Bassler
Hippodiplosia americana (Verrill)
Hippopodina feegeensis (Busk)
Hippopodinella adpressa (Busk)
Hippoporella gorgonensis Hastings
Hippoporidra edax (Busk)
H. senegambiensis (Carter)
Hippoporina pertusa (Esper)
H. porosa (Verrill)
Hippothoa divaricata var. *conferta* Hincks
Hippothoa hyalina (L.) [syn. *Celleporella hyalina* (L.)]
Margaretta triplex Harmer
Metrarabdotos unguiculatum Canu and Bassler
Microporella ciliata (Pallas)
Parasmittina trispinosa (Johnston)
Pentapora foliacea (Ellis and Solander)
Porella concinna (Busk)
Poricellaria rationiensis (Waters)
Reteporella graeffei (Kirchenpauer)
Rhynchozoon phrynoglossum Marcus
R. rostratum (Busk)
Schizomavella auriculata (Hassall)
S. immersa Powell
Schizoporella carvalhoi Marcus
S. errata (Waters)
S. unicornis (Johnston in Wood)
Smittina papulifera (MacGillivray)
Stylopoma duboisii (Audouin)
S. spongites (Pallas)
Trematooecia magnifica (Osburn)
T. turrita (Smitt)
Turbicellepora avicularis (Hincks)
Umbonula littoralis Hastings
Vittaticella contei (Audouin)
V. elegans (Busk)
Watersipora arcuata Banta
W. cucullata (Busk)

XII. References

Abbott, M. B. (1973). Intra- and intercolony variation in populations of *Hippoporina* Neviani (Bryozoa—Cheilostomata). *In* "Animal Colonies: Development and Function Through Time" (R. S. Boardman, A. H. Cheetham and W. A. Oliver, eds.), pp. 223–245. Dowden, Hutchinson and Ross, Stroudsburg.

Abbott, M. B. (1975). Relationship of temperature to patterns of sexual reproduction in some recent encrusting Cheilostomata. *In* "Bryozoa 1974" (S. Pouyet, ed.), Fasc. 1. *Documents des Laboratoires de Géologie de la Faculté des Sciences de Lyon*, Hors Série, **3,** 37–50.

Androsova, E. I. (1973). Bryozoa Cheilostomata (Anasca) of the Antarctic and Subantarctic. *In* "Living and Fossil Bryozoa" (G. P. Larwood, ed.), pp. 369–373. Academic Press, London and New York.

Annoscia, E. (Ed.) (1968). Proceedings of the first international conference on Bryozoa. *Atti della Società Italiana di Scienze Naturali, Milano*, **108,** 1–377.

Atkins, D. (1932). The ciliary feeding mechanism of the entoproct Polyzoa, and a comparison with that of the ectoproct Polyzoa. *Quarterly Journal of Microscopical Science*, **75,** 393–423.

Atkins, D. (1955*a*). The cyphonautes larvae of the Plymouth area and the metamorphosis of *Membranipora membranacea* (L.). *Journal of the Marine Biological Association of the United Kingdom*, **34,** 441–449.

Atkins, D. (1955b). The ciliary feeding mechanism of the cyphonautes larva (Polyzoa Ectoprocta). *Journal of the Marine Biological Association of the United Kingdom*, **34,** 451–466.

Aymes, Y. (1956). Croissance phototropique chez les bryozoaires du genre *Bugula*. *Comptes rendus hebdomadaire des Séances de l'Académie des Sciences*, **242,** 1237–1238.

Banta, W. C. (1968). The body wall of cheilostome Bryozoa, I. The ectocyst of *Watersipora nigra* (Canu and Bassler). *Journal of Morphology*, **125,** 497–508.

Banta, W. C. (1969). The body wall of cheilostome Bryozooa. II. Interzooidal communication organs. *Journal of Morphology*, **129,** 149–170.

Banta, W. C. (1970). The body wall of cheilostome Bryozoa, III. The frontal wall of *Watersipora arcuata* Banta, with a revision of the Cryptocystidea. *Journal of Morphology*, **131,** 37–56.

Banta, W. C. (1972). The body wall of cheilostome Bryozoa, V. Frontal budding in *Schizoporella unicornis floridana*. *Marine Biology*, **14,** 63–71.

Banta, W. C., McKinney, F. K. and Zimmer, R. L. (1974). Bryozoan monticules: excurrent water outlets? *Science, N.Y.*, **185,** 783–784.

Barrois, J. (1877). "Mémoire sur l'embryologie des Bryozoaires". 305 pp. Thèse de l'Université de Lille.

Bassindale, R., Ebling, F. J., Kitching, J. A. and Purchon, R. D. (1948). The ecology of the Lough Ine rapids with special reference to water currents. I. Introduction and hydrography. *Journal of Ecology*, **36,** 305–322.

Bishop, J. W. and Bahr, L. M. (1973). Effects of colony size on feeding by *Lophopodella carteri* (Hyatt). *In* "Animal Colonies: Development and Function Through Time" (R. S. Boardman, A. H. Cheetham and W. A. Oliver, eds.), pp. 433–437. Dowden, Hutchinson and Ross, Stroudsburg.

Boaden, P. J. S., O'Connor, R. J. and Seed, R. (1975). The composition and zonation of a *Fucus serratus* community in Strangford Lough, Co. Down. *Journal of Experimental Marine Biology and Ecology*, **17,** 111–136.

Boardman, R. S. (1968). Colony development and convergent evolution of budding pattern in "rhombotrypid" Bryozoa. *Atti della Società Italiana di Scienze Naturali, Milano*, **108,** 179–184.

Boardman, R. S. and Cheetham, A. H. (1969). Skeletal growth, intra-colony variation, and evolution in Bryozoa: a review. *Journal of Paleontology*, **43,** 205–233.

Boardman, R. S. and Cheetham, A. H. (1973). Degrees of colony dominance in stenolaemate and gymnolaemate Bryozoa. *In* "Animal Colonies: Development and Function Through Time (R. S. Boardman, A. H. Cheetham and W. A. Oliver, eds.), pp. 121–220. Dowden, Hutchinson and Ross, Stroudsburg.

Boardman, R. S., Cheetham, A. H. and Cook, P. L. (1970). Intracolony variation and the genus concept in Bryozoa. *Proceedings of the North American Paleontological Convention*, Part C, 294–320.

Bobin, G. (1968). Morphogenèse du termen et des épines dans les zoécies d'*Electra verticillata* (Ellis et Solander) (Bryozoaire Chilostome, Anasca). *Cahiers de Biologie marine*, **9,** 53–68.

Bobin, G. (1971). Histophysiologie du système rosettes-funicule de *Bowerbankia imbricata* (Adams) (Bryozoaire Ctenostome). Les lipides. *Archives de Zoologie expérimentale et générale*, **112,** 771–792.

Bobin, G. (1976). Interzooecial communications and the funicular systems. *In* "The Biology of Bryozoans" (R. M. Woollacott and R. L. Zimmer, eds.). Academic Press, New York.

Bobin, G. and Prenant, M. (1952). Structure et histogénèse du gésier des Vésicularines. *Archives de Zoologie expérimentale et générale*, **89,** 175–201.

Bobin, G. and Prenant, M. (1968). Sur le calcaire des parois autozoéciales d'*Electra verticillata* (Ell. et Sol.), Bryozoaire chilostome, Anasca. Notions préliminaires. *Archives de Zoologie expérimentale et générale*, **109**, 157–191.

Borg, F. (1926). Studies on Recent cyclostomatous Bryozoa. *Zoologiska Bidrag från Uppsala*, **10**, 181–507.

Brien, P. (1960). Classe des Bryozoaires. *In* "Traité de Zoologie" Tom. 5, Fasc. 2 (P.-P. Grassé, ed.), pp. 1054–1335. Masson et Cie, Paris.

Briggs, J. C. (1974). "Marine zoogeography", 475 pp. McGraw-Hill, New York.

Bronstein, G. (1937). Etude du système nerveux de quelques Bryozoaires Gymnolémides. *Travaux de la Station biologique de Roscoff*, **15**, 155–174.

Bullivant, J. S. (1967). Release of sperm by Bryozoa. *Ophelia*, **4**, 139–142.

Bullivant, J. S. (1968a). The rate of feeding of the bryozoan, *Zoobotryon verticillatum*. *New Zealand Journal of Marine and Freshwater Research*, **2**, 111–134.

Bullivant, J. S. (1968b). The method of feeding of lophophorates (Bryozoa, Phoronida, Brachiopoda.) *New Zealand Journal of Marine and Freshwater Research*, **2**, 135–146.

Bullivant, J. S. (1969). Distribution of selected groups of marine invertebrates in waters south of 35°S latitude: Bryozoa. Folio eleven, Antarctic map folio series, pp. 22–23. American Geographical Society.

Bullivant, J. S. and Bils, R. F. (1968). The pharyngeal cells of *Zoobotryon verticillatum* (Della Chiaje), a gymnolaemate bryozoan. *New Zealand Journal of Marine and Freshwater Research*, **2**, 438–446.

Bullock, T. H. and Horridge, G. A. (1965). "Structure and Function in the Nervous Systems of Invertebrates", 2 vols, 1719 pp. W. H. Freeman, San Francisco.

Bushnell, J. H. (1966). Environmental relations of Michigan Ectoprocta, and the dynamics of natural populations of *Plumatella repens*. *Ecological Monographs*, **36**, 95–123.

Calvet, L. (1900). Contributions à l'histoire naturelle des Bryozoaires ectoproctes marins. *Travaux de l'Institut de zoologie de l'Université de Montpellier et de la Station zoologique de Cette*, **8**, 1–488.

Cerame-Vivas, M. J. and Gray, I. E. (1966). The distributional pattern of benthic invertebrates on the continental shelf off North Carolina. *Ecology*, **47**, 260–270.

Cheetham, A. H. (1968). Morphology and systematics of the bryozoan genus *Metrarabdotos*. *Smithsonian Miscellaneous Collections*, **153**, 1–121.

Clark, P. J. and Evans, F. C. (1954). Distance to nearest neighbor as a measure of spatial relationships in populations. *Ecology*, **35**, 445–453.

Cook, P. L. (1960). The development of *Electra crustulenta* (Pallas) (Polyzoa, Ectoprocta). *Essex Naturalist*, **30**, 258–266.

Cook, P. L. (1962). The early development of *Membranipora seurati* (Canu) and *Electra crustulenta* (Pallas), Polyzoa. *Cahiers de Biologie marine*, **3**, 57–60.

Cook, P. L. (1968a). Bryozoa (Polyzoa) from the coasts of tropical West Africa. *Atlantide Report*, **10**, 115–262.

Cook, P. L. (1968b). Observations on living Bryozoa. *Atti della Società Italiana di Scienze Naturali, Milano*, **108**, 155–160.

Cook, P. L. (1973). Settlement and early colony development in some Cheilostomata. *In* "Living and Fossil Bryozoa", (G. P. Larwood, ed.), pp. 65–71. Academic Press, London and New York.

Cook, P. L. (1976). Colony-wide water currents in living Bryozoa. *Cahiers de Biologie marine*, in the press.

Cook, P. L. and Hayward, P. J. (1966). The development of *Conopeum seurati* (Canu), and some other species of membraniporine Polyzoa. *Cahiers de Biologie marine*, **7**, 437–443.

Cori, C. J. (1941). Bryozoa. *Handbuch der Zoologie*. Bd. **3**(5), 263–502.

Corrêa, D. D. (1948). A embryologia de *Bugula flabellata* (J. V. Thompson) (Bryozoa Ectoprocta). *Boletim da Faculdade de Filosofia, Ciências e Letras*, (Zool.), **13**, 7–71.

Crisp, D. J. (1961). Territorial behaviour in barnacle settlement. *Journal of Experimental Biology*, **38**, 419–446.

Crisp, D. J. (1973). Mechanisms of adhesion of fouling organisms. *In* "Proceedings of the 3rd International Congress on Marine Corrosion and Fouling" (R. F. Acker *et al.*, eds.), pp. 691–699. National Bureau of Standards, Gaithersburg.

Crisp, D. J. (1974). Factors influencing the settlement of marine invertebrate larvae. *In* "Chemoreception in Marine Organisms" (P. T. Grant and A. M. Mackie, eds.), pp. 177–265. Academic Press, London and New York.

Crisp, D. J. (1976). The role of the pelagic larva. *In* "Perspectives in Experimental Biology", Vol. 1, (P. Spencer Davies, ed.) pp. 145–155. Pergamon Press, Oxford.

Crisp, D. J. and Ryland, J. S. (1960). Influence of filming and of surface texture on the settlement of marine organisms. *Nature, London*, **185**, 119 only.

Crisp, D. J. and Williams, G. B. (1960). Effect of extracts from fucoids in promoting settlement of epiphytic Polyzoa. *Nature, London*, **188**, 1206–1207.

Cuffey, R. J. (1973). Bryozoan distribution in the modern reefs of Eniwetok Atoll and the Bermuda Platform. *Pacific Geology*, **6**, 25–50.

Cumings, E. R. (1904). Development of some Paleozoic Bryozoa. *American Journal of Science*, **17**, 49–78.

Doak, W. (1971). "Beneath New Zealand Seas", 113 pp. A. H. and A. W. Reed, Wellington.

Dudley, J. E. (1973). Observations on the reproduction, early larval development, and colony astogeny of *Conopeum tenuissimum* (Canu). *Chesapeake Science*, **14**, 270–278.

Dudley, J. W. (1970). Differential utilization of phytoplankton food resources by marine ectoprocts. *Biological Bulletin, Marine Biological Laboratory, Woods Hole*, **139**, 420 only.

Eakin, R. M. (1968). Evolution of photoreceptors. *Evolutionary Biology*, **2**, 194–242.

Ebling, F. J., Kitching, J. A., Purchon, R. D. and Bassindale, R. (1948). The ecology of the Lough Ine rapids with special reference to water currents. II. The fauna of the *Saccorhiza* canopy. *Journal of Animal Ecology*, **17**, 223–244.

Edmunds, M. (1975). An eolid nudibranch feeding on Bryozoa. *The Veliger*, **17**, 269–270.

Eggleston, D. (1970). Embryo colour in Manx ectoprocts. *Report of the Marine Biological Station, Port Erin*, **82**, 39–42.

Eggleston, D. (1971). Synchronization between moulting in *Calocaris macandreae* (Decapoda) and reproduction in its epibiont *Tricella* [sic] *koreni* (Polyzoa Ectoprocta). *Journal of the Marine Biological Association of the United Kingdom*, **51,** 409–410.

Eggleston, D. (1972a). Patterns of reproduction in marine Ectoprocta of the Isle of Man. *Journal of Natural History*, **6,** 31–38.

Eggleston, D. (1972b). Factors influencing the distribution of sub-littoral ectoprocts off the south of the Isle of Man (Irish Sea). *Journal of Natural History*, **6,** 247–260.

Eitan, G. (1972). Types of metamorphosis and early astogeny in *Hippopodina feegeensis* (Busk) (Bryozoa: Ascophora). *Journal of Experimental Marine Biology and Ecology*, **8,** 27–30.

Elmhirst, R. (1922). Habits of *Echinus esculentus*. *Nature, London*, **110,** 667.

Fiala-Médioni, A. (1974). Les peuplements sessiles des fonds rocheux de la région de Banyuls-sur-Mer: Ascidies—Bryozoaires. (2e partie et fin). *Vie et Milieu*, B, **21,** 143–182.

Fisher, R. A. and Yates, F. (1963). "Statistical Tables for Biological, Agricultural and Medical Research", 6th Edn. Oliver and Boyd, Edinburgh and London.

Fraenkel, G. S. and Gunn, D. L. (1940). "The Orientation of Animals". Oxford University Press, London.

Franzén, Å. (1956). On spermiogenesis, morphology of the spermatozoon, and biology of fertilization among invertebrates. *Zoologiska Bidrag från Uppsala*, **31,** 356–482.

Franzén, Å. (1960). *Monobryozoon limicola* n. sp., a ctenostomatous bryozoan from the detritus layer on soft sediment. *Zoologiska Bidrag från Uppsala*, **33,** 135–147.

Franzén, Å. (1976). Gametogenesis of bryozoans. *In* "The Biology of Bryozoans" (R. M. Woollacott and R. L. Zimmer, eds.). Academic Press, New York and London.

Fry, W. G. (1965). The feeding mechanisms and preferred foods of three species of Pycnogonida. *Bulletin of the British Museum* (*Natural History*), (Zoology), **12,** 197–223.

Gautier, Y. (1954). Sur l'*Electra pilosa* des feuilles de Posidonies. *Vie et Milieu*, **5,** 66–70.

Gautier, Y. V. (1962). Recherches écologiques sur les Bryozoaires Chilostomes en Mediterranée occidentale. *Recueil des Travaux de la Station marine d'Endoume*, **38,** 1–434.

Geraci, S. (1976). Infralittoral bryozoans settled on artificial substrata in the Ligurian Sea. *In* "Bryozoa 1974" (S. Pouyet, ed.) Fasc. 2. *Documents des Laboratoires de Géologie de la Faculté des Sciences de Lyon*, Hors Série **3,** 335–346.

Geraci, S. and Relini, G. (1970). Osservazioni sistematico-ecologiche sui Briozoi del fouling portuale di Genova. *Bolletino dei Musei e degli Istituti Biologici dell'Università di Genova*, **38,** 103–139.

Gerwerzhagen, A. (1913). Untersuchungen an Bryozoen. *Sitzungsberichte der Heidelbergen Akademie der Wissenschaften*, B, **1913** (9).

Gordon, D. P. (1968). Zooidal dimorphism in the polyzoan *Hippopodinella adpressa* (Busk). *Nature, London*, **219,** 633–634.

Gordon, D. P. (1970). Reproductive ecology of some northern New Zealand Bryozoa. *Cahiers de Biologie marine*, **11,** 307–323.

Gordon, D. P. (1971a). Colony formation in the cheilostomatous bryozoan *Fenestrulina malusii* var. *thyreophora*. *New Zealand Journal of Marine and Freshwater Research*, **5,** 342–351.

Gordon, D. P. (1971b). Zooidal budding in the cheilostomatous bryozoan *Fenestrulina malusii* var. *thyreophora*. *New Zealand Journal of Marine and Freshwater Research*, **5,** 453–460.

Gordon, D. P. (1972). Biological relationships of an intertidal bryozoan population. *Journal of Natural History*, **6,** 503–514.

Gordon, D. P. (1973). A fine-structure study of brown bodies in the Gymnolaemate *Cryptosula pallasiana* (Moll). *In* "Living and Fossil Bryozoa" (G. P. Larwood, ed.), pp. 275–286. Academic Press, London.

Gordon, D. P. (1974). Microarchitecture and function of the lophophore of a marine bryozoan. *Marine Biology*, **27,** 147–163.

Gordon, D. P. (1975a). Ultrastructure and function of the gut of a marine bryozoan. *Cahiers de Biologie marine*, **16,** 367–382.

Gordon, D. P. (1975b). The resemblance of bryozoan gizzard teeth to "annelid-like" setae. *Acta zoologica*, **56,** 283–289.

Gordon, D. P. (1975c). The occurrence of a gizzard in a bryozoan of the order Cheilostomata. *Acta zoologica*, **56,** 279–282.

Gordon, D. P. (1975d). Ultrastructure of communication pore areas in two bryozoans. *In* "Bryozoa 1974" (S. Pouyet, ed.), Fasc. 1. *Documents des Laboratoires de Géologie de la Faculté des Sciences de Lyon*, Hors Série, **3,** 187–192.

Gordon, D. P. (1976). The ageing process in bryozoans. *In* "The Biology of Bryozoans" (R. F. Woollacott and R. L. Zimmer, eds.). Academic Press, New York and London.

Gordon, D. P. and Hastings, A. B. (in prep.) Interzooidal communications in *Hippothoa* (Bryozoa, Cheilostomata).

Graupner, H. (1930). Zur Kenntnis der feineren Anatomie der Bryozoen. *Zeitschrift für wissenschaftliche Zoologie*, **136,** 38–97.

Grave, B. H. (1930). The natural history of *Bugula flabellata* at Woods Hole, Massachusetts, including the behavior and attachment of the larva. *Journal of Morphology*, **49,** 355–383.

Gray, J. S. (1971). Occurrence of the aberrant bryozoan *Monobryozoon ambulans* Remane, off the Yorkshire coast. *Journal of Natural History*, **5,** 113–117.

Haderlie, E. C. (1970). Marine fouling and boring organisms in Monterey Harbor. II. Second year of investigation. *The Veliger*, **12,** 182–192.

Harmelin, J. G. (1969a). Bryozoaires des grottes sous-marins obscures de la région marseillaise: faunistique et écologie. *Tethys*, **1,** 793–806.

Harmelin, J. G. (1969b). Bryozoaires récoltés au cours de la campagne du *Jean Charcot* en Mediterranée orientale (Août–Septembre 1967). I. Dragages. *Bulletin de Museum d'Histoire naturelle, Paris*, **40,** 1179–1208; **41,** 295–311.

Harmelin, J. G. (1973a). Les Bryozoaires des peuplements sciaphiles de Mediterranée: le genre *Crassimarginatella* Canu (Chilostomes Anasca). *Cahiers de Biologie marine*, **14,** 471–492.

Harmelin, J. G. (1973b). Bryozoaires de l'herbier de Poisdonies de l'île de Port-Cros. *Rapports de la Commission internationale de la Mer Mediterranée*, **21,** 675–677.

Harmelin, J. G. (1973c). Morphological variations and ecology of the Recent cyclostome bryozoan "*Idmonea*" *atlantica* from the Mediterranean. *In* "Living and Fossil Bryozoa" (G. P. Larwood, ed.), pp. 95–106. Academic Press, London and New York.

Harmelin, J. G. (1974). Les bryozoaires cyclostomes de Méditerranée : écologie et systemique. *Thèse de l'Université d'Aix-Marseille.*

Harmelin, J. G. (1976). Relations entre la forme zoariale et l'habitat chez les Bryozoaires cyclostomes. Consequences toxonomiques. *In* " Bryozoa 1974 " (S. Pouyet, ed.), Fasc. 2. *Documents des Laboratories de Géologie de la Faculté des Sciences de Lyon*, Hors Série **3,** 369–384.

Harmer, S. F. (1891). On the nature of the excretory processes in marine Polyzoa. *Quarterly Journal of Microscopical Science*, **33,** 123–167.

Harmer, S. F. (1893). On the occurrence of embryonic fission in cyclostomatous Polyzoa. *Quarterly Journal of Microscopical Science*, **34,** 199–241.

Harvey, P. H. Ryland, J. S. and Hayward, P. J. (1976a). Niche breadth in bryozoans. *Nature, London* **260,** 77 only.

Harvey, P. H., Ryland, J. S. and Hayward, P. J. (1976b). Pattern analysis in bryozoan and spirorbid communities. II. Distance sampling methods. *Journal of Experimental Marine Biology and Ecology*, **21,** 99–108.

Hastings, A. B. (1927). Zoological results of the Cambridge expedition to the Suez Canal, 1924. No. 20, Report on the Polyzoa. *Transactions of the Zoological Society of London*, **22,** 321–354.

Hayward, P. J. (1973). Preliminary observations on settlement and growth in populations of *Alcyonidium hirsutum* (Fleming). *In* "Living and Fossil Bryozoa" (G. P. Larwood, ed.), pp. 107–113. Academic Press, London and New York.

Hayward, P. J. (1974). Studies on the cheilostome bryozoan fauna of the Aegean island of Chios. *Journal of Natural History*, **8,** 369–402.

Hayward, P. J. (1976). Observations on the bryozoan epiphytes of *Posidonia oceanica* from the island of Chios (Aegean Sea). *In* "Bryozoa 1974" (S. Pouyet, ed.), Fasc. 2. *Documents des Laboratoires de Géologie de la Faculté des Sciences de Lyon*, Hors Série **3,** 347–356.

Hayward, P. J. and Harvey, P. H. (1974a). The distribution of settled larvae of the bryozoans *Alcyonidium hirsutum* (Fleming) and *Alcyonidium polyoum* (Hassall) on *Fucus serratus* L. *Journal of the Marine Biological Association of the United Kingdom*, **54,** 665–676.

Hayward, P. J. and Harvey, P. H. (1974b). Growth and mortality of the bryozoan *Alcyonidium hirsutum* (Fleming) on *Fucus serratus* L. *Journal of the Marine Biological Association of the United Kingdom*, **54,** 677–684.

Hayward, P. J. and Ryland, J. S. (1975). Growth, reproduction and larval dispersal in *Alcyonidium hirsutum* (Fleming) and some other Bryozoa. *Pubblicazioni della Stazione zoologica di Napoli*, **39,** Suppl. 1, 226–241.

Herring, P. J. (1972). Observations on the distribution and feeding habits of some littoral echinoids from Zanzibar. *Journal of Natural History*, **6,** 169–175.

Hincks, T. (1880). "A History of the British Marine Polyzoa". Vols 1 and 2. Van Voorst, London.

Hiller, S. (1939). The so-called "colonial nervous system" in Bryozoa. *Nature London*, **143,** 1069–1970.

Hondt, J. L. d' (1972). Metamorphose de la larve d'*Alcyonidium polyoum* (Hassall), Bryozoaire Cténostome. *Compte rendu hebdomadaire des Séances de l'Académie des Sciences, Paris*, Ser. D, **275,** 767–770.

Hondt, J. L. d' (1973). Etude anatomique et cytologique de la larve d'*Alcyonidium polyoum* (Hassall, 1841), Bryozoaire Cténostome. *Archives de zoologie expérimentale et générale*, **114,** 537–602.

Hondt, J. L. d' (1974). La métamorphose larvaire et la formation du "cystide" chez *Alcyonidium polyoum* (Hassall, 1841), Bryozoaire Cténostome. *Archives de Zoologie expérimentale et générale*, **115,** 577–605.

Hondt, J. L. d' (1975). Etude anatomique et cytologique comparée de quelques larves de Bryozoaires Cténostomes. *In* "Bryozoa 1974" (S. Pouyet, ed.), Fasc. 1. *Documents des Laboratoires de Géologie de la Faculté des Sciences de Lyon*, Hors Série **3,** 125–134.

Hondt, J. L. d' (1976). Bryozoaires cténostomes bathyaux et abyssaux de l'Atlantique nord. *In* "Bryozoa 1974" (S. Pouyet, ed.), Fasc. 2. *Documents des Laboratoires de Géoloqie de la Faculté des Sciences de Lyon*, Hors Serie **3,** 311–333.

Hornsey, I. S. and Hide, D. (1974). The production of antimicrobial compounds by British marine Algae, I. Antibiotic-producing marine Algae. *British Phycological Journal*, **9,** 353–361.

Hornsey, I. S. and Hide, D. (1976). The production of antimicrobial compounds by British marine Algae, III. Distribution of antimicrobial activity within the algal thallus. *British Phycological Journal*, **11,** 175–181.

Humphries, E. M. (1973). Seasonal settlement of bryozoans in Rehoboth Bay, Delaware, U.S.A. *In* "Living and Fossil Bryozoa" (G. P. Larwood, ed.), pp. 115–128. Academic Press, London and New York.

Hunt, O. D. (1925). Food of the bottom fauna of the Plymouth fishing grounds. *Journal of the Marine Biological Association of the United Kingdom*, **13,** 560–599.

Hyman, L. H. (1959). "The Invertebrates: Smaller Coelomate Groups". Vol. 5. McGraw-Hill, New York.

Jackson, J. B. C. and Buss, L. (1975). Allelopathy and spatial competition among coral reef invertebrates. *Proceedings of the National Academy of Sciences*, **72,** 5160–5163.

Jägersten, G. (1972). "Evolution of the Metazoan Life Cycle". Academic Press, London and New York.

Jebram, D. (1968). A cultivation method for saltwater Bryozoa and an example for experimental biology. *Atti della Società Italiana di Scienze Naturali, Milano*, **108,** 119–128.

Jebram, D. (1970). Preliminary experiments with Bryozoa in a simple apparatus for producing continuous water currents. *Helgoländer wissenschaftliche Meeresuntersuchungen*, **20,** 278–292.

Jebram, D. (1973). Preliminary observations on the influences of food and other factors on the growth of Bryozoa. *Kieler Meeresforschungen*, **29,** 50–57.

Jebram, D. (1975). Effect of different foods on *Conopeum seurati* (Canu) (Bryozoa Cheilostomata) and *Bowerbankia gracilis* Leidy (Bryozoa Ctenostomata). *Documents des Laboratoires de Géologie de la Faculté des Sciences de Lyon*, Hors Série **3,** 97–108.

Joliet, L. (1877). Contributions à l'histoire naturelle des bryozoaires des côtes de France. *Archives de Zoologie expérimentale et générale*, **6,** 193–304.

Kaufmann, K. W. (1971). The form and functions of the avicularia of *Bugula* (phylum Ectoprocta). *Postilla*, **151,** 1–26.

Kerneis, A. (1960). Contribution à l'étude faunistique et écologique des herbiers de Posidonies de la région de Banyuls. *Vie et Milieu*, **11,** 145–187.

Kitching, J. A. and Ebling, F. J. (1967). Ecological studies at Lough Ine. *Advances in Ecological Research*, **4,** 197–291.

Knight-Jones, E. W. (1954). Relations between metachronism and the direction of ciliary beat in Metazoa. *Quarterly Journal of Microscopical Science*, **95,** 503–521.

Knight-Jones, E. W. and Moyse, J. (1961). Intraspecific competition in sedentary marine animals. *Symposia of the Society for Experimental Biology*, **15,** 72–95.

Kupelwieser, H. (1906). Untersuchungen über den feineren Bau und die Metamorphose des Cyphonautes. *Zoologica, Stuttgart*, **47,** 1–50.

Lagaaij, R. (1963). *Cupuladria canariensis* (Busk)—portrait of a bryozoan. *Palaeontology*, **6,** 172–217.

Lang, J. (1973). Interspecific aggression by scleractinian corals. 2. Why the race is not only to the swift. *Bulletin of Marine Science*, **23,** 260–279.

Larwood, G. P. (ed.) (1973). "Living and fossil Bryozoa: recent advances in research". Academic Press, London and New York.

Levinsen, G. M. R. (1909). "Morphological and systematic studies on the cheilostomatous Bryozoa". Nationale Forfatterers Forlag, Copenhagen.

Lilly, S. J., Sloane, J. F., Bassindale, R., Ebling, F. J. and Kitching, J. A. (1953). The ecology of the Lough Ine rapids with special reference to water currents. IV. The sedentary fauna of sublittoral boulders. *Journal of Animal Ecology*, **22,** 87–122.

Lutaud, G. (1953). Progression de la calcification au cours de la métamorphose de la larve chez *Escharoides coccinea* Abildgaard, Bryozoaire Chilostome. *Archives de Zoologie expérimentale et générale*, **91,** 36–50.

Lutaud, G. (1955). Sur la ciliature du tentacule chez les bryozoaires chilostomes. *Archives de Zoologie expérimentale et générale*, **92,** 13–19.

Lutaud, G. (1961). Contribution a l'étude du bourgeonnement et de la croissance des colonies chez *Membranipora membranacea* (Linné), Bryozoaire Chilostome. *Annales de la Societé royale zoologique de Belgique*, **91,** 157–300.

Lutaud, G. (1969). Le "plexus" pariétal de Hiller et la coloration du système nerveux par le bleu de méthylène chez quelques Bryozoaires Chilostomes. *Zeitschrift für Zellforschung und mikroskopische Anatomie*, **99,** 302–314.

Lutaud, G. (1973a). L'innervation du lophophore chez le Bryozoaire chilostome *Electra pilosa* (L.). *Zeitschrift für Zellforschung und mikroskopische Anatomie*, **140,** 217–234.

Lutaud, G. (1973b). The great tentacle sheath nerve as the path of an innervation of the frontal wall structures in the cheilostome *Electra pilosa* (Linné). *In* "Living and Fossil Bryozoa" (G. P. Larwood, ed.), pp. 317–326. Academic Press, London and New York.

Lutaud, G. (1974). Le plexus pariétal des Cténostomes chez *Bowerbankia gracilis* Leydi [sic] (Vésicularines). *Cahiers de Biologie marine*, **15,** 403–408.

Lynch, W. F. (1947). The behavior and metamorphosis of the larvae of *Bugula neritina* (Linnaeus): experimental modification of the length of the free-swimming period and the responses of the larvae to light and gravity. *Biological Bulletin, Marine Biological Laboratory, Woods Hole*, **92,** 115–150.

Marcus, E. (1926). Beobachtungen und Versuche an lebenden Meeresbryozoen. *Zoologische Jahrbücher* (*Systematik*), **52,** 1–102.

Marcus, E. (1938). Bryozoarios marinhos Brasileiros, II. *Boletim da Faculdada de Filosofia, Ciências e Letras, São Paulo,* (Zoologia), **2,** 1–196.

Marcus, E. (1939). Bryozoarios marinhos brasileiros, III. *Boletim da Faculdada de Filosofia, Ciências e Letras, São Paulo,* (Zoologia), **3,** 111–354.

Marcus, E. (1941). Sobre os Briozoa do Brasil. *Boletim da Faculdada de Filosofia, Ciências e Letras, São Paulo,* (Zoologia), **5,** 3–208.

Matricon, I. (1973). Quelques données ultrastructurales sur un myoépithelium: le pharynx d'un Bryozoaire. *Zeitschrift für Zellforschung und mikrosckopische Anatomie,* **136,** 569–578.

Maturo, F. J. S. (1968). The distributional pattern of the Bryozoa of the east coast of the United States exclusive of New England. *Atti della Società Italiana di Scienze Naturali, Milano,* **108,** 261–284.

Mawatari, S. (1947). On the attachment of the larvae of *Tubulipora pulchra* MacGillivray. *Zoological Magazine, Tokyo,* **57,** 49–50. (In Japanese.)

Mawatari, S. (1948). On the metamorphosis of *Tubulipora misakiensis* Okada. *Zoological Magazine, Tokyo,* **58,** 27–28. (In Japanese.)

Mawatari, S. (1951). The natural history of a common fouling bryozoan, *Bugula neritina* (Linnaeus). *Miscellaneous Reports of the Research Institute for Natural Resources, Tokyo,* **20,** 47–54.

Mawatari, S. and Mawatari, S. F. (1975). Development and metamorphosis of the cyphonautes larva of *Membranipora serrilamella* Osburn. *Documents des Laboratoires de Géologie de la Faculté des Sciences de Lyon,* Hors Serie **3,** 13–18.

Mawatari, S. F. and Itô, T. (1972). The morphology of the cyphonautes larva of *Membranipora serrilamella* Osburn from Hokkaido. *Journal of the Faculty of Science, Hokkaido University,* (VI, Zoology), **18,** 400–405.

McDougall, K. D. (1943). Sessile marine invertebrates at Beaufort, North Carolina. *Ecological Monographs,* **13,** 321–374.

Meadows, P. S. (1969). Sublittoral fouling communities on northern coasts of Britain. *Hydrobiologia: Acta hydrobiologia, hydrographica et protistologica,* **34,** 273–294.

Meadows, P. S. and Campbell, J. I. (1972). Habitat selection by aquatic invertebrates. *Advances in Marine Biology,* **10,** 271–382.

Médioni, A. (1972). Les peuplements sessile des fonds rocheux de la région de Banyuls-sur-Mer: Ascidies–Bryozoaires (Première partie). *Vie et Milieu,* B, **21,** 591–656.

Menon, N. R. (1972a). Heat tolerance, growth and regeneration in three North Sea bryozoans exposed to different constant temperatures. *Marine Biology,* **15,** 1–11.

Menon, N. R. (1972b). The growth rates of four species of intertidal bryozoans in Cochin backwaters. *Proceedings of the Indian National Science Academy,* **38B,** 397–402.

Menon, N. R. (1973). Vertical and horizontal distribution of fouling bryozoans in Cochin backwaters, southwest coast of India. *In* "Living and Fossil Bryozoa" (G. P. Larwood, ed.), pp. 153–164. Academic Press, London and New York.

Menon, N. R. (1974). Clearance rates of food suspension and food passage rates as a function of temperature in two North Sea bryozoans. *Marine Biology*, **24,** 65–67.

Menon, N. R. and Nair, N. B. (1971). Ecology of fouling bryozoans in Cochin backwaters. *Marine Biology*, **8,** 280–307.

Miller, M. A. (1946). Toxic effects of copper on the attachment and growth of *Bugula neritina*. *Biological Bulletin, Marine Biological Laboratory, Woods Hole*, **90,** 122–140.

Miller, M. C. (1961). Distribution and food of the nudibranchiate Mollusca of the south of the Isle of Man. *Journal of Animal Ecology*, **30,** 95–116.

Miller, M. C. (1970). Two new species of the genus *Caldukia* Burn and Miller, 1969 (Mollusca: Gastropoda: Opisthobranchia) from New Zealand waters. *The Veliger*, **12,** 279–289.

Milligan, H. N. (1916). Observations on the feeding habits of the purple-tipped sea-urchin. *The Zoologist*, **20,** 81–99.

Moore, P. G. (1973). Bryozoa as a community component on the north-east coast of Britain. *In* "Living and Fossil Bryozoa" (G. P. Larwood, ed.), pp. 21–36. Academic Press, London and New York.

Moore, P. G. (1975). The role of habitat selection in determining the local distribution of animals in the sea. *Marine Behaviour and Physiology*, **3.** 97–100.

Morton, J. (1960). The functions of the gut in ciliary feeders. *Biological Reviews of the Cambridge Philosophical Society*, **35,** 92–140.

Morton, J. and Chapman, V. J. (1968). "Rocky Shore Ecology of the Leigh Area, North Auckland". University of Auckland Press, Auckland.

Morton, J. and Miller, M. (1968). "The New Zealand Sea Shore". Collins, London and Auckland.

Moyano, G., H.I. (1967). Sobre la fusión de dos colonias de *Membranipora hyadesi* Jullien, 1888. *Noticiario mensual del Museo Nacional de Historia Natural, Santiago*, **11** (126), 1–4.

Moyano, G., H.I. and Bustos, H. E. (1974). Distribucion vertical de briozoos sobre algas des genero *Macrocystis* en el Golfo de Arauco. *Boletín de la Sociedad de biologia de Concepción*, **47,** 171–179.

Nielsen, C. (1970). On metamorphosis and ancestrula formation in cyclostomatous bryozoans. *Ophelia*, **8,** 217–256.

Nielsen, C. (1971). Entoproct life-cycles and the entoproct-ectoproct relationship. *Ophelia*, **9,** 209–341.

Nilsson, L. and Jägersten, G. (1959). "Liv i hav". Tidens Förlag, Stockholm.

Nitsche, H. (1869). Beitrage zur Kenntniss der Bryozoen, I Heft. *Zeitschrift für wissenschaftliche Zoologie*, **20** (1), 1–36.

Norton, T. A. (1973). Orientated growth of *Membranipora membranacea* (L.) on the thallus of *Saccorhiza polyschides* (Lightf.) Batt. *Journal of Experimental Marine Biology and Ecology*, **13,** 91–95.

Norton, T. A., Ebling, F. J. and Kitching, J. A. (1971). Light and the distribution of organisms in a sea cave. *In* "Fourth European Marine Biology Symposium" (D. J. Crisp, ed.), pp. 409–432. Cambridge University Press, London.

O'Connor, R. J., Boaden, P. J. S. and Seed, R. (1975). Niche breadth in Bryozoa as a test of competition theory. *Nature, London*, **256,** 307–309.

O'Donoghue, C. H. (1926). Observations on the early development of *Membranipora villosa* Hincks. *Contributions to Canadian Biology and Fisheries*, **3,** 249–263.

Oka, H. (1970). Colony specificity in compound ascidians. *In* "Profiles of Japanese Science and Scientists" (H. Yukawa, ed.), pp. 196–206. Kodansha, Tokyo.

Osburn, R. C. (1921). Bryozoa as food for other animals. *Science, New York*, **53,** 451–453.

Pérès, J. M. (1967). The Mediterranean benthos. *Oceanography and Marine Biology, an Annual Review*, **5,** 449–533.

Pérez, R. (1969). Croissance de *Laminaria digitata* (L.) Lamouroux étudiée sur trois années consecutives. *In* "Proceedings of the 6th International Seaweed Symposium" (R. Margalef, ed.), pp. 329–344. Madrid.

Pielou, E. C. (1962). The use of plant-to-neighbour distances for the detection of competition. *Journal of Ecology*, **50,** 357–367.

Pinter, P. (1969). Bryozoan-algal associations in southern California waters. *Bulletin of the Southern California Academy of Sciences*, **68,** 199–218.

Pouyet, S. (1971). *Schizoporella violacea* (Canu et Bassler, 1930) (Bryozoa Cheilostomata): variations et croissance zoariale. *Geobios*, **4,** 184–197.

Pouyet, S. (ed.) (1975). "Bryozoa 1974", Fasc. 1. *Documents des Laboratoires de Géologie de la Faculté des Sciences de Lyon*, Hors Série, **3,** 1–256.

Pouyet, S. (ed.) (1976). "Bryozoa 1974", Fasc. 2. *Documents des Laboratoires de Géologie de la Faculté des Sciences de Lyon*, Hors Série, **3,** 257–690.

Powell, N. A. (1967). Sexual dwarfism in *Cribrilina annulata* (Cribrilinidae–Bryozoa). *Journal of the Fisheries Research Board of Canada*, **24,** 1905–1010.

Powell, N. A. (1969). Indo-Pacific Bryozoa new to the Mediterranean coast of Israel. *Israel Journal of Zoology*, **18,** 157–168.

Prenant, M. (1927). Notes éthologiques sur la faune marine sessile des environs de Roscoff, II. *Travaux de la Station biologique de Roscoff*, **6,** 5–58.

Prenant, M. and Teissier, G. (1924). Notes éthologiques sur la faune marine sessile des environs de Roscoff: Cirripèdes, Bryozoaires, Hydraires. *Travaux de la Station biologique de Roscoff*, **2,** 3–49.

Qasim, S. Z. (1957). The biology of *Blennius pholis* L. (Teleostei). *Proceedings of the Zoological Society of London*, **128,** 161–208.

Reger, J. F. (1971). A fine structure study on spermiogenesis in the ectoproct, *Bugula* sp. *Journal of submicroscopical Cytology*, **3,** 193–200.

Rey, P. (1927). Observations sur le corps brun des Bryozoaires Ectoproctes. *Bulletin de la société zoologique de France*, **52,** 367–386.

Riedl, R. (1966). "Biologie der Meereshöhlen". Paul Parey, Hamburg and Berlin.

Rogick, M. D. (1956). Bryozoa of the United States navy's 1947–1948 Antarctic expedition, I–IV. *Proceedings of the United States National Museum*, **105,** 221–317.

Rogick, M. D. (1965). Bryozoa of the Antarctic. *In* "Biogeography and Ecology in Antarctica (P. van Oye and J. van Miegham, eds.). *Monographiae biologicae*, **15,** 401–413.

Rogick, M. D. and Croasdale, H. (1949). Studies on marine Bryozoa, III. Woods Hole region Bryozoa associated with algae. *Biological Bulletin, Marine Biological Laboratory, Woods Hole*, **96,** 32–69.

Ross, J. R. P. (1974). Reef associated Ectoprocta from central region, Great Barrier Reef. *In* "Proceedings of the 2nd International Congress on Coral Reefs" (A. M. Cameron *et al.*, eds.), Vol. 1, pp. 349–352. Brisbane.

Ryland, J. S. (1958). Embryo colour as a diagnostic character in Polyzoa. *Annals and Magazine of Natural History*, Ser. 13, **1,** 552–556.

Ryland, J. S. (1959a). The settlement of Polyzoa larvae on algae. *In* "Proceedings of the 15th International Congress of Zoology", pp. 236–239.

Ryland, J. S. (1959b). Experiments on the selection of algal substrates by polyzoan larvae. *Journal of Experimental Biology*, **36,** 613–631.

Ryland, J. S. (1960). Experiments on the influence of light on the behaviour of polyzoan larvae. *Journal of Experimental Biology*, **37,** 783–800.

Ryland, J. S. (1962a). The association between Polyzoa and algal substrata. *Journal of Animal Ecology*, **31,** 331–338.

Ryland, J. S. (1962b). The effect of temperature on the photic responses of polyzoan larvae. *Sarsia*, **6,** 41–48.

Ryland, J. S. (1963a). The species of *Haplopoma* (Polyzoa). *Sarsia*, **10,** 9–18.

Ryland, J. S. (1963b). Systematic and biological studies on Polyzoa (Bryozoa) from western Norway. *Sarsia*, **14,** 1–59.

Ryland, J. S. (1964). The identity of some cyphonautes larvae (Polyzoa). *Journal of the Marine Biological Association of the United Kingdom*, **44,** 645–654.

Ryland, J. S. (1965). Polyzoa (Bryozoa): Order Cheilostomata: cyphonautes larvae. *Fiches d'Identification du Zooplancton*, **107,** 1–6.

Ryland, J. S. (1967a). Polyzoa. *Oceanography and Marine Biology, an Annual Review*, **5,** 343–369.

Ryland, J. S. (1967b). Respiration in Polyzoa (Ectoprocta). *Nature, London*, **216,** 1040–1041.

Ryland, J. S. (1968). Terminological problems in Bryozoa. *Atti della Società Italiana di Scienze Naturali, Milano*, **108,** 225–236.

Ryland, J. S. (1970). "Bryozoans". Hutchinson University Library, London.

Ryland, J. S. (1972). The analysis of pattern in communities of Bryozoa, I. Discrete sampling methods. *Journal of Experimental Marine Biology and Ecology*, **8,** 277–297.

Ryland, J. S. (1973). The analysis of spatial distribution patterns. *In* "Living and Fossil Bryozoa" (G. P. Larwood, ed.), pp. 165–172. Academic Press, London.

Ryland, J. S. (1974a). Observations on some epibionts of gulfweed, *Sargassum natans* (L.) Meyen. *Journal of Experimental Marine Biology and Ecology*, **14,** 17–25.

Ryland, J. S. (1974b). Bryozoa in the Great Barrier Reef province. *In* "Proceedings of the 2nd International Congress on Coral Reefs" (A. M. Cameron *et al.*, eds.), Vol. 1, pp. 341–348. Brisbane.

Ryland, J. S. (1975). Parameters of the lophophore in relation to population structure in a bryozoan community. *In* "Proceedings of the 9th European Marine Biology Symposium" (H. Barnes, ed.), pp. 363–393. Aberdeen University Press, Aberdeen.

Ryland, J. S. (1976a). Behaviour, settlement and metamorphosis of bryozoan larvae: a review. Proceedings of a Symposium on Marine Invertebrate Larvae, 1973. *Thalassia jugoslavica*, **10**, 239–262.

Ryland, J. S. (1976b). Taxes and tropisms. *In* "The Biology of Bryozoans" (R. M. Woollacott and R. L. Zimmer, eds.). Academic Press, New York and London, in the press.

Ryland, J. S. and Gordon, D. P. (1976). Some New Zealand and British species of *Hippothoa*. *Journal of the Royal Society of New Zealand*, in the press.

Ryland, J. S. and Nelson-Smith, A. (1975). Littoral and benthic investigations on the west coast of Ireland –IV. (Section A: faunistic and ecological studies.) Some shores in counties Clare and Galway. *Proceedings of the Royal Irish Academy*, B, **75,** 245–266.

Ryland, J. S. and Stebbing, A. R. D. (1971). Settlement and orientated growth in epiphytic and epizoic bryozoans. *In* " Fourth European Marine Biology Symposium " (D. J. Crisp, ed.), pp. 105–123. Cambridge University Press.

Scheer, B. T. (1945). The development of marine fouling communities. *Biological Bulletin, Marine Biological Laboratory, Woods Hole*, **89,** 103–121.

Schneider, D. (1955). Phototropische Wachstum der Zoide und Rhizoide von *Bugula avicularia*. *Naturwissenschaften*, **42,** 48–49.

Schneider, D. (1957). Orientiertes Wachstum von Calcit-Kristallen in der Cuticula mariner Bryozoen. *Verhandlungen der Deutschen zoologischen Gesellschaft*, 1957, (*Zoologischer Anzeiger*, Supp. 21) 250–255.

Schneider, D. (1959). Der Aufbau der *Bugula*-Tierstöcke und seine Beeinflussung durch Aussenfaktoren. *Biologisches Zentralblatt*, **78,** 250–283.

Schneider, D. (1960). Über den mechanismus des phototropischen knospenwachstums bei marinen Bryozoen. *Verhandlungen der Deutschen zoologischen Gesellschaft*, 1959, 238–247.

Schneider, D. (1963). Normal and phototropic growth reactions in the marine bryozoan *Bugula avicularia*. *In* " The Lower Metazoa: Comparative Biology and Physiology " (E. C. Dougherty, ed.). pp. 357–371. University of California Press, Berkeley and Los Angeles.

Schopf, T. J. M. (1967). The literature of the phylum Ectoprocta: 1555–1963. *Systematic Zoology*, **16,** 318–327.

Schopf, T. J. M. (1973). Population genetics of ectoprocts: status as of January, 1972. *In* " Living and Fossil Bryozoa " (G. P. Larwood, ed.), pp. 585–592. Academic Press, London and New York.

Schopf, T. J. M. (1974). Ectoprocts as associates of coral reefs: St. Croix, U.S. Virgin Islands. *In* " Proceedings of the 2nd International Congress on Coral Reefs " (A. M. Cameron *et al.*, eds.), Vol. 1, pp. 353–356. Brisbane.

Sevilla, J. Z. (1961). On the food preferences of *Strongylocentrotus drobachiensis*. Unpubl. report, Friday Harbor Lab., Univ. Washington, 14 pp.

Silén, L. (1943). Notes on Swedish marine Bryozoa. *Arkiv för Zoologi*, **35**A (7), 1–16.

Silén, L. (1944a). On the division and movements of the alimentary canal of the Bryozoa. *Arkiv för Zoologi*, **35**A, No. 12, 1–40.

Silén, L. (1944b). On the formation of the interzoidal communications of the Bryozoa. *Zoologiska Bidrag från Uppsala*, **22,** 433–488.

Silén, L. (1945). The main features of the development of the ovum, embryo and ooecium in the ooeciferous Bryozoa Gymnolaemata. *Arkiv för Zoologi*, **35**A, No. 17, 1–34.

Silén, L. (1951). Notes on Swedish marine Bryozoa, II. *Arkiv för Zoologi*, Ser 2, **2**: 569–573.

Silén, L. (1966). On the fertilization problem in the gymnolaematous Bryozoa. *Ophelia*, **3**, 113–140.

Silén, L. (1972). Fertilization in the Bryozoa. *Ophelia*, 10, 27–34.

Silén, L. (1976). Polymorphism in marine bryozoans. *In* "The Biology of Bryozoans" (R. M. Woollacott and R. L. Zimmer, eds.), Academic Press, New York and London.

Silén, L. and Harmelin, J. G. (1976). *Haplopoma sciaphilum* sp. nov.—a cave-dwelling bryozoan from the Skagerrak and the Mediterranean. *Zoologica Scripta*, in the press.

Silén, L. and Jansson, B.-O. (1972). Occurrence of *Electra crustulenta* (Bryozoa) in relation to light. *Oikos*, **23**, 59–62.

Skellam, J. G. (1952). Studies in statistical ecology. I. Spatial pattern. *Biometrika*, **39**, 346–362.

Smith, L. W. (1973). Ultrastructure of the tentacles of *Flustrellidra hispida* (Fabricius). *In* "Living and Fossil Bryozoa" (G. P. Larwood, ed.), pp. 335–342. Academic Press, London and New York.

Soule, D. F. (1973). Morphogenesis of giant avicularia and ovicells in some Pacific Smittinidae. *In* "Living and Fossil Bryozoa (G. P. Larwood, ed.), pp. 485–495. Academic Press, London and New York.

Soule, D. F. and Soule, J. D. (1972). Ancestrulae and body wall morphogenesis of some Hawaiian and eastern Pacific Smittinidae (Bryozoa). *Transactions of the American Microscopical Society*, **91**, 251–260.

Soule, J. D. (1973). Histological and histochemical studies on the bryozoan-substrate interface. *In* "Living and Fossil Bryozoa" (G. P. Larwood, ed.), pp. 343–347. Academic Press, London and New York.

Soule, J. D. and Soule, D. F. (1969). Systematics and biogeography of burrowing bryozoans. *American Zoologist*, **9**, 791–802.

Soule, J. D. and Soule, D. F. (1974). The bryozoan-coral interface on coral and coral reefs. *In* "Proceedings of the 2nd International Congress on Coral Reefs" (A. M. Cameron *et al.*, eds.), Vol. 1, pp. 335–340. Brisbane.

Soule, J. D. and Soule, D. F. (1975). *Spathipora*, its anatomy and phylogenetic affinities. *Documents des Laboratoires de Géologie de la Faculté des Sciences de Lyon*, Hors Série, **3**, 247–253.

Stebbing, A. R. D. (1970). Studies on *Rhabdopleura* and the Bryozoa. Thesis, University of Wales.

Stebbing, A. R. D. (1971a). Growth of *Flustra foliacea* (Bryozoa). *Marine Biology*, **9**, 267–272.

Stebbing, A. R. D. (1971b). The epizoic fauna of *Flustra foliacea* (Bryozoa). *Journal of the Marine Biological Association of the United Kingdom*, **51**, 283–300.

Stebbing, A. R. D. (1972). Preferential settlement of a bryozoan and serpulid larvae on the younger parts of *Laminaria* fronds. *Journal of the Marine Biological Association of the United Kingdom*, **52**, 765–772.

Stebbing, A. R. D. (1973a). Competition for space between the epiphytes of *Fucus serratus* L. *Journal of the Marine Biological Association of the United Kingdom*, **53**, 247–261.

Stebbing, A. R. D. (1973b). Observations on colony overgrowth and spatial competition. *In* "Living and Fossil Bryozoa" (G. P. Larwood, ed.), pp. 173–183. Academic Press, London and New York.

Strathmann, R. R. (1971). The feeding behaviour of planktotrophic echinoderm larvae: mechanisms, regulation and rates of suspension feeding. *Journal of Experimental Biology and Ecology*, **6,** 109–160.

Strathmann, R. R. (1973). Function of lateral cilia in suspension feeding of lophophorates (Brachiopoda, Phoronida, Ectoprocta). *Marine Biology*, **23,** 129–136.

Ström, R. (1969). Sexual reproduction in a stoloniferous bryozoan, *Triticella koreni* (G. O. Sars). *Zoologiska Bidrag från Uppsala*, **38,** 113–128.

Tavener-Smith, R. and Williams, A. (1972). The secretion and structure of the skeleton of living and fossil Bryozoa. *Philosophical Transactions of the Royal Society*, B **264,** 97–159.

Thompson, T. E. (1958). The natural history, embryology, larval biology and post-larval development of *Adalaria proxima* (Alder and Hancock) (Gastropoda Opisthobranchia). *Philosophical Transactions of the Royal Society*, B **242,** 1–58.

Thompson, T. E. (1962). Grazing and the life cycles of British nudibranchs. *In* "Grazing in Terrestrial and Marine Environments" (D. J. Crisp, ed.), pp. 275–297. Blackwell, Oxford.

Thorpe, J. P. (1975). Behaviour and colonial activity in *Membranipora membranacea* (L.). *Documents des Laboratoires de Géologie de la Faculté des Sciences de Lyon*, Hors Série, **3,** 115–121.

Thorpe, J. P., Shelton, G. A. B. and Laverack, M. S. (1975). Electrophysiology and co-ordinated responses in the colonial bryozoan *Membranipora membranacea* (L.). *Journal of Experimental Biology*, **62,** 389–404.

Thorson, G. (1946). Reproduction and larval development of Danish marine bottom invertebrates, with special reference to the planktonic larvae in the Sound (Øresund). *Meddeleser fra Kommissionen for Havundersøgelser*, Bd. 4, No. 1, 1–523.

Wisely, B. (1958). The settling and some experimental reactions of a bryozoan larva, *Watersipora cucullata* (Busk). *Australian Journal of Marine and Freshwater Research*, **9,** 362–371.

Wisely, B. (1960). Observations on the settling behaviour of larvae of the tubeworm *Spirorbis borealis* Daudin (Polychaeta). *Australian Journal of Marine and Freshwater Research*, **11,** 55–72.

Wisely, B. (1962). Attachment of marine invertebrates and algae to polytetrafluorethylene (PTFE) surfaces. *Australian Journal of Science*, **24,** 389 only.

Withers, R. G., Farnham, W. F., Lewey, S., Jephson, N. A., Haythorn, J. M. and Gray, P. W. G. (1975). The epibionts of *Sargassum muticum* in British waters. *Marine Biology*, **31,** 79–86.

Wood, T. S. (1973). Colony development in species of *Plumatella* and *Fredericella* (Ectoprocta: Phylactolaemata). *In* "Animal Colonies: Development and Function Through Time" (R. S. Boardman, A. H. Cheetham and W. A. Oliver, eds.), pp. 395–432. Dowden, Hutchinson and Ross, Stroudsberg.

Wood, T. S. (1974). Food, feeding and lophophore morphology in freshwater Bryozoa. Unpublished paper given at the 3rd International Bryozoology Conference, Lyon, France.

Woollacott, R. M. and North, W. J. (1971). Bryozoans of California and northern Mexico kelp beds. *In* "The Biology of Giant Kelp Beds (*Macrocystis*)" (W. J. North, ed.). *Beihefte zur Nova Hedwigia*, **32,** 600 pp.

Woollacott, R. M. and Zimmer, R. L. (1971). Attachment and metamorphosis of the cheilo-ctenostome bryozoan *Bugula neritina* (Linné). *Journal of Morphology*, **134,** 351–382.

Woollacott, R. M. and Zimmer, R. L. (1972a). Origin and structure of the brood chamber in *Bugula neritina* (Bryozoa). *Marine Biology*, **16,** 165–170.

Woollacott, R. M. and Zimmer, R. L. (1972b). A simplified placenta-like brooding system in *Bugula neritina* (Bryozoa). *30th Annual Proceedings of the Electron Microscope Society of America*, 130–131.

Woollacott, R. M. and Zimmer, R. L. (1972c). Fine structure of a potential photoreceptor organ in the larva of *Bugula neritina. Zeitschrift für Zellforschung und mikroskopische Anatomie*, **123,** 458–469.

Woollacott, R. M. and Zimmer, R. L. (1975). A simplified placenta-like system for the transport of extraembryonic nutrients during embryogenesis of *Bugula neritina* (Bryozoa). *Journal of Morphology*, **147**, 355–378.

Wyer, D. W. and King, P. E. (1973). Relationships between some British littoral and sublittoral bryozoans and pycnogonids. *In* "Living and Fossil Bryozoa" (G. P. Larwood, ed.), pp. 199–207. Academic Press, London and New York.

Yamaguchi, M. (1975). Growth and reproductive cycles of the marine fouling ascidians *Ciona intestinalis*, *Styela plicata*, *Botrylloides violaceus* and *Leptoclinum mitsukurii* at Aburatsubo-Moroiso Inlet (Central Japan). *Marine Biology*, **29,** 253—259.

Zimmer, R. L. (1975). Morphological and developmental affinities of the lophophorates. *In* "Living and Fossil Bryozoa" (G. P. Larwood, ed.), pp. 593–597. Academic Press, London and New York.

Taxonomic Index

A

B

D

H

I

J

K

R

S

Subject Index

D

E

K

L

N

O

R

S

T

U

Cumulative Index of Titles

Cumulative Index of Authors